McDONALD INSTITUTE MONOGRAPHS

Neanderthals and modern humans in the European landscape during the last glaciation:
archaeological results of the Stage 3 Project

Edited by Tjeerd H. van Andel* & William Davies

with contributions by
Leslie C. Aiello, Judy R.M. Allen, Eric J. Barron, Robert A. Foley, Clive S. Gamble, Piers Gollop, Brian Huntley, Olaf Jöris, Marta Mirazón Lahr, Anastasia Markova, Rudolf Musil, Heiko Pälike, David Pollard, Cheryl Ross, John R. Stewart, Chris Stringer, Paul Valdes, Thijs van Kolfschoten, Bernhard Weninger & Peter Wheeler

* Co-ordinator of the Stage 3 Project

Published by:

McDonald Institute for Archaeological Research
University of Cambridge
Downing Street
Cambridge
CB2 3ER
UK
+44-(0)-1223-339336 (Publications); +44-(0)1223-333538 (Institute Main Office); +44-(0)1223-333536 (FAX)

Distributed by Oxbow Books
  United Kingdom: Oxbow Books, Park End Place, Oxford, OX1 1HN, UK.
   Tel: +44-(0)-1865-241249; Fax: +44-(0)-1865-794449; http://www.oxbowbooks.com/
  USA: The David Brown Book Company, P.O. Box 511, Oakville, CT 06779, USA.
   Tel: 860-945-9329; FAX: 860-945-9468

ISBN: 1-902937-21-X
ISSN: 1363-1349

Edited for the Institute by Chris Scarre (*Series Editor*) and Dora A. Kemp (*Production Editor*).

Cover illustrations: Neanderthal and *Homo sapiens sapiens.* (Drawings by Rudy Zallinger; courtesy of the
National Science Foundation, *Mosaic* 11(6), 1980.)

Printed and bound by Short Run Press, Bittern Rd, Sowton Industrial Estate, Exeter, EX2 7LW, UK.

In memory of

Dr Olga Marie van Andel (1925–2002)
Professor of Botany (emeritus), the University of Utrecht
who supported the Stage 3 Project in many ways.

CONTENTS

CONTRIBUTORS

LESLIE C. AIELLO
Department of Anthropology, University College London, Gower Street, London, WX1E 6BT, UK.

JUDY R.M. ALLEN
Environmental Research Centre, School of Biological and Biomedical Sciences, University of Durham, South Road, Durham, DH1 3LE, UK.

ERIC J. BARRON
College of Earth and Mineral Sciences, The Pennsylvania State University, 116 Deike Building, University Park, PA 16802, USA.

WILLIAM DAVIES
Centre of the Archaeology of Human Origins, Department of Archaeology, University of Southampton, Avenue Campus, Southampton, SO17 1BJ, UK.

ROBERT A. FOLEY
Department of Biological Anthropology, University of Cambridge, Pembroke Street, Cambridge, CB2 3DZ, UK.

CLIVE S. GAMBLE
Centre of the Archaeology of Human Origins, Department of Archaeology, University of Southampton, Avenue Campus, Southampton, SO17 1BJ, UK.

PIERS GOLLOP
14 Wellington Street, Hertford, SG14 3AW, UK.

BRIAN HUNTLEY
School of Biological and Biomedical Sciences, University of Durham, South Road, Durham, DH1 3LE, UK.

OLAF JÖRIS
Forschungsbereich Altsteinzeit des Römisch-Germanischen Zentralmuseums Mainz, Schloss Monrepos, D-56567 Monrepos-Segendorf, Germany.

MARTA MIRAZÓN LAHR
Department of Biological Anthropology, University of Cambridge, Pembroke Street, Cambridge, CB2 3DZ, UK.

ANASTASIA MARKOVA
Laboratory of Biogeography, Institute of Geography, Russian Academy of Sciences, Staromonetny 29, Moscow 109017, Russia.

RUDOLF MUSIL
Department of Geology and Paleontology, Masaryk University, Kotlarska 2, 61137 Brno, Czech Republic.

HEIKO PÄLIKE
Department of Geology and Geochemistry, Stockholm University, S-10691 Stockholm, Sweden & Godwin Institute of Quaternary Research, Department of Earth Sciences, Cambridge University, Downing Street, Cambridge, CB2 3EQ, UK

DAVID POLLARD
Earth Systems Science Center, Pennsylvania State University, University Park, PA 16802, USA.

CHERYL ROSS
Centre of the Archaeology of Human Origins, Department of Archaeology, University of Southampton, Avenue Campus, Southampton, SO17 1BJ, UK.

JOHN R. STEWART
Department of Anthropology, University College London, Gower Street, London, WX1E 6BT, UK.

CHRIS STRINGER
Vertebrates and Anthropology Division, Department of Palaeontology, Natural History Museum, Cromwell Road, London SW7 5BD, UK.

PAUL VALDES
School of Geographical Sciences, University of Bristol, University Road, Bristol, BS8 1SS, UK.

TJEERD H. VAN ANDEL
Department of Earth Sciences & Godwin Institute of Quaternary Research, University of Cambridge, Downing Street, Cambridge, CB2 3EQ, UK.

THIJS VAN KOLFSCHOTEN
Faculty of Archaeology, Leiden University, P.O. Box 9515, 2300-RA, Leiden, The Netherlands.

BERNHARD WENINGER
Institut für Ur- und Frühgeschichte, Universität zu Köln, Weyertal 125, D-50923 Köln, Germany.

PETER WHEELER
Liverpool John Moores University, Byron Street, Liverpool, L3 3AF, UK.

Figures

Tables

Appendices

Acknowledgements[1]

The Stage 3 Project is a team effort that has involved at its peak 34 members, experts in more than a dozen subjects. Although not all of them are amongst the authors of this book, each has contributed to its content by their research and experience, their discussions, their planning, their critique and their moral support. Without them the simulations and data bases which have allowed us (and now also everyone else who desires it) to think in new ways and greater depth about glacial times and their human populations would not have been possible. As Co-ordinator of the Stage 3 Project I am deeply indebted to them all, not only because they have made this experiment in multi-disciplinary research work so well and provided it with so many new and valuable insights, but also because of the cheerful harmony that prevailed throughout the eight years of the Project.

I am deeply indebted to my sister Professor O.M. van Andel, now deceased; her strong interest in the project often sustained mine and, remembering that research results must eventually be published, she also left me with some of the means to make this possible.

We are most grateful to the Leverhulme Trust Ltd of London and the Director and Deputy Director of the McDonald Institute for Archaeological Research of the University of Cambridge for their generous financial support of the archaeological and mammalian components of the research. The McDonald Institute also provided hospitality for several years for two post-docs. We are honoured that the Institute have accepted this book, our main archaeological publication, for their series of distinguished monographs.

A large and complex enterprise like ours with its widely-scattered membership and periodic gatherings needs modest but critical financial resources for the programme as a whole, for its many meetings, for travel and other forms of communication, for data processing and so forth. These not so glamorous items are rarely of interest to major funding agencies, but are nonetheless essential. In our case we owe our survival during the publication phase to numerous donors from amongst the alumni of Stanford University in California where the Stage 3

co-ordinator once happily taught, and especially to Nancy and Bob Phelps. At a critical moment late in the game when we faced a serious under-estimate of the skills, time and expense required for state-of-the-art data processing we were saved by generous donations from the Jon B. Jolly family and Mark Baerwaldt, both from Seattle WA, USA. Our gratitude to all of them is difficult to express adequately.

Others helped in various ways. Sir Nicholas J. Shackleton provided the wisdom based on his long experience in the subject of palaeoclimatology and with multi-disciplinary projects for which in particular the co-ordinator is grateful. A large set of late Pleistocene mammalian data from eastern Europe was donated by Dr Anastasia Markova and her colleagues at the Biogeography Laboratory, Institute of Geography, The Russian Academy of Sciences, in Moscow. Professor Rudolf Musil of the Masaryk University of Brno (Czech Republic) provided us with a floral and faunal data base for central Europe that language barriers would otherwise have kept beyond our reach. As the archaeological and animal data bases grew, mapping their contents would have overwhelmed us, had not Dr Alan Smith and Lawrence Rush written software for a friendly map-plotting program. Piers Gollop's efforts to compile in readily accessible form the information needed for the climate geography underpin much of our results, in particular Chapter 11. For the geological background of the late Quaternary, we owe much to the quietest of our colleagues, Dr Ko van Huissteden for his readiness to share his deep understanding of the environments of the Weichselian Glaciation.

It is appropriate that I thank on a separate line our editors from the McDonald Institute, Chris Scarre who did much to arrange publication of this monograph and Dora Kemp who edited our manuscripts and typesetted them with skill, dispatch and a large dose of patient understanding.

The GISP2 $\delta^{18}O$ record on which the graphs of the changing climate in many chapters rest was obtained from The Greenland Icesheet Project 2 summary web site, http://www.gisp2.sr.unh.edu/GISP2/.

We thank the members of that project for making them so freely available. Climate and vegetation simulations used for figures in Chapters 4, 5, 6 and 8 were generously provided by Eric Barron and David Pollard of the Earth Systems Science Center, Pennsylvania State University, University Park PA 16802, USA. The cover drawings are the work of Rudy Zallinger; they appeared in the magazine *Mosaic* vol. 11(6) (Nov./Dec. 1980) of the The National Science Foundation, Washington DC, USA.

Finally, the co-ordinator and members of the Stage 3 Project will never forget the superb organization by Margaret Johnston, Administrator of the Department of Earth Sciences at the University of Cambridge, of our meetings and workshops with all the attendant problems of housing and feeding members. Without her, nor would our annual project dinners, presented each time with style in one or another of the Cambridge Colleges, have become such memorable traditional events.

Tjeerd H. van Andel
Co-ordinator of the Stage 3 Project

Note

1. Acknowledgements specific to individual chapters are expressed at the end of each chapter.

Foreword

Clive Gamble

Interdisciplinarity is easy to say, difficult to achieve. We are encouraged to break down disciplinary boundaries and the rewards, we hope, are grant success and international recognition for research which achieves new insights into established questions. In practice such large-scale endeavours are often multi- rather than inter-disciplinary. They tend to be asymmetrical in effort, dominated by a lead subject with others tagging along behind. Integration is the key to overcoming such problems.

The very best interdisciplinary studies pose new questions and change the subject entirely. I can think of two shining examples which certainly set the agenda for much of what I have been researching in the past thirty years. First there was the groundbreaking *Man the Hunter* symposium (Lee & DeVore 1968) that brought together anthropologists, demographers, biologists and archaeologists. The result provided a global view of how small-world societies were structured and how they had changed. Pattern was not only perceived, where previously it had not, but the cross-cultural regularities in group size, mobility and the variations in adaptive solutions now had a framework for explanation.

Second was the creation of a continuous record of global climate change. This project combined the work of oceanographers, physicists, chemists, quaternary scientists and statisticians (Imbrie & Imbrie 1979). When allied with climate modellers the results were far-reaching (CLIMAP 1976). The maps of ice age earth pieced together from proxy records such as pollen and fauna were now supplanted by predictive models. These used such measures as seasurface temperature and combined them with a more sophisticated understanding of how the earth's weather systems worked.

Spurred on by such examples archaeologists tried their own global overviews of significant timeslices (Gamble & Soffer 1990; Soffer & Gamble 1990), although their first priority was to assemble the data from a single discipline rather than draw in specialists from others. In his contribution to these volumes, however, Tjeerd van Andel (1990), who was the moving force behind CLIMAP and would become the architect and shepherd of the Stage 3 project, had this to say for the future.

> When correlated with the global palaeoenvironmental geography, the (archaeological) results should permit a much more rigorous evaluation than is possible now of site patterns, population densities, extraction territories, exploitation strategies, migration routes, and of the role played by environmental opportunities and constraints in the flow of Late Palaeolithic human economic and social activity (van Andel 1990, 24)

What we had to avoid, he reminded us, was the 'potential Babylonian confusion among members of many disciplines' (van Andel 1990, 31). CLIMAP had managed this feat and the fusion of Palaeolithic archaeology and Palaeoecology could learn from their experience. His warning was well taken.

The Stage 3 Project is a model of interdisciplinarity. It has met those goals set out by van Andel in 1990 and quoted above. Furthermore it has achieved all this in under ten years. The key decisions which have led to its spectacular success are worth re-iterating (see Chapter 1).

- Palaeoclimate modelling formed the framework for the integration of all the terrestrial archives.
- The spatial scale, all of Europe, was deliberately large to study the impact of climate change on all biota and to escape archaeological parochialism.
- Europe was chosen because it had sufficient data and internal diversity to provide answers to key questions.
- The purpose of the project was to examine how humans adapted within a well-defined palaeoecological and palaeoclimatic framework.

The Stage 3 Project has therefore addressed big questions at a continental scale. In particular it has clarified a horribly muddled part of the last interglacial/glacial cycle. The monolithic view of cold interspersed with a few short-lived interstadials, as indicated from pollen profiles, has been shattered. In its place is a complex temporal mosaic of cold and warm phases which the climatic modelling then transposes into a spatial dimension. Moreover, the varying scales of climate change are built into the framework from the full cycle to the short-lived Dansgaard/Oeschger (D/O) interstadial oscillations associated predominantly with the warm phases of Stage 3 (see Chapter 2, Fig. 2.1).

What follows from the papers in this volume will mean that the interdisciplinary study of the so-called Ice Age will never be the same again. The gauntlet is thrown down to many of the proxy palaeo-ecological records concerning their traditional assessments of tree cover, faunal communities and the conditions under which aeolian activity is found. The palaeoclimate modelling provides estimates of relevant variables such as snow cover and wind speed as well as the more expected ones of temperature and precipitation. We are suddenly faced with a multitude of variables for studying such complex problems as the extinction of the Neanderthals or the response of populations to general climate trends. Explanations for changes in the distribution of plants and animals based on a single environmental variable can no longer be accepted. The door is finally opened to multi-causal and historically contingent accounts of what were obviously very complex processes.

Finally, I must mention the Neanderthals for whom Stage 3 was their last hurrah. The Project has shown just what can be done using radiocarbon dates as data to demonstrate, as Chris Stringer once put it, that Neanderthals went out with a whimper not a bang. Chapter 3 should be compulsory reading for all archaeologists. Even allowing for the vagaries of calibration at such a time-range, and the small samples, the patterns are very revealing. The Stage 4 refuge (Fig. 4.3a) emerges with much more clarity than I could have hoped to expect. But equally the patterns of settlement in the Stage 3 warm phase and the overlap in the transitional phase with the earliest Upper Palaeolithic point to rather different patterns of continental land use. If these are indeed the geographical signatures of Neanderthals and Crô-Magnons then they suggest a resident population of the former engaged in small-scale ebb and flow. This is in contrast to a repeated pattern of demic expansion for the rather incoherent archaeological entity known as the earliest Upper Palaeolithic. As both Davies (2001) and Conard & Bolus (2003) have argued the Crô-Magnons take the fast-track route along the Danube right into the heart of Europe. Moreover, they took the same train several times. In the future we can contrast two populations of hominids

with very different social networks. The Neanderthals, with short-distance adjustments to climate and biotic changes and a local approach to their social worlds, and the Crô-Magnons, whose social success is expressed in their ability to cope with linear environments, but one which comes at a price. Their extended social 'supply lines' were frequently cut by climatic changes such as the D/O interstadial oscillations when temperatures were 5–7°C warmer than the longer duration intervening cold phases (Lowe & Walker 1997, 340). It is not the temperature itself, however, which is important but the very marked amplitude changes over 500–1000 years. A factor discussed by Potts (1998) in terms of variability selection on hominid behaviour. In the context of Crô-Magnon population expansion these high-amplitude, high-frequency changes led to the rolling back of their entire network with further expansion along the same line occurring at a later date. In the colder phase of Late Stage 3 and into Stage 2 this variability is less marked. It is tempting to see the appearance, sometime after 33,000 years ago (Gamble 1999), of the first archaeologically coherent, widespread Upper Palaeolithic entity, the Gravettian, as a signature of these less variable conditions. Furthermore the Gravettian above 45°N coincides with the demic contraction of the Neanderthals into an Iberian refuge last seen in Stage 4 (see Chapter 4). The possibility for their historically expected pattern of later population expansion was therefore compromised. Instead extinction became a possibility since part of their traditional range response to long-term climate change had altered with the settled arrival of Crô-Magnons.

These are just my readings of one of the patterns that the Stage 3 project lays before us. There are many more in this encyclopaedia of insights, challenges and paths for future research. Interdisciplinarity indeed!

References

CLIMAP, 1976. The surface of ice age earth. *Science* 191, 1131–7.

Conard, N. & M. Bolus, 2003. Radiocarbon dating the appearance of modern humans and timing of cultural innovations in Europe: new results and new challenges. *Journal of Human Evolution* 44, 331–71.

Davies, S.W.G., 2001. A very model of a modern human industry: new perspectives on the origins and spread of the Aurignacian in Europe. *Proceedings of the Prehistoric Society* 67, 195–217.

Gamble, C.S., 1999. *The Palaeolithic Societies of Europe*. Cambridge: Cambridge University Press.

Gamble, C.S. & O. Soffer (eds.), 1990. *The World at 18,000 BP*, vol. 2: *Low Latitudes*. London: Unwin Hyman.

Imbrie, J. & K.P. Imbrie, 1979. *Ice Ages: Solving the Mystery*. London: Macmillan.

Lee, R.B. & I. DeVore (eds.), 1968. *Man the Hunter*. Chicago (IL): Aldine.

Lowe, J.J. & M.J.C. Walker, 1997. *Reconstructing Quaternary Environments*. Harlow: Longman.

Potts, R., 1998. Variability selection in hominid evolution. *Evolutionary Anthropology* 7, 81–96.

Soffer, O. & C. Gamble (eds.), 1990. *The World at 18,000 BP*, vol. 1: *High Latitudes*. London: Unwin Hyman.

van Andel, T.H., 1990. Living in the last high glacial — an interdisciplinary challenge, in *The World at 18,000 BP*, vol. 1: *High Latitudes*, eds. O. Soffer & C. Gamble. London: Unwin Hyman, 24–38.

Prologue

Tjeerd H. van Andel

In the final quarter of the twentieth century the amount of human skeletal and archaeological material available for the study of human evolution has increased enormously. The result has been a striking rise of the vigour of human palaeontology/anthropology that has turned human ancestry into a subject of lively public interest. At about the same time, human genetics has used with growing success, but not without controversy, female-line mitochondria and male-line Y chromosomes to track evolutionary relationships. The resulting lines of human ancestry and chronology and their patterns of migration in time and space have become an independent challenge to hypotheses based on archaeological and skeletal data. This, in turn, has led to a greater awareness of the potential role of the palaeoenvironment and especially the palaeoclimate in the human past.

Not surprisingly, these developments have led to a convergence of the interests of human palaeontologists and those students of the Quaternary who explore the changing landscapes, climates and oceans of our not so remote past. The meeting of those two fields of inquiry has produced a truly important step forward that promised major progress in several directions, perhaps even breakthroughs, and so it has.

Identifiable relationships between evolving humans and their ever-changing environments are likely to be spotted most easily at the branching points and branch ends of the hominid evolutionary tree. The arrival of the first hominids is one of those, the spread of *Homo erectus* from Africa northward into Europe and eastward into Asia another. How, why, and under what conditions those events took place is currently the subject of much optimistic speculation, but the palaeoenvironmental data needed to underpin and test those speculations is still sparse and not specific enough. Moreover, the state of the chronology of both environmental and human event sequences remains much too insecure to lend credibility to claims of synchronicity, not to mention causality — of human and environmental event histories and their mutual interactions prior to the last 100,000 years or so.

Compared to the lively activity in human palaeobiology and Quaternary earth history, the archaeology of pre-Holocene humans has maintained a more conservative stance with a strong local focus on archaeological sites and their contents. Whether this betrays a lingering reluctance left from the heyday of archaeological theory to take seriously any form of environmental determinism, or is caused by the somewhat disappointing results of promising anthropological studies of present hunter-gatherers is hard to say. Perhaps it is simply a consequence of the strong provincialism that has so long been a mark of European Palaeolithic research.

Yet the arguments for studying human evolution in a context of changing climates and environments are clear. In fact they become more compelling as we dig farther back into the past at about the same rate that our palaeoenvironmental knowledge decays and the volume of the evidence decreases. On the palaeoenvironmental history of Europe we are better informed than thirty years ago, but well-documented images of climates and landscapes of the African human homeland remain patchy except where chance preservation of botanical and geological remains allows local reconstructions of the environment. Along both archaeological and environmental lines of research hard data is sparse and the chronology lacks the temporal resolution needed to illuminate human evolution under rapid environmental changes.

Only when we come to the final days of multispecies human history when anatomically modern humans entered Europe followed by and, as many believe, causing the demise of their Neanderthal predecessors, does our ability to consider human changes in an environmental context reach a more promising level. From *c*. 50,000 years ago to the present, human events, although by no means free of chronological controversy, are fairly well dated and so is the record of climatic change.

Oddly, received wisdom continues to paint the

arrival of modern humans and the demise of the Neanderthals against a harsh background of fully glacial landscapes where much of the time the reality was very different. The recent Weichsel/Würm Glaciation, began in earnest about 65,000 years ago with a modest glacial maximum and ended with a much more severe and extensive one that started about 25,000 years ago. The intervening 40,000 years were from time to time surprisingly mild, the landscapes less barren and hostile and the Fennoscandian ice-sheets far smaller than usually assumed, until a gradual cooling began some 35,000 years ago that brought on the Last Glacial Maximum 10,000 years later. It may be meaningful that, not even until halfway through that decline and well before that final maximum, the Neanderthals — who had flourished after the first glacial maximum — became extinct.

The assumption of a continuously severe glacial climate has persisted in the face of doubts raised by Quaternary geologists for many years. It inhibits a realistic appreciation of the world as it was when modern humans entered Europe and when the Neanderthals left it for good. Whether climate and environment played a role in those key events — and how much of a role — is a question crucial for our understanding of hominid extinctions and why the world has now a single human species.

The dialogue about Palaeolithic climates and landscapes on one side and human history on the other has thus far involved mostly archaeologists. Being alone, their efforts in assembling and evaluating climatic and landscape information have been limited and until very recently not very enlightening. Thus the time seemed ripe several years ago for an integrated interdisciplinary study covering all of Europe and most of the last glaciation. The study, planned and executed jointly by archaeologists, palaeoclimatologists, geologists and several other specialists was initiated in 1996 by the Godwin Institute of Quaternary Research at the University of Cambridge. It was named *The Stage 3 Project* after Oxygen Isotope Stage 3 (OIS-3) because of its focus on the middle part of the last glaciation, and addressed two major questions.

1. Can we, using mainly existing data, construct a reasonable facsimile of the changing climates and landscapes of the interval between the two maxima of the last glaciation?
2. Do Middle and Upper Palaeolithic human events as recorded at archaeological sites reflect the history of the glacial climate as we understand it at this time and if so, to what degree?

In this book we direct our extensive environmental research toward the first question, and published articles in journals dealing with the Quaternary, relating to the issues raised in the second one. The chapters on palaeo-environmental results focus on archaeological issues and have been written for an archaeological readership. The historical analysis uses a chrono-archaeological data base as a proxy for the human presence in time and space with a continent-wide perspective, taking in the interval of 60,000–25,000 years before present.

The chapters proceed from the objectives, methods and data bases of the first phase of the Stage 3 Project by way of simulations of the climate changes and landscapes of Europe between 60,000 and 20,000 years ago to comparisons between simulated palaeo-environments and contemporaneous human history.

Since the birth of the Stage 3 Project seven years ago, interest in examining the evolution of hominoids, hominids and humans in different palaeoenvironmental contexts at different points in their evolution has spread, finally drawing enough attention to persuade some funding agencies to make modest grants available to pursue this further. Perhaps our experience in what is possible today now and what is not combined with our experience of the complexities and priorities of interdisciplinary research, may be useful to those who intend to spend some of their years and a share of the shrinking, ever-shrinking promised millions on this same subject.

A great and lasting pleasure of the Stage 3 Project has been the happy collaboration between its members who together represented a dozen disciplines; a pleasure that we hope our readers will be able to share. We also hope that, armed with insights in the pitfalls and surprises of seven years of interdisciplinary dialogue, we have avoided, at least partly, that bane of much interdisciplinary research, the failure of putting clearly and readably on paper for everyone what we wrote so easily for our own peers.

Chapter 1

The Stage 3 Project: Initiation, Objectives, Approaches

Tjeerd H. van Andel

Birth and purpose of the Stage 3 Project

Conception and initiation
The Stage 3 Project was conceived late in 1995 by Sir N.J. Shackleton and T.H. van Andel of the Godwin Institute for Quaternary Research at the University of Cambridge and saw the light of day at a Godwin Conference on Oxygen Isotope Stage 3 (OIS-3) convened in 1996. At its first plenary meeting in September 1997 new, high-resolution climate models developed by the Earth Systems Center at Pennsylvania State University, USA, were reviewed. The results were encouraging, the technique was adopted, a basic strategy agreed, detailed plans developed and the Stage 3 Project was underway (van Andel 2002; van Andel & Tzedakis 1998).

The Project drew its members and research methodology from two independent fields of inquiry. In the late 1980s the prehistory of the Middle and Upper Palaeolithic and especially of the Neanderthals, their demise and replacement by the ancestors of modern humans was attracting wide interest among professionals as well as the lay public (*viz.* Trinkaus & Shipman 1993; Stiner 1994; Tattersall 1995; Kuhn 1995; Mellars 1996; Roebroeks & Gamble 1999). The enthusiasm continues today, feeding numerous television programmes which, although sometimes of dubious quality, have raised the popularity of Neanderthals and Anatomically Modern Humans alike. At about the same time but from an entirely independent direction, the discovery of hitherto unknown, frequent, major climatic changes recorded in Greenland ice cores of the last glacial period (Dansgaard *et al.* 1993; Grootes *et al.* 1993) created a great stir amongst Quaternary palaeoclimatologists. Known as Dansgaard/Oeschger (D/O) oscillations, the unstable, strikingly bipolar climate contrasted sharply with the conventional image of a long, slowly cooling but stable mid-glacial interval. The cause(s?) of the instability are not yet understood and con-

tinue to present an irresistible challenge to palaeoclimatologists and palaeoceanographers alike.

Two objectives and two principles
The Stage 3 Project, borrowing its objectives and methods from the convergence of those two strands, one palaeoclimatological, the other archaeological, set out to address the following questions for OIS-3 (60,000–24,000 years ago), the middle pleni-glacial period.
1. What was the climate of Europe like during OIS-3 and to what degree did the drastic changes displayed by Greenland ice cores influence the European landscape and its flora and fauna?
2. Do the human events of the Middle and early Upper Palaeolithic reflect the OIS-3 climatic and environmental history and in what way and to what degree?

At their first meeting the members of the Project agreed on two basic principles of action: a) to use only existing data to evaluate the current state of our knowledge; and b) to aim specifically at formulating better questions, while regarding better answers as a bonus. The second principle has had interesting consequences for our research procedures, especially when applied to the very broad second question. Of course, 'better questions' may also be read as 'more specific explanations', or 'multiple testable hypotheses'.

The members of the Stage 3 Project
Except for two post-docs, the Stage 3 Project operated exclusively with academic volunteers, many of them seduced by the opportunity to study the details of the glacial climate from many perspectives or tempted by the opportunity offered by high-resolution models and large data bases and/or simply by the multi-disciplinary context. Specialists, including several Quaternary geologists, ranged from palaeoclimatologists and palaeoceanographers to practitioners of chronology, archaeology, anthropology and

1

human physiology. During the first four years actively contributing members, based in ten countries on three continents (see Appendix 1.1), numbered about 30. The members operated through specialist panels that furnished input for the models and took part in the model evaluation. Some 15 observers also attended the annual meetings where they provided depth and breadth and widened the disciplinary range (for instance human physiology).

The principal reward offered to all team members was free access to the entire body of Stage 3 data bases and climate simulations (Barron *et al.* 2003) and vegetation studies (Huntley & Allen 2003). The membership was held together by an e-mail network centred on the Project Co-ordinator at the University of Cambridge. Small, special-purpose workshops dealt with specific subjects such as vegetation modelling, the ecology of the mammalian fauna, or human dispersal and adaptation. What is now so oddly called 'membership bonding' was achieved at annual plenary meetings attended by most members and observers. There the results of previous model runs were reviewed, the level of data collection evaluated and plans for next year's experiments worked out in detail. A happy side-effect of those annual meetings was a substantial lowering of professional language barriers between the numerous specialists. Twice-yearly hefty newsletters appeared offering detailed minutes of meetings, reports on progress of the project, and various news items. Together those 12 newsletters now constitute a rare historical record of a major research project of somewhat unusual design. Since mid-2000 the Project has also had a modest Stage 3 website.[1]

Stage 3 research strategy
The goals of the project were certainly ambitious and, if pursued in a traditional manner by the large-scale acquisition of new data, would have required much time and a great expenditure of energy and human resources that would have to be initially invested in the tedious, time-consuming struggle for major funding. Instead, our almost total reliance on existing data brought, amongst many other advantages, the opportunity to use that energy and time to better purposes. Our reliance on volunteers who possessed the full range of expertise required to construct data bases and models and carry out their evaluation and interpretation constituted a distinguished, diverse and highly motivated team. With goodwill on all sides, we rarely found it difficult to match the personal priorities of the members to the Project requirements.

At the first plenary meeting the following tasks, in rough order of execution, were specified:
- develop a comprehensive body of existing palaeoclimatic and some other palaeoenvironmental data, to be synthesized by various means including computer-based, medium-resolution climate and vegetation models;
- concentrate on the interval 45–25 ka BP[2] which includes the arrival of modern humans and the demise of the Neanderthals; the range was later widened to 60–20 ka BP;[2]
- construct a chrono-archaeological data base of all dated Neanderthal and Modern Human sites in Europe published up to the end of the year 2000;
- construct a comprehensive, ecology-oriented data base of the large mammals of the late Pleistocene to supplement existing data bases which tend towards taxonomic and stratigraphic issues. Besides its ecological value, this data base provided much needed insight into the availability of prey animals as a function of climate and landscape (Stewart *et al.* 2003; Musil 2003);
- integrate the environmental and human results and construct as many hypotheses as seem practical regarding the impact of climate on human affairs and resources, and on the extinction of the Neanderthals and the successes of modern humans.

A provisional palaeoenvironmental study (van Andel & Tzedakis 1996; 1998) had shown that existing data were capable of yielding generalized but challenging results using the traditional mode of judging them by standards of 'geological common sense'. However, methods capable of yielding spatially detailed, semi-quantitative results amenable to validation by sets of reserved data would be preferable over the intuitive skills of geologists. This meant medium-resolution computer climate modelling because of its capacity for testing simulations for their sensitivity to a range of input variables and evaluating them with independent observations.

A two-phase programme
The objectives listed above demanded a two-phase strategy. Phase One, the first order of business, would develop one or more syntheses of all information relevant to Stage 3 palaeoenvironments, with an emphasis on climate and climate changes. The syntheses should, as far as possible, be tested against sets of evidence specifically reserved for this purpose. Besides being the key to the archaeological goal, Phase One had a considerable methodological and palaeoclimatological potential of its own and so attracted outstanding volunteers.

In Phase Two our chrono-archaeological data base would be used as a set of proxies for the human activity in space and time that would enable us to construct the shifting patterns of the human presence in Europe from about 60 to 20 cal ka BP. The mammalian data base would allow us to address important survival issues such as the possibility that the climatic preferences of human beings might be guided by the climate tolerance of food animals rather than directly by the climate impact upon themselves. The human dispersal in space and time and the ecology of prey mammals, when combined with the environmental results of Phase One, might permit us to refine and divide broad questions of the kind stated above, replacing them with more specific ones that would encourage further research.

Phase One: climate and landscape

Higher-resolution climate modelling

The ability of palaeoclimate models to perform the functions we required depends on their spatial resolution. Conventional global circulation models (GCMs) have an output grid-spacing of the order of 4.5° latitude by 7.5° longitude. This would, for example, provide a mere four or five output squares (pixels) for Great Britain as a whole, a number obviously too small to reflect the complex climate of the British Isles. Also, GCMs smooth high-relief terrain like that of the Alps, Apennines or Pyrenees so severely that orographic patterns are largely suppressed. Because other variables critical in our quest also vary on a regional or even local scale, GCMs would fail to construct the details of climate, landscape and human resource models needed for our purpose.

Fortunately, a novel approach allowing higher-resolution models by means of a nested strategy was made available to us by Eric Barron at Pennsylvania State University (USA). The nested strategy begins by constructing a GCM based on minimal essential input variables such as insolation (as modulated by orbital parameters), dimensions of polar ice sheets, atmospheric CO_2 levels, sea-surface temperatures and sea-ice limits. As the GCM simulation steps through the seasonal cycle for many years, it supplies the boundary conditions for a regional higher-resolution circulation model (RegCM2) that are essential for the construction of simulations on a 60×60 km output grid. For the United Kingdom the RegCM2 raises the number of output points to about 60, quite enough for an acceptably detailed image of even the British climate.

The RegCM2 is supplied with its own regional input variables that include relief, shorelines and the dimensions of regional ice sheets, all isostatically compensated for changes in ice volume and extent. Also essential are sea-surface temperatures and sea-ice boundaries for the adjacent seas and oceans. The regional programme generates maps of over fifty output variables, such as winter and summer mean temperatures and precipitation, diurnal temperature variations, wind-chill, snow thickness and number of snow days per year, seasonal wind patterns at several altitudes and so forth. For a complete description of the Stage 3 climate modelling programme (Fig. 1.1) and the simulations see Chapter 5 (Barron *et al.* 2003; Barron & Pollard 2002; Pollard & Barron 2003).[3]

An evolving methodology

Our initially envisaged *modus operandi* was simplicity itself, being directly derived from the approach used with so much success by the CLIMAP Project (CLIMAP 1981). Our approach differed from theirs only because instead of obtaining new climatic data we based our modelling experiments on existing evidence which was abundant and diverse but not always ideal or sufficient for the purpose. We then tested the output simulations for their sensitivity to a range of values for key inputs such as ice sheet dimensions. Ultimately, we would validate the simulations with data kept in reserve especially for that purpose.

In the course of the first two years it became clear that a far simpler and more effective approach was to reduce the number of input variables to an absolutely essential minimum and reserve all other available data to test and validate the output.

Because each suite of combined GCM and RegCM2 experiments took months of supercomputer time at Penn State, we ran only a single sequence of simulations per year, labelled Phases 2, 3 and 4 of the Project. At annual plenary meetings the modelling approach itself and its results were thoroughly examined, additional compilations of input data organized, procedural improvements for the next phase devised and the next scheme of simulation experiments designed. This practice also allowed us to bring our members together annually without overburdening them or interfering too much with their own duties.

Modelling the plant cover

The most striking example of the change in design referred to above is the vegetation modelling programme. From many points of view, archaeological as well as climatological, the plant cover is an important component of the system. It shapes the land-

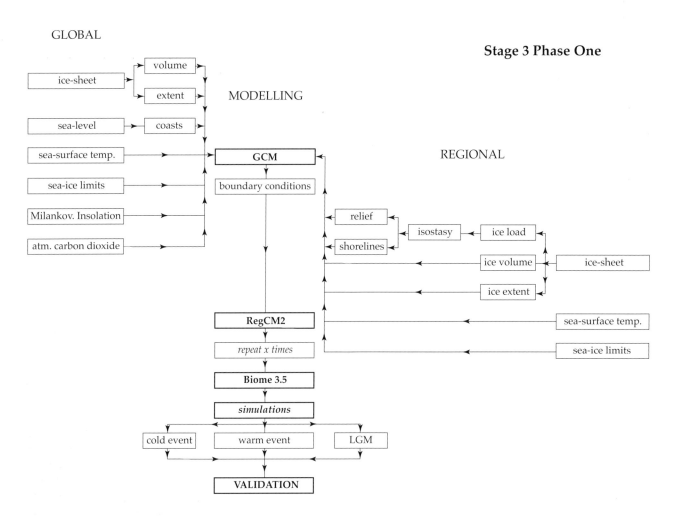

Figure 1.1. *Diagram of the climate and vegetation modelling programme of Phase One of the Stage 3 Project. Left: input variables for the global circulation models (GCM). Right: additional inputs for the inserted higher-resolution regional models (RegCM2). The climatic end members of typical cold and typical warm D/O events and the last glacial maximum (LGM) were constructed as in separate experiment runs.*

scape, provides shelter and fuel and is an essential food source for herbivores, the main hunted mammalian category, as well as an important resource for humans. Plant cover and climate also interact strongly with each other by altering many characteristics of the land surface such as roughness and albedo (reflectivity) that affect local and regional precipitation and evaporation and modify seasonal climate contrasts. A provisional synthesis based on long pollen cores (van Andel & Tzedakis 1996; 1998) suggested striking differences in vegetation between long milder and cold D/O events, and hinted at plant biomes that differed from their modern equivalents (Davies *et al.* 2000). Their results were, however, only qualitative and did not suffice for the objectives of the Stage 3 Project. What was really needed was a set of simulated biomes constructed independently

of their presumed modern counterparts and thereby appropriate for testing the result.

Therefore, instead of using the pollen data base as input for climate/biome modelling, the higher-resolution climate output of the RegCM2 model was coupled to a bio-dynamic model (BIOME 3.5; Prentice *et al.* 1992; Haxeltine & Prentice 1996) that simulated vegetation patterns by means of functional units (Huntley *et al.* 2003). The final simulations were verified with pollen spectra that had also been converted to functional types.

Testing the climate models
From the outset it was obvious that it would be necessary to test the climate and vegetation models against as many independent palaeoenvironmental data sets as could reasonably be assembled. The ob-

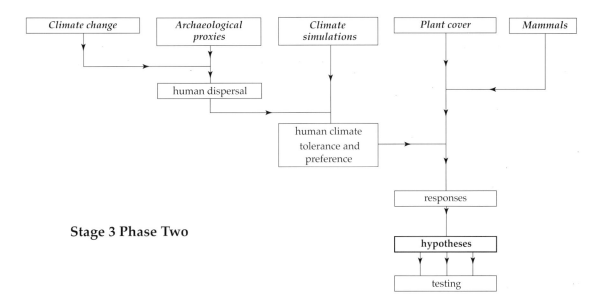

Stage 3 Phase Two

Figure 1.2. *Conceptual diagram of the Stage 3 Project, Phase Two, analysis of the Neanderthal and modern human history in the context of Stage 3 climate changes, simulations of climate end members and the fauna and flora.*

vious choice was the palaeoflora in the form of pollen records which are available for much of the Mediterranean and western Europe, but rather scarce for central, eastern and northeastern Europe (Huntley *et al*. 2003). Another would be dated and identified charcoal analyses, but unfortunately, this material proved rare except for parts of central Europe (Musil 2003).

Another potential source was geological evidence such as permafrost features, loess and other wind-blown deposits and river sediments. A special Stage 3 panel took advantage of a data base compiled at the Free University in Amsterdam to prepare syntheses of permafrost features and river and wind deposits (van Huissteden *et al*. 2003; van Huissteden & Pollard 2003) for comparison with the climate simulations. The results were used together with pollen data by Alfano *et al*. (2003) to test the Stage 3 cold event simulation. We had hoped to use for verification microfaunal data such as micromolluscs and coleoptera to obtain semi-quantitative estimates of precipitation and temperature, but except for coleoptera from maritime western Europe (Coope 2002) the existing data and their dating proved insufficient.

Early on we adopted the fundamental tenet that the very high-resolution Greenland ice-core record would at least qualitatively reflect the climate changes of Greenland, the North Atlantic Ocean and probably the maritime parts of Europe. Whether it would also provide a guide for climate changes in the Medi-

terranean and central and eastern Europe was in 1996 an open question which has now been confirmed for the western Mediterranean (van Andel 2003). That eastern and central Europe underwent glacial climate changes of similar magnitude is likely and has been partly confirmed by sparsely dated loess studies, but so far no robust direct correlation with Greenland has been established. That temporal climate changes on a multi-millennial time-scale have marked central and eastern Europe during Stage 3 is clear (Musil 2003), but their synchronicity remains hard to prove (van Andel 2003).

Phase Two: the Palaeolithic in a climatic context

The ultimate goal of the Stage 3 Project, the analysis of the Middle and Early Upper Palaeolithic in Europe in the context of rapidly changing climates and environments was the task of Phase Two when climate and environmental contexts developed in Phase One would be compared with the assembled human record.

First, a chrono-archaeological data base was compiled that held all dates published up to the end of the year 2000 for archaeological cave, abri and open-air sites in the interval from 60 to 20 cal ka BP except for papers, mainly in Slavic languages, that were inaccessible to us. The overwhelming majority (more than 80 per cent) of the dates had been obtained by radiocarbon analysis and was therefore expressed in ^{14}C years, but the Greenland ice-core base record is calendrical. Therefore all ^{14}C dates had to be converted to calendar years (van Andel 1998). This was accomplished with CalPal (Jöris & Weninger 1998; 2000), a broad-based software-engine designed to perform ^{14}C age calibrations with any available chronological data sets.[4]

Using dated archaeological sites as proxies for the human presence, spatio-temporal models of the changes of Neanderthal and modern human site distributions were to be constructed that could be compared with climate changes recorded in Greenland ice cores and then searched for non-random distribution patterns.

Figure 1.2 shows the conceptual flow of the Phase Two analysis, but crucial questions might be raised about the evaluation and interpretation of our results. Are our input data sufficient in number and meaningful or are they not? Do our models simulate past reality, if not perfectly, at least usefully when we compare them with independent proxies? Mismatches are bound to occur between our simulations and the output proxies; do they derive from flawed proxies, from faulty models or from both? Can insights gained from flawed climate simulations nonetheless have something useful to say about relations between the mid-glacial climate of Europe and its human inhabitants?

No hard and fast answers exist for most of those questions and the underpinnings of Phase Two can be truly evaluated only by the quality of the hypotheses they raise. Ultimately, the broad question whether any observed interactions between humans and climate might have affected human successes and failures the Stage 3 Project can only begin to address and then only tentatively.

Notes

1. The Stage 3 website, now open to all visitors but limited to all data bases and the results of climate and vegetation simulations is located at: http://www.esc.esc.cam.ac.uk/oistage3/Details/Homepage/html. A website established at the University of Southampton houses an enlarged chrono-archaeological data base which includes in addition to the Stage 3 chrono-archaeological data base the Near and Middle East and North Africa with an extended time range to approximately 8000 ka BP.
2. Because our chronological standard, the Greenland ice core GISP2 (see van Andel 2003), measures time in calendrical years, all other dates had to be converted to calendrical units. The notation xxxx ka BP denotes an age in thousands (ka) of calendar years before present for ice core, U-Th, TL, OSL, ESR and calibrated ^{14}C dates; xxxx ka uncal b.p. denotes uncalibrated ^{14}C ka years. Calibrated and uncalibrated

[14]C years can be found in the chrono-archaeological data base at the Stage 3 website; see note 1 for the address.
3. As the Project advanced and accumulating evidence and targets did gain clarity, the chronological limits of the Project widened from Stage 3 *s.s.* to the entire interval from the start of the first glacial maximum (OIS-4) to the peak of the final one (OIS-2) at approximately 20–18 ka BP.
4. The CalPal ^{14}C calibration programme is available at its website: http://www.calpal.de.html.

References

Alfano, M.J., E. Barron, D. Pollard, B. Huntley & J.R.M. Allen, 2003. Comparison of climate model results with European vegetation and permafrost during Oxygen Isotope Stage 3. *Quaternary Research* 59, 97–107.

Barron, E. & D. Pollard, 2002. High-resolution climate simulations of Oxygen Isotope Stage 3 in Europe. *Quaternary Research* 58, 296–309.

Barron, E., T.H. van Andel & D. Pollard, 2003. Glacial environments II: reconstructing the climate of Europe in the last glaciation, in *Neanderthals and Modern Humans in the European Landscape during the Last Glaciation*, Chapter 5, eds. T.H. van Andel & W. Davies. (McDonald Institute Monographs.) Cambridge: McDonald Institute for Archaeological Research, 57–78.

CLIMAP (Climate: Long-range Investigation, Mapping and Prediction), 1981. *Seasonal Reconstructions of the Earth's Surface at the Last Glacial Maximum.* Boulder (CO): Geological Society of America.

Coope, R., 2002. Changes in the thermal climate of northwest Europe during Oxygen Isotope Stage 3 estimated from fossil insect assemblages. *Quaternary Research* 57, 401–8.

Dansgaard, W., S.J. Johnsen, H.B. Clausen, D. Dahl-Jensen, N.S. Gundestrup, C.U. Hammer, C.S. Hvidberg, J.P. Steffensen, H. Sveinbjörnsdottir, J. Jouzel & G. Bond, 1993. Evidence for general instability of past climate from a 250-kyr ice-core record. *Nature* 364, 218–20.

Davies, W., J. Stewart & T.H. van Andel, 2000. Neanderthal landscapes: a preview, in *Neanderthals on the Edge*, eds. C. Stringer, R.N.E. Barton & J.C. Finlayson. Oxford: Oxbow Books, 1–8.

Grootes, P.M., M.J. Stuiver, J.W. White, S. Johnsen & J. Jouzel, 1993. Comparison of oxygen isotope records from the GISP2 and GRIP Greenland ice cores. *Nature* 366, 552–4.

Haxeltine, A. & I.C. Prentice, 1996. An equilibrium terrestrial biosphere model based on ecophysical constraints, resource availability and competition among plant functional types. *Global Geochemical Cycles* 10, 693–709.

Huntley, B. & J.R.M. Allen, 2003. Glacial environments III: palaeo-vegetation patterns in Late Glacial Europe, in *Neanderthals and Modern Humans in the European Landscape during the Last Glaciation*, Chapter 6, eds.

T.H. van Andel & W. Davies. (McDonald Institute Monographs.) Cambridge: McDonald Institute for Archaeological Research, 79–102.

Huntley, B., M.J. Alfano, J.R.M. Allen, D. Pollard, P.C. Tzedakis, J.L. de Beaulieu, E. Grüger & W.S. Watts, 2003. European vegetation during Marine Oxygen Isotope Stage 3. *Quaternary Research* 59, 195–212.

Jöris, O. & B. Weninger, 1998. Extension of the ¹⁴C calibration curve to *c.* 40,000 cal BC by synchronising Greenland ¹⁸O/¹⁶O ice-core records and North Atlantic Foraminifera profiles: a comparison with U/Th coral data. *Radiocarbon* 40, 495–504.

Jöris, O. & B. Weninger, 2000. Radiocarbon calibration and the absolute chronology of the Late Glacial, in L'Europe Centrale et Septentrionale au Tardiglaciaire, Table Ronde de Nemours 13–16 mai 1997. *Mémoires de la Musée de Préhistoire de l'Ile de France* 7, 19–54.

Kuhn, S.L., 1995. *Mousterian Lithic Technology: an Ecological Perspective.* Princeton (NJ): Princeton University Press.

Mellars, P.A., 1996. *The Neanderthal Legacy: an Archaeological Perspective from Western Europe.* Princeton (NJ): Princeton University Press.

Musil, R., 2003. The Middle and Upper Palaeolithic game suite in central and southeastern Europe, in *Neanderthals and Modern Humans in the European Landscape during the Last Glaciation,* Chapter 10, eds. T.H van Andel & W. Davies. (McDonald Institute Monographs.) Cambridge: McDonald Institute for Archaeological Research, 167–90.

Pollard, D. & E.J. Barron, 2003. Causes of model-data discrepancies in European climate during oxygen isotope stage 3 with insights from the Last Glacial Maximum. *Quaternary Research* 59, 108–13.

Prentice, I.C., W. Cramer, S.P. Harrison, R. Leemans, R.A. Monserud & A.M. Solomon, 1992. A global biome model based on plant physiology and dominance, soil properties and climate. *Journal of Biogeography* 19, 117–34.

Roebroeks, W. & C. Gamble (eds.), 1999. *The Middle Palaeolithic Occupation of Europe.* Leiden: University of Leiden.

Stewart, J.S., T. van Kolfschoten, A. Markova & R. Musil, 2003. The mammalian faunas of Europe during Oxygen Isotope Stage Three, in *Neanderthals and Modern Humans in the European Landscape during the Last Glaciation,* Chapter 7, eds. T.H van Andel & W. Davies. (McDonald Institute Monographs.) Cambridge: McDonald Institute for Archaeological Research, 103–30.

Stiner, M.C., 1994. *Honor Among Thieves: a Zooarchaeological Study of Neandertal Ecology.* Princeton (NJ): Princeton University Press.

Tattersall, I., 1995. *The Last Neanderthal: the Rise, Success and Mysterious Extinction of our Closest Human Relatives.* New York (NY): MacMillan.

Trinkaus, E. & P. Shipman, 1993. *The Neandertals: Changing the Image of Mankind.* London: Pimlico Press.

van Andel, T.H., 1998. Middle and Upper Palaeolithic environments and the calibration of ¹⁴C dates beyond 10,000 BP. *Antiquity* 72, 27–33.

van Andel, T.H., 2002. The Stage 3 Project and the climate and landscape of the middle part of the Weichselian Glaciation in Europe. *Quaternary Research* 57, 2–8.

van Andel, T.H., 2003. Glacial environments I: the Weichselian climate in Europe between the end of the OIS-5 interglacial and the Last Glacial Maximum, in *Neanderthals and Modern Humans in the European Landscape during the Last Glaciation,* Chapter 2, eds. T.H van Andel & W. Davies. (McDonald Institute Monographs.) Cambridge: McDonald Institute for Archaeological Research, 9–20.

van Andel, T.H. & P.C. Tzedakis, 1996. Palaeolithic landscapes of Europe and environs, 150,000–25,000 years ago. *Quaternary Science Reviews* 15, 481–500.

van Andel, T.H. & P.C. Tzedakis, 1998. Priority and opportunity: reconstructing the European Middle Palaeolithic climate and landscape, in *Science in Archaeology: an Agenda for the Future,* ed. J. Bailey. London: English Heritage, 37–46.

van Huissteden, K. & D. Pollard, 2003. Oxygen isotope Stage 3 fluvial and aeolian successions in Europe compared with climate model results. *Quaternary Research* 59, 223–33.

van Huissteden, K., J. Vandenberghe & D. Pollard, 2003. Palaeotemperature reconstructions of the European permafrost zone during Marine Oxygen Isotope Stage 3 compared with climate model results. *Journal of Quaternary Research* 18, 453–64.

Appendix 1.1. *Contributing members of the Stage 3 Project.*

PROFESSOR DR TJEERD H. VAN ANDEL
Department of Earth Sciences and Godwin Institute of Quaternary Research, University of Cambridge, Cambridge, UK.

PROFESSOR DR LESLIE C. AIELLO
Department of Anthropology, University College London, London, UK.

DR MARY JO ALFANO
American Geological Institute, Alexandria VA, USA.

DR JUDY R.M. ALLEN
Environmental Research Centre, School of Biological Sciences and Biomedical Sciences, University of Durham, Durham, UK.

DR NEIL ARNOLD
Scott Polar Research Institute, University of Cambridge, Cambridge, UK.

PROFESSOR DR ERIC J. BARRON
College of Earth and Mineral Sciences, Pennsylvania State University, University Park, PA, USA.

Dr Jean-Pierre Bocquet
CNRS, EP 1781, Paris, France.

Dr Mark Chapman
School of Environmental Sciences, University of East Anglia, Norwich, UK.

Professor Dr Russell G. Coope
Centre for Quaternary Research, Royal Holloway College, University of London, Egham, UK.

Dr William Davies
Department of Archaeology, University of Southampton, Southampton, UK.

Professor Dr Jacques-Louis de Beaulieu
Institut Méditerranéen d'Ecologie et de Paleoécologie, Faculté des Sciences St-Jerome, Marseille, France.

Dr Robert A. Foley
Department of Biological Anthropology, University of Cambridge, Cambridge, UK.

Professor Dr Clive S. Gamble
Department of Archaeology, University of Southampton, Southampton, UK.

Dr Eberhard Grüger
Institute für Palynologie und Quartärwissenschaften, Universität Göttingen, Göttingen, Germany.

Professor Dr Brian Huntley
School of Biological Sciences and Biomedical Sciences, University of Durham, Durham, UK.

Dr Olaf Jöris
Forschungsbereich Altsteinzeit des Römisch-Germanischen Zentralmuseums, Schloss Monrepos, Neuwied, Germany.

Dr Marta Mirazón Lahr
Department of Biological Anthropology, University of Cambridge, Cambridge, UK.

Professor Dr Kurt Lambeck
Research School of Earth Sciences, Australian National University, Canberra, Australia.

Dr Anastasia Markova
Laboratory of Biogeography, Institute of Geography, Russian Academy of Sciences, Moscow, Russia.

Professor Dr Rudolf Musil
Department of Geology and Paleontology, Masaryk University, Brno, Czech Republic.

Dr Heiko Pälike
Department of Geology and Geochemistry, Stockholm University, Stockholm, Sweden.

Dr Uwe Pflaumann
Institut für Geowissenschaften, Universität Kiel, Kiel, Germany.

Dr David Pollard
Earth Systems Science Center, Pennsylvania State University, University Park, PA, USA.

Cheryl Ross
Centre of the Archaeology of Human Origins, Department of Archaeology, University of Southampton, Southampton, UK.

Professor Dr Michael Sarnthein
Institut für Geowissenschaften, Universität Kiel, Kiel, Germany.

Professor Dr Sir Nicholas J. Shackleton
The Godwin Laboratory, University of Cambridge, Cambridge, UK.

Dr John R. Stewart
Department of Anthropology, University College London, London, UK.

Professor Dr Chris Stringer
Department of Palaeontology, Natural History Museum, London, UK.

Dr Chronis Tzedakis
School of Geography, Leeds University, Leeds, UK.

Professor Dr Paul Valdes
School of Geographical Sciences, University of Bristol, University Road, Bristol, UK.

Dr J. van Huissteden
Faculty of Earth Sciences, Vrije Universiteit, Amsterdam, Netherlands.

Dr T. van Kolfschoten
Faculty of Archaeology, Leiden University, Leiden, The Netherlands.

Peter Wheeler
Liverpool John Moores University, Liverpool, UK.

Professor Dr William A. Watts
Department of Botany, Trinity College, Dublin University, Dublin, Ireland.

Dr Bernhard Weninger
Institut für Ur- und Frühgeschichte, Universität zu Köln, Germany.

Chapter 2

Glacial Environments I:
the Weichselian Climate in Europe between the End of the OIS-5 Interglacial and the Last Glacial Maximum

Tjeerd H. van Andel

Not so dreadfully cold?

The last glacial interval, known in Europe as the Weichsel-Würm Glaciation, is widely regarded among archaeologists as having had a climate of such extreme severity that it was at its best marginally fit for human survival. North of the Pyrenean, Alpine and Carpathian mountain ranges, glacial conditions were thought to be at or beyond the limit of human endurance (cf. Sykes 2001, 9). Some archae-ologists are persuaded that the glacial climate was fatal to the Neanderthals and might have driven even modern humans to the brink of disaster, but others feel that, environmental determinism being out of fashion, the severe climatic stress of the last glaciation played little if any part in the human history of the middle and late Palaeolithic.

Actually, both sides may be wrong! Stephen Porter (1989) suspected that the traditional view of the Pleistocene as a series of long, cold glaciations alternating with short warmer interglacials was too simple. Instead, the glacial periods themselves contained long, milder intervals (Fig. 2.1: left). Just two climate variables suffice to illustrate the temporal and spatial diversity of the European climate during the last glaciation: the variation in size of the Fennoscandian ice sheet and the position of the warmest mid-summer isotherm between 65 and 20 ka BP[1] (Fig. 2.2). During the 45,000 years that separate the first major ice advance in OIS-4 from the deglaciation late in OIS-2, the ice sheet at its maximum may have existed for less than one third of the total time (Arnold *et al.* 2002). At other times the European climate was considerably milder.

Still, the image of unrelentingly severe arctic conditions in Europe with an ice sheet extend-

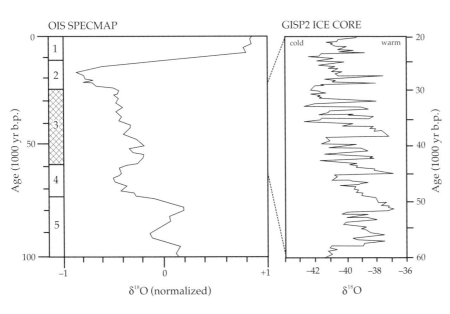

Figure 2.1. *Climate history of the last glacial/interglacial cycle.*
Left: SPECMAP oxygen isotope record displaying a stable mild climate during OIS-3 (after Martinson et al. 1987). Right: Stage 3 climate with high-frequency Dansgaard/Oeschger (D/O: Dansgaard et al. 1993) climate oscillations (from GISP2 Greenland ice core; after Meese et al. 1997; Johnsen et al. 2001; Stuiver & Grootes 2000).

Figure 2.2. *Four stages of the Weichselian Glaciation showing the extent of the Fennoscandian ice sheet and the mean June-July-August 15°C (dashed) and 20°C (solid) isotherms. Isotherms generalized from Stage 3 climate simulations (Barron & Pollard 2002; Pollard & Barron 2003). No acceptable simulation exists for the late Stage 3 cold event. The OIS-4 ice sheet derives from van Andel & Tzedakis (1996); the minimal OIS-3 ice sheet from Arnold et al. (2002); and the Late Glacial Maximum (LGM) ice sheet from Donner (1995). Modern coastlines given only for orientation.*

ing across the Baltic throughout the Middle and Upper Palaeolithic refuses to die (e.g. Stringer & Gamble 1993, fig. 1; Svoboda *et al.* 1996, fig. 1.11). The gap between the glacial climate as perceived by many archaeologists and the evidence widened further when early in the 1990s Greenland ice cores revealed many brief, high-frequency climate oscillations of a surprisingly large amplitude (Fig. 2.1: right). During the pleni-glacial interval some 15–20 such events occurred that were up to 7°C warmer than the intervening cold spells and at times only 2°C cooler than the local Holocene average (GRIP 1993; Bond *et al.* 1993; McManus *et al.* 1994).

These *third order* climatic events, called Dansgaard/Oeschger (D/O) oscillations (Dansgaard *et al.* 1993; GRIP 1993; Grootes *et al.* 1993), were common throughout the middle half of the Weichselian Glaciation but seem to have been nearly absent during the initial and terminal glacial maxima.

The discovery of the D/O oscillations delivered a startling message to students of the middle and later Palaeolithic in Europe, because under conditions of such dramatic instability a large impact of the climate on human beings and their evolution would seem almost inevitable (van Andel & Tzedakis 1998). The human scene which since the previous interglacial had been dominated solely by Neanderthals was altered to an unknown degree by the arrival of Anatomically Modern Humans (AMH) around 45 ka BP. The subsequent 15 millennia during which the two species co-existed span a time of major climatic instability until the demise of the Neanderthals in roughly 30 ka BP made the AMH the sole owners of the earth. Explanations for this last crucial event in human history range from genocide committed by our direct ancestors to a lack of cognitive ability among Neanderthals that kept them from adapting to the extreme environmental conditions, but the role of the Weichselian climate has not been thoroughly and systematically examined so far. Syntheses of the late glacial hominid response to the environment do of course exist, e.g. Svoboda *et al.* (1996 *passim*), Mellars & Stringer (1989), Mellars (1996) and Roebroeks *et al.* (1999 *passim*), but are rare and tend to be regional in approach. A noteworthy exception to this limitation are Gamble's two synopses (1986; 1999).

A brief history of the Weichselian climate in Europe

The penultimate Eemian interglacial (OIS-5e) ended 115 ka ago when a slow deterioration of the hitherto warm climate began, a decline briefly interrupted twice by major warm interstadials which for a few millennial returned the world to near-interglacial conditions (van Andel & Tzedakis 1996). The long cooling trend was accompanied by a large fall in sea level caused by the gradual accumulation of a huge continental ice volume mainly on North America (Lambeck & Chappell 2001, fig. 3A), but there is no direct evidence for the growth of an extensive Fennoscandian ice sheet (Mangerud 1991) until well into OIS-4 (74–60 ka BP). Not until late in OIS-4 an extensive open tundra or cold steppe indicating high-arctic conditions replaced all or most of the boreal woodland and taiga that covered Europe north of the Pyrenean and Alpine ranges (van Andel & Tzedakis 1996). However, this first peak of the last glaciation lasted only 5000–6000 years and the ice margin probably failed to reach the southern and western shores of the Baltic Sea (Donner 1995, 66–7). The Alps and Pyrenees also appear to have been free of large ice caps (van Andel & Tzedakis 1996).

With the onset of Stage 3 a sudden warming initiated a sequence of long, fairly mild D/O events that were only occasionally interrupted by brief, not yet very cold events (Fig. 2.1: right). At Les Echets in the Massif Central of France (Guiot *et al.* 1989) the warm moist events each lasting several millennia reached an annual mean of 7°C (the mean today is 11°C) and the annual precipitation was 500–600 mm (Fig. 2.3), drier than today but far from cold aridity. Annual mean temperatures of the intervening cold events were 0–2°C, close to those of the late OIS-4 glacial maximum, but conditions were drier. At Grande Pile in the Vosges (northeastern France), two warm events at 37 and 40–43 ka BP had mean July temperatures of 20–22°C (based on pollen data) or 16–18°C (estimated from combined pollen and coleopteran data: Guiot *et al.* 1993, figs. 9 & 10). Coope (2002), summarizing Coleoptera-based palaeotemperatures from northwestern Europe for the interval 45–25 ^{14}C ka b.p., has shown that during the only warm event in this interval, perhaps the one at about 45 ka BP known from Greenland ice cores, conditions prevailed that were fairly close to those existing today.

The early warm period lasted until *c.* 45 ka BP when the climate began to deteriorate and between 42 and 38 ka BP cold D/O events with very low minimum temperatures began to appear in close-spaced clusters (Fig. 2.1: right). A final relatively

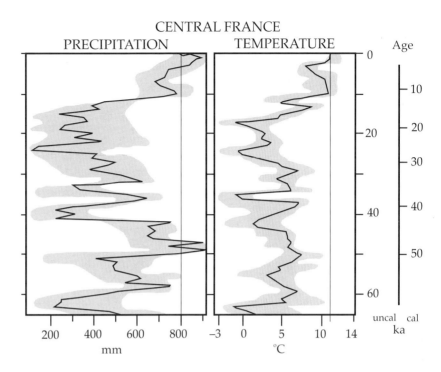

Figure 2.3. *Precipitation and temperature history for OIS-3 derived from a long pollen record at Les Echets in the eastern Massif Central, France (Guiot et al. 1989, fig. 3). Shading indicates confidence limits of the estimates. The original time-scale in ^{14}C years was non-linear; reference points in calendrical time have been added on the right. Location on Figure 2.4.*

warm event — but shorter and cooler than its predecessors — ended 37 ka ago (Fig. 2.1: right). Subsequent conditions were generally close to those of the Late Glacial Maximum (LGM = OIS-2) with lowest temperatures that equalled or were even colder than those of the LGM. Still, much of Scandinavia below an elevation of 1500 m may have remained ice-free until about 25–26 ka ago when the main ice advance towards northern Germany and Denmark gained speed (Arnold *et al.* 2002; Olsen 1997; 1998; Olsen *et al.* 1996; 2001a,b). As regards temperatures, the LGM may thus have begun about 35 cal ka ago (Fig. 2.1: right), but sensitivity tests with climate models show that even large variations in the dimensions of the Fennoscandian ice sheet would have had little impact on the European climate south of the Baltic Sea (Barron & Pollard 2002).

The main warm events of OIS-3 in Europe north of the Pyrenean and Alpine barrier, the Oerel, Glinde, Hengelo, Moershoofd and Denekamp 'interstadials'[2]), are marked by arboreal pollen excursions also seen but more clearly south of the Alps. Other, more ephemeral warm events are recorded only in Mediterranean pollen cores as is evident (Figs. 2.4 & 2.5) when we compare continuous pollen sequences from

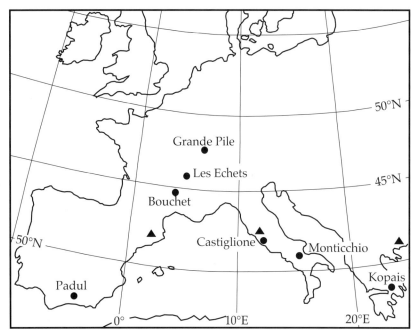

Figure 2.4. *Locations of sites of the long palynological and climatic records that span the interval from OIS-4 through OIS-2 (c. 70–20 ka BP) and are mentioned in the text.*

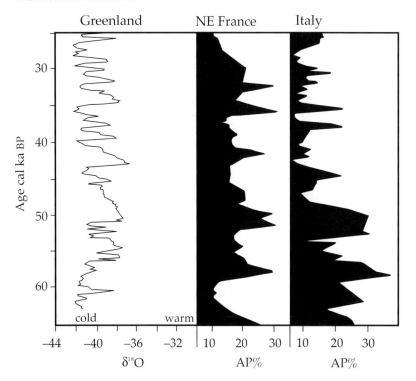

Figure 2.5. *OIS-3 climate oscillations displayed by the δ¹⁸O record of the GRIP ice core (Greenland: GRIP Members 1993) compared with arboreal pollen percentages from Grande Pile (Vosges, France: de Beaulieu & Reille 1992) and Valle di Castiglione (Italy: Follieri* et al. *1988). Calendrical time scale. After Davies* et al. *2000 (Fig. 6). AP% = per cent arboreal pollen. For locations see Figure 2.4.*

France at Grande Pile and the Massif Central (Reille & de Beaulieu 1990) with those of Valle di Castiglione in Italy (Follieri *et al.* 1988).

The Greenland ice core record and the glacial climate in Europe

Two high-resolution climate records from Greenland span the entire Weichselian Glaciation: the GRIP (GRIP Members 1993) and the GISP2 (Grootes *et al.* 1993; Meese *et al.* 1997; Stuiver & Grootes 2000) ice cores. Together with four other Greenland cores (Johnsen *et al.* 2001), they provide a coherent picture of the last 100,000 years of climate changes as seen from the Greenland ice cap. Ice core oxygen isotope (δ¹⁸O) values reflect the local air temperature; after correction for the constant difference between continental ice sheet δ¹⁸O and oceanic δ¹⁸O (the 'Dole effect'), they constitute a useful proxy for the local air temperature history.

Detailed studies of dated high-resolution ocean sediment cores (e.g. Appenzeller *et al.* 1998; Bond *et al.* 1993; Grootes *et al.* 1993; Jöris & Weninger 1998) confirm the continuity of Greenland D/O climate oscillations across the North Atlantic as far south as 38°N latitude (de Abreu 2000; Roucoux *et al.* 2001; Sánchez-Goñi *et al.* 2000). In addition, six major ice-rafted deposits dispersed all over the North Atlantic Ocean between OIS-5a and the end of OIS-2 mark collapses of the North American — and at times also the Fennoscandian ice sheet — which generated vast fleets of icebergs reaching across the Atlantic to the east and southeast of North America, and as far as the coast of Portugal. These so-called Heinrich events are key elements in the detailed correlation of the Greenland D/O climate sequence with North Atlantic sediment records (Bond *et al.* 1992; MacAyeal 1993; Roucoux *et al.* 2001; Sánchez-Goñi *et al.* 2000;

2002; Völker *et al.* 1998).

Not surprisingly, the Greenland ice cores (Johnson *et al.* 2001) have become a standard for the chronology of the European climate during the Weichselian Glaciation. For different intervals the time-scales of the various cores differ in some ways, but within the limits of confidence of our climatic and archaeological data those differences are of minor importance. For the interval of our main interest, from about 65 to 20 ka BP, the GISP2 core has a near-annual resolution and a very good age model (Meese *et al.* 1997) allowing us to use it as our climate age model.

Before we do so, however, we must consider whether the broad features of the Greenland climate record adequately parallel European climate changes between OIS-4 and OIS-2. Detailed correlations have indeed been achieved for Iberia, Italy and southern France, i.e. for the Atlantic and Mediterranean maritime regions; the same is likely to be true for northwestern Europe, but long pollen cores are rare there. In central, eastern and southeastern Europe sparse records, mainly loess deposits and accumulations of faunal and macro-plant remains (Musil 2003) suggest significant climate changes on a multi-millennial time-scale, but the long high-resolution records needed for direct correlations on appropriate time-scales like those available in the maritime zone are lacking. Moreover, too little is known about interactions between major regional climate systems or across the gradient between maritime and continental climate zones to suggest on theoretical grounds alone that the high-resolution climate changes displayed in Greenland ice cores are capable of transmission beyond the zone of North Atlantic influence.

A pre-Stage 3 Project summary of the glacial climate changes in Europe contained in a synthesis by van Andel & Tzedakis (1996) was derived from data obtained in the 1980s and early 1990s. The summary showed that brief but strong climate changes similar to the Greenland D/O oscillations marked western European and Mediterranean regions. The chronology of the long pollen records on which this synthesis was built, however, had a time resolution insufficient to support claims of synchronicity for millennial-scale warm D/O events in Greenland and the North Atlantic with the arboreal pollen peaks of Europe and the Mediterranean. Furthermore, the long cores available north of the Pyrenean-Alpine-Carpathian mountain ranges came all from the maritime climate zone west of the 10°E meridian and did not represent the continental climate of central and eastern Europe (Barron *et al.* 2003).

What is the state of our knowledge today? The most important step forward has probably been the publication of the cores from volcanic Lago Grande di Monticchio in central Italy (Watts *et al.* 1996; Allen *et al.* 1999; 2000). The Monticchio record (Fig. 2.6) consists of long sections where the chronology is based on counts of thin, annually deposited strata (varves: Zolitschka 1998) and dates of volcanic ash beds pin the chronological record down. The Monticchio results are for now Europe's only completely dated, high-resolution pollen record for the entire last glacial cycle. Chronologically it is comparable with the Greenland ice cores and, spanning more than 100,000 years, it enables us to match the Weichselian climate of Greenland with the climate of southern Europe in terms of botanical responses to glacial climate changes (Allen *et al.* 2000; Allen & Huntley 2000). The correspondence between the Monticchio arboreal pollen stratigraphy and broad features of the GISP2 glacial climate record (Fig. 2.6) is remarkably good as the climate evolved from the pre-70 ka warm phase through the cold OIS-4 glacial maximum and subsequent long mild phase to its slow decline between 40 and 25 ka BP to the final glacial maximum.

Two recently published high-resolution marine cores from the Portuguese continental margin (MD95-20390: Roucoux *et al.* 2001; MD95-2042: Sánchez-Goñi *et al.* 2000) support the idea of a series of climate changes that simultaneously affected the northern hemisphere from Greenland and the North Atlantic Ocean to maritime Europe. A core from the Alboran Sea in the western Mediterranean (MD95-2043: Cacho *et al.* 1999; Sánchez-Goñi *et al.* 2002) carries the same message. Another core (MD95-20390), located 180 km west of the Portuguese coast at about 40°N, has a rapid sedimentation rate and a well-preserved high-resolution pollen record only slightly reduced in diversity by long-distance river and sea transport (Roucoux *et al.* 2001). This pollen record closely parallels the vegetation history on the adjacent land at Padul in southern Spain (Pons & Reille 1988). Its marine oxygen isotope record (Fig. 2.7) reasonably matches the GRIP ice core in Greenland. The same is true for other cores (MD95-2042 & MD95-2043) along the Portuguese margin. As at Monticchio, the correspondence between arboreal pollen fluctuations and major climatic features is good (Allen *et al.* 2000, fig. 7, table 5) and even some millennial scale secondary events — though by no means all — can be identified in all records with boundary discrepancies of a few thousand years.

Clearly, the Monticchio record, with its unique

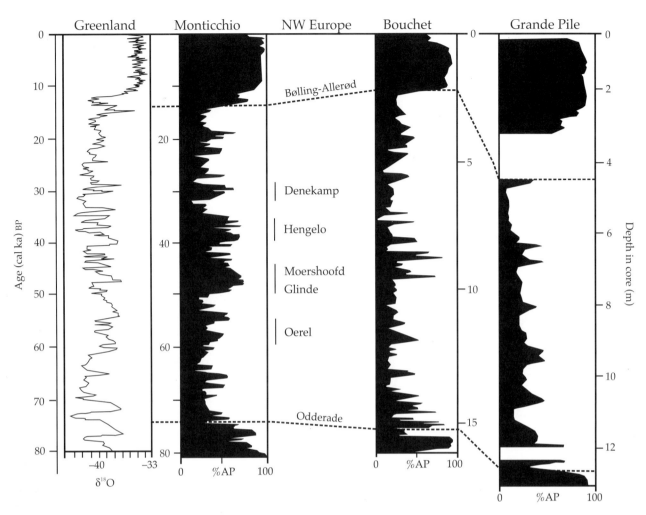

Figure 2.6. *The GISP2 Greenland ice-core record (Meese* et al. *1997) and a pollen record from Italy (Lago Grande di Monticchio in Italy: Allen* et al. *2000; Allen & Huntley 2000), both with calendrical time scales. They are compared here with pollen records from Lac du Bouchet in the southern Massif Central, France (Reille & de Beaulieu 1990) and Grande Pile in the Vosges, France (Woillard 1978, redrawn by Allen & Huntley (2000, fig. 3) and plotted on a depth-in-hole scale). Centre panel: the main Stage 3 interstadials of northern Europe (after Allen & Huntley 2000; Behre 1989). AP% = per cent arboreal pollen. Locations on Figure 2.4.*

high-resolution chronology and rich flora, together with the key land–sea correlation bridges of the cores mentioned above, furnishes solid evidence that first-order multi-millennial features of the glacial history of southern Europe and the Mediterranean are linked to first-order climate changes generated in the North Atlantic. Some millennial-scale climate features can also be identified in cores from elsewhere in the Mediterranean and southern Europe (Fig. 2.6) such as Lac du Bouchet in southern France and the Grande Pile in the Vosges (Allen & Huntley 2000, fig. 3), Valle de Castiglione in Italy (Follieri *et al.* 1988), Padul in southern Iberia (Pons & Reille 1988) and Kopaïs in Greece (Tzedakis 1999).

If we consider individual D/O oscillations at

the centennial to millennial time-scale, however, visual correlation begins to fail (Allen & Huntley 2000). Since Monticchio and the IMAGES marine cores of Cacho *et al.* (1999) and Roucoux *et al.* (2001) are at this time the only records with a robust chronology, their correlation with events recorded at other sites in the Mediterranean and southwestern Europe rests only on visual inspection with at best spottily dated pollen-stratigraphic records. Consequently, all correlations of millennial or shorter time-scales in western and Mediterranean Europe are in essence hypothetical as are all assumptions of synchronicity (Allen & Huntley 2000).

When we compare the Mediterranean climate history with that from Europe north of the Pyrenean-

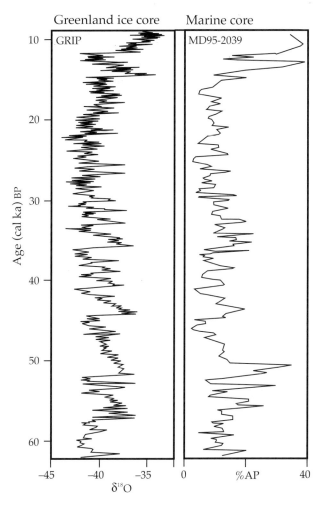

Figure 2.7. *Arboreal pollen record of a late Quaternary high-resolution marine sediment core taken at latitude 40°34'N, c. 180 km off the Portuguese Atlantic coast, compared with the δ¹⁸O climatic record of the GRIP ice core (after Roucoux* et al. *2001, fig. 2). AP% = per cent arboreal pollen.*

correlation, the rates of change, duration and synchronicity of climatic events can not be determined. In the Mediterranean region itself recovery after cold events is swift, probably due to the presence of refugia in the vicinity (Allen *et al.* 2000; Allen & Huntley 2000; Roucoux *et al.* 2001; Tzedakis 1999).

A wholly independent source of climate information is furnished by stratigraphic sections of the Weichselian loess that mantles much of Europe including the Russian plain. The sections show cycles of deposition of fresh loess produced by the barren lands and high winds of cold stadials, alternating with milder climates, landscape stabilization and soil formation during interstadials. Palaeosols also mark the interstadial glacio-fluviatile deposits of Stage 3 age in Norway (Olsen 1997; 1998; Olsen *et al.* 1996; 2001a,b). And sequences of palaeosols in northwestern France (Antoine *et al.* 1999, fig. 5) and the Rhine region (Schirmer 2000, fig. 10) confirm that the pleniglacial climate of western Europe was subject to periods of warming that stopped loess accumulation and lasted long enough for palaeosols to form. Unfortunately, so far the loess data fall short of yielding a chronology capable matching the Greenland D/O record.

The loess chronology depends almost entirely on sediment dates with luminescence methods (e.g. Frechen *et al.* 1999), which are potentially powerful, but dated sequences of sufficient length to match the ice-core chronology remain elusive and the dating technique itself has not yet achieved the precision needed to compare loess cycles with even the multi-millennial ice-core record. Nonetheless, the loess record, unambiguous unlike pollen records in the matter of climate oscillations of secondary magnitude in Europe, will be increasingly useful, especially in the continental regions.

Adopting a type record for 60,000 years of climate change in the North Atlantic, Europe and the Mediterranean

The long-sustained climate deterioration towards the Late Glacial Maximum displayed by Greenland ice cores is clearly reflected in the European climate north of the Pyrenean-Alpine-Carpathian mountain ranges and the northern Mediterranean region. The stratigraphic comparison between northern Europe and the northern Mediterranean, however, rest mainly on visual correlation. Consequently, all that can be safely said at this time is that the climate of western Europe appears to have oscillated between colder and milder events on a roughly millennial

Alpine ranges (Fig. 2.7) to correlate the time-based Monticchio record with the depth-based stratigraphy of Lac du Bouchet and Grande Pile, we find that the numbers, amplitudes and durations of the warm events are much attenuated in the North. Lac du Bouchet, the southernmost site, shows some arboreal pollen peaks, but at Grande Pile, several degrees farther north, only a few peaks can be seen and some of those may be compounded of shorter events. Although it seems plausible that the arboreal peaks of the northern pollen records mark some major events portrayed at Monticchio, in the absence of robust dating it is not possible to say which of the many features displayed in the south are expressed in northern Europe too. Also, given the limits of visual

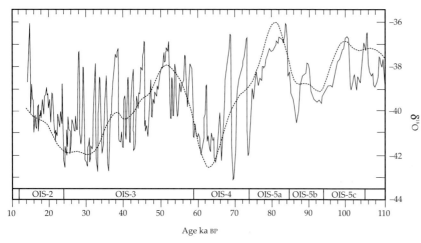

Figure 2.8. *First-order climate changes between the end of the penultimate interglacial (OIS-5d) and the onset of the Holocene (OIS-1) as reflected in the δ¹⁸O climate record from the GISP2 ice core (Meese et al. 1997). The heavy dashed line depicts the general trend of the climate changes for the interval; it is the first reconstructed component of a singular spectral analysis (SSA; Vautard & Ghil 1989; Vautard et al. 1992; Yiou et al. 1996). The curve captures a large percentage of the total variance and provides a much better estimate of the trend than any linear curve (courtesy of H. Pälike).*

time-scale similar to that of Greenland and the North Atlantic. When, how fast and especially where the European climate changed is difficult to say, however. Furthermore, because of the marked climatic changes from maritime to continental and arctic to subtropical conditions even if a single large climate system prevailed across the Atlantic and all of Europe, those gradients might have greatly modified the amplitude and duration of its oscillatory changes and their expression in terms of seasonal temperature ranges, precipitation and many other variables.

The point is illustrated in Chapter 10 where Musil (2003) describes drastic regional changes in the mammalian fauna of east central Europe and the Balkans during the period 26–20 ka BP. Those changes brought to southern parts of the region an early northward movement of thermophile species and the disappearance of truly arctic species. The early warming that this requires ill fits the North Atlantic maritime scheme (Svoboda *et al.* 1996, 135–6).

Today, other major regional climate systems such as the Northwest Indian Ocean and Atlantic Ocean Monsoons are known to interact with the North Atlantic System at the margins of the Atlantic-European-Mediterranean region to a degree that sometimes varies in a major way. It is obvious that this must have happened in the past as well (Roberts & Wright 1993), but for the time being neither the

available climatic data nor our basic grasp of the dynamics of the great global climatic systems suffice to assess those interactions effectively (van Andel 2000).

Over the long-term, the (δ¹⁸O) air temperature trend of the Greenland ice cores (Fig. 2.8) agrees reasonably well with pollen-based climate records from many parts of Europe, but it would not be surprising if the main ice core D/O oscillations would have expressed themselves quite differently in continental pollen and loess records (Sánchez-Goñi *et al.* 2000; 2002).

These matters are unlikely to be resolved soon. The Palaeolithic chronology of the Stage 3 Project has a resolution of a few thousand years (van Andel *et al.* 2003), and so for the moment and for our purpose it suffices to have shown that multi-millennial-scale climate changes (Fig. 2.8) did occur in many parts of Europe at approximately the same times as they do in ice cores. Of course, the nature of the events may have varied between different regions.

Therefore we have adopted the GISP2 ice core (Fig. 2.8) as an acceptable qualitative representation of the medium-term (heavy dashed line) and some short-term (D/O events) climate changes across Europe during the Weichselian Glaciation. The standard should be useful to compare the human history with snapshot climate simulations generated for medium-term warm and cold climate events (phases[3]) superimposed on the overall trend (Barron *et al.* 2003).

Acknowledgements

The subject of this chapter is complex and controversial and I am in debt to Nick Shackleton, Isabel Cacho and several others who guided me gently through this booby-trapped field. The illustrations come from a wide range of sources and I am most grateful to Sharon Copan who turned them into the present neat, pleasant and uniform style.

Notes

1. Calendar dates (ice, TL, OSL, ESR and U-Th methods) and calibrated ¹⁴C dates (van Andel *et al.* 2003) are used throughout. The phrase 'xxxx ka BP' refers by

definition to calibrated ^{14}C and other calendrical dates. Calibration was by CalPal.1998 (Jöris & Weniger 2000; see also van Andel *et al.* 2003). The raw versions of the ^{14}C dates can be found in the Stage 3 Project's chrono-archaeological data base at its website or in references cited there. The Stage 3 Web site can be consulted at: http://www.esc.esc.cam.ac.uk/oistage3/Details/Homepage/html.

2. In recent years the terms stadial and interstadial, originally narrowly defined, have been used for stratigraphic events on quite different time-scales. Not wishing to take a stand on the proper use of the terms, we shall simply refer to the cold/dry and warm/moist intervals within OIS 3 as *events*.

3. The term *climate zone* will be used in the geographic sense, while units of climate change will be labelled *climate phases*.

References

Allen, J.R.M. & B. Huntley, 2000. Weichselian palynological records from southern Europe: correlation and chronology. *Quaternary International* 73–4, 111–26.

Allen, J.R.M., U. Brandt, A. Brauer, H.-W. Hubberten, B. Huntley, J. Keller, M. Kraml, A. Mackensen, J. Mingram, J.F. Negendank, N.R. Nowackzyk, H. Oberhänsli, W. Watts, S. Wulf & B. Zolitschka, 1999. Rapid environmental changes in southern Europe during the last glacial period. *Nature* 400, 740–43.

Allen, J.R.M., W.A. Watts & B. Huntley, 2000. Weichselian palynostratigraphy, palaeovegetation and palaeoenvironment: the record from Lago Grande di Monticchio. *Quaternary International* 73/74, 91–110.

Antoine, P., D.D. Rousseau, J.P. Lautridou & C. Hatté. 1999. Last interglacial-glacial climatic cycle in loess-palaeosol successions of north-western France. *Boreas* 28, 551–63.

Appenzeller, C., T.F. Stocker & M. Anklin, 1998. North Atlantic oscillation dynamics recorded in Greenland ice cores. *Science* 282, 446–9.

Arnold, A.S., T.H. van Andel & V. Valen, 2002. Extent and dynamics of the Scandinavian ice sheet during Oxygen Isotope Stage 3 (60,000–25,000 yr BP). *Quaternary Research* 57, 38–48.

Barron, E.J. & D. Pollard, 2002. High-resolution climate simulations of Oxygen Isotope Stage 3 in Europe. *Quaternary Research* 58, 296–309.

Barron, E., T.H. van Andel & D. Pollard, 2003. Glacial environments II: reconstructing the climate of Europe in the Last Glaciation, in *Neanderthals and Modern Humans in the European Landscape during the Last Glaciation*, Chapter 5, eds. T.H. van Andel & W. Davies. (McDonald Institute Monographs.) Cambridge: McDonald Institute for Archaeological Research, 57–78.

Behre, K.-E. 1989. Biostratigraphy of the last glacial period in Europe. *Quaternary Science Reviews* 8, 25–44.

Bond, G.C., H. Heinrich, W.S. Broecker, L. Labeyrie, J. McManus, J.T. Andrews, S. Huon, R. Jantschik, S. Clasen, C. Simet, K. Tedesco, M. Klas, G. Bonani & S. Ivy, 1992. Evidence for massive discharge of icebergs into the North Atlantic Ocean during the last glacial period. *Nature* 360, 245–9.

Bond, G.C., W.D. Broecker, S. Johnsen, J. McManus, L. Labeyrie, J. Jouzel & G. Bonani, 1993. Correlations between climate records from North Atlantic sediments and Greenland ice. *Nature* 365, 143–7.

Cacho, I., J.O. Grimalt, C. Pelejero, M. Canals, F.J. Sierro, J.A. Flores & N.J. Shackleton, 1999. Dansgaard-Oeschger and Heinrich event imprints in Alboran Sea palaeotemperatures. *Paleoceanography* 14, 698–705.

Coope, G.R., 2002. Changes in the thermal climate in North-western Europe during marine oxygen isotope stage 3, estimated from fossil insect assemblages. *Quaternary Research* 57, 401–8.

Dansgaard, W., S.J. Johnsen, H.B. Clausen, D. Dahl-Jensen, N.S. Gundestrup, C.U. Hammer, C.S. Hvidberg, J.P. Steffensen, H. Sveinbjörnsdottir, J. Jouzel & G. Bond, 1993. Evidence for general instability of past climate from a 250-kyr ice-core record. *Nature* 364, 218–20.

Davies, W., J. Stewart & T.H. van Andel, 2000. Neandertal landscapes — a preview, in *Neanderthals on the Edge*, eds. C. Stringer, R.N.E. Barton & J.C. Finlayson. Oxford: Oxbow Books, 1–8.

de Abreu, L., 2000. High-resolution Palaeoceanography off Portugal during the Last Two Glacial Cycles. Unpublished PhD dissertation, University of Cambridge, Cambridge.

de Beaulieu, J.L. & M. Reille, 1992. The climatic cycle at Grande Pile (Vosges, France): a new pollen profile. *Quaternary Science Reviews* 11, 431–8.

Donner, J., 1995. *The Quaternary History of Scandinavia.* Cambridge: Cambridge University Press.

Follieri, M., D. Magri & L. Sadori, 1988. 250,000-year pollen record from Valle di Castiglione (Roma). *Pollen et Spores* 30, 329–56.

Frechen, M., A. Zander, V. Cilek & V. Lozek, 1999. Loess chronology of the last Interglacial/Glacial cycle in Bohemia and Moravia. *Quaternary Science Reviews* 18, 1467–94.

Gamble, C.S., 1986. *The Palaeolithic Settlement of Europe.* Cambridge: Cambridge University Press.

Gamble, C.S., 1999. *The Palaeolithic Societies of Europe.* Cambridge: Cambridge University Press.

GRIP (Greenland Ice-core Project) Members, 1993. Climate instability during the last interglacial period recorded in the GRIP ice core. *Nature* 364, 203–7.

Grootes, P.M., M. Stuiver, J.W. White, S. Johnsen & J. Jouzel, 1993. Comparison of oxygen isotope records from the GISP2 and GRIP Greenland ice cores. *Nature* 366, 552–4.

Guiot, J A., J.L. Pons, J.L. de Beaulieu & M. Reille, 1989. A 140,000-year continental climate reconstruction from two European pollen records. *Nature* 338, 309–13.

Guiot, J., J.L. de Beaulieu, R. Cheddadi, F. David, P. Ponel, & M. Reille, 1993. The climate in western Europe during the last glacial/interglacial cycle derived from pollen and insect remains. *Palaeogeography, Palaeo-climatology, Palaeoecology* 103, 73–94.

Johnsen, S.J., D. Dahl-Jensen, N. Gundestrup, J.P. Steffensen, H.B. Clausen, H. Miller, V. Masson-Delmotte, A.E. Sveinbjörnsdottir & J. White, 2001. Oxygen isotope and paleotemperature records from six Greenland ice-core stations, Camp Century, Dye-3, GRIP, GISP2, Renland and North GRIP. *Journal of Quaternary Science* 16, 299–308.

Jöris, O. & B. Weninger, 1998. Extension of the 14-C calibration curve to *c.* 40,000 cal BC by synchronising Greenland $^{18}O/^{16}O$ ice-core records and North Atlantic Foraminifera profiles: a comparison with U/Th coral data. *Radiocarbon* 40, 495–504.

Jöris, O. & B. Weninger, 2000. Calendric age-conversion of glacial radiocarbon dates at the transition from the Middle to the Upper Palaeolithic in Europe. *Bulletin de la Société Préhistorique Luxembourgeoise* 18, 43–55.

Lambeck, K. & J. Chappell, 2001. Sea level changes through the last glacial cycle. *Science* 292, 679–86.

MacAyeal, D.R., 1993. Growth/purge oscillations of the Laurentide icesheet as a cause of the North Atlantic Heinrich events. *Paleoceanography* 8, 775–84.

McManus, J.F., G.C. Bond, W.S. Broecker, S. Johnsen, L. Labeyrie & S. Higgins, 1994. High-resolution climate records from the North Atlantic during the last interglacial. *Nature* 371, 326–9.

Mangerud, J., 1991. The last ice age in Scandinavia. *Striae* 34, 15–30.

Martinson, D., N.G. Pisias, J.D. Hays, J. Imbrie, T.C. Moore Jr. & N.J. Shackleton, 1987. Age dating and the orbital theory of the Ice Ages: development of a high-resolution 0–300,000-year chronostratigraphy. *Quaternary Research* 27, 1–29.

Meese, D.A., A.J. Gow, R.B. Alley, G.A. Zielinsky, P.M. Grootes, M. Ram, K.C. Taylor, P.A. Mayewski & J.F. Bolzan, 1997. The Greenland Ice Sheet Project 2 depth-age scale: methods and results. *Journal of Geophysical Research* 102, 26,411–23.

Mellars, P., 1996. *The Neanderthal Legacy: an Archaeological Perspective from Western Europe.* Princeton (NJ): Princeton University Press.

Mellars, P. & C. Stringer (eds.), 1989. *The Human Revolution.* Edinburgh: Edinburgh University Press.

Musil, R., 2003. The Middle and Upper Palaeolithic game suite in central and southeastern Europe, in *Neanderthals and Modern Humans in the European Landscape during the Last Glaciation,* Chapter 10, eds. T.H. van Andel & W. Davies. (McDonald Institute Monographs.) Cambridge: McDonald Institute for Archaeological Research, 167–90.

Olsen, L., 1997. Rapid shifts in glacial extension characterise a new conceptual model for glacial variations during the Mid and Late Weichselian in Norway. *Norges Geologiske Undersøgelse Bulletin* 433, 54–5.

Olsen, L., 1998. Pleistocene paleosols in Norway: implications for past climate and glacial erosion. *Catena* 34, 75–103.

Olsen, L., V. Mejdahl, & S.F. Selvik, 1996. Middle and late Pleistocene stratigraphy, chronology and glacial history in Finnmark, North Norway. *Norges Geologisk Undersøgelse Bulletin* 429, 93–118.

Olsen, L., K. van der Borg, B. Bergstrøm, H. Sveian, S.-E. Lauritzen & G. Hansen, 2001a. AMS radiocarbon dating of glacigenic sediments with low organic content — an important tool for reconstructing the history of glacial variations in Norway. *Norsk Geologisk Tidsskrift* 81, 59–92.

Olsen, L., H. Sveian & B. Bergstrøm, 2001b. Rapid adjustments of the western part of the Scandinavian ice-sheet during the mid- and Late Weichselian — a new model. *Norsk Geologisk Tidsskrift* 81, 93–118.

Pollard, D. & E.J. Barron, 2003. Causes of model-data discrepancies in European climate during oxygen isotope stage 3 with insights from the Last Glacial Maximum. *Quaternary Research* 59, 108–13.

Pons, A. & M. Reille, 1988. The Holocene and upper Pleistocene pollen record from Padul near Granada, Spain: a new study. *Palaeogeography, Palaeoclimatology, Palaeoecology* 66, 243–63.

Porter, S.C., 1989. Some geological implications of average Quaternary glacial conditions. *Quaternary Research* 32, 245–61.

Reille, M. & J.-L. de Beaulieu, 1990. Pollen analysis of a long upper Pleistocene continental sequence in a Velay Maar (Massif Central, France). *Palaeogeography, Palaeoclimatology, Palaeoecology* 80, 35–48.

Roberts, N. & H.E. Wright Jr, 1993. Vegetational, lake-level and climatic history of the Near East and Southwest Asia, in *Global Climates Since the Last Glacial Maximum,* eds. H.E. Wright Jr, J.E. Kutzbach, T. Webb III, W.F. Ruddiman, F.A. Street-Perrott & P.J. Bartlein. Minneapolis (MN): University of Minneapolis Press, 194–220.

Roebroeks, W., M. Mussi, J. Svoboda & K. Fennema (eds.), 1999. *Hunters of the Golden Age: the Mid Upper Palaeolithic of Eurasia 30,000–20,000 BP.* Leiden: University of Leiden.

Roucoux, K.H., N.J. Shackleton & L. de Abreu, 2001. Combined marine proxy and pollen analyses reveal rapid Iberian vegetation response to North Atlantic millennial-scale climate oscillation. *Quaternary Research* 56, 128–32.

Sánchez-Goñi, M.F., J.-L. Turon, F. Eynaud & S. Gendreau, 2000. European climatic response to millennial-scale changes in the atmosphere-ocean system during the last glacial period. *Quaternary Research* 54, 394–403.

Sánchez-Goñi, M.F., I. Cacho, J.-L. Turon, J. Guiot, F.J. Sierro, J.-P. Peypouquet, J.O. Grimalt & N.J. Shackleton, 2002. Synchronicity between marine and terrestrial responses to millennial-scale climatic variability during the last glacial period in the Mediterranean region. *Climate Dynamics* 19, 95–105.

Schirmer, W., 2000. Rhein loess, ice cores and deep-sea cores during MIS 2-5. *Zeitschrift der deutsche geologische Gesellschaft* 151, 309–32.

Stringer, C. & C. Gamble, 1993. *In Search of the Neanderthals.* London: Thames & Hudson.

Stuiver, M. & P. Grootes, 2000. GISP2 oxygen isotope ratios. *Quaternary Research* 53, 277–84.

Svoboda, J., V. Lozek & E. Vlcek, 1996. *Hunters between East and West: the Palaeolithic of Moravia.* New York (NY): Plenum Press.

Sykes, B., 2001. *The Seven Daughters of Eve.* London: Bantam Press.

Tzedakis, P.C., 1999. The last climatic cycle at Kopaïs, Greece. *Journal of the Geological Society of London* 155, 425–34.

van Andel, T.H., 2000. Where received wisdom fails: the mid-Palaeolithic and Early Neolithic climates, in *Archaeogenetics: DNA and the Population Prehistory of Europe,* eds. C. Renfrew & K. Boyle. (McDonald Institute Monographs.) Cambridge: McDonald Institute for Archaeological Research, 31–9.

van Andel, T.H. & P.C. Tzedakis, 1996. Palaeolithic landscapes of Europe and Environs, 150,000–25,000 years ago. *Quaternary Science Reviews* 15, 481–500.

van Andel, T.H. & P.C. Tzedakis, 1998. Priority and opportunity: Reconstructing the European Middle Palaeolithic climate and landscape, in *Science in Archaeology: an Agenda for the Future,* ed. J. Bayley. London: English Heritage, 37–46.

van Andel, T.H., W. Davies, B. Weninger & O. Jöris, 2003. Archaeological dates as proxies for the spatial and temporal human presence in Europe: a discourse on the method, in *Neanderthals and Modern Humans in the European Landscape during the Last Glaciation,* Chapter 3, eds. T.H. van Andel & W. Davies. (McDonald Institute Monographs.) Cambridge: McDonald Institute for Archaeological Research, 21–30.

Vautard, R. & M. Ghil, 1989. Singular spectrum analysis in non-linear dynamics with applications to paleoclimatic time series. *Physica D* 35, 395–424.

Vautard, R., P. Yiou & M. Ghil, 1992. Singular spectrum analysis; a toolkit for short, noisy chaotic signals. *Physica D* 58, 95–126.

Völker, A.H., M. Sarnthcin, P.M. Grootes, H. Erlenkeuser, C. Laj, A. Mazaud, M.-J. Nadeau & M. Schleicher, 1998. Correlation of marine ^{14}C ages from the Nordic Seas with the GISP2 isotope record: Implications for radiocarbon calibration beyond 25 ka BP. *Radiocarbon* 40, 517–34.

Watts, W.A., J.R.M. Allen & B. Huntley, 1996. Vegetation history and palaeoclimate of the last glacial period at Lago Grande di Monticchio, southern Italy. *Quaternary Science Reviews* 15, 133–53.

Woillard, G.M., 1978. Grande Pile peat bog: a continuous pollen record for the last 140,000 years. *Quaternary Research* 9, 1–21.

Yiou, P., E. Baert & M.F. Loutré, 1996. Spectral analysis of climate data. *Surveys in Geophysics* 17, 619–63.

Zolitschka, B., 1998. Paläoklimatische Bedeutung laminierter Sediment — Holzmaar (Eiffel, Deutschland), Lake C (Northwest Territorien, Kanada) und Lago Grande di Monticchio (Basilikate, Italien). *Relief, Böden, Paläoklima* 13, 1–176.

Chapter 3

Archaeological Dates as Proxies for the Spatial and Temporal Human Presence in Europe: a Discourse on the Method

Tjeerd H. van Andel, William Davies, Bernhard Weninger & Olaf Jöris

Linking the human history of the Last Glaciation to its changing climate

The Stage 3 Project set itself two principal and sequential tasks: 1) to reconstruct the changing climate and environments of Europe for the last (Weichselian) Glaciation; and 2) to analyze the human history of the same period. As proxy for the history of the glacial climate the Greenland ice core GISP2 (Fig. 2.8) was adopted. The second task, formulated *ab initio* as a question, *viz.* 'Was human history during the last glacial period influenced by climatic and environmental changes' required a chronological record of the human movements across Europe throughout the glacial period. For this purposes (van Andel 2002) a chrono-archaeological data base)[1] was compiled by William Davies. To be able to use the two time-series in combination the following conditions must be satisfied:

- That the Greenland ice core reflects at least qualitatively the main features of the European climate changes from 70 to 20 ka BP[2]) and that at least some Greenland Dansgaard/Oeschger events have parallels amongst short-term European climatic events.
- That the chronology of both climatic and human histories is based upon a common time unit. The Greenland ice core records the climate in calendar years and therefore the dates (which constitute the bulk of the archaeological record, i.e. 87 per cent) have been converted into calendar years.
- That the sites of the chrono-archaeological data base reflect the European human distribution in time and space.

Our first assumption has been examined in Chapter 2 (van Andel 2003) where we have shown that broad features of the Greenland glacial climate can be traced across the North Atlantic into western and Mediterranean Europe to justify its use as a qualitatively proportional climate guide (Fig. 2.8), although the assumption is somewhat questionable in central and eastern Europe.

The second assumption rules out the use of the ^{14}C time-scale as a base for comparisons between climate changes and human affairs because of the large variations in the production of ^{14}C in the upper atmosphere during the glacial period (Mazaud *et al.* 1991; Tric *et al.* 1992; Thouvény *et al.* 1993; Laj *et al.* 1996). Other dating methods, such thermal (TL) and optical luminescence (OSL), electron spin resonance (ESR) and the uranium/thorium series (U/Th) do yield calendrical dates. Although they only constitute 13 per cent of the total number of dates, for the time prior to *c.* 45 cal ka BP this group forms by far the largest part of the chrono-archaeological data base. The conversion of ^{14}C dates to calendar years is discussed in the third section of this chapter.

Archaeological dates as proxies for the human presence in time and space

Of the three conditions stated above the third one does require full discussion here. The use of dates as proxies for human geography was pioneered by Ammerman & Cavalli Sforza (1971; 1984) while developing models for the immigration of early Anatolian farmers into Europe. On a smaller scale ^{14}C dates were similarly used by Rick (1987) in Peru, Holdaway & Porch (1995) in Tasmania, Housley *et al.* (1997) for the recolonization of north-central Europe and Bocquet-Appel & Demars (2000) in demographic studies of Middle Palaeolithic Europe.

This use of the chrono-archaeological data base differs both in principle and in practice from the archaeological tradition which gives pride of place to archaeological sites as locations of human occupation and activity and as the resting place of their artefacts. In that context archaeological dates tend to be relegated to the service of stratigraphy. The emphasis is on individual dates in well-established stratigraphies of known precision and an assumed high estimated accuracy.

Our approach dispenses with a rigorous *a priori* assessment of each individual date, using instead the aggregate set of dates to generate potentially meaningful temporal and spatial patterns of human presence. Non-random site distributions that seem interesting are adopted as working hypotheses worthy of further evaluation. Eventually, when sufficiently rigorous questions or predictions have been formulated, a more quantitative analysis may be attempted.

This approach may seem irresponsible given the common view that the scientific process *par excellence*, the trademark of *real science*, consists of constructing hypotheses that yield predictions testable by means of experiments. This **prediction-driven** approach contrasts with **question-driven** *historical science* which infers past causes from their consequences. Unable to test its hypotheses by experiment, historical science is seen as sadly speculative and therefore not a science at all (e.g. Gee 1999, 5, 8). Yet geology, stratigraphy, palaeontology, historical geology and other variants of the earth and environmental sciences perforce operate in the historical mode. For explanations they rely on a narrative logic that makes sense of the observations within a broad understanding of a region, event sequence or process rather than an evaluation resting on a framework of predictive general laws secured by experiments (Frodeman 1995).

Experimental science evaluates its predictions in two ways. The first is inductive and accepts a hypothesis when confirmed by a 'reasonable' (whatever reasonable means) number of experiments. This ignores the fact that no matter how many experiments are positive, negative results can never be excluded. The alternative approach accepts that experiments can not confirm predictions but that they can be rejected with even a single failure. At first sight this seems reasonable enough, but in particular in the earth or environmental sciences predictions tend to involve many underlying secondary and tertiary assumptions. The presence (or lack) of any of those may cause even a true prediction *s.s.* to yield false negatives or fail. Absolute confirmation can only be achieved if all possible lower-rank variables

and their interactions have been tested and found to be positive, but this can rarely if ever be achieved.

Although historical hypotheses are not amenable to empirical testing, the common availability of alternatives allows them to be ranked according to their probability, an approach proposed long ago by T.C. Chamberlin (1897) as the 'method of multiple working hypotheses'. Appropriate statistical methods may sometimes be helpful here, but in the early stages of research they rarely repay the effort. Also, statistics being an enemy of serendipity, they may tempt one to ignore apparently unimportant anomalies which in a complex system may well play a key role at some point in time or at an early stage. We state explicitly here that we regard the questions and hypotheses derived from our data too imprecise and the risk of discarding information by premature use of statistical methods too great to justify an early quantitative approach.

Instead, we shall, as many have done before us, rely only on eye and mind before proceeding in more interpretative chapters to compare our speculations with a range of independent palaeo-environmental evidence. Only then shall we feel justified to explore whether a quantitative analysis might be productive and statistical method might fit the properties of our data.

The contrast between experimental and historical science is the result of an asymmetry of causation (Lewis 1991; Cleland 2001). To predict an event we must be confident that no process, even one unrecognized or not preserved, was present to prevent the event from happening. In contrast, when seeking the causes of an observed event such as human migration in the context of climate change, numerous consequences are often evident that may allow us to determine the relative probabilities of a range of inferred causes. In the matter of creating and testing hypotheses, experimental science can not claim universal superiority over historical science.

Besides its heuristic merit, our approach has the advantage of conserving information, because decisions concerning the robustness of our sets of dates and the meaning of their patterns are deferred until they have generated hypotheses that may (or may not) make contextual sense and can be evaluated with independent evidence.

Our use of archaeological dates as proxies for the evaluation of temporo-spatial patterns is unlikely to please everyone because it flies in the face of a widely held view that each date must be tested for its worthiness almost to its destruction before it may be used for any purpose whatsoever by anyone. Indeed, some members of the ^{14}C dating community have recently begun to act as arbiters in this respect, berating users of calibrated ^{14}C dates for their carelessness (or impertinence?) before a consensus has been established by experts regarding the best method of calibration for the interval 25–45 ka BP. Authors of offending works have been warned away from the subject until the time that ^{14}C experts have approved a calibration method and authorized its use (e.g. van der Plicht 1999). A curious example is the patronizing attack, neither thoughtful nor constructive, of Pettitt & Pike (2001) on a statistical study based on ^{14}C dates for the analysis of Neanderthal demography and extinction (Bocquet-Appel & Demars 2000). Bocquet-Appel's method, in some ways analogous to ours, places a heavier burden on the dates than we do because of their statistical approach, but this hardly justifies Pettitt and Pike's high-handed dictate that '. . . chronological reconstructions of the palaeodemography of Neanderthal extinction . . . are premature at best and possibly seriously misleading!' In just one sentence this astonishing view sweeps away a fundamental, time-honoured and widely accepted research principle stating that any approach with the potential of leading to a useful question or tentative explanation may be legitimately employed (van Andel 1998).

In sum, we regard our attempt at clarifying the demise of the Neanderthals and the survival of modern humans in Europe by means of the application of methods of historical science as a complementary procedure rather than one that competes with other broad regional studies. In excellent syntheses Svoboda *et al.* (1996), Mellars (1996), Gamble (1986; 1999) and Roebroeks *et al.* (1999 *passim*) have approached successfully problems similar to ours from other, quite different perspectives. Here we simply offer another way of reviewing vast amounts of existing data available for a large region and over a long time in order to discover what they might tell us.

Converting radiocarbon dates into calendar years: the CalPal ^{14}C Calibration Curve

The need for radiocarbon age-calibration
It is easy to forget that the primary product of radiocarbon dating is not an age but a dimensionless logarithmic ratio (Mook & van der Plicht 1999) scaled to the AD 1950 ^{14}C/^{12}C count rate of an arbitrary beetroot sample known as NBS-oxalic acid. To convert that number into an age we must perform a ^{14}C age calibration if we want a proper calendrical date. If

no calendrically dated sequence suitable for the calibration exists, one is forced to use a simple model, i.e. that the production rate of ^{14}C in the upper atmosphere has been invariant over time. This yields the linear calibration that was used in the early decades of ^{14}C dating to obtain actual dates, although the term calibration was not then in fashion.

The discovery that ^{14}C was produced at rates that widely varied over time forced the introduction of an age calibration known for the Holocene as the 'dendrochronological ^{14}C calibration'. Deprived of this highly successful approach by lack of trees in deposits of the last Glaciation, a more general 'calendric age conversion' has arisen which uses a range of calendrical time-series for calibration. The method is considered to be mathematically cumbersome and indeed the calibration of, for instance, the Stage 3 chrono-archaeological ^{14}C data required a rather complex, non-linear curve-folding analysis, but those basic mathematical and statistical problems were solved long ago from a variety of different logico-mathematical view-points. Nonetheless, the question whether or not to calibrate ^{14}C dates of the last glaciation appears to some (e.g. van der Plicht 1999) to be fraught with uncertainty. The answer to the question, however, is a simple YES; we do actually have no choice, since all time-series displaying climatic changes operate on a calendrical or astronomic time-scale with units of equal length.

An equally compelling reason for ^{14}C calibration is that age-converted ^{14}C data are much more age-precise than uncalibrated ^{14}C ages left to hover as ghostly dimensionless entities above the ^{14}C scale. In fact, contrary to what many assume, it is not possible to quantify the differences between ^{14}C years and calendric years. One might say that the difference between ^{14}C and calendric time-scales ranges from a few centuries in the Holocene to a few millennia in the last glaciation, but in saying this we assume that the ^{14}C years can be subtracted from calendric years, an erroneous operation because it scrambles scaling dimensions. Comparing calendric and conventional radiocarbon years is analogous to subtracting grams from centimetres.

Consequently, only calibrated ^{14}C ages allow us to perform a complete quantitative analysis of the dating errors of the ^{14}C method. Moreover only on the calendric time-scale can we sensibly perform comparisons of radiocarbon dates with the results of other dating methods such as luminescence, electron spin and uranium-series or the climatic history recorded in Greenland ice cores, for example in terms of rates of change.

Table 3.1. *Some of the time-series available in CalPal for comparison in radiocarbon calibrations of sediments of the Last Glaciation.*

CalAge Range BP	Data set	Method	N = Pairs	References
0–11,885	INTCAL98	Tree-rings	N = 1156 (1156)	Stuiver *et al.* 1998
11,887–12,927	Lake Gosciar	Lake varves	N = 48 (107)	Goslar *et al.* 2000
11,906–12,852	Lake Perespilno	Lake varves	N = 15 (26)	Goslar *et al.* 2000
11,900–12,440	Vanuatu	U/Th-coral	N = 34 (39)	Burr *et al.* 1998
11,885–14,659	Cariaco	Marine varves	N = 218 (349)	Hughen *et al.* 2000
11,887–37,404	Lake Suigetsu	Lake varves	N = 214 (270)	Kitagawa & van der Plicht 2000
12,544–30,173	V23-81	GISP2	N = 20 (21)	Bond *et al.* 1993; Jöris & Weninger 1998
12,650–30,505	DSDP-609	GISP2	N = 17 (17)	Bond *et al.* 1993; Jöris & Weninger 1998
15,720–53,261	PS2644	GISP2	N = 90 (91)	Völker *et al.* 2000
12,260–30,230	Barbados	U/Th-coral	N = 12 (20)	Bard *et al.* 1998
11,930–13,850	Tahiti	U/Th-coral	N = 14 (27)	Bard *et al.* 1998
15,585–23,510	Mururoa	U/Th-coral	N = 4 (4)	Bard *et al.* 1998
41,100–	New Guinea	U/Th-coral	N = 1 (1)	Bard *et al.* 1998; Edwards *et al.* 1993

Calibration of older ^{14}C ages is thus not intended to identify differences between uncalibrated and calibrated ^{14}C ages since such differences have no physical meaning. Instead, it is an attempt to identify and minimize differences between calibrated ^{14}C ages and other time-scales such as the Greenland ice-core scale. In a typical Quaternary application, for instance, the initial age-scale will have been constructed with reference to Greenland and Antarctic ice cores in order to provide proxy data that are of climatological interest to Palaeolithic archaeologists. Differences between the time-scales of climate proxies used for calibration can be displayed by simple but effective graphic measures, but as the number of sets of climate data grows those methods become ever more cumbersome and inconvenient for updating age-models. An approach that integrates various age-scales as does the CalPal programme seems more appropriate; it is also more convenient for the use of the ^{14}C method.

The palaeoclimatological framework of ^{14}C-age calibration

Radiocarbon dates can be calibrated by means of any calendrical data set of appropriate length and resolution from pharaonic kings' lists to tree-ring counts. An argument used against ^{14}C age conversion on the 10–45 ka BP time range is that supporting data of sufficient precision and temporal resolution do not currently exist. This is simply not true; Table 3.1 lists fourteen sequences well-suited for the construction of a 10–45 ka ^{14}C calibration curve. Some of those record climate changes directly; others can be correlated to climate records, for example by using the chain GISP2 ice core ⇨ North Atlantic ocean sediments (Völker *et al.* 1998) ⇨ Lago di Monticchio core (Allen & Huntley 2000) and several variants (Chapter 2, van Andel 2003).

Seen from this wider perspective, what renders the inclusion of archaeological data in the chain to some degree problematic — and hence interesting — is the fact that the process of radiocarbon age calibration itself is only one step in the road from sample to date. In isolation, each of the steps may appear to be simple, but since there may well be many steps, the process of ^{14}C calibration becomes quickly complicated. Assume for a moment that a serviceable archaeological ^{14}C data base exist together with a basic grasp of the statistical and archaeological properties of the data. Subsequent processes often involve numerous tests, filters and corrections that, although important, are boring and time-consuming before we reach the final research phase and fulfilment of our hope for new discoveries. Perhaps the most important aspect of the final phase of ^{14}C age calibration is not the acquired *dating precision* (although this helps), but rather the *speed and technical efficiency* with which we can perform the analysis. For a heuristic approach when the potential gain of knowledge is but loosely defined, the speed with which the entire process can be iterated and completed under systematically varying conditions is of great importance.

An example illustrates the process. Assume that we expect interesting results from an all-European calibrated Palaeolithic ^{14}C-chronology by first using all dates and then building the same chronology using only charcoal data, omitting bone material that is suspected of a bone-specific preservation bias. Next we consider the data by climatically different regions, perhaps to test whether over time the Aurignacians in Spain had reacted differently from the contemporaneous cultures in South France. Finally, we might wish to compare the regional chronologies in the west with the climate signal from the North Atlantic and relate the Crimean ^{14}C chronology to the Vostok ice core in the Antarctic because it might better reflect the continental climate of eastern Europe and beyond.

The following procedures — which may appear in any order — are basic to most Palaeolithic studies involving ^{14}C data: 1) selecting and critical filtering the data; and 2) constructing an appropriate up-dated working ^{14}C calibration curve for the period. The result may be enriched by ^{14}C-age conversions using different methods and comparisons with graphic representations of the chronological results, finishing the chain with some elementary mapping procedures for all studies involving spatial-temporal relationships.

Clearly, what we really propose when talking of radiocarbon calibration in archaeological studies is not a fairly simple change in time-scales or a switch to another numbering system, but an often much more complicated and frequently extensive sequence of technical procedures and corresponding application of analytical and graphic methods, any or all of which may be crucial to the quality of the desired scientific output. Experts in those matters are more often found amongst the users than amongst the producers of ^{14}C dates.

Construction of a glacial-period ^{14}C calibration curve
The design and construction of glacial calibration curves should not be constrained by the availability of one or more calibration data sets, nor should it be ruled by a desire for perfect matches between some calibration data sets. Inevitably quantitative differences between data sets will always exist, but established statistical procedures allow us to quantify those differences. Therefore, it is of greatest importance that a working calibration curve provides realistic dating errors on both the ^{14}C and calendric time-scales.

In summary, a working glacial calibration curve is one which allows application to a range of calibration data sets. It is also one that admits the use of one or more of the standardized and widely applied archaeological calibration methods, such as archaeological wiggle matching, dispersion, culture group calibration or Bayesian calibration.

The method itself allows for differences between data, and the corresponding differences in research opinions can be dealt with by ample and free discussion.

The CalPal computer programme package
The development of ^{14}C analytical methods and matching software has traditionally been performed by the community of ^{14}C data producers. What we present below is the independent construction of a working calibration curve for the last glacial period as an analytical *archaeological* tool. To support the research envisaged by the Stage 3 Project by providing glacial ^{14}C calibrations we have developed a computer program called *CalPal*, an acronym for 'The Köln Radiocarbon Calibration and Palaeoclimate Research Package'. Programs included in the CalPal package provide a highly automated calendric age conversion of large archaeological ^{14}C data sets by the several methods listed above. The programming language is Fortran.9 and the CalPal package makes substantial use of the Winteracter 4.0 GUI-libraries (Graphics User Interface) as well as Canaima F90SQL (Structured Query Language) for data-import through the ODBC (Open Database Connectivity) interface under Windows. CalPal incorporates several archaeological databases containing ^{14}C data from the Palaeolithic, Mesolithic and Neolithic periods in Europe and the Near East. These databases are in an EXCEL format that facilitates the import/export from other platforms.

To support the plotting of site distributions we have provided CalPal with a user-friendly interface to the PanMap mapping program that uses high-resolution topographic data covering most of the earth's surface supplied by the National Geodetic Data Centre in Boulder, Colorado. The palaeoclimate data base incorporated in CalPal covers about 60 different climate proxies with emphasis on polar and equatorial ice cores. Typical among ice core data are oxygen isotope, methane and carbon dioxide contents, ice accumulation rates and ^{10}Be production.

A special feature of CalPal is the 'Climate Composer', a menu allowing completely independent, simultaneous graphic displays of any two selected climate data sets. User-friendly facilities support the linear shifting and expansion of the underlying time-scales and proxy axes. Given these facilities, often no more than half a dozen mouse-clicks are required to produce a plot showing for example a large Stage 3 Project set of calibrated dated sites of the Palaeolithic in a graphic context based on such climate data as GISP2 compared to VOSTOK. Other selections supply maps with the geographic distribution of all radiocarbon-dated sites.

The chrono-archaeological data base

The Stage 3 chrono-archaeological data base — which has other than chronological uses as well — was constructed first and foremost to place European

Table 3.2. *Statistics of the edited chrono-archaeological data base by lithic industries. Total dates before first edit 1896; bold numbers are after editing.*

		Total dates	Aurignacian	Gravettian	Mousterian	Middle Pal.	E. Upper Pal.	Upper Pal.
Dates	before edit	1666	425	388	452	14	133	109
	after edit	1363	354	374	303	13	99	95
Sites	before edit	454	101	96	135	5	29	31
	after edit	372	88	89	97	5	20	24
Cave sites	before edit	267	60	41	90	5	13	17
	after edit	217	48	39	65	5	13	13
Abri sites	before edit	69	21	16	20	0	1	0
	after edit	61	20	14	15	0	1	0
Open-air sites	before edit	119	22	39	25	0	16	10
	after edit	92	15	36	17	0	11	7
Method (after edit)	^{14}C all	1186	332	372	160	13	99	94
	^{14}C AMS	407	115	103	59	5	66	29
	U-series	15	1	0	14	0	0	0
	TL, OSL	51	15	2	27	0	0	0
	ESR	106	2	0	103	0	0	0
	?	6	4	0	0	0	0	1
Sample (after edit)	bone	779	160	202	178	11	78	77
	charcoal	350	120	131	44	2	16	8
	other	70	20	3	35	0	2	0
	unknown	166	54	38	46	0	3	10

		Bachokirian	Bohunician	Châtelperronian	Magdalenian	Solutrean	Szeletian	Uluzzian
Dates	before edit	4	7	46	28	35	18	7
	after edit	3	7	36	25	34	15	5
Sites	before edit	1	2	18	10	16	7	3
	after edit	1	2	12	9	15	7	3
Cave sites	before edit	1	0	13	6	11	5	5
	after edit	1	0	9	5	11	5	3
Abri sites	before edit	0	0	4	4	3	0	0
	after edit	0	0	4	4	3	0	0
Open-air sites	before edit	0	2	1	0	2	2	0
	after edit	0	2	1	0	1	2	0
Method (after edit)	^{14}C all	3	7	28	24	34	15	5
	^{14}C AMS	3	0	10	6	9	0	2
	U-series	0	0	0	0	0	0	0
	TL, OSL	0	0	7	0	0	0	0
	ESR	0	0	1	0	0	0	0
	?	0	0	0	1	0	0	0
Sample (after edit)	bone	2	0	25	21	20	2	3
	charcoal	1	7	2	2	12	4	1
	other	0	0	7	2	0	0	1
	unknown	0	0	4	0	2	9	0

Middle and Upper Palaeolithic sites in a time-and-space framework. Currently it contains just under 1900 *published* dates from excavated sites covering the interval from *c.* 100 ka BP[2]) (late OIS-5) to 20 ka BP at the peak of the Late Glacial Maximum (LGM). The sites are from all over Europe as far east as the great Russian Plain and from the European Mediterranean. To the best of our knowledge, the data base holds most dates published before the end of the year 2000, the most important exceptions being dates published solely in Slavic languages and not accessible to us (see Musil 2003), and a major compendium of Russian dates (Sinitsyn & Praslov 1997) which came only recently to our attention.

Table 3.2 lists 1896 raw dates but of these 230 were rejected because they were taken from sterile strata or faunal elements, or lacked industrial affiliations. This left 1666 dates obtained from 454 sites. After further screening for dates regarded as suspect for other reasons (Table 3.3), 1363 dates for 372 sites remained for use. Amongst those are many dates indicating a 'greater than' age which have little value in our approach and have mainly been ignored.

A large majority (*c.* 80 per cent) of the data base consists of raw ^{14}C dates with standard deviations (SD$_{lab}$), the rest are calendrical U/Th, TL and OSL luminescence and ESR dates. All sources are cited. The ^{14}C dates were calibrated with the 1998 version of CalPal and are given with their calendrical calibration ranges (SD$_{cal}$). Analytical standard deviations

Table 3.3. *Criteria used in the first quality screening of the chrono-archaeological data base.*

Samples
Unless clearly associated with artifacts, reject:
1. Sediments and palaeosols unless clearly associated with artefacts.
2. ^{14}C dates on flowstone, stalagmites, stalactites, shell, travertine *etc.* unless corrected for dead CO_2.

Industries
1. Reject mixed suites, e.g. Aurignacian/Gravettian.
2. Reject question-marked industry identifications, e.g. Mousterian?.
3. Reject industrial labels marked pre-xxxx ka or post-xxxx ka.
4. Ignore industrial sub-types of industries, such as Dufour Aurignacian, Denticulate Mousterian, Ancient Solutrean, Late Gravettian etc.

Processing of dates
1. Reject all dates labeled >xxxx (older than).
2. Round standard deviations of calibrated ^{14}C and all TL/OSL, ESR and U series dates to the nearest 100 years.
3. Round all dates to the nearest 1000 years (ka).
4. Reject all dates with standard deviations >2.5 ka except for the interval before *c.* 45 ka BP.
5. Ignore standard deviations for the zone >45 ka.

Table 3.4. *Site numbers and SD_{cal} distribution by major lithic complexes and broad climate phases; na = not applicable.*

Lithic industry	N dates	Dates by SD_{cal} classes				
		<1000	<1500	<2000	<2500	>2500
Mousterian						
20–26 cal ka BP	11	7	0	0	0	0
27–36 cal ka BP	56	23	14	16	3	0
37–44 cal ka BP	74	13	9	22	10	20
>44 cal ka BP	150					
Châtelperronian						
20–26 cal ka BP	1	1	0	0	0	0
27–36 cal ka BP	18	5	4	6	2	1
37–44 cal ka BP	6	0	0	3	1	2
Aurignacian						
20–26 cal ka BP	43	25	15	2	1	0
27–36 cal ka BP	143	61	35	42	3	2
37–44 cal ka BP	37	11	14	5	3	4
>44 cal ka BP	5	na	na	na	na	na
Early Upper Palaeolithic						
20–26 cal ka BP	15	8	5	2	0	0
27–36 cal ka BP	65	34	13	12	3	3
37–44 cal ka BP	6	1	1	3	0	1
>44 cal ka BP	6	0	0	0	0	6
Magdalenian						
<20–26 cal ka BP	4	4	0	0	0	0
27–36 cal ka BP	5	4	1	0	0	0
37–44 cal ka BP	2	2	0	0	0	0
Gravettian						
20–26 cal ka BP	136	107	27	2	0	0
27–36 cal ka BP	60	43	14	3	0	0
Solutrean						
20–26 cal ka BP	27	25	0	0	2	2
27–36 cal ka BP	1	1	0	0	0	0
Upper Palaeolithic						
20–26 cal ka BP	58	52	4	0	2	0
27–36 cal ka BP	31	24	5	0	1	1

of calendrical U/Th, TL, OSL and ESR dates are for convenience also labelled SD_{cal}. Laboratory numbers, analytical methods and dated substances are included where available. Because our standard age model is derived from calendrical dates, only calibrated ^{14}C dates can be used; the pre-calibrate ^{14}C values are not listed but can be found in the data base.

Because of its chronological focus, the data base limits site characteristics to basic categories such as cave, abri and open-air, but where available the dates have been supplemented with as many excavation records and other information that might be relevant to the stratigraphic context as could be extracted from the sources listed at the end of each date line.

Up to an age of 38 ka BP, we have been able to rely mainly on dates with SD_{lab} and SD_{cal} ranges below ±1000 years (Table 3.4); in this age range ^{14}C dates dominate so strongly that all other kinds (U/Th, OSL, TL and ESR) carry little weight. Beyond *c.* 38 ka the ratio between calibrated and primary calendar dates changes rapidly and ^{14}C dates with SD_{cal} between ±1000 and ±2000 years had to be used increasingly. After about 40 ka BP they constitute the majority and up to 45 ka BP ^{14}C calibration values continue to be in line with the SD_{cal} of primary calendar dates and SD_{cal} values beyond ±2500 years are few. This changes drastically at around 45 ka BP when the difference between raw and calibrated dates is so

much in excess of ^{14}C dates younger than 45 ka BP that serious calibration problems are implied for this interval. The ^{14}C SD_{cal} ranges are very wide, mostly ±4 to ±8 ka, and the primary calendrical dates have similar ranges. Clearly, beyond about 45 ka BP even our 150 dates will only broadly arrange occupation of most Mousterian sites between the end of Stage 4 (60 ka BP) and about 45 ka, but allows little detail.

The data base is vulnerable to several biases. More excavations have taken place in France than in Spain or Greece. More prosperous archaeologists have been able to buy more dates per site than their poorer colleagues, and sites that were excavated many years ago tend to be less well dated than those that were studied more recently. As regards site types, the 278 cave and abri sites (Table 3.2) seem fairly representative for their geographic and temporal distribution. The same can not be said with as much confidence for the 92 open-air sites, because relatively few such sites have been published and even fewer properly dated. Still, the fact that about one third of all our dated sites are in the open-air category eases somewhat our initial concern that Middle and Upper Palaeolithic site maps of Europe might merely reflect a sketchy pattern of the distribution of limestone. No hard and fast remedies exist for these biases, but they should be kept firmly in mind.

Recently, the validity of bone, horn and antler dates has been questioned by some with various stated consequences that range from the possibility that they may be systematically too young to the claim that even their ages might vary from site to site and hence might be basically worthless. Because bone dates constitute about half of the total data set and are twice as numerous as charcoal-based dates (Table 3.2), these claims, if correct, would leave us with a very small, in effect probably insufficient data set. However, no robust evidence has been presented so far that any of these claims are correct and we shall therefore ignore them here and in Chapter 4 (van Andel *et al.* 2003).

In subsequent chapters we have used the edited data set for various main purposes.
- to construct time/site frequency curves for each major lithic complex in Europe and the individual sub-regions defined by Gamble (1999, fig. 5.1);
- to map geographically the archaeological site distributions by lithic complexes, by time slices and by broad temporal climate phases;
- to identify sites of long occupation;
- to determine the preferences and tolerances of each major lithic industry for a range of modelled summer and winter temperatures, wind-chill values, snow days/year and snow depth.

Notes

1. Construction details and properties of the chrono-archaeological data base are presented in the Preamble of the data base which can be found on the Stage 3 website at: http://www.esc.esc.cam.ac.uk/oistage3/Details/Homepage.html.
2. To avoid confusion in this and the next chapter the notation 'xxxx ka BP' denotes primary calendric dates (ice-core, U-Th, TL and OSL and ESR) and calibrated ^{14}C dates in millennia Before Present. The range of the

data base is currently being extended chronologically to the early Holocene and geographically to North Africa and the Near East by William Davies of the Department of Archaeology, University of Southampton.
3. The CalPal ^{14}C calibration programme is available on its website at: http://www.calpal.de.html.

References

Allen, J.R.M. & B. Huntley, 2000. Weichselian palynological records from southern Europe: correlation and chronology. *Quaternary International* 73–4, 111–26.
Ammerman, A.J. & L.L. Cavalli-Sforza, 1971. Measuring the rate of spread of early farming in Europe. *Man* 6, 674–88.
Ammerman, A.J. & L.L. Cavalli-Sforza, 1984. *The Neolithic Transition and the Genetics of Population in Europe.* Princeton (NJ): Cambridge University Press.
Bard, E., M. Arnold, B. Hamelin, N. Tisnerat-Laborde & G. Cabioch, 1998. Radiocarbon calibration by means of mass spectrometric [230]Th/[234]U and ^{14}C ages of corals: An updated database including samples from Barbados, Mururoa and Tahiti. *Radiocarbon* 40, 1085–92.
Bocquet-Appel, J.-P. & P.Y. Demars, 2000. Neanderthal contraction and modern human colonization in Europe. *Antiquity* 74, 544–52.
Bond, G., W. Broecker, S. Johnsen, J. McManus, L. Labeyrie, J. Jouzel & G. Bonani, 1993. Correlations between climate records from North Atlantic sediments and Greenland ice. *Nature* 365, 143–6.
Burr, G.S., J.W. Beck, F.W. Taylor, J. Récy, R.L. Edwards, G. Cabioch, T. Corrège, D.J. Donahue & J.M. O'Malley, 1998. A high resolution radiocarbon calibration between 11,700 and 12,400 calender years BP derived from [230]Th ages of corals from Espiritu Santo Island, Vanuatu. *Radiocarbon* 40, 3–105.
Chamberlin, T.C., 1897. The method of multiple working hypotheses. *Journal of Geology* 5, 837–48; republished 1995 in *Journal of Geology* 103, 349–54.
Cleland, C.E., 2001. Historical science, experimental science and the scientific method. *Geology* 29, 897–990.
Edwards, R.L., J.W. Beck, G.S. Burr, D.J. Donahue, J.M.A. Chappell, A.L. Bloom, E.R.M. Druffel & F.W. Taylor, 1993. A large drop in atmospheric ^{14}C/^{12}C and reduced melting in the Younger Dryas, documented with [230]Th ages of corals. *Science* 260, 962–8.
Frodeman, R., 1995. Geologic reasoning: geology as an interpretative and historical science. *Geological Society of America, Bulletin* 107, 960–68.
Gamble, C.S., 1986. *The Palaeolithic Settlement of Europe.* Cambridge: Cambridge University Press.
Gamble, C.S., 1999. *The Palaeolithic Societies of Europe.* Cambridge: Cambridge University Press.
Gee, H., 1999. *In Search of Deep Time.* New York (NY): The Free Press.
Goslar, T., M. Arnold, N. Tisnerat-Laborde, C. Hatté, M. Paterne & M. Ralska-Jasiewiczowa, 2000. Radiocarbon calibration by means of varve versus ^{14}C ages of

terrestrial macrofossils from Lake Gościąż and Lake Perespilno, Poland. *Radiocarbon* 42, 335–48.

Holdaway, S. & N. Porch, 1995. Cyclical patterns in the Pleistocene human occupation of Southwest Tasmania. *Archaeology in Oceania* 30, 74–82.

Housley, R.A., C.S. Gamble, M. Street & P. Pettitt, 1997. Radiocarbon evidence for the late glacial human recolonisation of Northern Europe. *Proceedings of the Prehistoric Society* 63, 25–54.

Hughen, K.A., J.R. Southon, S.J. Lehman & J.T. Overpeck, 2000. Synchronous radiocarbon and climate shift during the Last Deglaciation. *Science* 290, 1951–4.

Jöris, O. & B. Weninger, 1998. Extension of the 14-C calibration curve to *c.* 40,000 cal BC by synchronising Greenland $^{18}O/^{16}O$ ice-core records and North Atlantic Foraminifera profiles: a comparison with U/Th coral data. *Radiocarbon* 40, 495–504.

Kitagawa, H. & J. van der Plicht, 2000. Atmospheric radiocarbon variation beyond 11,900 cal BP from Lake Suigetsu. *Radiocarbon* 42, 369–80.

Laj, C., A. Mazaud & J.C. Duplessy, 1996. Geomagnetic intensity and ^{14}C abundance in the atmosphere and ocean during the past 50 kyr. *Geophysical Research Letters* 23, 2045–8.

Lewis, D., 1991. Counterfactual dependence and time's arrow, in *Conditionals*, ed. E. Jackson. Oxford: Oxford University Press, 46–75.

Mazaud, A., C. Laj, E. Bard, M. Arnold & E. Tric, 1991. Geomagnetic field control of ^{14}C production over the last 80 ky; Implication for the radiocarbon time scale. *Geophysical Research Letters* 18, 1885–8.

Mellars, P.A., 1996. *The Neanderthal Legacy: an Archaeological Perspective from Western Europe*. Princeton (NJ): Princeton University Press.

Mook, W.G. & J. van der Plicht, 1999. Reporting ^{14}C activities and concentrations. *Radiocarbon* 41, 227–39.

Musil, R., 2003. The Middle and Upper Palaeolithic game suite in central and southeastern Europe, in *Neanderthals and Modern Humans in the European Landscape during the Last Glaciation,* Chapter 10, eds. T.H. van Andel & W. Davies. (McDonald Institute Monographs.) Cambridge: McDonald Institute for Archaeological Research, 167–90.

Pettitt, P.B. & A.W.G. Pike, 2001. Blind in a cloud of dates: problems with the chronology of Neanderthal extinction and anatomically modern human expansion. *Antiquity* 75, 415–17.

Rick, J.W., 1987. Dates as data: an examination of the Peruvian Pre-ceramic radiocarbon record. *American Antiquity* 52, 55–73.

Roebroeks, W., M. Mussi, J. Svoboda & K. Fennema (eds.), 1999. *Hunters of the Golden Age: the Mid Upper Palaeolithic of Eurasia 30,000–20,000 BP.* (Analecta Praehistorica Leidensia 31.) Leiden: University of Leiden.

Sinitsyn, A.A. & N.D. Praslov (eds.), 1997. *Radiocarbon Chronology of the Palaeolithic of Eastern Europe and Northern Asia: Problems and Perspectives.* Saint Petersburg: Russian Academy of Sciences, Institute of the History of Material Culture.

Stuiver, M., P.J. Reimer, E. Bard, J.W. Beck, G.S. Burr, K.A. Hughen, B. Kromer, B.G. McCormac & J. van der Plicht (eds.), 1998. INTCAL98 Radiocarbon age calibration, 24,000–0 cal BP. *Radiocarbon* 40, 1041–84.

Svoboda, J., V. Lozel & E. Vlcek, 1996. *Hunters between East and West: the Palaeolithic of Moravia.* New York (NY): Plenum Press.

Thouvény, N., K.M. Creer & D. Williamson, 1993. Geomagnetic moment variations in the last 70,000 years: Impact on the production of cosmogenic isotopes. *Global and Planetary Change* 7, 157–67.

Tric, E., J.P. Valet, P. Tucholka, L. Labeyrie, F. Guichard, L. Tauxe & M. Fontugne, 1992. Palaeo-intensity of the geomagnetic field during the last 80,000 years. *Journal of Geophysical Research* 97, 9337–51.

van Andel, T.H., 1998. Middle and Upper Palaeolithic environments and the calibration of ^{14}C dates beyond 10,000 BP. *Antiquity* 72, 27–33.

van Andel, T.H., 2002. Reconstructing climate and landscape of the middle part of the last glaciation in Europe : the Stage 3 Project. *Quaternary Research* 57, 2–8.

van Andel, T.H., 2003. Glacial environments I: the Weichselian climate in Europe between the end of the OIS-5 interglacial and the Last Glacial Maximum, in *Neanderthals and Modern Humans in the European Landscape during the Last Glaciation,* Chapter 2, eds. T.H. van Andel & W. Davies. (McDonald Institute Monographs.) Cambridge: McDonald Institute for Archaeological Research, 9–20.

van Andel, T.H., W. Davies & B. Weninger, 2003. The human presence in Europe during the last glacial period I: human migrations and the changing climate, in *Neanderthals and Modern Humans in the European Landscape during the Last Glaciation,* Chapter 4, eds. T.H. van Andel & W. Davies. (McDonald Institute Monographs.) Cambridge: McDonald Institute for Archaeological Research, 31–56.

van der Plicht, J., 1999. Radiocarbon calibration for the Middle/Upper Palaeolithic: a comment. *Antiquity* 73, 119–23.

Völker, A.H., M. Sarnthein, P.M. Grootes, H. Erlenkeuser, C. Laj, A. Mazaud, M.-J. Nadeau & M. Schleicher, 1998. Correlation of marine ^{14}C ages from the Nordic Seas with the GISP2 isotope record: implications for radiocarbon calibration beyond 25 ka BP. *Radiocarbon* 40, 517–34.

Völker, A.H.L. & workshop participants, 2002. Global distribution of centennial-scale records for Marine Isotope Stage (MIS) 3: a data base. *Quaternary Science Reviews* 21, 1185–212.

Chapter 4

The Human Presence in Europe during the Last Glacial Period I: Human Migrations and the Changing Climate

Tjeerd H. van Andel, William Davies & Bernard Weninger

Tracking Neanderthals and Anatomically Modern Humans

The Stage 3 chrono-archaeological data base lists most of the human skeletal material dates published up to December 2000. The finds derive from only a few sites (Table 4.1), but their pattern is consistent. With few exceptions, Neanderthal remains (*Homo sapiens neanderthalensis*) are associated with Mousterian and Châtelperronian lithic industries, whilst Anatomically Modern Human bones (*Homo sapiens sapiens*) are accompanied by Aurignacian, Gravettian and other Early Upper and Upper Palaeolithic artefacts. A few somewhat doubtful *Homo neanderthalensis?* bones are not associated with lithics but, ignoring those, only three exceptions exist: *Homo neanderthalensis* is associated with an Upper Palaeolithic assemblage at Trou de l'Abîme and a few certain and some doubtful Neanderthal bones have been retrieved from Aurignacian strata at Vindija Cave and Bacho Kiro (Table 4.1).

For our purposes the chrono-archaeological data base has provided time and space co-ordinates for two human species; the Neanderthals are represented by the Mousterian technocomplex, while the Aurignacian and Gravettian represent Anatomically Modern Humans (AMH). To the Aurignacian we must add the many sites labelled Early Upper Palaeolithic (EUP) and to the Gravettian Upper Palaeolithic (UP) as their chronologically equivalents. We are conscious of the many distinctions involved in those categories (van Andel *et al.* 2003), but in the broad overview of space and time we are aiming for here we must regard those distinctions as of secondary interest. Anyway, many of the lesser units have too few dated sites to be useful given our *modus operandi* which depends on large data sets.

For convenience we use the term Aurignacian for the Aurignacian + Early Upper Palaeolithic and Gravettian for the Gravettian + Upper Palaeolithic techno-complexes. Inevitably, we shall at times speak of Mousterians when we mean Neanderthals and carelessly call the earliest arriving modern humans Aurignacians and those of later date Gravettians.

An anonymous reviewer of an earlier version of this chapter asked why we cast our study in terms of human species, thereby adding another speculative dimension to our work. The answer is that we are concerned with the behaviour of human beings belonging to two species and their responses to the ever-changing climate and landscape of the last glaciation; their stone tools are merely a means of labelling them. If we accept Mousterian lithics as proxies for the presence of the Neanderthals whose fate we wish to trace, we can draw on a much larger data set than if we limit ourselves to Neanderthal skeletal finds (cf. Tables 4.1 & 4.2). The Aurignacian, Gravettian and other, later lithic complexes reliably carry the flag for the newly arrived Anatomically Modern Humans. But in the end, when we consider the final question of the Stage 3 Project: 'Why did the Neanderthals perish while Anatomically Modern Humans flourished?' we shall propose hypotheses and explanations about people, not industrial complexes.

We recognize that our tripartite scheme does not cover the full breadth and depth of human life in Europe during the last glaciation, but prefer not to participate here in discussions about the meaning and validity of the many subdivisions of each major industry that grace the columns of our data base. We do not question the cultural and historical significance of those sub-units, but in the context of this study with its long time and vast space to cover, their role is secondary and best left to subsequent studies.

Table 4.1. *List of sites with dated human skeletal material.*

Age range, cal ka BP	Total sites*	with *H. neander.*	with *H. neander.*?
H. neanderthalensis			
24–37	36	12	2
38–44	22	9	
45–59	50	8	2
60–66	8	3	1
67–74	7	3	1
H. sapiens sapiens	Total sites**	with *H. sapiens*	with *H. sapiens*?
16–23	93	6	3
24–36	106	31	13
37–45	53	9	6

NOTES on questionable *H. neanderthalensis*

Site	Industry	Species	Ages, cal ka BP
Salemas [algar]	Mousterian	6 *H. neander.*?	24, 28, 28, 28, 31, 34
Columbeira, G. Nova	Mousterian	6 *H. neander.*?	30, 33, 36, 54, 61, 101
Banyoles	?	1 *H. neander.*?	45
Jaurens	?	4 *H. neander.*	33, 33, 34, 37
Vergisson, La Falaise	FAUNA/?Neanderthal	1 *H. neander.*?	33
Trou de l'Abime	Middle Palaeolithic	2 *H. neander.*	29, 57
Trou de l'Abime	Upper Palaeolithic	1 *H. neander.*	30
Feldhofer cave	?	2 *H. Neander.*	45, 48
Vindija Cave	Aurignacian	1 *H. neander.*?	30, 31
Vindija Cave	Aurignacian?	2 *H. neander.*?	30, 38, 45
Vindija Cave	?	4 *H. neander.*	32, 33, 46, 51
Ohapa Ponor Cave	Mousterian	1 *H. neander.*?	44
Bacho Kiro	Bachokirian	4 *H. neander.*?	39, 42, 43, 59
Bacho Kiro	Aurignacian	1 *H. neander.*?	33

*   Mousterian and Chatelperronian sites.
** Aurignacian, Gravettian, Early Upper and Upper Palaeolithic sites.

Table 4.2. *Summary of number of dates and sites with their age ranges[1] (after Table 3.2); industries used in this chapter are shown in **bold italics**. Questionable limits shown with ?.*

Lithic industries	N dates	N sites	Age range* ka** BP
Aurignacian	361	88	23,000–47,000
Bachokirian	5	3	41,000–42,000
Bohunician	7	2	43,000–47,000
Châtelperronian	36	13	28,000–43,000
Early Upper Palaeolithic	99	20	26,000–41,000
Gravettian	386	93	21,000–38,000
Magdalenian	25	19	20,000***–28,000
Middle Palaeolithic	13	5	30,000–45,000
Mousterian	303	98	23,000?–108,000
Solutrean	31	12	21,000–26,000
Szeletian	15	7	37,000–45,000
Uluzzian	5	3	31,000–37,000
Upper Palaeolithic	90	24	24,000–31,000

*   Ranges from the Stage 3 chrono-archaeological data base.
** In calendar ka (1000) BP.
*** Lower limit of data set is 20 ka BP.

Zoning[1] the record of Stage 3 climate changes

In this chapter we track human migrations through time and across Europe, using time-slice maps and time graphs to determine the degree to which their wanderings may have been direct or indirect responses to the changing climate. For this purpose we compare the ice-core climate record (Meese *et al.* 1997; Johnsen *et al.* 2001; Stuiver & Grootes 2000) with dated sites from the chronological data base. Because calibration ranges (SD_{cal}) of the archaeological dates extend from less than ±500 to more than ±2500 years (Table 3.4), matching Palaeolithic time-slices with Dansgaard/Oeschger oscillations is not possible. Instead, we have subdivided the GISP2 ice-core record into climate sub-units (Fig. 4.1; Table 4.3) of longer durations that enable us to exploit archaeological time-slices of the same order of magnitude.

Relief and palaeoclimate

For its size Europe has a remarkably diverse geography dominated by west–east trending trans-European mountain ranges that start with the Pyrenees and Alps and at about 15°E longitude divide into a northern Carpathian branch and the southeasterly trending Dinarides extending into western Greece. Between them the two branches embrace the Hungarian lowland traversed by the Danube which, having passed the Iron Gates, enters a large alluvial plain before it reaches the Black Sea. It was almost certainly one or even the main age-old highway from the Near East to western Europe. The trans-European ranges have elevations of up to two or three thousand metres and form a major climatic barrier shielding the Mediterranean region with its intricate arrays of islands, peninsulas and mountains from the nordic weather.

Reduced to such simplicity the geography of Europe is familiar to all, but the main features should be kept in mind lest we fall, albeit only mentally, in the trap that has caught many a modeller whose techniques were suited best to a table-top surface (e.g. Bocquet-Appel & Demars 2000a,b; Zubrow

1992). In Chapter 11 (Davies *et al.* 2003) we return in more detail to the relations between topography, climate and early human settlement.

North of the trans-European mountain range and bordering the Fennoscandian ice margin is the North European Plain which in glacial times began on the dry bed of the fully emerged North Sea. The Low Countries are part of this plain and so is northern Germany where it gradually widens as it forms the lowlands of Poland and the Russian plain, ending at the Urals. Its northern boundary is often thought to have been the edge of the Fennoscandian ice sheet, located south of the southern Baltic shore throughout the entire last glaciation. The ice sheet probably reached the northern Baltic shore during its first maximum (OIS-4) and again during the Late Glacial Maximum (LGM) when it extended deep into northern Germany, the Baltic countries and northernmost Poland. For the 35,000 years between the two glacial maxima, however, the ice sheet was much reduced and during OIS-3 may have been limited to local ice-caps covering the highest peaks of southern Norway (Arnold *et al.* 2002; Olsen 1997; Olsen *et al.* 1996; 2001a,b; Sveian *et al.* 2001). In northeastern Russia during the Briansk Interstade (33–24 uncal [14]C b.p. or about 36–26 ka BP[2]) forest tundra, periglacial forest-tundra and periglacial forest-steppe landscapes occupied the region between 55° and 65°N lat. rather then an ice sheet. This region contains many Palaeolithic sites which render it likely — but not yet prove (Larsson 1999) — that in Scandinavia too humans hunted at higher latitudes than the usually accepted 55°N limit.

The late Pleistocene climate of Europe was marked by two continental-scale trends, the north–south gradient from arctic to cold temperate conditions north of the trans-European mountain barrier and the west–east transition from the maritime Atlantic to the continental climate of the plains of easternmost Europe (Barron *et al.* 2003). The trans-European mountain barrier was the protector of the milder climate in the northern Mediterranean region from eastern Iberia to Turkey. Dominated by sharp contrasts between high country and the sea, the region displays a complex mosaic of climate zones that

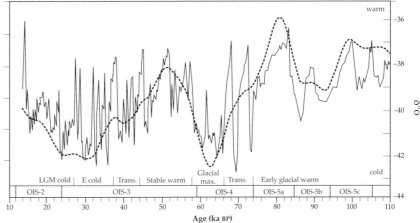

Figure 4.1. *Second- and third-order (Dansgaard/Oescher) climate phases between the late penultimate interglacial (OIS-5d) and the onset of the Holocene (OIS-1) as reflected in the δ18O climate record from the GISP2 ice core (Meese et al. 1997; Johnsen et al. 2001; Stuiver & Grootes 2000). Climate phases as in Table 4.3. The broken heavy line depicts the general trend of the OIS-3/OIS-2 climate changes.*

Table 4.3. *Second-order climate phases of the Weichselian Glaciation between OIS-5a and OIS-1, based on the GISP2 Greenland ice core and used to place middle and late Palaeolithic human history in a climatic context.*

SPECMAP	Climate phase	Age (ka BP)
OIS-5a	Early Glacial Warm Phase	>74
OIS-4	Transitional Phase	74–66
OIS-4	First Glacial Maximum	66–59
OIS-3	Stable Warm Phase	59–44
OIS-3	Transitional Phase	44–37
OIS-3	Early Cold Phase	37–27
OIS-2	Last Glacial Maximum	27–16

is largely a function of topography (Barron *et al.* 2003).

Shifting patterns of human occupation: 70–20 ka BP[2]

The Neanderthals in Europe during the last glaciation
The temporal and spatial record of the three main techno-complexes on which our review of the decline and extinction of the Neanderthals and the entry and expansion of modern humans rests reveals clear parallels but also some striking differences.

The fairly continuous Mousterian record (Fig. 4.2) begins more than 70,000 years ago in early OIS-4 (*c.* 74 ka BP), well before the onset of the first ice advance of the last glaciation at ~65 ka BP. The early part of the record is dominated by U/Th series, TL/OSL and ESR and some questionable ^{14}C dates, all with large SD$_{cal}$ ranges, and has a low resolution.

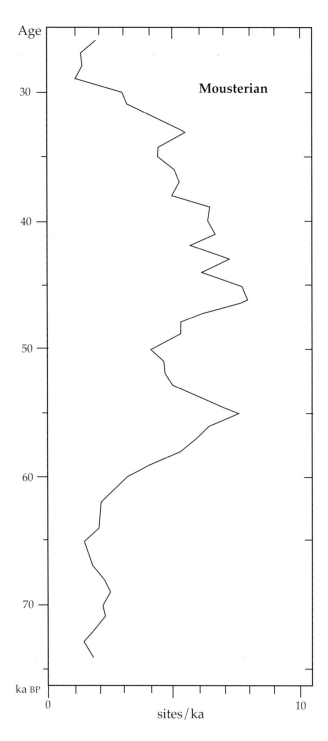

Figure 4.2. *The Neanderthal (Mousterian) temporal record in Europe, 70–25 ka BP, presented as five-point running average of the number of sites per millennium plotted against their age in thousands of years (ka) BP.*

✳ Even so it parallels the broad trends of the climatic record (Fig. 4.1).

The OIS-4 glacial maximum, lasting a mere 5000

years, features 12 dated sites, all situated south of 45°N latitude (Fig. 4.3a:top) in the Mediterranean and southern France of which six go back another 10–20 ka BP; the 67 ka BP date from Temnata is questionable. Time-slices of the Mousterian site distribution in Europe (Figs. 4.3a–d) give considerable detail about the Mousterian geography. See Appendices 4.1–4.3 for site lists by industry and time-slice.

The human presence as expressed by dated sites remained low until an abrupt warming around 59 ka ✳ ago started the longest warm period within the entire Weichselian glaciation. The recovery from the ✳ OIS-4 glacial advance initiated a rich final Neanderthal *floruit* during the long mild interval of early OIS-3 time, and human activity, as measured by the number of sites per millennium, expanded drastically. A brief decline around 53–49 ka BP may reflect temporary cold-dry steppe conditions which are recorded at Monticchio in Italy (Allen *et al.* 2000) and elsewhere, but the climate curve soon rose again.

The time-slices of 59–48 ka (Fig. 4.3a:bottom) and 47–43 ka BP (Fig. 4.3b:top) which together encompass the Stable Warm Phase display a settlement pattern very different from that of OIS-4. In the northern Mediterranean, the number of sites quickly doubled and a vigorous expansion in western, central and eastern Europe and the Crimea and into northern latitudes is clear, but few dated site lie north of 50°N (Pavlov & Indrelid 1999; Pavlov *et al.* 2001). The pattern (Fig. 4.3a:bottom) suggests that the re-colonization of central and western Europe may have come from the east up the great river systems of the Danube, Main and Rhine, a pattern also implied during the 47–43 ka BP time-slice (Fig. 4.3b:bottom). There is also a suggestion that the post-OIS-4 colonization of north-central and eastern Europe came from points farther east — perhaps from the Crimea and the northeastern Black Sea coast — and moved along the lower and middle courses of the large Ukrainian rivers. Here too few sites are near or beyond the 50°N parallel, the apparent northern limit of Neanderthal expansion in Europe until *c.* 40 ka BP.

At about 42 ka BP, the transition towards the Early Cold Phase began with a few thousand years of unstable climate conditions, but there was hardly any decrease in the number of Neanderthal sites. The 42–38 ka time-slice (Fig. 4.3b:bottom) shows a human retreat away from the North European Plain towards southern France, as well as an eastward retreat to the Ukraine and to southeastern Russia where Neanderthals still flourished. A single open-air site at 53°N, 34°E (Betovo: Soffer 1989, 724) points

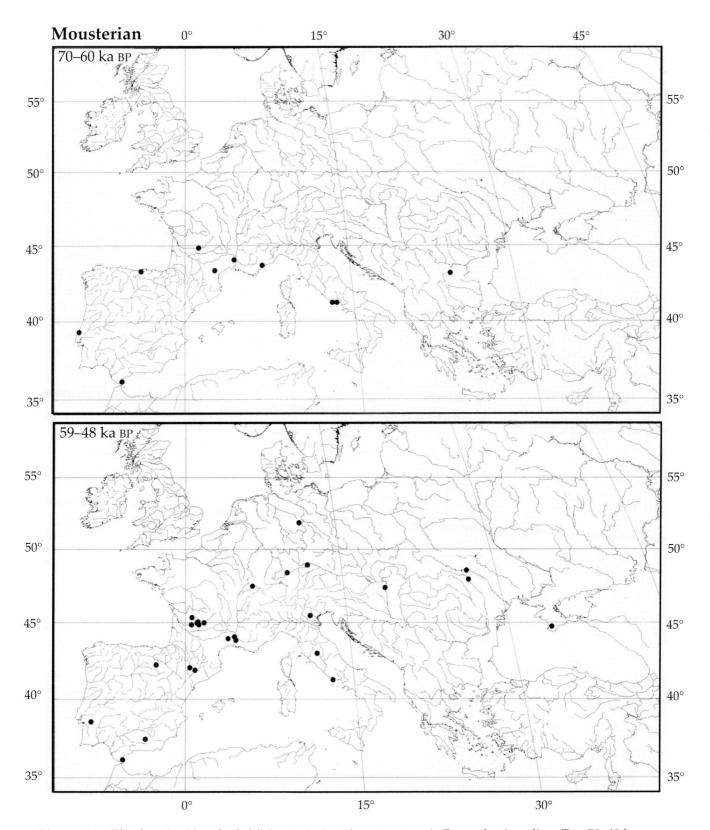

Figure 4.3a. *The changing Neanderthal (Mousterian) settlement pattern in Europe by time-slices. Top: 70–60 ka BP; bottom 59–47 ka BP. Sites are marked with black dots. Due to the map scale (diameter of black dots = ≈28 km) not all sites may show separately.*

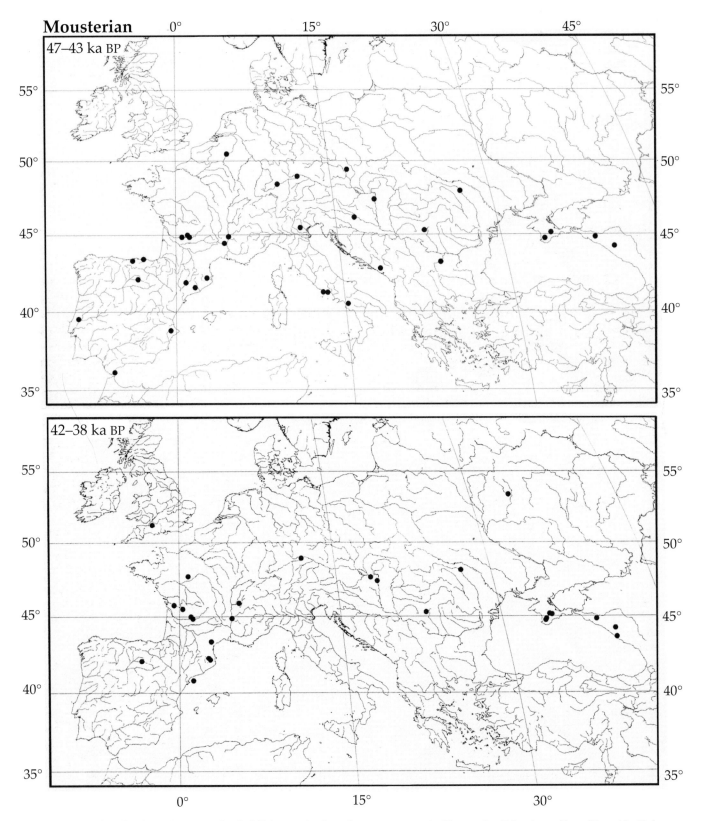

Figure 4.3b. *The changing Neanderthal (Mousterian) settlement pattern in Europe by 5-ka time-slices. Top: 46–43 ka* BP; *bottom: 42–38 ka* BP. *Sites are marked with black dots. Due to the map scale (diameter of black dots = ≈28 km) not all sites may show separately.*

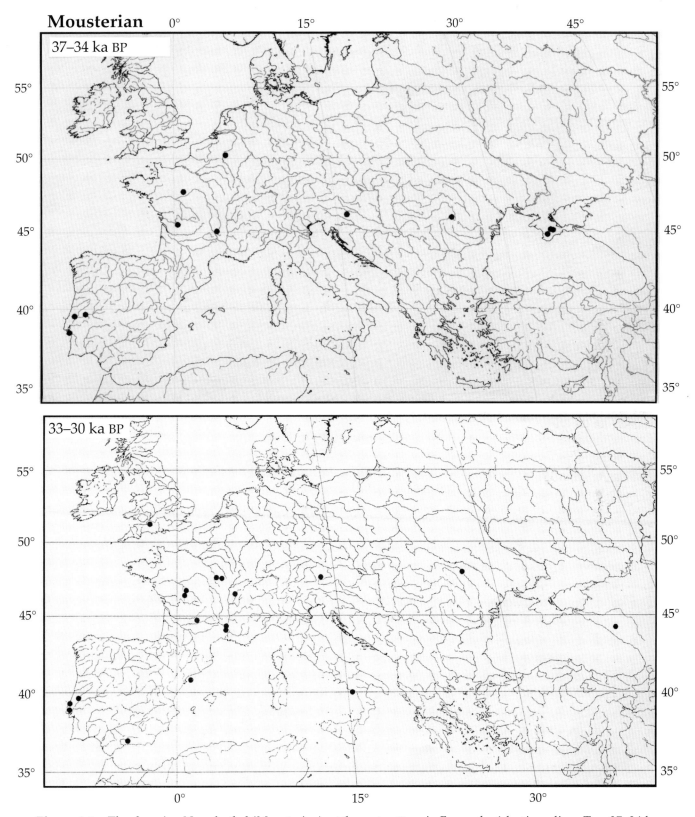

Figure 4.3c. *The changing Neanderthal (Mousterian) settlement pattern in Europe by 4-ka time-slices. Top: 37–34 ka BP; bottom: 33–30 ka BP. Sites are marked with black dots. Due to map scale (diameter of black dots = ≈28 km) not all sites may show separately.*

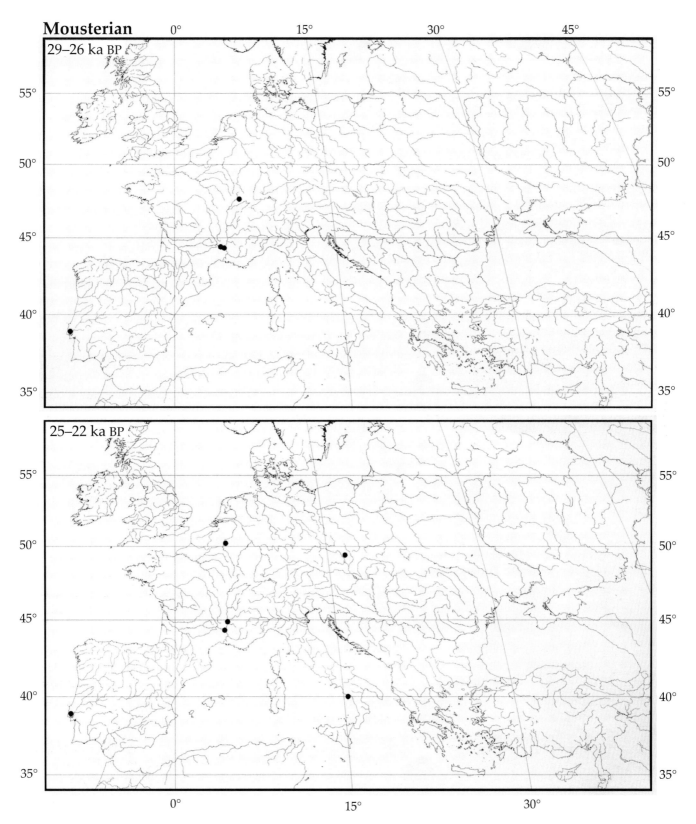

Figure 4.3d. *The changing Neanderthal (Mousterian) settlement pattern in Europe by 4-ka time-slices. Top: 29–26 ka* BP*; bottom: 25–22 ka* BP*. Sites are marked with black dots. Due to the map scale (diameter of black dots = ≈28 km) not all sites may show separately.*

to continued interest in the arctic. In central Europe north of the Alps only a few long-occupied sites remained and many sites there and in eastern France were abandoned.

The steady deterioration of the climate towards the Last Glacial Maximum (LGM) which began about 37 ka ago was accompanied by a gradual contraction of the Neanderthal presence (Fig. 4.3c:top) and the slow withdrawal from central Europe became total a little later (Fig. 4.3c:bottom). The Neanderthal realm was thus reduced to two disconnected regions, a western one in France with a small northernmost foot-hold in the Ardennes, and an eastern one in the Cri-mea and along the northeastern Black Sea shore; 4–6 sites continued to be occupied there until 30 ka ago.

The 29–26 and 25–22 ka BP maps (Fig. 4.3d) show that the whole of Europe east of *c.* 10°E as well as the Black Sea region were deserted. For a while a few sites continued to be occupied in southern and south-central France and on the Atlantic coast of Portugal; but then it was over. A few 'Mousterian' dates range down to 25 ka BP, but whether those sites were occupied by Neanderthals or reflect errors in cultural assignment or dating is unclear. Thus the final demise of the Neanderthals, between 32 and perhaps 28 ka ago, came well before the onset of the peak of the LGM around 25–23 ka BP, suggesting that its cause lay elsewhere than with the severely cold climate.

[handwritten marginal note: JH - but may've begun - ic/ demise with the previous cold spell c.34 ??]

Points of note *[handwritten: NB]*
- A clear parallelism between Neanderthal move-ments and climate changes. Evidence:
  - A rapid spread across the Mediterranean and throughout Europe south of the 50°N parallel when the long Stable Warm Phase began.
  - A two-pronged withdrawal westward to the Atlantic shores and southeastward to the Black Sea; perhaps a response to the final climate deterioration after 37 ka BP.
- A Neanderthal re-entry into central Europe from the southeast up the Danube–Main–Rhine sys-tem is suggested by the site pattern, but the dates do not suffice to imply a point of origin and the direction.
- Sparse archaeological exploration may be the cause of the apparent southeastward Neander-thal withdrawal, but might also be due to dete-riorating climate conditions and dwindling food resources on the Russian plain.

Arrival and early dispersal of modern humans
The temporal history of Anatomically Modern Hu-

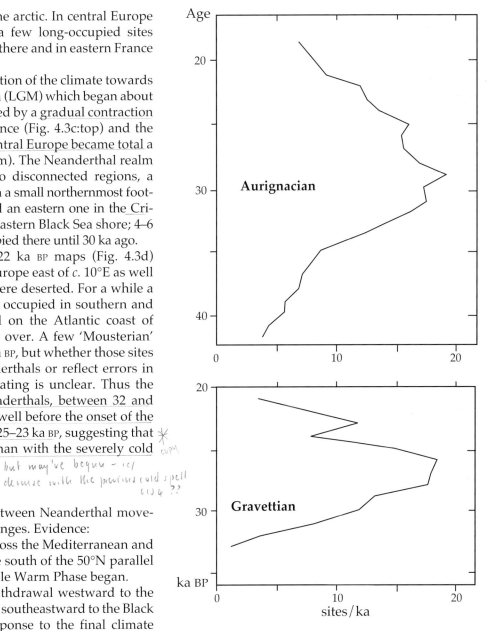

Figure 4.4. *The temporal records of Anatomically Mod-ern Humans (Aurignacian + EUP and Gravettian + UP) in Europe, 45–20 ka BP, presented as 5-point running averages of the number of sites per millennium plotted against age in ka BP.*

mans (Fig. 4.4) differs from that of the Neanderthals, but their spatial pattern is rather similar (Roebroeks *et al.* 1992). The rising part of the Aurignacian tem-poral curve could well describe an unconfined popu-lation increase which, starting slowly before 40 ka BP, accelerated until it levelled off around 30 ka BP, perhaps due to external forces such as pressure from other populations, resource limitations and/or envi-

Table 4.4. *The earliest settlements of Anatomically Modern Humans in Europe with oldest dates. Only dates with $SD_{cal} <\pm2500$ years are listed.*

		Age ka BP					
Map*	Site	43	44	45	46	47	>50
1.	Castillo	x		x	x		
2.	Abric Romani	x					
3.	L'Arbreda		x	x			
4.	Reclau viver			x			
5.	Trou Magritte				x		
6.	Trou al Wesse			x			
7.	Das Geissenklösterle		x	xx			
8.	Willendorf II	x	x				51
9.	Grotte di Paine				x		
10.	Abri Fumane			x			
11.	Istallósko			x		x	54
12.	Temnata	x		x	x		
13.	Kostienki I (EUP)			x			

* Map number (differs from site numbers from the archaeo-chronological data base.

Note: Oldest dates of sites 5 and 6 are separated from later ones by several thousand years and hence questionable.

Table 4.5. *Time-slices by techno-complexes: numbers of sites: actual and on maps (Figs. 4.4, 4.7 & 4.9).*

Time-slice ka BP	Mousterian		Aurignacian+EUP		Gravettian+UP	
	actual	map	actual	map	actual	map
25–22	6	5	7	7	52	41
29–26	4	4	34	30	66	54
33–30	18	17	48	40	39	30
37–34	13	37	36	31	6	5
42–38	27	26	32	29		
46–43	32	32	13	13		
59–47	31	26				
70–60	12	11				

ronmental stress. Where and when did this growth start and how and why did the newly arrived species spread across Europe until it stabilized?

Twenty-two oldest dates are available for the Aurignacian and Early Upper Palaeolithic techno-complex if we discard those with SD_{cal} ranges of $>\pm2500$ ka as too vague and ages above 50 ka as improbable (Table 4.4). The 22 dates, derived from 13 sites (Fig. 4.5a) suggest that the entry of modern humans began as early as 47 ka BP. Those early sites spread thinly across Europe and until the 33–30 ka BP time-slice their increase was slow (Fig. 4.5b).

In the tradition of *Ex Oriente Lux* the origin of European modern humans has generally been sought in the Near and Middle East (Garrod 1936; Mellars & Stringer 1989; Mellars 1992; Davies 2001). This implies entry across European Turkey or from the Black Sea region and six of the earliest sites (Fig. 4.5a:top), Temnata (#12), Istallósko (#11), Willendorf (#8),

Geissenklösterle (#7) and on the Meuse River Trou Magritte (#5) and Trou al'Wesse (#6) indeed mark a plausible route up the Danube river system and across by way of the River Main valley to the Rhine and thence into northwestern Europe. This is the same path that palaeontologists have widely accepted as been taken by hippopotamus and other eastern warm-climate animals when repeatedly repopulating the Thames valley during successive interglacials. Far away from this trail across central Europe are four sites isolated in coastal Spain of which two, Castillo (#1) and l'Arbreda (#3), have long date sequences. Iberia was readily accessible across the Straits of Gibraltar which was only about 20 km wide when the sea stood at –70 m to –80 m during Stage 3 (Shackleton *et al.* 1984). Thus, from a strictly geographic point of view, a western addition to the traditional entry from the southeast seems worth considering (cf. Straus 1996, 212). More dates from northern Africa and earlier dates from southern Iberia — there the earliest Aurignacoid assemblages, such as Pego do Diabo and Gorhams Cave, only date to *c.* 30 ka BP — are needed in order to take the hypothesis of a western entry further.

The Aurignacian expansion began slowly around 42 ka BP (Fig. 4.5a:bottom). The result was a widely scattered, open site pattern with clusters in a few areas such as the Dordogne and the Ardennes and along the north coast of Spain. During the next two time-slices (Fig. 4.6b) the number of sites increased considerably (Table 4.5), but the pattern did hardly change until the Aurignacian reached its peak around 33–30 ka BP (Fig. 4.6b:bottom) with 48 sites. At the same time, the number of sites in the Mediterranean increased by half (Fig. 4.b:bottom), perhaps in an initial response to the worsening climate.

Also at this time a fine-scale clustering, almost entirely absent earlier, is revealed by the fact that the number of sites visible on the maps is significantly smaller than the total number recorded for the time-slice (Table 4.5), because the dots marking the sites have a diameter of *c.* 28 km and so can conceal more than one sites. Although several clusters are open-air sites where excavation practice raise the number of closely-spaced but separately-named sites, most clusters appears to be real and deliberate. This raises the question whether these dense site complexes might have exhausted the supply of food and thereby contributed to the decline in numbers after 30 ka BP.

Whatever was the reason, the Aurignacian withdrawal to southern France and sites in Iberia is reminiscent of the earlier response by Mousterians to an increasingly severe climate (Fig. 4.4b). The number

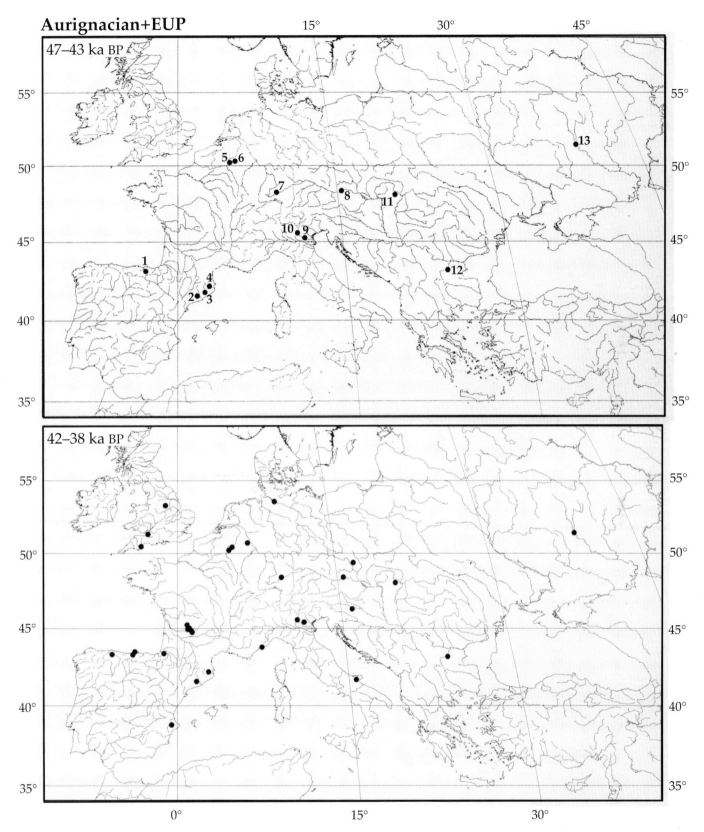

Figure 4.5a. *The changing early modern human (Aurignacian+EUP) settlement pattern in Europe shown by 5-ka time-slices. Top: 47–43 ka BP; bottom: 42–38 ka BP. Top (oldest) sites are marked with numbered black dots; see Table 4.4 for names and ages. Due to the map scale (diameter of black dots = ≈28 km) not all sites may show separately.*

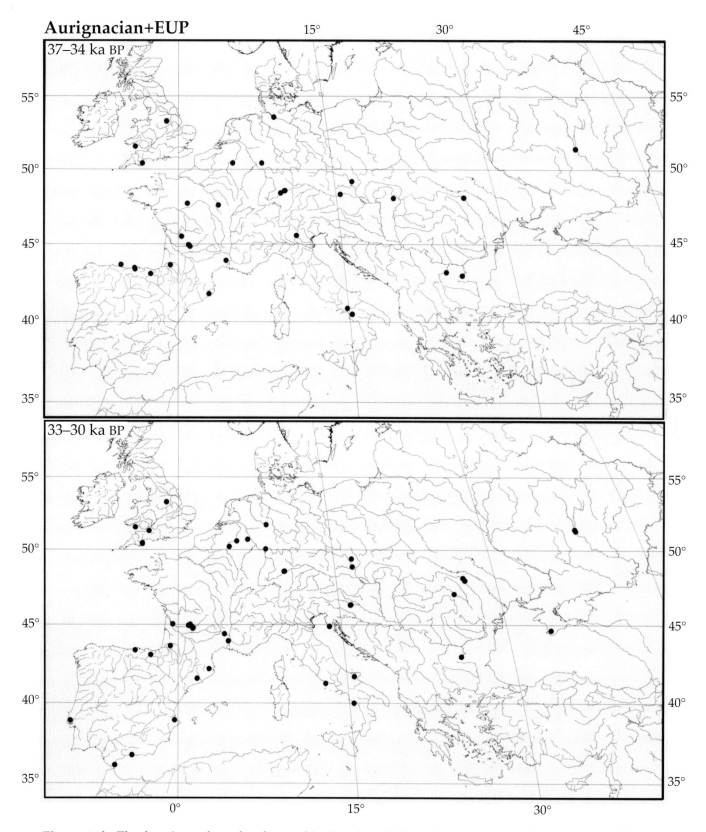

Figure 4.5b. *The changing early modern human (Aurignacian+EUP) settlement pattern in Europe shown by 4-ka time-slices. Top: 37–34 ka BP; bottom: 32–30 ka BP. Sites are marked with black dots. Due to the map scale (diameter of black dots = ≈28 km) not all sites may show separately.*

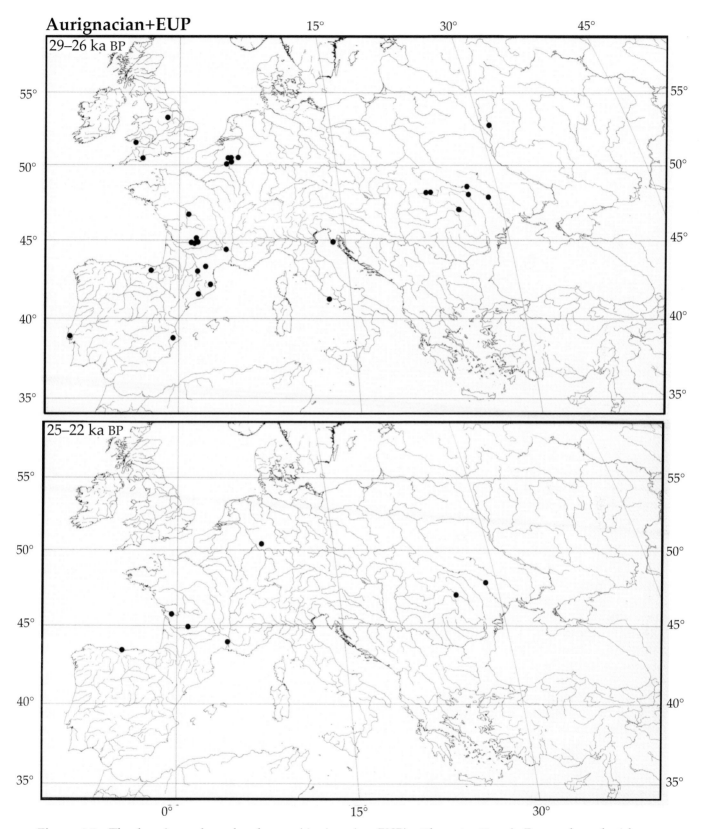

Figure 4.5c. *The changing early modern human (Aurignacian+EUP) settlement pattern in Europe shown by 4-ka time-slices. Top: 29–26 ka BP; bottom: 25–22 ka BP. Sites are marked with black dots. Due to the map scale (diameter of black dots = ≈28 km) not all sites may show separately.*

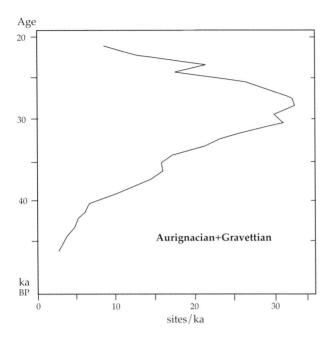

Figure 4.6. *Aurignacian and Gravettian temporal records combined. Number of sites per millennium plotted against age in ka BP. A five-point running average was applied to smooth minor irregularities.*

of sites began to decrease during the 29–26 ka BP interval and a population divide developed between western and eastern Europe (Fig. 4.5c:top). The now empty region between 5°E and 20°E separated two strikingly different terrains. To the west lie the coastal plains of the Low Countries, the rolling hills of northern France, the Massif Central and the rocky coast of northern Iberia, all having a marked maritime climate (Barron *et al.* 2003). In contrast, the central and northeastern region, north to the Baltic shore and southeast to the milder coastal zone of the Black Sea, was a vast plain with a continental climate.

The apparent southward withdrawal of the Aurignacians (Fig. 4.5c:top) has two exceptions. A small number of sites in the Ardennes which may have served as a winter refuge (Davies *et al.* 2000) long remained occupied in an otherwise empty region. A similar cluster persisted between about 20°E and 30°E just south of the 50th parallel in the Ukrainian and Russian plain.

The final time-slice (25–22 ka BP) covers the early part of the Last Glacial Maximum Phase when the edge of the Fennoscandian ice sheet had crossed the Baltic and was advancing into the North European Plain until it halted at about 52°–53°N. The seven dated Aurignacian sites in the plain may have housed direct descendants of the earliest wave of modern

humans who entered Europe about 20,000 years earlier, but the evidence is ambiguous. At any rate, whether the Aurignacian came to an end 30,000 years ago or during the 22–25 ka time-slice, it concluded a very long period of slow but steady development.

Points of note
- The dispersal of earliest Aurignacian sites up the Danube River into northwestern Europe supports the view that modern humans entered Europe from the east and probably migrated up the Danube and down the Main and Rhine.
- Four of the oldest Aurignacian sites are in Spain and, being distant from the Danube trail across Europe, their founders may have entered across the Straits of Gibraltar, a possibility worth evaluating.
- The migrations of early modern humans and late Neanderthals during the late 'Stable Warm' and 'Transitional' climate phases (45–37 ka BP) show very similar preferences in terms of climate zones.
- This raises the question whether Neanderthals and early modern humans both were highly adapted to temperate or boreal resources with essentially sedentary animal resources and thereby became incapable of adjusting to a subarctic or arctic mode of living with its forced seasonal mobility. *Bang go Nanumuik!*

A high-glacial focus on north-central and northeastern Europe

The time curve of the later modern humans (Gravettian+UP: Fig. 4.4) is based on 368 ^{14}C and AMS dates with SD_{cal} ranges of <±1500 years. The curve is truncated at the limit of our data base at 20 ka BP, but its nearly linear rising limb also looks curiously artificial. If we sum the Gravettian+UP and Aurignacian+EUP curves (Fig. 4.6), however, rapid population growth appears to begin again after the shoulder that marks the stabilization of the early Aurignacian expansion at 36–34 ka BP. Contrary to the Mousterian and Aurignacian temporal curves which declined as the climate deteriorated, the Gravettian rise led to a second plateau at 30–25 ka BP and reached its maximum at *c.* 25 ka BP (Fig. 4.1) during the severe climate of the early LGM before it comes to its artificial cut-off.

Two explanations come to mind. The Gravettian expansion may have been due to indigenous development: adoption of a new technology, new social systems or hunting strategies which singly or in combination compensated for the rising climatic stress by encouraging appropriate adaptations. Hunting a full-glacial arctic fauna (Stewart *et al.* 2003; Musil 2003) may have opened up new opportunities and new territories. Alternatively, the expansion may have been due to the arrival of vigorous newcomers (Richards *et al.* 2000; Gibbons 2000) from elsewhere who, equipped with major technical and/or social advantages provided new adaptations to the high-glacial conditions.

The six earliest Gravettian sites, Höhle Fels and Dolní Vestonice in central Europe, Temnata in Bulgaria and Mezmaiskaya and two Kostienki sites in Russia (Fig. 4.7a:top) range in age from 38 to 35 ka BP. Their distribution tells us little concerning the provenance of the techno-complex except that it might have emerged north and northeast of the trans-European mountain barrier amongst a swarm of Aurignacian sites (Fig. 4.5b) or it might have had a more distant eastern provenance, perhaps in the region of the Kostienki and Mezmaiskaya sites in Russia. The next time-slice (33–30 ka BP: Fig. 4.7a:bottom) shows 39 Gravettian sites north of the 45th parallel scattered amongst 48 Aurignacian sites (Table 4.5; Fig. 4.5b:bottom) from the Atlantic shores to the Russian plain, with only four dated human sites in northern Iberia and two in Italy. This pattern suggests a local origin from Aurignacian roots in central Europe, perhaps in Moravia (Svoboda *et al.* 1996; 1999).

In marked contrast to the preference for mid-latitudinal and southern climates of the Neanderthals and early modern humans, the Gravettian and Upper Palaeolithic populations, while not wholly averse to lands south and east of the Bay of Biscay or Italy, clearly had mastered life in the foreland north of the Pyrenees and Alps from east the Ardennes to the 15°E meridian and in the Northern European Plain. In Russia six sites west of the Urals at locations beyond 50°N (Pavlov & Indrelid 1999, fig. 1; Pavlov *et al.* 2001) span the Early Cold Phase when there was probably no major Fennoscandian ice sheet yet (Arnold *et al.* 2002; Olsen 1997). These sites suggest continuous occupation from 40 ka BP at the Mamontovaya site (66°N: 37,360±970 ^{14}C yr b.p.) to 27 ka BP at Byzovaya (62°N: 25,540±380 ^{14}C yr b.p.) with a gap at the acme of the LGM. Settlement in this region resumed from 20 ka BP at the Medvezhya site (62°N: 20,072±180 ^{14}C yr b.p.) to late sites at Pymva Shore I around 12 ka BP (67°N: 11,460±80 ^{14}C yr b.p.). The sites are found in a forest-tundra-steppe landscape (Markova *et al.* 2002). In Scandinavia mammalian skeletal material from later Stage 3 suggests similar landscapes, but no human sites are (yet?) known there (Larsson 1999).

Between 29 and 26 ka BP the number of Gravettian+UP sites nearly doubled to 66 (Fig. 4.7b:top) but except for a few sites in Iberia and Italy and two in the Balkans the techno-complex is confined to Europe north of the trans-European mountain barrier even during OIS-2, the Last Glacial Maximum (Fig. 4.7b:bottom). On the maps the sites appear to be widely spaced except for a cluster in the Dordogne and another in the Ukraine. Table 4.5 shows that about forty per cent of all sites must be hidden underneath the black dots within 30 km of one or more others. No such concealment has afflicted the Mousterian and Aurignacian+EUP sites.

This is particularly striking in central Europe north of the mountains where the three major sites of Pavlov, Milovice and Dolní Vêstonice, often regarded as permanent settlements with sizeable populations, are in such close proximity that it seems unlikely that the local fauna could have sustained them for more than a very short time. Perhaps they were permanent winter quarters to which summer populations of sites farther north in the tundra seasonally returned or might it be that the conservation practices of, for instance, the Northwest Pacific Indians who rotated their exploitation between a few years of hunting and many years of recovery has deep roots in time?

Unlike the Aurignacian abandonment of Europe north of the trans-European mountain barrier after 30 ka BP in favour of lands in the southwest of France and the Black Sea region, the Gravettians, although liking Atlantic France, show a much stronger and more persistent preference for the North European and Russian plains. While roughly limited to the 50th parallel in western and central Europe, well south of the Fennoscandian ice sheet margin, their sites reach up to 67°N in the great Russian plain. The central European alpine foreland, at one time thinly populated by Neanderthals and/or Aurignacians, now was well settled, mainly along river courses and in particular in the region of Moravia.

Our view that Gravettian humans preferred the north and largely abandoned the Mediterranean and the Balkans, conflicts with an elegant paper by Margherita Mussi (1999). Mussi's goal was in principle the same as ours, *viz.* to set Gravettian history in the context of severe climatic deterioration, but in this case in Italy. However, while we rely on a climatic and human history limited strictly to space and time data, her approach is traditional, exploiting with considerable success a rich hoard of climatic, chronological, stratigraphic, artefactual and cultural information. A brief comparison of the two

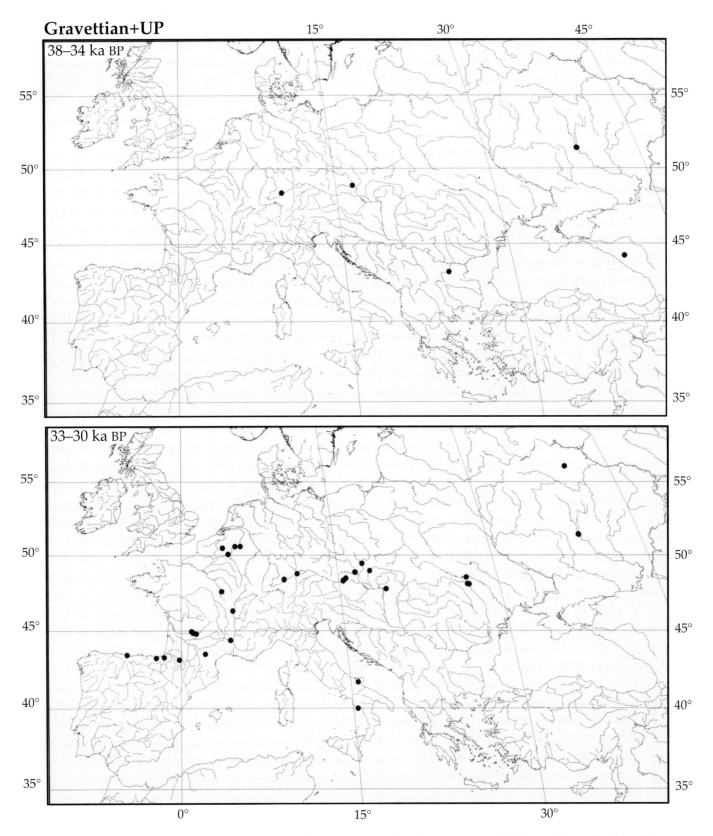

Figure 4.7a. *The changing later modern human (Gravettian+UP) settlement pattern in Europe shown by time-slices. Top: 5-ka time-slice, 38–34 ka BP; bottom: 4-ka time-slice, 33–30 ka BP. Due to the map scale (diameter of black dots = ≈28 km) not all sites may show separately.*

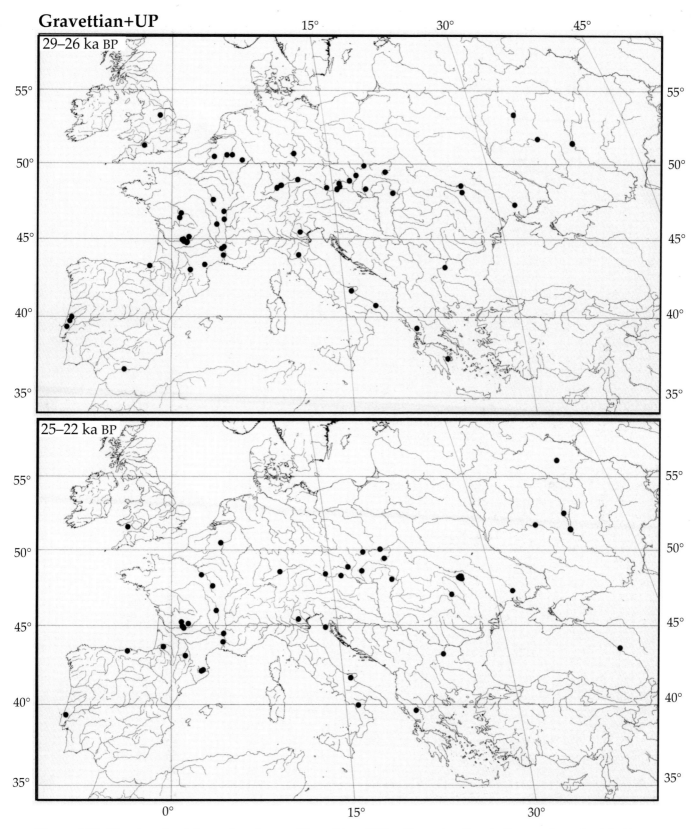

Figure 4.7b. *The changing later modern human (Gravettian+UP) settlement pattern in Europe shown by 4-ka time-slices. Top: 29–26 ka BP; bottom: 25–22 ka BP. Due to the map scale (diameter of black dots = ≈28 km) not all sites may show separately.*

ways of attacking the problem reveals the strengths and weaknesses of each approach.

For the interval 30,000–20,000 ^{14}C years Mussi lists 30 Italian sites, including four clusters of two to six sites while the others are widely scattered across the country (Mussi 1999, fig. 1). Her set of sites includes nine for which no dates are given (do dates exist?), three dated sites missing from our own data base (Arene candide with four dates and Fosso di Pagliano and Cala delle Ossa with one date each) and five sites which our own data base regards as Aurignacian. For the same region and interval we have seven *dated* sites.

Converting Mussi's ^{14}C years into calendar dates enables us to sort the combined set of dated Italian sites into three time-slices with the following result (numbers in () are from Appendix 3): 33–30 ka BP: 2 (2); 29–26 ka BP: 6 (5) and 25–22 ka BP: 6 (3) sites. Comparison with the total number of European Gravettian sites in each of the time-slices of 39, 66 and 52 sites makes Mussi's proposed exodus to the south seem modest indeed.

This does not detract from the value of Mussi's paper, because the differences between Mussi's way and our own in placing the Gravettian in a climatic context are interesting. Having chosen an interval long enough to include undated sites, Mussi has woven a rich history of human retreat southward that demonstrates the strength of a broad-based regional overview, but fails to capture the continental Gravettian decline as the climate grew severe. Our ability to define changes in settlement pattern in response to millennial-scale climate changes for the whole of Europe, on the other hand, provides a useful perspective to judge the importance of a regional exodus to Italy. Other studies that would allow similar comparisons are Hahn (1999), Scheer (1999) and Bosinski (1999) which stress the importance of a riverine component. Street & Terberger's (2000) high-resolution study of the German Upper Palaeolithic is similar in intend but introduces a level of analysis well beyond that attempted here.

Points of note
- The earliest dated Gravettian sites are thinly scattered across Europe, offering no hint to their point of origin.
- A striking feature of the 33–20 ka BP interval, equivalent to half the Early Cold Phase and LGM combined, is the large increase of the number of sites north of the trans-European mountain barrier, especially in the centre of the region. In contrast, dated upper Palaeolithic sites are rare in

southeastern Europe and the Mediterranean.
- This raises the question whether the Gravettian expansion was due to local technical and social developments suitable as adaptations to specialised arctic hunting or, alternatively, was introduced by immigrants from the East.
- A closely related question asks why the population density in Iberia, Italy and Greece was so low during the interval 30–20 ka BP compared to the earlier Stage 3, since dating or exploration biases can probably not be held responsible in this case.

Summary and perspective

Two characteristics of the *modus operandi* of Palaeolithic archaeology that continue to the present day are the local or modestly regional focus and the use of largely intuitive palaeo-environmental settings. Sedimentology and palynology have exerted a disproportionate influence on studies concerned with a local or regional environmental context (e.g. Laville *et al.* 1980, passim), but a severe shortage of absolute dates continues to be a major handicap. Other attempts have suffered from an injudicious mixture of old and new concepts. While some authors have tried to marry the traditional 'Weichsel/Würm' stadials and interstadials to the oxygen isotope SPECMAP sequence (e.g. Rigaud 1999; Churchill & Smith 2000), others, like Guthrie & van Kolfschoten (1999), Street & Terberger (1999) or Sánchez-Goñi & d'Errico (2001) have used SPECMAP by itself, with varying levels of inference about the relevance of Greenland ice cores with regard to regions of Europe distant from the western maritime zone (van Andel 2003).

The Stage 3 Project has *ab initio* claimed the whole of Europe as its territory, the time from the first glacial advance (OIS-4) to the LGM as its interval and as its aim the human occupation in time and space to the degree that they are represented by dated archaeological sites. Unlike the thoughtful reviews of the European middle and late Palaeolithic, for instance by Mellars (1996), Mellars & Stringer (1989) or Svoboda *et al.* (1996) and papers in Roebroeks *et al.* (1999 *passim*), our approach has armed us with coherent geographic and climatic targets which, on a macro-scale, possess broadly defined boundaries. Below we list some of our observations that we believe have merit and would benefit from further study in the light of independent palaeoenvironmental data and climate simulations assembled by the Stage 3 Project and others.

We do not present our work as a new or necessarily better way of doing things, but as an alternative overview that will add to the sum of our archaeological understanding. We do not pretend that all our questions are novel or profound, nor that our tentative explanations are all correct, but hope that some of both will be stimulating at least. We also recognize that our approach, as everything in life, has its price; as alternative explanations are devised and refined until rigorous multiple working hypotheses can be constructed, meso-scale data and meso-scale studies must increasingly take over the primary role from our macro-scale approach. That task and its rewards are beyond the scope of the Stage 3 Project and rest in the hands of others.

Questions, the evidence and some proposed explanations
Origins and migration paths
- Did the re-entry of the Neanderthals into Europe north of the 45°N parallel after OIS-4 come from southeastern Europe or from adjacent Asia?
- Can the suggestion of the site patterns in eastern and central Europe that the modern human migration to northwestern Europe came by way of the Danube and Rhine drainages be confirmed by independent data?
- The Danube–Main–Rhine–Thames path is widely thought to have carried repeated invasions of the interglacial large-mammal fauna, each followed by adaptation and eventual extinction. Might this be a suitable model for the history of middle to late Pleistocene hominids also?

Relations between human dispersal patterns and climate
- Question: Is there a relation between Neanderthal migration patterns and climate changes?
  - Evidence:
    - The parallelism between Neanderthal movements over time and climatic changes.
    - The rapid spread across Europe after OIS-4 during the next 20 millennia of mild stable climate.
    - The westward withdrawal to maritime France and southeasterly retreat to the Black Sea region after 40 ka BP when serious climate deterioration began.
- Question: Is there a relation between the preferred environments of early modern humans and climate changes?
  - Evidence:
    - Aurignacian withdrawal west- and southeastward from *c.* 30 ka BP.

- The great increase of Gravettian sites north of the trans-European mountain barrier as well as the ultimately near-total withdrawal from southeastern Europe and the Mediterranean.

Variations in human responses to the changing climate
- Question: Are the responses to climate changes of Neanderthals and early modern humans similar or different?
  - Evidence:
    - In terms of relief, latitude and proximity to the sea Mousterian and Aurignacian dispersal patterns seem similar.
    - The moves of Neanderthals and early modern humans during the late 'Stable Warm' and 'Transitional' climate phases are near-identical.
    - Southward withdrawal patterns of Neanderthals and modern humans when the Late Glacial Maximum approached are similar.
    - Alternatively, given the small number of sites involved might the apparent southeastward withdrawal be a function of inadequate exploration?

Direct and indirect impact of climate changes
- Question: Are observed parallels between climate changes and human migration patterns due to direct impact of climate changes on humans or did the changes mainly affect the resources on which humans depended for survival?

The Neanderthal extinction
- Question: Were Neanderthals and early modern humans both adapted to temperate or boreal food resources and incapable of adjusting to a subarctic or arctic mode of living?
- Question: Given the many similarities listed above, why did Neanderthals and early modern humans not both perish near the LGM?

Advanced adaptation
- Question: Was the survival of Gravettian and other upper Palaeolithic humans due to successful adaptation to the climate deterioration towards the LGM, perhaps similar to the mode of arctic living akin still practised today.
- Was this an internal regional development in the cold tundra and steppe between the Trans-European barrier and the advancing ice margin, or

did it arrive with new immigrants from the east as suggested, for instance, by Richards *et al.* (2000).

Filling or not filling space
- Question: Why are Mousterian and Aurignacian occupations of southeastern Europe and the Balkans so scattered, leaving regions such as the Hungarian and Danube plains or the southern Balkans so thinly occupied?
- Question: Why was the population density in Iberia, Italy and Greece so low during the LGM compared to earlier times during Stage 3, or is this merely an artefact of inadequate exploration?
- Question: While Mousterian and Aurignacian sites generally seem to have maintained a substantial (>30 km) distance from each other, why did the Upper Palaeolithic develop such strong clustering that one third of all sites were spaced less than a few tens of kilometre apart?

Acknowledgements

This chapter depends heavily on its numerous maps and we owe a debt of gratitude to Sharon Copan who patiently and painstakingly drew them and drew them again when we rememberd omissions and spotted mislocations.

Notes

1. The term *zone*, as in *'climate zone'*, is used in its geographic sense, while units of climate change are labelled *climate phases*.
2. Calendar dates (van Andel *et al.* 2003) are used throughout. The phrase 'xxxx ka BP' refers to calibrated ^{14}C dates (calibration by CalPal.v1998: Jöris & Weninger 2000; see also van Andel *et al.* 2003) and other calendrical dates. Raw versions of ^{14}C dates can be found in the Stage 3 Project's chrono-archaeological data base at its website or in references cited there. The Stage 3 Web site can be consulted at: http://www.esc.esc.cam.ac.uk/oistage3/Details/Home page/html.

References

Allen, J.R.M., W.A. Watts & B. Huntley, 2000. Weichselian palynostratigraphy, palaeovegetation and palaeoenvironment: the record from Lago Grande di Monticchio. *Quaternary International* 73/74, 91–110.

Arnold, N.S., T.H. van Andel & V. Valen, 2002. Extent and dynamics of the Scandinavian ice sheet during Oxygen Isotope Stage 3 (60,000 to 30,000 ka BP). *Quaternary Research* 57, 38–48.

Barron, E., T.H. van Andel & D. Pollard, 2003. Glacial environments II: reconstructing the climate of Europe in the Last Glaciation, in *Neanderthals and Modern Humans in the European Landscape during the Last Glaciation*, Chapter 5, eds. T.H. van Andel & W. Davies. (McDonald Institute Monographs.) Cambridge: McDonald Institute for Archaeological Research, 57–78.

Bocquet-Appel, J.-P. & P.-Y. Demars, 2000a. Neanderthal contraction and modern human colonization of Europe. *Antiquity* 74, 544–52.

Bocquet-Appel, J.-R. & P.-Y. Demars, 2000b. Population kinetics in the Upper Palaeolithic in Western Europe. *Journal of Archaeological Science* 27, 551–70.

Bosinski, G. 1999. The period 30,000–20,000 bp in the Rhineland, in *Hunters of the Golden Age: the Mid Upper Palaeolithic of Eurasia 30,000–20,000 BP*, eds. W. Roebroeks, M. Mussi, J. Svoboda & K. Fennema. Leiden: *Analecta Praehistorica Leidensia* 31, 271–80.

Churchill, S.E. & F.H. Smith, 2000. Makers of the Early Aurignacian of Europe. *Yearbook of Physical Anthropology* 4, 61–115.

Davies, W., 2001. A very model of a modern human industry: new perspectives on the origins and spread of the Aurignacian in Europe. *Proceedings of the Prehistoric Society* 67, 195–217.

Davies, W., J. Stewart & T.H. van Andel, 2000. Neandertal landscapes: a preview, in *Neanderthals on the Edge*, eds. C. Stringer, R.N.E. Barton & J.C. Finlayson. Oxford: Oxbow Books, 1–8.

Davies, W., P. Valdes, C. Ross & T.H. van Andel, 2003. The human presence in Europe during the Last Glacial Period III: site clusters, regional climates and resource attractions, in *Neanderthals and Modern Humans in the European Landscape during the Last Glaciation*, Chapter 11, eds. T.H. van Andel & W. Davies. (McDonald Institute Monographs.) Cambridge: McDonald Institute for Archaeological Research, 191–220.

Garrod, D., 1936. The Upper Palaeolithic in the light of recent discovery. *Report of the British Association for the Advancement of Science, Presidential Address*, Section H (Blackpool), 155–72.

Gibbons, A., 2000. Evolutionary genetics, Europeans trace ancestry to Paleolithic people. *Science* 290, 1080–81.

Guthrie, D. & T. van Kolfschoten, 1999. Neither warm and moist, nor cold and arid, the ecology of the Mid Upper Palaeolithic, in *Hunters of the Golden Age: the Mid Upper Palaeolithic of Eurasia 30,000–20,000 BP*, eds. W. Roebroeks, M. Mussi, J. Svoboda & K. Fennema. Leiden: *Analecta Praehistorica Leidensia* 31, 13–20.

Hahn, J., 1999. The Gravettian in southern Germany: environment and economy, in *Hunters of the Golden Age: the Mid Upper Palaeolithic of Eurasia 30,000–20,000 BP*, eds. W. Roebroeks, M. Mussi, J. Svoboda & K. Fennema. Leiden: *Analecta Praehistorica Leidensia* 31, 249–56.

Johnsen, S., D. Dahl-Jensen, N. Gundestrup, J. Steffensen, H. Clausen, H. Miller, V. Mason-Delmotte, A.

Sveinbjörnsdottir & J. White, 2001. Oxygen isotope and palaeotemperature records from six Greenland ice-core stations, Camp Century, Dye-3, GRIP, GISP2, Renland and NorthGRIP. *Journal of Quaternary Science* 16, 299–307.

Jöris, O. & B. Weninger, 2000. Calendric age-conversion of glacial radiocarbon dates at the transition from the Middle to the Upper Palaeolithic in Europe. *Bulletin de la Société Préhistorique Luxembourgeoise* 18, 43–55.

Larsson, L., 1999. Plenty of mammoths but no humans? Scandinavia during the Middle Weichselian, in Roebroeks *et al.* (eds.), 155–64.

Laville, H., J.-P. Rigaud & J. Sackett, 1980. *Rockshelters of the Perigord.* New York (NY): Academic Press.

Markova, A.K., A.N. Simakova, A.Y. Puzachenko & L.M. Kitaev, 2002. Environments of the Russian Plain during the Middle Valdai Briansk Interstade (33,000–24,000 yr BP) indicated by fossil mammals and plants. *Quaternary Research* 57, 391–400.

Meese, D.A., A.J. Gow, R.B. Alley, G.A. Zielinski, P.M. Grootes, M. Ram, K.C. Taylor, P.A. Mayewski & J.F. Bolzan, 1997. The Greenland Ice sheet Project 2 depth-age scale: methods and results. *Journal of Geophysical Research* 102, 26,411–23.

Mellars, P.A., 1992. Archaeology and modern human origins in Europe. *Proceedings of the British Academy* 82, 1–35.

Mellars, P.A., 1996. *The Neanderthal Legacy: an Archaeological Perspective from Western Europe.* Princeton (NJ): Princeton University Press.

Mellars, P.A. & C. Stringer (eds.), 1989. *The Human Revolution.* Edinburgh: Edinburgh University Press.

Musil, R., 2003. The Middle and Upper Palaeolithic game suite in central and southeastern Europe, in *Neanderthals and Modern Humans in the European Landscape during the Last Glaciation,* Chapter 10, eds. T.H. van Andel & W. Davies. (McDonald Institute Monographs.) Cambridge: McDonald Institute for Archaeological Research, 167–90.

Mussi, M., 1999. Heading south, the Gravettian colonisation of Italy, in *Hunters of the Golden Age: the Mid Upper Palaeolithic of Eurasia 30,000–20,000 BP,* eds. W. Roebroeks, M. Mussi, J. Svoboda & K. Fennema. Leiden: *Analecta Praehistorica Leidensia* 31, 355–74.

Olsen, L., 1997. Rapid shifts in glacial extension characterise a new conceptual model for glacial variations during the Mid and Late Weichselian in Norway. *Norges Geologiske Undersøgelse, Bulletin* 433, 54–5.

Olsen, L., V. Mejdahl & S.F. Selvik, 1996. Middle and Late Pleistocene stratigraphy, chronology and glacial history in Finnmark, North Norway. *Norges Geologisk Undersøgelse Bulletin* 429.

Olsen, L., K. van der Borg, B. Bergstrøm, H. Svein, S.-E. Lauritzen & G. Hansen, 2001a. AMS radiocarbon dating of glacigenic sediments with low organic content: an important tool for reconstructing the history of glacial variations in Norway. *Norsk Geologisk Tidsskrift* 81, 59–92.

Olsen, L., H. Svein & B. Bergstrøm, 2001b. Rapid adjustments of the western part of the Scandinavian ice-sheet during the mid- and Late Weichselian: a new model. *Norsk Geologisk Tidsskrift* 81, 93–118.

Pavlov, P. & S. Indrelid, 1999. Human occupation in northeastern Europe during the period 35,000–18,000 BP. in *Hunters of the Golden Age: the Mid Upper Palaeolithic of Eurasia 30,000–20,000 BP,* eds. W. Roebroeks, M. Mussi, J. Svoboda & K. Fennema. Leiden: *Analecta Praehistorica Leidensia* 31, 165–72.

Pavlov, P., J.I. Svendsen & S. Indrelid, 2001. Human presence in the European Arctic nearly 40,000 years ago. *Nature* 413, 64–7.

Richards, M., V. Macaulay, E. Hickey, E. Vega, B. Sykes, V. Guida, C. Rengo, D. Sellitto, F. Cruciani, T. Kivisild, R. Villems, M. Thomas, S. Rychkov, O. Rychkov, Y. Rychkov, M. Golge, D. Dimitrov, E. Hill, D. Bradley, V. Romano, F. Cali, G. Vona, A. Demaine, S. Papiha, C. Triantaphyllidis, G. Stefanescu, J. Hatina, M. Belledi, A. di Rienzo, A. Novelletto, A. Oppenheim, S. Nørby, N. Al-Zaheri, S. Santachiara-Benerecetti, R. Scozari, A. Torroni & H.-J. Bandelt, 2000. Tracing European founder lineages in the Near Eastern mtDNA pool. *American Journal of Human Genetics* 67, 1251–76.

Rigaud, J.-P., 1999. Human adaptation to the climatic deterioration of the last Pleniglacial in southwestern France (30,000–20,000 bp), in *Hunters of the Golden Age: the Mid Upper Palaeolithic of Eurasia 30,000–20,000 BP,* eds. W. Roebroeks, M. Mussi, J. Svoboda & K. Fennema. Leiden: *Analecta Praehistorica Leidensia* 31, 325–36.

Roebroeks, W., N.J. Conard & T. van Kolfschoten, 1992. Dense forests, cold steppes and the Palaeolithic settlement of northern Europe. *Current Anthropology* 33, 551–86.

Roebroeks, W., M. Mussi, J. Svoboda & K. Fennema (eds.), 1999. *Hunters of the Golden Age: the Mid Upper Palaeolithic of Eurasia 30,000–20,000 BP.* Leiden: *Analecta Praehistorica Leidensia* 31.

Sánchez-Goñi, M.F. & F. d'Errico, 2001. New evidence on the chronology and climatic framework of the Middle–Upper Palaeolithic transition, in *Neanderthals and Modern Humans in Late Pleistocene Eurasia (Abstracts, Calpe 2001 Conference, Gibraltar, 16–19 August 2001),* ed. C. Finlayson. Gibraltar: Gibraltar Museum.

Scheer, A., 1999. The Gravettian in Southwest Germany, stylistic features, raw material resources and settlement patterns, in *Hunters of the Golden Age: the Mid Upper Palaeolithic of Eurasia 30,000–20,000 BP,* eds. W. Roebroeks, M. Mussi, J. Svoboda & K. Fennema. Leiden: *Analecta Praehistorica Leidensia* 31, 257–70.

Shackleton, J.C., T.H. van Andel & C.N. Runnels, 1984. Coastal paleogeography of the central and western Mediterranean during the last 125,000 years and its archaeological implications. *Journal of Field Archaeology* 11, 307–14.

Soffer, O., 1989. The Middle to Upper Palaeolithic Transition on the Russian Plain, in *The Human Revolution,* eds. P.A. Mellars & C. Stringer. Edinburgh: Edinburgh University Press, 714–42.

Stewart, J.S., T. van Kolfschoten, A. Markova & R. Musil, 2003. The mammalian faunas of Europe during Oxygen Isotope Stage Three, in *Neanderthals and Modern Humans in the European Landscape during the Last Glaciation*, Chapter 7, eds. T.H. van Andel & W. Davies. (McDonald Institute Monographs.) Cambridge: McDonald Institute for Archaeological Research, 103–30.

Straus, L.G., 1996. A tale about the human diversity, in *The Last Neandertals, the First Anatomically Modern Humans: Cultural Change and Human Evolution, the Crisis at 40 ka BP*, eds. E. Carbonell & M. Vaquero. Tarragona: Universitat Rovira i Virgili, 203–18.

Street, M. & T. Terberger, 1999. The German Upper Palaeolithic, 35,000–15,000 bp. New dates and insights with emphasis on the Rhineland, in *Hunters of the Golden Age: the Mid Upper Palaeolithic of Eurasia 30,000–20,000 BP*, eds. W. Roebroeks, M. Mussi, J. Svoboda & K. Fennema. Leiden: *Analecta Praehistorica Leidensia* 31, 280–91.

Stuiver, M. & P. Grootes, 2000. GISP2 oxygen isotope ratios. *Quaternary Research* 53, 277–84.

Sveian, O.L., H. Bergstrøm, B. Selvik, S.F. Lauritzen, S.-E. Stokland & K. Grøsfjeld, 2001. Methods and stratigraphies used to reconstruct Mid- and Late Weichselian palaeoenvironmental and palaeoclimatic changes in Norway. *Norges geologiske undersøkelse Bulletin* 438, 21–46.

Svoboda, J., V. Lozel & E. Vlcek, 1996. *Hunters Between East and West: the Palaeolithic of Moravia*. New York (NY): Plenum Press.

Svoboda, J., B. Klíma, L. Jarosová & P. Skrdla, 1999. The Gravettian in Moravia: climate, behaviour and technological complexity, in *Hunters of the Golden Age: the Mid Upper Palaeolithic of Eurasia 30,000–20,000 BP*, eds. W. Roebroeks, M. Mussi, J. Svoboda & K. Fennema. Leiden: *Analecta Praehistorica Leidensia* 31, 197–218.

van Andel, T.H., 2003. Glacial environments I: the Weichselian climate in Europe between the end of the OIS-5 interglacial and the Last Glacial Maximum, in *Neanderthals and Modern Humans in the European Landscape during the Last Glaciation*, Chapter 2, eds. T.H. van Andel & W. Davies. (McDonald Institute Monographs.) Cambridge: McDonald Institute for Archaeological Research, 9–20.

van Andel, T.H., W. Davies, B. Weninger & O. Jöris, 2003. Archaeological dates as proxies for the spatial and temporal human presence in Europe: a discourse on the method, in *Neanderthals and Modern Humans in the European Landscape during the Last Glaciation*, Chapter 3, eds. T.H. van Andel & W. Davies. (McDonald Institute Monographs.) Cambridge: McDonald Institute for Archaeological Research, 21–30.

Zubrow, E.B.W., 1992. An interactive growth model applied to the expansion of Upper Palaeolithic populations, in *The Origins of Human Behaviour*, ed. R.A. Foley. Cambridge: Cambridge University Press, 82–96.

Appendix 4.1. *Sites used in this chapter: Mousterian.*

MOUSTERIAN

Longitude	Latitude	ID #	Site #	Site
25–22 ka				**6 sites**
–9.192	38.894	38	11	Salemas [algar]
4.837	44.889	584	115	A. Moula [Soyons]
4.542	44.3308	764	160	Gr. St-Marcel [d'Ardeche] [Bidon]
4.971	50.214	1004	203	Trou Magrite
15.381	40.001	1292	251	Gr. La Cala
16.7372	49.4001	1449	304	Kulna Cave
29–26 ka				**4 sites**
–9.192	38.894	39	11	Salemas [algar]
4.213	44.412	478	84	Les Pecheurs [Casteljau]
6.139	47.608	760	159	Gr. d'Echenoz-la-Meline [La Baume]
4.542	44.3308	762	160	Gr. St-Marcel [d'Ardeche] [Bidon]
33–30 ka				**19 sites**
–8.4648	39.6357	13	5	Caldeirao Cave
–9.192	38.894	41	11	Salemas [sima]
–9.1932	39.2998	57	17	Columbeira, Gruta Nova
–9.1932	39.2998	54	17	Columbeira, Gruta Nova
–4.1267	36.951	167	42	Zafarraya Cave
3.769	47.595	407	76	Grotte du Renne, Arcy-sur-Cure
1.223	40.8084	532	98	Combe Grenal [Domme]
1.848	44.716	542	101	A du Mas Viel [St-Simon]
5.4796	46.4757	566	107	Gr. de La Baume [Gigny sur Suran]
4.484	44.059	679	120	A. Brugas [?O/A] [Vallabrix]
0.7167	46.3833	706	138	L'Ermitage [Lussac-les-Chateaux]
4.542	44.3308	763	160	Gr. St-Marcel [d'Ardeche] [Bidon]
0.872	46.708	767	162	A. Sabourin [Dousse]
4.2833	47.533	789	169	Montagne de Girault [Genay]
–2.6761	51.2263	946	188	Hyaena Den
13.696	47.611	1169	236	Salzofenhohle
15.381	40.001	1291	251	Gr. La Cala
27.14	47.943	1618	326	Ripiceni-Izvor
40.0034	44.2217	1887	379	Mezmaiskaya
37–34 ka				**13 sites**
–7.6431	39.6611	8	3	Foz do Enxarrique
–8.969	38.486	9	4	Figueira Brava Cave
–8.5898	39.5307	23	6	Oliveira Cave [Almonda cave system]
–8.5898	39.5307	26	7	Almonda [EVS]
0.851	47.7	422	78	Les Cottes [St. Pierre de Maille]
0.303	45.504	452	81	La Quina Y-Z [Villebois la Valette]
3.8685	45.0615	700	135	Les Rivaux [Espaly-St-Marcel]
4.971	50.214	1003	203	Trou Magrite
15.875	46.169	1381	280	Krapina
25.514	45.977	1641	331	Gura Cheii-Risnov
34.396	45.16	1700	338	Buran-Kaya III
34.0006	44.8479	1711	340	Kabazi II
34.5987	45.1224	1717	341	Zaskal'naya VI
42–38 ka				**27 sites**
2.747	42.161	154	37	L'Arbreda
2.611	42.265	163	40	Ermitons Cave
–3.4256	42.0735	194	49	Cueva Millan
–0.506	45.747	390	73	Roche a Pierrot [St.-Cesaire]
0.851	47.7	423	78	Les Cottes [St. Pierre de Maille]
0.303	45.504	453	81	La Quina Y-Z [Villebois la Valette]
1.223	44.8084	533	98	Combe Grenal [Domme]
4.837	44.889	583	115	A. Moula [Soyons]
1.249	44.8673	606	117	Pech de l'Aze II [Carsac]
5.551	45.885	719	146	Gr. de la Chenelaz [Hostias]
2.8818	43.3409	749	153	Gr. Tournal Grande Grotte de Bize)
1.07	45.002	797	170	Le Moustier
1.07	45.002	798	170	Le Moustier
–2.6761	51.2263	944	188	Hyaena Den
11.794	48.931	1145	230	Sesselfelsgrotte
18.334	47.633	1414	297	Tata
18.917	47.379	1418	298	Erd
27.161	48.125	1553	324	Buzdujeni I Cave
23.06	45.293	1646	333	Cioarei/Borosteni
34.396	45.16	1699	338	Buran-Kaya III
33.893	44.744	1705	339	Starosel'e
34.0006	44.8479	1710	340	Kabazi II
34.5987	45.1224	1715	341	Zaskal'naya VI
34.026	53.354	1757	356	Betovo
38.542	44.839	1876	375	Il'skaya
39.899	43.626	1885	378	Vorontsov Cave
40.0034	44.2217	1889	379	Mezmaiskaya
47–43 ka				**33 sites**
–8.5898	39.5307	24	6	Oliveira Cave [Almonda cave system]
–3.9555	43.2918	128	35	Castillo
1.676	41.539	134	36	Abric Romani
2.747	42.161	156	37	L'Arbreda
–0.491	38.778	161	39	Cova Beneito
0.8699	41.866	178	45	Roca dels Bous
–3.4256	42.0735	193	49	Cueva Millan
–2.958	43.402	197	52	Kurtzia
–5.3	36.133	253	67	Vanguard Cave
1.223	44.8084	534	98	Combe Grenal [Domme]
1.249	44.8673	604	117	Pech de l'Aze II [Carsac]
4.836	44.891	686	126	[Gr.] Neron [Soyons]
0.5584	44.866	731	149	Barbas III [Creysse]
1.07	45.002	796	170	Le Moustier
4.4295	44.4442	803	172	Abri du Ranc de l'Arc [Lagorce]
5.046	50.488	1032	209	Sclayn Cave
9.772	48.401	1049	215	Das Geissenklosterle
11.794	48.931	1147	230	Sesselfelsgrotte
15.237	40.495	1236	247	Castelcivita
11.565	45.466	1307	258	Gr. del Broion
13.1	41.232	1330	266	Gr. Guattari
13.508	41.219	1339	267	Gr. di Sant'Agostino
16.767	46.159	1361	275	Divje Babe
18.502	42.775	1388	285	Crvena Stijena
18.917	47.379	1416	298	Erd
16.7372	49.4001	1448	304	Kulna Cave
27.14	47.943	1615	326	Ripiceni-Izvor
23.06	45.293	1648	333	Cioarei/Borosteni
23.983	43.196	1697	337	Samuilitsa
33.893	44.744	1706	339	Starosel'e
34.5987	45.1224	1716	341	Zaskal'naya VI
38.542	44.839	1877	375	Il'skaya
40.0034	44.2217	1888	379	Mezmaiskaya
59–47 ka				**31 sites**
–8.137	38.5345	35	10	Gruta do Escoural
1.249	44.8673	59	17	Columbeira, Gruta Nova
0.8699	41.866	177	45	Roca dels Bous
–3.432	37.438	184	47	Cariguela
–2.619	42.2089	206	54	Pena Miel 1
0.3989	42.0189	213	59	Los Moros I [Gabasa]
–5.3	36.133	243	65	Gorham's Cave
–5.3	36.133	251	67	Vanguard Cave
1.17	45.055	541	100	Regourdou [Montignac]
6.3833	47.4667	570	108	Gr. aux Ours [Gondenans les Moulins]
1.728	44.995	587	116	La Chapelle-aux-Saints
1.249	44.8673	656	117	Pech de l'Aze II [Carsac]
4.621	43.832	670	118	Ioton [Beaucaire]
0.6095	45.341	672	119	Fonseigner [Bourdeilles]
4.484	44.059	676	120	A. Brugas [?O/A] [Vallabrix]
3.903	43.943	680	121	La Roquette II [Conquerac]
0.5584	44.866	732	149	Barbas III [Creysse]
1.07	45.002	791	170	Le Moustier
9.772	48.401	1052	215	Das Geissenklosterle
11.409	51.828	1133	225	Konigsaue
11.794	48.931	1143	230	Sesselfelsgrotte
1.249	44.8673	1293	251	Gr. La Cala
1.249	44.8673	1306	258	Gr. del Broion
11.88	42.966	1310	259	Gr. di Gosto
13.1	41.232	1326	266	Gr. Guattari
11.565	45.466	1342	267	Gr. di Sant'Agostino
18.917	47.379	1415	298	Erd
1.249	44.8673	1447	304	Kulna Cave
27.14	47.943	1617	326	Ripiceni-Izvor

Appendix 4.1. *(cont.)*

MOUSTERIAN

Longitude	Latitude	ID #	Site #	Site	Longitude	Latitude	ID #	Site #	Site
33.893	44.744	1704	339	Starosel'e	−5.3	36.133	250	67	Vanguard Cave
27.1533	48.5613	1856	372	Korman' IV	1.249	44.8673	663	117	Pech de l'Aze II [Carsac]
					4.484	44.059	674	120	A. Brugas [?O/A] [Vallabrix]
70–60 ka				*12 sites*	2.7005	43.3534	685	125	[Gr.] Aldene [Cesseras]
−9.1932	39.2998	55	17	Columbeira, Gruta Nova	7.049	43.705	688	127	Pie[d] Lombard [Tourrettes-sur-Loup]
−3.9555	43.2918	132	35	Castillo	13.1	41.232	1334	266	Gr. Guattari
−5.3	36.133	246	65	Gorham's Cave	13.485	41.236	1346	268	Gr. dei Moscerini
1.223	44.8084	528	98	Combe Grenal [Domme]	24.088	43.175	1668	336	Temnata Cave

Appendix 4.2. *Sites used in this chapter: Aurignacian and Early Upper Palaeolithic.*

AURIGNACIAN + EUP

Longitude	Latitude	ID #	Site #	Site	Longitude	Latitude	ID #	Site #	Site
25–22 ka				*7 sites*	1.257	44.876	447	80	Abri Caminade [Caneda]
−4.837	43.42	203	53	La Riera	1.366	44.797	460	82	Le Piage [Fajoles]
1	44.933	272	69	Abri Pataud	4.542	43.95	465	83	La Salpetriere [Remoulins]
−0.506	45.747	380	73	Roche a Pierrot [St-Cesaire]	4.213	44.412	481	84	Les Pecheurs [Casteljau]
4.542	43.95	470	83	La Salpetriere [Remoulins]	−0.492	45.044	486	86	Roc de Marcamps [Prignac-et-Marcamps]
8.137	50.421	1130	224	Wildscheuer	−0.692	43.638	513	91	Gr. de Hyenes, Brassempouy
28.97	47.847	1537	321	Climautsy II	1.062	44.978	734	151	Abri du Facteur
25.936	47.022	1634	328	Ceahlau-Cetatica II	1.1016	45.0158	743	152	La Rochette [St Leon sur Vezere]
					−3.4987	50.4625	832	174	Kent's Cavern
29–26 ka				*33 sites*	−1.1957	53.2643	858	177	Pin Hole Cave
−9.217	38.901	5	2	Pego do Diabo	−4.2419	51.548	910	180	Paviland Cave [Goat's Hole]
−2.458	43.066	85	25	Labeko Koba	−3.492	50.4005	986	199	Bench Quarry 'Tunnel' cavern
1.676	41.539	139	36	Abric Romani	−2.8778	51.2908	990	200	Picken's Hole, Layer 3
2.747	42.161	145	37	L'Arbreda	4.971	50.214	1001	203	Trou Magrite
−0.491	38.778	160	39	Cova Beneito	5.722	50.589	1013	204	Trou Walou
1.0833	44.85	328	70	Le Flageolet I [Bezenac]	10.2	48.56	1096	216	Vogelherd Cave
1.366	44.797	458	82	Le Piage [Fajoles]	6.804	50.7	1109	218	Lommersum
4.213	44.412	482	84	Les Pecheurs [Casteljau]	10.146	48.554	1111	219	Bockstein-Torle
2.335	43.314	487	87	Canecaude I [Villardonel]	8.747	51.713	1119	222	Paderborn
1.5394	45.143	751	154	Le Raysse [Brive-la-Gaillarde]	8.5069	50.0521	1122	223	Kelsterbach
1.6285	43.0104	754	156	Tuto de Camalhot [St-Jean de Verges]	15.6264	41.6671	1247	248	Gr. Paglicci
0.855	46.703	765	161	Fontenioux [St Pierre de Maille]	15.381	40.001	1288	251	Gr. La Cala
1.673	44.877	773	165	Gr des Fieux [Miers]	13.051	41.229	1350	270	Gr. Barbara
−3.4987	50.4625	839	174	Kent's Cavern	16.088	46.305	1371	276	Vindija Cave
−1.1957	53.2643	857	177	Pin Hole Cave	16.016	46.291	1373	277	Velika Pecina 2
−4.2419	51.548	919	180	Paviland Cave [Goat's Hole]	13.8856	44.9012	1375	278	Sandalja II
4.971	50.214	1002	203	Trou Magrite	16.6995	48.8538	1420	300	Milovice I
4.972	50.483	1018	205	Gr. de la Princesse [Marche-les-Dames]	16.723	49.389	1437	302	Pod Hradem Cave A
4.995	50.215	1019	206	Trou du Renard	27.036	48.111	1563	325	Mitoc Malul Galben
4.513	50.06	1020	207	Trou de l'Abime, Couvin	27.14	47.943	1622	326	Ripiceni-Izvor
4.674	50.478	1041	212	Gr du Spy	25.882	47.028	1623	327	Bistricioara-Lutarie
5.662	50.507	1044	213	Gr. du Haleux [Sprimont]	25.421	42.944	1667	335	Bacho Kiro
13.083	41.226	1301	254	Gr. del Fossellone	33.858	44.642	1731	344	Siuren I
13.8856	44.9012	1376	278	Sandalja II	39.0452	51.3985	1765	357	Kostienki I
28.97	47.847	1538	321	Climautsy II	39.0713	51.293703	1797	359	Kostienki VIII [Tel'manskaya site]
25.882	47.028	1624	327	Bistricioara-Lutarie	39.0553	51.3948	1827	365	Kostienki XVII [Spitsyn site]
25.936	47.022	1633	328	Ceahlau-Cetatica I					
27.136	48.033	1729	343	Korpach	*37–34 ka*				*36 sites*
30.931	52.75	1753	351	Berdyzh	−3.8834	43.4398	77	24	Ruso [I]
23.129	48.157	1847	369	Korolevo I	−2.458	43.066	84	25	Labeko Koba
57.2	65.1	1853	370	Byzovaya	−5.117	43.657	98	30	La Guelga
23.57	48.171	1855	371	Molochnyi Kamen'	−3.84	43.361	106	31	Cueva Morin
27.1533	48.5613	1858	372	Korman' IV	2.711	41.776	196	51	Cal Coix
					1	44.933	269	69	Abri Pataud
33–30 ka				*48 sites*	1.0833	44.85	323	70	Le Flageolet I [Bezenac]
−9.217	38.901	3	2	Pego do Diabo	0.941	44.955	347	71	La Ferrassie
−2.458	43.066	82	25	Labeko Koba	4.3259	43.9296	376	72	Esquicho-Grapaou
−3.84	43.361	104	31	Cueva Morin	3.769	47.595	418	76	Grotte du Renne, Arcy-sur-Cure
−0.3	38.926	111	33	Mallaetes Cave	0.851	47.7	429	78	Les Cottes [St Pierre de Maille]
1.676	41.539	138	36	Abric Romani	0.303	45.504	455	81	La Quina Y-Z [Villebois la Valette]
2.747	42.161	146	37	L'Arbreda	−0.692	43.638	511	91	Gr. de Hyenes, Brassempouy
2.747	42.161	166	41	Reclau Viver	−1.1916	53.266	849	175	Robin Hood's Cave
−3.892	36.748	179	46	Nerja	−1.1957	53.2643	873	177	Pin Hole Cave
−5.3	36.133	236	65	Gorham's Cave	−4.2419	51.548	900	180	Paviland Cave [Goat's Hole]
1	44.933	270	69	Abri Pataud	−3.492	50.4005	987	199	Bench Quarry 'Tunnel' cavern
0.941	44.955	355	71	La Ferrassie	5.294	50.421	1045	214	Trou Al'Wesse
1.333	44.767	436	79	Roc de Combe [Nadaillac]					

54

Appendix 4.2. *(cont.)*

AURIGNACIAN + EUP

Longitude	Latitude	ID #	Site #	Site	Longitude	Latitude	ID #	Site #	Site
9.772	48.401	1075	215	Das Geissenklosterle	0.8836	45.2307	724	147	A. Combe Sauniere [Sarliac-sur-l'Isle]–
10.2	48.56	1094	216	Vogelherd Cave	3.4987	50.4625	831	174	Kent's Cavern
10.168	48.551	1100	217	Hohlenstein-Stadel IV	–1.1916	53.266	848	175	Robin Hood's Cave
10.146	48.554	1112	219	Bockstein-Torle	–1.1957	53.2643	862	177	Pin Hole Cave
9.7626	53.548	1118	221	Hahnofersand	–1.1957	53.2643	872	177	Pin Hole Cave
8.137	50.421	1127	224	Wildscheuer	–2.8778	51.2908	989	200	Picken's Hole, Layer 3
15.399	48.323	1186	240	Willendorf II	4.971	50.214	998	203	Trou Magrite
15.237	40.495	1238	247	Castelcivita	5.294	50.421	1047	214	Trou Al'Wesse
10.902	45.57	1274	250	Abri Fumane	9.772	48.401	1070	215	Das Geissenklosterle
14.86	40.858	1302	255	Serino	6.804	50.7	1101	218	Lommersum
20.4108	48.0654	1402	290	Istallosko cave	9.7626	53.548	1117	221	Hahnofersand
16.6758	49.1845	1429	301	Stranska-skala IIa	15.587	48.413	1212	241	Krems-Hundssteig
27.036	48.111	1554	325	Mitoc Malul Galben	15.6264	41.6671	1246	248	Gr. Paglicci
25.421	42.944	1664	335	Bacho Kiro	11.49	45.418	1266	249	Gr. di Paina
24.088	43.175	1678	336	Temnata Cave	10.902	45.57	1277	250	Abri Fumane
39.0452	51.3985	1763	357	Kostienki I	7.5475	43.7894	1296	252	Riparo Mochi
39.0524	51.3957	1804	361	Kostienki XII [Volkovskaya]	16.016	46.291	1372	277	Velika Pecina 2
39.0553	51.3948	1832	365	Kostienki XVII [Spitsyn site]	20.4264	48.0468	1403	291	Pesko cave
					16.723	49.389	1434	302	Pod Hradem Cave A
42–38 ka				*34 sites*	24.088	43.175	1675	336	Temnata Cave
–3.776	43.481	88	26	Arenillas	39.0553	51.3948	1831	365	Kostienki XVII [Spitsyn site]
–5.8055	43.314	92	28	La Vina					
–3.9555	43.2918	124	35	Castillo	*47–43 ka*				*8 sites*
1.676	41.539	143	36	Abric Romani	1.676	41.539	144	36	Abric Romani
2.747	42.161	152	37	L'Arbreda	2.747	42.161	150	37	L'Arbreda
2.747	42.161	158	38	Mollet Cave	4.971	50.214	995	203	Trou Magrite
–0.491	38.778	159	39	Cova Beneito	15.399	48.323	1182	240	Willendorf II
0.941	44.955	344	71	La Ferrassie	10.902	45.57	1276	250	Abri Fumane
1.333	44.767	433	79	Roc de Combe [Nadaillac]	20.4108	48.0654	1398	290	Istallosko cave
1.257	44.876	449	80	Abri Caminade [Caneda]	39.0452	51.3985	1764	357	Kostienki I
–1.196	43.371	708	139	Isturitz [Isturits]	24.088	43.175	1674	366	Temnata Cave
1.0995	45.0068	712	141	A. Castanet [Sergeac]					

Appendix 4.3. *Sites used in this chapter: Gravettian and Upper Palaeolithic.*

GRAVETTIAN + UP

Longitude	Latitude	ID #	Site #	Site	Longitude	Latitude	ID #	Site #	Site
25–22 ka				*54 sites*	17.8802	48.6095	1517	313	Moravany-Zakovska
–8.9844	39.3541	47	13	Terra do Manuel	17.8863	48.6087	1518	314	Moravany-Lopata II
–3.84	43.361	107	31	Cueva Morin	19.924	50.053	1524	317	Spadzista St. A
2.747	42.161	157	37	L'Arbreda	19.93	50.054	1529	318	Krakow
2.578	42.104	176	44	Roc de la Melca	20.1563	49.443	1531	320	Oblazowa 1
1	44.933	280	69	Abri Pataud	27.042	48.243	1539	322	Ciuntu Cave
1.0833	44.85	339	70	Le Flageolet I [Bezenac]	27.167	48.25	1541	322	Ciuntu Cave
3.769	47.595	420	76	Grotte du Renne, Arcy-sur-Cure	27.171	48.076	1549	323	Brinzeni Cave I
4.542	43.95	474	83	La Salpetriere [Remoulins]	27.171	48.076	1551	323	Brinzeni Cave I
–0.692	43.638	508	91	Brassempouy [Grande Galerie 2]	27.036	48.111	1612	325	Mitoc Malul Galben
0.95	44.967	535	99	Laugerie-Haute Est	25.882	47.028	1630	327	Bistricioara-Lutarie
2.756	48.341	684	124	Montigny[-sur-Loing]	26.896	48.176	1642	332	Coto Miculinti
1.487	45.125	705	137	Puy-Jarrige II [Brive-La-Gaillarde]	24.088	43.175	1691	336	Temnata Cave
4.631	44.499	718	145	Bouzil [Saint-Thome]	35.7886	51.6891	1737	346	Avdeevo
0.8836	45.2307	725	147	A. Combe Sauniere [Sarliac-sur-l'Isle]	31.722	47.282	1749	349	Leski
1.199	43.057	775	166	Gr. d'Enlene [Montesquieu-Avantes]	39.0452	51.3985	1767	357	Kostienki I
4.0286	45.9822	777	167	La Vigne Brun [St-Maurice-sur-Loire]	39.0524	51.3957	1809	361	Kostienki XII [Volkovskaya]
–4.2419	51.548	889	180	Paviland Cave [Goat's Iole]	39.0669	51.3739	1824	363	Kostienki XV [Gorodtsov site]
4.674	50.478	1042	212	Gr. du Spy	39.0466	51.4112	1838	366	Kostienki XXI [Gmelinskaya]
10.146	48.554	1113	219	Bockstein-Torle	40.479	56.134	1870	374	Sungir' [Vladimir]
15.8795	48.2943	1171	237	Langmannersdorf A	38.9954	52.4957	1879	376	Gagarino
14.428	48.4	1174	239	Alberndorf [in der Riedmark]	38.9954	52.4957	1881	376	Gagarino
15.6264	41.6671	1261	248	Gr. Paglicci	39.9898	43.5583	1884	377	Akhchtyr Cave
11.49	45.418	1267	249	Gr. di Paina					
16.036	39.963	1321	265	Gr. del Romito	*29–26 ka*				*64 sites*
40.479	56.134	1869	274	Sungir' [Vladimir]	–8.923	39.376	27	8	Cabeco de Porto Marinho III
13.8856	44.9012	1379	278	Sandalja [III]	–8.541	39.988	52	15	Buraca Escura
20.9165	39.6301	1396	289	Kastritsa	–8.696	39.745	63	19	Abrigo do Lagar Velho
20.5324	48.0517	1408	293	Balla cave	–3.892	36.748	182	46	Nerja Vestibulo
16.634	48.875	1479	308	Dolni Vestonice I	–1.9044	43.2654	224	61	Aitzbitarte III
16.641	48.8666	1491	308	Dolni Vestonice II	1.0833	44.85	334	70	Le Flageolet I [Bezenac]
16.634	48.875	1478	308	Dolni Vestonice I	0.941	44.955	364	71	La Ferrassie
18.262	49.865	1513	310	Petrkovice	3.769	47.595	393	74	Arcy-sur-Cure [Grande Grotte?]

Appendix 4.3. *(cont.)*

GRAVETTIAN + UP

Longitude	Latitude	ID #	Site #	Site
3.7694	47.595	395	74	Grande Grotte, Arcy-sur-Cure
1.333	44.767	442	79	Roc de Combe [Nadaillac]
1.366	44.797	463	82	Le Piage [Fajoles]
4.542	43.95	475	83	La Salpetriere [Remoulins]
4.424	44.387	501	88	Grotte Chauvet
4.72	46.818	549	103	Les Vignes [St-Martin sous Montaigu]
4.726	46.299	558	106	Solutre [O/A]
4.726	46.299	555	106	Solutre [O/A]
0.71667	46.4	574	110	Gr. de Laraux
4.631	44.499	717	145	Bouzil [Saint-Thome]
1.062	44.978	736	151	Abri du Facteur
2.8818	43.3409	750	153	Gr. Tournal (or Grande Grotte de Bize) [Bize-Minervois]
1.5394	45.143	752	154	Le Raysse [Brive-la-Gaillarde]
1.6285	43.0104	755	156	Tuto de Camalhot [St-Jean de Verges]
0.855	46.703	766	161	Fontenioux [St Pierre de Maille]
4.0286	45.9822	781	167	La Vigne Brun [St-Maurice-sur-Loire]
−1.1916	53.266	846	175	Robin Hood's Cave
−2.6761	51.2263	949	188	Hyaena Den
5.722	50.589	1008	204	Trou Walou
3.987	50.473	1029	208	Maisieres-Canal
5.205	50.581	1038	210	L'Hermitage [Huccorgne]
9.772	48.401	1081	215	Das Geissenklosterle
10.2	48.56	1098	216	Vogelherd Cave
10.146	48.554	1114	219	Bockstein-Torle
11.669	50.681	1137	226	Kniegrotte
11.79	48.93	1138	227	Obere Klause
9.758	48.378	1153	231	Hohle[r] Fels
6.642	50.228	1160	232	Magdalenahohle
15.662	48.67	1173	238	Horn (Raberstrasse)
14.428	48.4	1175	239	Alberndorf [in der Riedmark]
15.399	48.323	1202	240	Willendorf II
15.693	48.475	1227	244	Langenlois
15.395	48.293	1233	245	Aggsbach
15.6264	41.6671	1254	248	Gr. Paglicci
11.565	45.466	1308	258	Gr. del Broion
11.259	43.976	1351	271	Bilancino
17.584	40.727	1353	272	Gr. di Santa Maria di Agnano
20.845	39.289	1391	286	Asprochaliko
23.137	37.4201	1393	288	Franchthi
20.5324	48.0517	1409	293	Balla cave
17.361	49.219	1424	300	Milovice I
16.678	48.872	1461	307	Pavlov I
18.262	49.865	1514	310	Petrkovice
18.084	48.322	1516	312	Nitra-Cerman
20.1563	49.443	1535	320	Oblazowa 1
27.036	48.111	1577	325	Mitoc Malul Galben
24.088	43.175	1690	336	Temnata Cave
27.055	48.547	1728	342	Molodova V [Kosoutsy]
35.7886	51.6891	1740	346	Avdeevo
31.722	47.282	1748	349	Leski
39.0481	51.3848	1799	360	Kostienki X
39.0524	51.3857	1802	361	Kostienki XI [Anosovka site 2]
39.0365	51.3931	1814	362	Kostienki XIV [Markina Gora]

Longitude	Latitude	ID #	Site #	Site
39.0321	51.3957	1826	364	Kostienki XVI [Uglianka]
39.0466	51.4112	1839	366	Kostienki XXI [Gmelinskaya]
34.1497	53.3253	1844	368	Khotylevo II
57.2	65.1	1854	370	Byzovaya VI
33–30 ka				*39 sites*
−1.506	43.26	212	58	Alkerdi
−2.2008	43.2274	216	60	Amalda Cave
−4.8375	43.4205	229	64	Cueto de la Mina
1	44.933	283	69	Abri Pataud
1.0833	44.85	330	70	Le Flageolet I [Bezenac]
0.941	44.955	369	71	La Ferrassie
3.7694	47.595	394	74	Grande Grotte, Arcy-sur-Cure
1.366	44.797	462	82	Le Piage [Fajoles]
4.424	44.387	502	88	Grotte Chauvet
4.726	46.299	557	106	Solutre [O/A]
2.15	43.4833	697	132	Gr. du Castellas [Dourgne]
−0.162	43.107	733	150	Trou du Rhinoceros [St-Pe-de-Bigorre]
−0.162	43.107	802	171	Grotte de Courau (Grotte Saucet) [St-Pe-de-Bigorre]
5.722	50.589	1009	204	Trou Walou
4.513	50.06	1022	207	Trou de l'Abime, Couvin
3.987	50.473	1024	208	Maisieres-Canal
5.205	50.581	1034	210	L'Hermitage [Huccorgne]
9.772	48.401	1085	215	Das Geissenklosterle
9.772	48.401	1092	215	Das Geissenklosterle
11.065	48.773	1139	228	Weinberghohlen [Mauern 2]
9.758	48.378	1150	231	Hohle[r] Fels
15.399	48.323	1191	240	Willendorf II
15.604	48.415	1213	242	Krems-Wachtberg
15.693	48.475	1230	244	Langenlois
15.395	48.293	1232	245	Aggsbach
15.6264	41.6671	1251	248	Gr. Paglicci
15.381	40.001	1286	251	Gr. La Cala
19.3813	47.7768	1411	295	Puspokhatvan
16.678	48.872	1462	307	Pavlov I
16.634	48.875	1467	308	Dolni Vestonice I
17.4395	49.4562	1510	309	Predmosti
18.089	48.966	1515	311	Nemsova
27.171	48.076	1548	323	Brinzeni Cave I
27.036	48.111	1574	325	Mitoc Malul Galben
27.055	48.547	1725	342	Molodova V [Kosoutsy]
39.0452	51.3985	1792	357	Kostienki I
39.0365	51.3931	1819	362	Kostienki XIV [Markina Gora]
39.0669	51.3739	1825	363	Kostienki XV [Gorodtsov site]
40.479	56.134	1866	374	Sungir' [Vladimir]
38–34 ka				*6 sites*
9.758	48.378	1159	231	Hohle[r] Fels
16.634	48.875	1472	308	Dolni Vestonice I
24.088	43.175	1686	336	Temnata Cave
39.0756	51.3712	1796	358	Kostienki VI
39.0365	51.3931	1821	362	Kostienki XIV [Markina Gora]
40.0034	44.2217	1892	379	Mezmaiskaya

Chapter 5

Glacial Environments II:
Reconstructing the Climate of Europe in the Last Glaciation

Eric Barron, Tjeerd H. van Andel & David Pollard

Archaeological objectives and the Middle and Upper Palaeolithic climate

The Eemian Interglacial ended with a long period of slow cooling, but no large Scandinavian ice sheet came into being until late in OIS-4. Between 66 and 60 ka BP[1]) the Fennoscandian ice sheet spread south- and southeastwards, but it lasted only a few thousand years and its margins probably did not reach the southern and western shores of the Baltic (Donner 1995). Around 59 ka BP the ice withdrew rapidly and OIS-3 brought moderate warming that lasted until c. 44 ka BP (Fig. 5.1). The climate then gradually deteriorated until around 37 ka BP it was as cold as the Last Glacial Maximum (LGM: 27–16 ka BP). Upon these long, multi-millennial climate changes were superimposed the high frequency, millennial-scale Dansgaard/Oeschger oscillations (Dansgaard et al. 1993; Grootes

et al. 1993) throughout most of OIS-3 (Fig. 5.1).

The Stage 3 climate modelling programme was designed with two objectives in mind. The first phase was to yield millennial-scale simulations for the generic extremes of the Dansgaard/Oeschger events to provide insight in the dynamics of the OIS-3 climate and its vegetation and environments. Secondly, those climate simulations would be used to illustrate the multi-millennial climate phases between 60 and 20 ka BP, i.e. the Stable Warm (59–45 ka BP), Transitional (44–37 ka BP), Early Cold (37–27 ka BP) and Last Glacial Maximum (27–16 ka BP) Phases[2]) (van Andel et al. 2003b, Fig. 4.1, Table 4.3). These multi-millennial climate phases are needed to study human responses to climate changes, because the calibration ranges (SD_{cal}) of archaeological dates lie between <±0.5 and ±2.5 ka (van Andel et al. 2003a, Table 3.4) and so can not resolve Dansgaard/Oeschger oscillations. They have been widely used to place human adaptations in a climatic context, e.g. this volume Chapters 4 (van Andel et al. 2003b), 8 (Davies & Gollop 2003), 9 (Aiello & Wheeler 2003), 12 (Stewart et al. 2003b) and 13 (Stringer et al. 2003).

In this chapter we examine the objectives, procedures and results of the OIS-3 modelling programme to the degree that this is directly relevant to archaeological applications. For a comprehensive treatment of the Stage 3 palaeoclimatic and palaeovegetation modelling experiments see Barron & Pollard (2002), Pollard & Barron (2003), Alfano et al. (2003) and (Huntley & Allen 2003; Huntley et al. 2003).

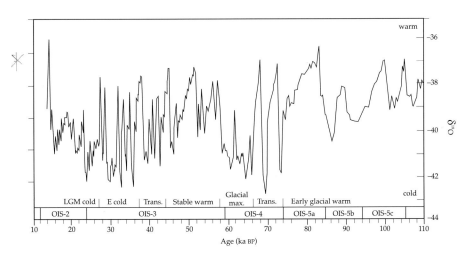

Figure 5.1. *The climate changes of the Weichselian Glaciation from the late penultimate interglacial (OIS-5d) to the onset of the Holocene as recorded by the δ^{18}O record of the GISP2 Greenland ice-core (Meese et al. 1997; Johnsen et al. 2001; Stuiver & Grootes 2000).*

Modelling strategy and modelling methods

Constructing palaeoclimate simulations for Stage 3
Even the capacity of supercomputers did not permit us to construct continuous models of the changing climate (Fig. 5.1) with the detail needed for archaeological purposes. Instead, we built 'snapshot' models of the generic end members of the Stage 3 climatic record (Pollard & Barron 2003; Barron & Pollard 2002), including the vegetation (Huntley & Allen 2003).

Conventional Global Circulation Models (GCMs) provide output grid-spacings (pixels) of the order of 4.5° lat. by 7.5° long. that are inadequate to display the spatial and temporal interactions between the palaeoclimate and human beings (van Andel *et al.* 2003b). For the entire United Kingdom, for example, the GCM scale yields only four or five pixels, a number insufficient to reveal even the larger features of its climate and the responses of human inhabitants.

To overcome this scale problem we have used a higher-resolution version of a regional climate model 'nested' within a GCM to simulate modern and OIS-3 climates and assess their sensitivity to a wide range of boundary conditions, a strategy pioneered by Dickinson *et al.* (1989). For the British Isles this 'meso-scale' model yields some sixty output pixels and provides an acceptably detailed image of the complex British climate.

The model we used is a 60×60 km grid-spacing regional model (RegCM2: Giorgi *et al.* 1993a,b) of Europe, embedded within the GENESIS 2.0 GCM model of Thompson & Pollard (1995a,b) and Pollard & Thompson 1994; 1995) and improved by Thompson & Pollard (1997). It is widely used in palaeoclimatic model studies.

Modelling procedures and model inputs
Meso-scale climate models are driven at their boundaries by a GCM linked to the regional modelling scheme. Marginal strips eight grid points wide exclude the inevitable spurious boundary effects from the output maps. Our simulations for the region between 20°W and 50°E long. and 35°N and 65°N lat. were plotted on a Lambert Conformal Projection with a 60 km grid spacing (Barron & Pollard 2002).

Vegetation is an important factor in shaping the climate and observed vegetation data provide an opportunity to verify climate models. The BIOME3.5 vegetation model (Haxeltine & Prentice 1996; Prentice *et al.* 1992) was run interactively with GENESIS and produced a global vegetation pattern consistent with the GCM results. The BIOME3.5 model, forced by the meso-scale climate simulation, was then run *a posteriori* in order to predict the plant cover for the meso-scale grid and compare the results with the reasonably good information on the OIS-3 vegetation existing for western and southern Europe.

The relief and shorelines of glacial Europe, continuously modified by isostatic compensation for the advances and retreats of global and Fennoscandian ice sheets (Lambeck 1995b,c), are complex and limit the palaeoclimatic uses of conventional GCMs. Even for the meso-scale models the extent and elevation of the ice sheets and the geometry of glacial shorelines presented a challenge. For Scandinavia conventional wisdom holds that during OIS-3 mountainous Norway was covered by a thick ice sheet with a western limit just inside the coastal fjords, while much of southern Sweden and Denmark was ice-free (Andersen & Mangerud 1989; Donner 1995). Recently, this concept of the OIS-3 ice sheet has been questioned (Arnold *et al.* 2002; Olsen *et al.* 1996; 2001a,b; Svendsen *et al.* 2001; Mangerud *et al.* 2003). Vegetation and fauna studies in northern Russia back this view (Markova *et al.* 2002). Therefore, our models (Fig. 5.2) were run with two ice sheet inputs, a small mountain-based one (Arnold *et al.* 2002) with an ice-free Sweden and Finland as in Porter (1989) and a larger ice sheet similar to that of the Younger Dryas to satisfy the traditionally minded.

Global eustatic sea-level curves (Chappell & Shackleton 1986; Chappell *et al.* 1996; Lambeck & Chappell 2001) indicate an average sea-level position of −80 m for the OIS-3 interval. The −80 m isobath is close to the shelf break and the steep upper slope; therefore the GCM glacial shoreline was, with few exceptions, adequately simulated by the −100 m isobath and did not require glacio-hydro-isostatic compensation. For the meso-scale model this was not everywhere true because the low-gradient, shallow floors of the North Sea, French Atlantic coast and areas of the Mediterranean such as the Adriatic responded to sea-level changes of only a few tens of metres with large shoreline shifts. For those areas we used shorelines from Lambeck (1995a,b,c; 1996) with supplementary data for the Mediterranean (van Andel & Shackleton 1982; Shackleton *et al.* 1984).

'Snapshot' simulations of glacial climate end members
The input data generated for the meso-scale experiments were chosen to allow simulations of three main *generic end members* of the glacial climate suite between 60 and 20 ka BP: a 'typical warm event' (OIS-3 'warm') for the warm Dansgaard/Oeschger events of the 45–42 ka BP interval and probably valid

Land Surface and Ice Sheet Boundary Conditions

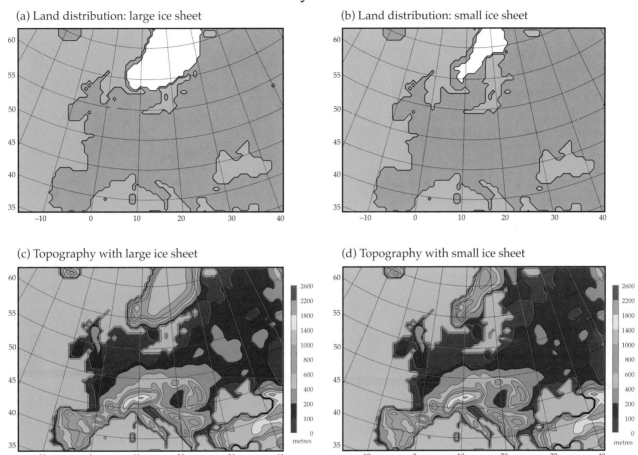

(a) Land distribution: large ice sheet

(b) Land distribution: small ice sheet

(c) Topography with large ice sheet

(d) Topography with small ice sheet

Figure 5.2. *Specifications of the meso-scale model for large (a) and small (b) Scandinavian ice sheets and isostatically compensated topography (c) and coastlines (d). Note the topographic barrier created by the trans-European mountain ranges of the Pyrenees, Alps and Carpathians and the southwest trending Dinarides of Yugoslavia and western Greece.*

for all 'warm' events from 60–45 ka BP, a 'typical cold event' (OIS-3 'cold') representing conditions around 30 ka BP, and the Last Glacial Maximum (LGM) 'cold' event. A 'Modern' simulation of the present European climate and vegetation was used to test the modelling process and provide a basis for comparison with the palaeoclimate simulations.

Of the four simulations the OIS-3 'cold' has turned out to be problematic, because it was judged too warm relative to observational data such as permafrost features (van Huissteden *et al.* 2003) and vegetation biomes based on pollen data (Alfano *et al.* 2003; Barron & Pollard 2002). To see how sensitive the OIS-3 'cold' simulation was to an array of temperatures and sea-ice limits, an *ad hoc* OIS-3 'cold' version was produced by increasing the annual sea-ice cover from 3 to 11 months and setting winter SSTs in the Bay of Biscay (where no data was avail-

able) to freezing. The resulting estimate of the extent to which SSTs and sea-ice duration control cooling helped us understand the dynamics of the OIS-3 'cold' event (Barron & Pollard 2002), but the conflict with observational data was not resolved. The pollen and permafrost arguments per se are less than very robust, however, and it is possible (see van Andel 2003) that the OIS-3 'cold' event has more merit than the palynologists, whose postulated plant cover it was unable to match (Alfano *et al.* 2003; Huntley *et al.* 2003) were willing to accept. Therefore we have retained the OIS-3 'cold' simulations for the purpose of making comparisons subject to the readers' judgement.

The 'Modern' simulation: evaluating the meso-scale model GENESIS global models tend to represent the present-day European climate reasonably well. For Europe

Modern Surface Temperatures

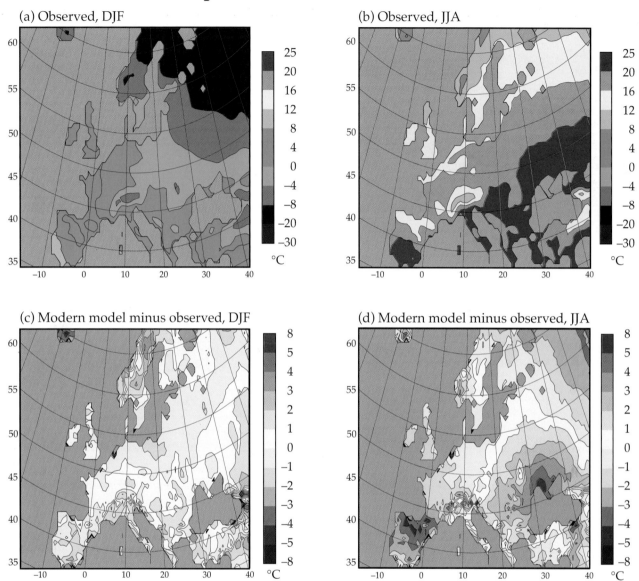

Figure 5.3. *Present-day observed European surface temperatures (°C) for winter (DJF) and summer (JJA) and the differences between the modern meso-scale simulation and the observed data (model minus observed).*

they show winter and summer surface temperatures that deviate by only ±2°C from observations, but deviations up to 2–5°C occur in winter in a band across southern Europe. GENESIS also captured the regional magnitude and areal variability of summer and winter precipitation within 2 mm/day of observations. Only in southern Europe does the GENESIS simulation exaggerate precipitation values.

The meso-scale simulations also place much of the continent within ±2°C of the observed temperature (Fig. 5.3:top) in winter and summer both. In winter larger differences of up to 4–8°C occur at high elevations, along the margins of the Black Sea

and in the far eastern part of the domain. In summer areas of high elevation show local conflicts between simulations and observations; the region north of the Black Sea, for example, is noticeably warmer than is observed. Precipitation patterns (Fig. 5.4:top), like those of temperature (Fig. 5.3), show much of Europe within 1–2 mm/day of modern observations. Exceptions are common in high-relief areas where large positive and negative precipitation deviations mark adjacent areas of alpine regions and along rugged coastlines.

In general this test of the global and regional models by means of present climate data inspires confidence in the ability of the GENESIS model to

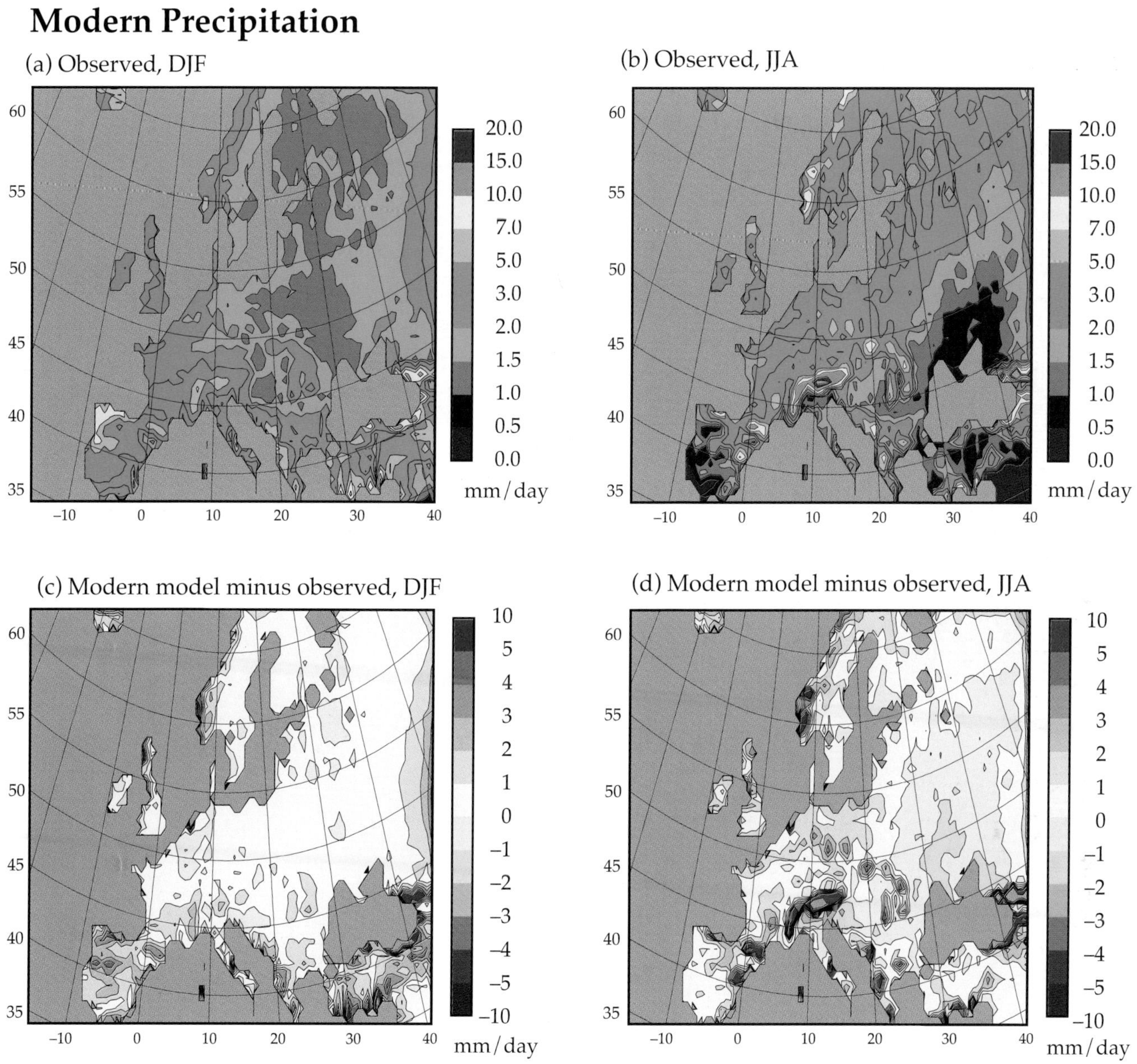

Modern Precipitation

(a) Observed, DJF

(b) Observed, JJA

(c) Modern model minus observed, DJF

(d) Modern model minus observed, JJA

Figure 5.4. *Present-day observed European precipitation (mm/day) for winter (DJF) and summer (JJA) and the differences between the modern meso-scale simulation and the observed data (model minus observed).*

simulate glacial climates realistically.

The 'LGM' experiment
The LGM experiment is based on the orbital parameters for 21 ka BP (Berger & Loutre 1991): 200 ppmv CO$_2$ (Barnola *et al.* 1987), a large ice sheet (ICE-4G of Peltier 1994), sea-level at –120 m and CLIMAP sea-surface temperatures (SSTs) modified with GLAMAP SSTs (Pollard & Barron 2003) that differ from those prescribed for OIS-3, but only in the North Atlantic.

The Greenland ice record for late OIS-3 (37–28 ka BP: Fig. 5.1) displays temperatures as cold as or colder than those of the LGM, but the LGM simu-

lations are much colder than all OIS-3 simulations. Since the sea-surface temperatures prescribed for the LGM and OIS-3 cold-phase simulations are quite similar, their role in creating the atmospheric temperature variations of the ice core must be minor. The same is true for the orbital configuration and for any variations in shoreline geography. So it seems that the configuration and huge size of the LGM ice sheet are the controlling factors, probably through their high solar reflectivity (albedo) and strong influence on air pressure patterns and winds at sea-level (Kageyama *et al.* 1999; Kageyama & Valdes 2000; Shinn & Barron 1989).

Modern and OIS-3 Sea Level Pressures (mb): DJF

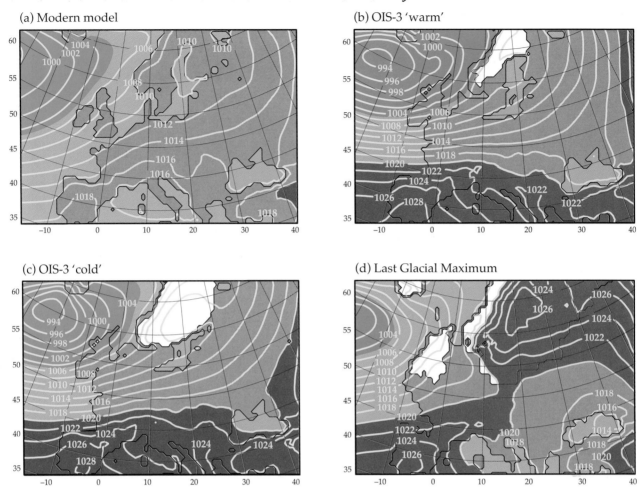

Figure 5.5. *Meso-scale simulations of winter (DJF) sea-level air pressure (SLP in mb) for the (a) Modern, (b) OIS-3 'warm', (c) OIS-3 'cold' (NB: for comparison only) and (d) LGM climate end members. Green = depression; red = high pressure.*

Climate end members of the last European glaciation

If we discard the OIS-3 'cold' simulation as too warm, two end member simulations remain to assess human response to the deteriorating glacial climate: the OIS-3 'warm' and LGM 'cold' events. Because ice-core temperatures between 37 and 27 ka BP differ little from the LGM condition (Fig. 5.1) we decided, although with some discomfort, to accept the LGM simulation as representative of the whole interval 37–20 ka BP. This substitution is not ideal; it exaggerates the severity of the Early Cold Phase (37–27 ka BP) and its atmospheric circulation may have differed from that of late OIS-3 time (Pollard & Barron 2003). Still, it allows us to speculate more confidently about the late Stage 3 climate than has been possible till now, and it enables us to ask

better questions about its impact on modern humans. To proceed beyond this, however, we shall need better input data and a deeper understanding of the dynamics of late OIS-3 conditions.

In toto, the modelling effort has yielded values for more than a hundred climate variables of which about half are relevant to palaeo-environments, resources and human beings (Appendix 5.1). Of those hundred only the following have so far attracted much attention: summer and winter mean temperatures, winter wind-chill factors, summer and winter precipitation, snow cover in days/year and mean snow thickness. Many others, such as the seasonal ranges of the above, day/night temperature ranges or cloud cover, all obviously relevant to the human existence, are readily available for future consideration.

Modern and OIS-3 Sea Level Pressures (mb): JJA

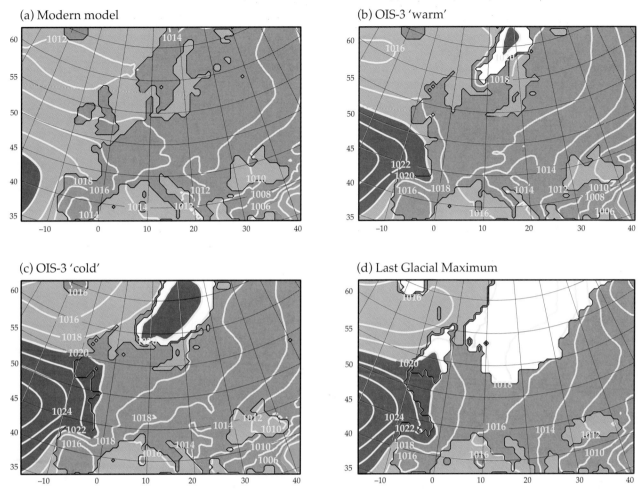

Figure 5.6. *Meso-scale simulations of summer (JJA) sea-level air pressure (SLP in mb) for the (a) Modern, (b) OIS-3 'warm', (c) OIS-3 'cold' (NB: for comparison only) and (d) LGM 'cold' end members.*

The glacial period: atmospheric circulation

Among the factors controlling the glacial climate, the atmospheric circulation system and the shifts in its position played a controlling role. Today the winter circulation has a deep low-pressure centre over the North Atlantic and a weak high-pressure ridge in the southern part of the model domain (Fig. 5.5a). The OIS-3 'warm' experiment (Fig. 5.5b) places this North Atlantic Low considerably farther east but only a little north of its present position. The 'warm' as well as well as the 'cold' OIS-3 experiments (Fig. 5.5:b & c) have a strongly developed high-pressure zone across the southern part of the domain. Its steeply north-dipping pressure gradient would have induced a vigorous westerly air flow over Europe creating strong zonal winds north of the transverse European mountain barrier.

During the LGM (Fig. 5.5d) the winter position of the North Atlantic Low was similar to that of the Modern simulation. A high-pressure belt extended from southwest to northeast across Europe including the Fennoscandian ice sheet and nearly all of northeastern Europe above the 53rd parallel. This belt caused a steep pressure gradient in the North Atlantic making for a cold northern Europe, but across the central part of Europe the LGM winds were weak. As a result the continental interior was dominated by radiative cooling rather than by cold air-flow from the central North Atlantic.

Today, the North Atlantic and European circulation systems in summer are dominated by a large high-pressure area offshore from Portugal, the Açores High, while the rest of the domain is almost featureless as far as air pressure is concerned (Fig. 5.5a). The OIS-3 'warm' event (Fig. 5.6b) had in summer two high-pressure centres, but they are too small to have had much impact on the atmospheric circulation except for maritime western Europe. The OIS-3 'cold' simulation has an 'Açores' high similar to that of the OIS-3 'warm' phase but penetrating a little farther into Europe. The fairly large high above the Fenno-scandian ice sheet is probably due to an ice sheet (Fig. 5.2b) considerably larger than the small Norwegian ice caps proposed by Arnold *et al.* (2002); if so, the OIS-3 'cold' air circulation might have been like the OIS-3 'warm' simulation.

The LGM summer air pressure pattern (Fig. 5.6d) is in principle similar to that of the 'warm' OIS-3 experiment (Fig. 5.6b), because with a weak jetstream air temperature and precipitation patterns are influenced mainly by the distribution of sea-surface temperatures. In winter when the jetstream was strong due to the large south-to-north temperature gradient, the ice sheet exerted a dominant influence on temperature conditions across northern Europe.

Snapshots as images of archaeological climate phases
How do the climatic end members discussed above relate to the issue of human survival in a glacial climate? Seeking an answer, we accept the Modern, OIS-3 'warm', and LGM 'cold' simulations as generalized images of the multi-millennial Modern, Stable Warm and LGM Cold Phases (Fig. 5.1) when considering past patterns of temperature, precipitation and snow conditions (Figs. 5.7, 5.8 & 5.9).

The European climate today: The European climate zones, present and past, are anchored by three continental-scale topographic features, the Atlantic Ocean in the west, the Russian plain in the east and the trans-European mountain barrier of Pyrenees, Alps and Carpathians and the southeast-trending Dinarides of Yugoslavia and northwestern Greece. Except for southernmost France, this barrier separates the subarctic and temperate climate zones of Europe from the hot, summer-dry, winter-wet climate of the Mediterranean about as far east as the 25°E meridian. Farther east the arctic climate zone changes through the subarctic and temperate zones more smoothly to the mild Mediterranean climate of the Black Sea coastal region (Fig. 5.7: top). In summer the north→south or, more precisely, the northwest→southeast temperature gradient dominates the climate; the extensive highlands of Europe have relatively little impact on temperature patterns. In winter, however, the dominant temperature change is west→east, with a strong gradient from maritime to continental conditions driven by the jetstream (Fig. 5.5a: Modern).

In winter the precipitation rapidly diminishes inland from 3–5 mm/day at the Atlantic coast to no more than 1–1.5 mm/day in western Russia (Fig. 5.8: top), while in summer the Atlantic precipitation zone is much more extensive and decreases to 2 mm/day much farther inland. Especially in winter the mountains exert a strong orographic control on precipitation, casting clear but complex rain-shadows to the east and southeast of the mountains. In the Mediterranean where summer rains are rare and mainly associated with thunderstorms, the winter precipitation tends to be more copious than in central and western Europe but the rains fall on fewer days.

Climates of the pleni-glacial
Surface temperatures: The Stable Warm Phase temperature pattern of the OIS-3 'warm' simulation (Fig. 5.7: centre) resemble that of the present, but it does cool 4–6°C eastwards, resulting in a distinct northwest→southeast trend for the main temperature gradient. In the west, mean summer values above 18°C extend no farther north than central Spain and the western Mediterranean, but farther east, beyond the 20°E meridian, Modern and Stable Warm temperature patterns are remarkably similar, *vide* the 20°C isotherm which extends northward almost to the 60°N parallel. Winter temperatures of –8°C to –20°C were found all over Scandinavia, but only beyond 30°E, in western Russia, did they reach south of the 50°N parallel.

LGM conditions (Fig. 5.7: bottom) were even in summer much more extreme than those discussed above; all isotherms had shifted south by about 5° latitude relative to the Warm Phase. Summer days above 18°C occurred only in the Mediterranean region and north of the trans-European barrier in a belt trending northeastward from 40°N at 20°E to 50°N at 50°E. In western Europe, summer temperatures barely reached 10°C, but in winter the Atlantic and Mediterranean shores of southern France, Iberia and western Italy were relatively mild with temperatures of 4°C–6°C. Elsewhere, winters were very cold with temperatures below –8°C, going down to –20°C near the ice sheet and to –30°C on the ice due to thermal forcing and reflection radiation and because an ice sheet acts as a physical barrier to atmospheric flow.

Precipitation: A meaningful visual analysis of the precipitation patterns (Fig. 5.8: top) is more difficult than for temperature maps, for two reasons. First, the differences between Modern, Stable Warm and LGM Phases precipitation, although quite small, are critical in terms of human tolerance and resource

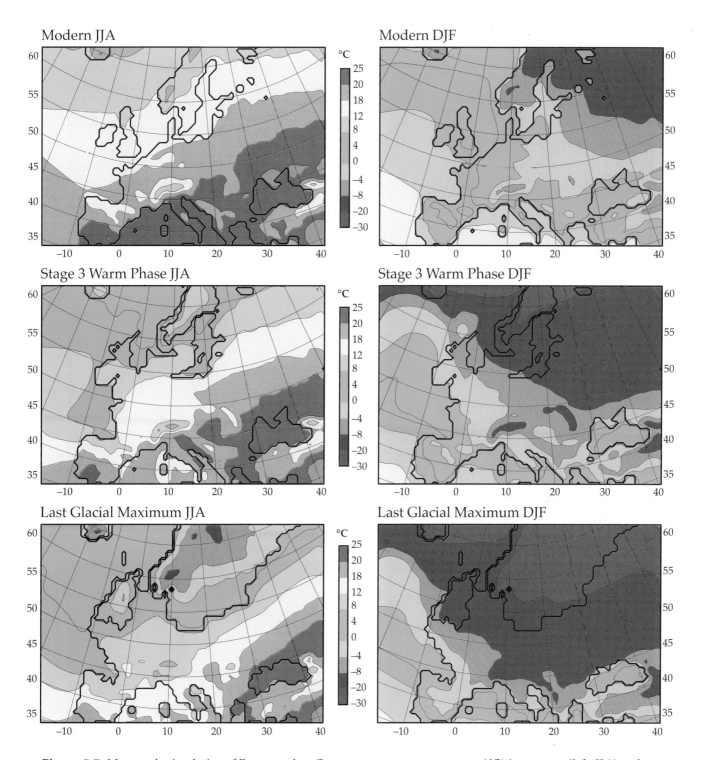

Figure 5.7. *Meso-scale simulation of European late Quaternary mean temperatures (°C) in summer (left, JJA) and winter (right, DJF). Top: Modern values; heavy line is today's coastline. Centre: OIS-3 'warm', illustrating warm events during the Stable Warm Phase and to 37 ka BP (Chapter 4, Table 4.1); heavy line is the mid-glacial coastline at −80 m with shores of the Baltic and Black Seas and a small ice sheet. Bottom: LGM 'cold' event showing prevailing conditions during the Last Glacial Maximum; heavy lines define the LGM coastline at −120 m, the shores of the Black Sea and the margins of the LGM ice sheet.*

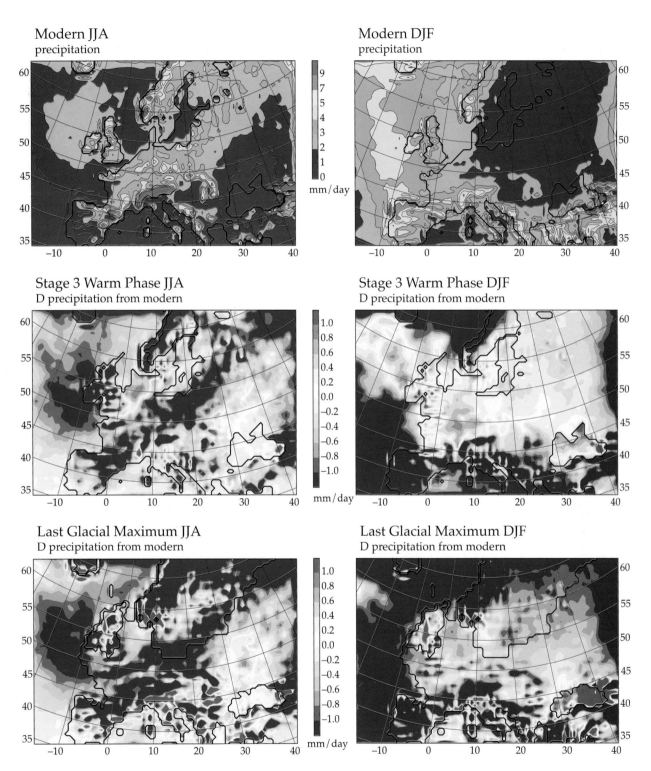

Figure 5.8. *Meso-scale simulation of European late Quaternary mean precipitation (mm/day) in summer (left, JJA) and winter (DJF, right). Top: Modern values; heavy line is today's coastline. Centre: difference between Modern and OIS-3 'warm' event values during the Stable Warm Phase and to c. 37 ka BP (Modern minus OIS-3 'warm'); heavy line is the mid-glacial coastline at −80 m with shores of the Baltic and Black Seas and a small ice sheet. Bottom: difference between Modern and LGM 'cold' values (Modern minus LGM) showing prevailing conditions during the Last Glacial Maximum; heavy lines define the LGM coastline at −120 m, the shores of the Black Sea and the margins of the LGM ice sheet.*

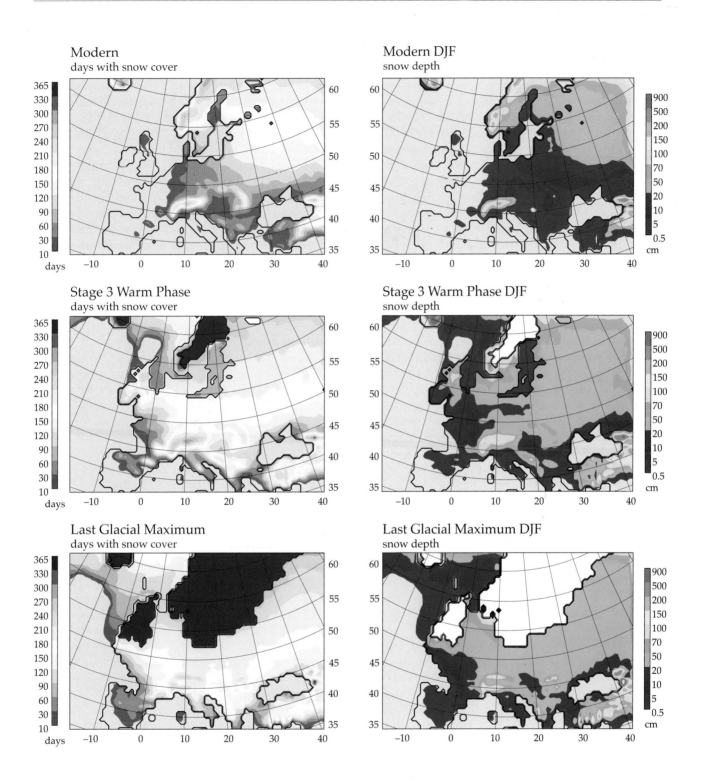

Figure 5.9. *Meso-scale simulations of late Quaternary snow conditions in winter (DJF). Left: days with snow cover (days/year). Right: actual snow depth in cm. Top: Modern; heavy line is today's coastline. Centre: OIS-3 'warm' event illustrating conditions between c. 59 and 37 ka BP; heavy line is the mid-glacial coastline at –80 m with the shores of the Baltic and Black Seas and small ice sheet. Bottom: LGM 'cold' event showing prevailing conditions during the Last Glacial Maximum; heavy line is the LGM coastline at –120 m with Black Sea and margins of the LGM ice sheet.*

availability as they switch the regime repeatedly from humid to arid and back again. Secondly, the topography of a large part of Europe is the cause of strong micro-scale orographic control on precipitation. For the interpretation of precipitation patterns in the Stable Warm and LGM Phases, maps displaying them as deviations from the modern values are much more convenient (Fig. 5.8: central and bottom). Single panel map copies at a larger scale (A4) of Figures 5.7, 5.8 and 5.9, available at the Stage 3), are also helpful.

Aridity is widely assumed to have been a dominant feature of glacial Europe north of the trans-European mountain barrier and in southeastern Europe and the Mediterranean, and its impact on fauna, flora and humans is being much discussed (e.g. Allen *et al.* 2000; Prentice *et al.* 1992; van Andel & Tzedakis 1996; 1998; Huntley *et al.* 2003; Guthrie & van Kolfschoten 1999). Two factors are mainly responsible for the difference between the Modern and the OIS-3 'warm' and LGM 'cold' precipitation (Figs. 5.4 & 5.8). Firstly, cooler SSTs and a wide-spread cover of sea-ice in glacial times much reduced evaporation, causing a sharply diminished moisture supply from the Atlantic and Mediterranean. Secondly, the OIS-3 atmospheric circulation differed from that of today, especially because of the strong westerly air flow induced in winter by the steep northward pressure gradient in the south (cf. Figs. 5.5 & 5.6). Thus the onshore air flow may have delivered as much or even slightly more moisture than today to western and central Europe north of the trans-European mountain barrier, especially west of 15°E and up to 55°N (Fig. 5.8). Small positive differences also mark the precipitation in large parts of Europe elsewhere, for instance in the Black Sea region where the difference between the Modern and OIS-3 'warm' precipitation is marginally positive.

In summer that situation was reversed; rainfall in the Atlantic region had a deficit of 3–4 mm/day relative to today and in eastern Europe the rainfall was generally lower by a few mm/day or at best equal to the modern simulation (Fig. 5.8). *Even so, the widely held view that during OIS-3 Europe was generally arid and hence poor in floral and faunal resources is not supported by our simulations.*

In southeastern Europe and the Mediterranean the winter months brought much lower precipitation — reduced by up to 4–8 mm/day — during the Stable Warm Phase and Last Glacial Maximum (Fig. 5.8), accompanied by a reduction in orographic rain.

Snow cover and snow depth: For humans to survive through the glacial winter in snow-covered areas,

snow depth and the time of the arrival and melting of the snow blanket were critical. Deep snow not only hampers the movements of humans and animals alike, so negatively affecting hunting and foraging, but snow depth and certain kinds of snow crusts control where grazers such as reindeer and mammoth, horse and muskox can forage and survive.

Snowfall patterns of the OIS-3 'warm' and LGM 'cold' simulations (Fig. 5.9: centre and bottom) differ greatly from the precipitation patterns of the modern winter (Fig. 5.8: right), and from each other. Although the amount of snow falling is a function of the precipitation rate, the annual accumulation of snow is modified by many factors such as the number of days when snow fell, oscillations of the temperature around freezing especially in the maritime zone, and long periods of clear inland skies that increase sublimation. The snow blanket, responding to temperature, precipitation and seasonality, neatly sums up the annual changes of state of the arctic and mid-latitude maritime and continental climate zones.

In western Europe, the modelled modern climate has a light, intermittent snow cover that, except at higher elevations, usually persists for only a few days, accumulating a snow depth that rarely exceeds 10–30 cm (Fig. 5.9: top). Farther east- and southeastward, the number of snow days rapidly rises to three to four months in central Europe and elsewhere in mountainous regions and to six to eight months in Scandinavia and northwestern Russia. Therefore, the snow depth, notwithstanding the declining precipitation rate, does increase east- and northward, but below the 50th to 55th parallel it does not exceed mean values of 20–50 cm (Fig. 8.9: top).

Even during the Stable Warm Phase and subsequent isolated warm events until *c.* 37 ka BP, the difference with the present snow cover is striking (Fig. 5.9: centre). In western coastal Europe snow was on the ground from two to six months and farther north (50°–60°N) 150–180 days of snow cover marked the region each year.

Snow depth, however, depends on wind drift and freezing as much as or more than on the precipitation rate. It was modest in western Europe, reaching 70 cm or more only in distant northern and eastern areas during the many months of the long winter. During the LGM snow was on the ground for two or three months almost everywhere including much of the Mediterranean except in Iberia, and increased to nine months and more in northern Russia. However, the cold ocean provided so little moisture that except along the Atlantic coast simulated snow depths rarely exceeded 20 cm.

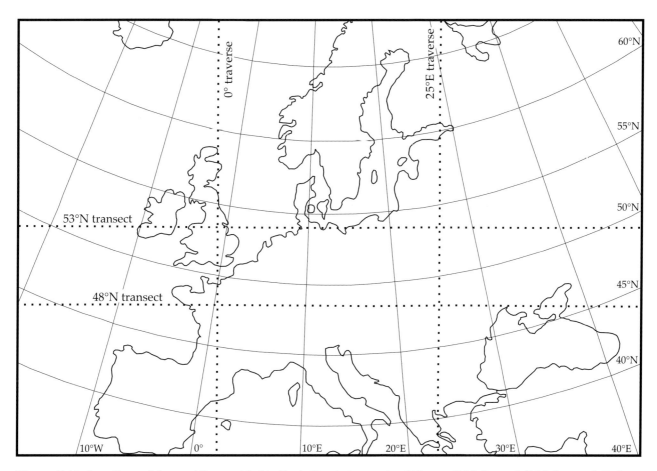

Figure 5.10. *Locations of the meridian and latitudinal climate transects of Figures 5.11 through 5.16 (heavy dotted lines ····).*

Major climatic gradients: maritime to continental and arctic to Mediterranean

Continental-scale climate features

The two main features of the European climate are the north→south arctic-to-temperate gradient north of the trans-European mountain barrier and the west→east maritime-to-continental gradient extending eastward from the Atlantic region to the vast Russian plain. The northern Mediterranean climate zone which extends from eastern Iberia to Turkey, a third major feature, is shielded from the others by the trans-European mountain barrier and dominated by sharp contrasts between sea and mountain. It consists of a complex mosaic of climate zones that can be understood mainly in topographic terms. Of the three features, it is the Atlantic maritime zone which has attracted most attention in the study of the relations between the glacial climate and the human past.

Figures 5.7, 5.8 and 5.9 display a panorama where even a single cursory glance contrasts sum-

mer with winter, maritime conditions with continental ones and the present with the Stable Warm and LGM Cold climate end members. For a first overview the maps are very helpful, but details and transitions are not easily grasped and important features such as the seasonality which controls key aspects such as the early arrival or late melting of the snow blanket are not conveyed. Attempts to superimpose images of the human presence in time and space are defeated by problems of localization and scale. Large-scale maps of simulations of the many individual parameters can be obtained from the Stage 3 website,[3] but comparisons of human distribution patterns (van Andel *et al.* 2003b) with sets of climate maps are cumbersome.

Other types of records may be better suited to human concerns. The Stage 3 models generate their output as numerical tables that list the values of all calculated climate variables (see Appendix 5.1); the tables can be downloaded from our website for manipulation with a wide range of graphic methods. Below we present a few examples that emphasize

Temperature

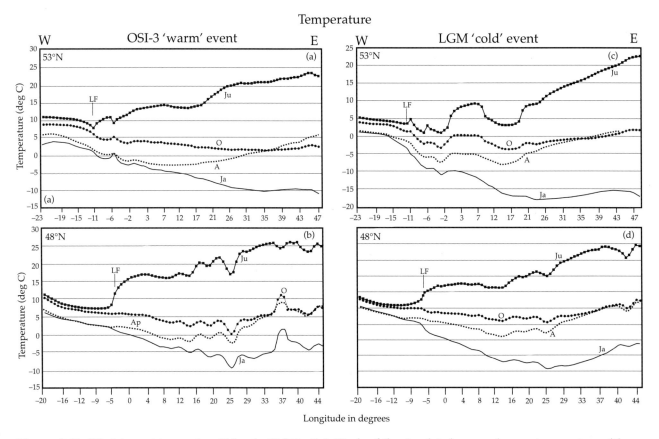

Figure 5.11. *West-to-east transects ~48° and ~53° North latitude of the simulated seasonal mean temperatures (°) for mid-winter (Ja = January), spring (A = April), mid-summer (Ju = July) and autumn (O = October). Left: OIS-3 'warm' event; right: LGM 'cold' event/ Locations on Figure 5.10. LF = landfall. Base line: Longitude in degrees west (negative) and east (positive).*

the main climate gradients by means of latitudinal and meridional transects (Fig. 5.10). The chosen variables are the mean surface temperature and precipitation in mid-summer (July), mid-autumn (October), mid-winter (January) and mid-spring (April) and their seasonal gradients and the same for snow depth in winter. The seasonal climate changes in particular are of vital interest for the human adaptation to climate (Davies & Gollop 2003; Aiello & Wheeler 2003), and for an understanding of the varying availability of prey animals (Stewart *et al.* 2003a; Musil 2003). Snow depth is a main factor controlling plant cover and limits the opportunity for grazers to survive the winter.

Regional temperature changes

Figure 5.11 shows west-to-east seasonal temperature changes for the OIS-3 'warm' (5.11:a,b) and LGM 'cold' simulations (5.11:c,d). The Atlantic climate zone is characterized by a narrow annual temperature range at both latitudes and in both climate phases. At 53°N the maritime zone was wide (cf. Fig. 5.10),

including most of the British Isles, but farther south at 48°N it only covered a narrow coastal part of France. During the OIS-3 'warm' phase the annual maritime temperature range extended from just below freezing in January to 10°±2°C in mid-summer with little latitudinal change (Fig. 5.11:a,b). To the east the curve rose in summer inland to a plateau of 15°±2°C at about 7°E on the 53°N and 4°E on the 48°N transect. Farther east a second rise took the average summer temperature to *c.* 20°C on the 53°N (5.11a) and 20°C–25°C on the 48°N (Fig. 5.11b) transects; the boundary between the maritime and continental regimes was at *c.* 20°E on the northern and 15°E on the southern transect.

The spring thaw (April) arrived early in the east but much later in the west (Fig. 5.11:a,b), while the autumn (October) had about the same temperature across the whole region at both latitudes.

On the LGM 'cold' simulation (Fig. 5.11:c,d) the maritime zone in central and southern Britain and westernmost France had nearly the same pattern as above. At 48°N (Fig. 5.11d) the coastal temperature

70

Temperature

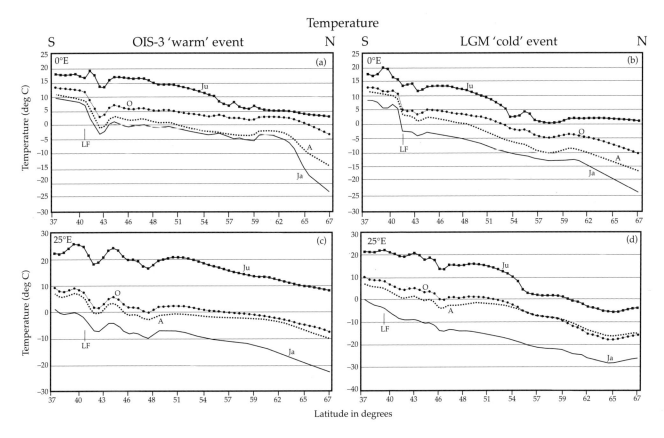

Figure 5.12. *North-to-south transects at ~0° and ~25° East longitude of the simulated seasonal mean temperatures (°) for mid-winter (Ja = January), spring (A = April), mid-summer (Ju = July) and autumn (O = October). Left: OIS-3 'warm' event; right: LGM 'cold' event. Locations on Figure 5.10. LF = landfall. Base line: Latitude in degrees north.*

ranged annually from a few degrees below freezing to a cool summer maximum of ~5°C–10°C; then it rose to a temperature plateau between 15°C and 20°C at ~12°E. From the western edge of the continental zone the temperature rose to 25°C at the eastern end of the transect. The 53°N transect (Fig. 5.11:c) shows essentially the same pattern, but the simple slope to the inland temperature plateau is modified by a ~5°C rise over the low, dry North Sea floor and a 6°–7°C drop over the southern Baltic Sea, then a large lake.

As in the case of the OIS-3 'warm' simulation, in neither autumn nor spring was there a stepwise gradient from the milder western to the more severe eastern region, but spring (April) remained long cold in the Atlantic region and warmed up slowly. In contrast, the mid-autumn temperatures (Fig. 11:c,d) were remarkably uniform across the whole of central Europe.

Compared to the west-to-east temperature changes, the north–south transects (Fig. 5.12) display a simple pattern of northern cold and southern warmth. Between the Pyrenees and 55°N lat.

the OIS-3 'warm' simulation shows a range of 10°–15°C in summer and slightly more in winter. The 0°E transect (Fig. 5.12a) except for a short leg over the Mediterranean Sea, passed mainly through the Atlantic maritime climate region except when crossing the Pyrenees, where OIS-3 'warm' and LGM 'cold' simulations both dip 5°C in all seasons. On the 0°E transect the winter (January) and spring (April) temperature curves differ little; regional and seasonal changes are small in the entire Atlantic climate zone, but compared to the continental region at 25°E (5.12c) where the spring warmed much earlier, the summer in the Atlantic was much delayed.

If we ignore the extensions of the traverses north of 55°N, the meridional LGM transects are very similar (Fig. 5.12:b,d), changing from –10°C in winter to 15°C in summer in the Balkans and afflicted by severe –30°C winters and chilly –5°C summers beyond 60°N latitude. Autumn and spring values were close, but the range between summer and winter temperatures was of the order of 25° to 35°C.

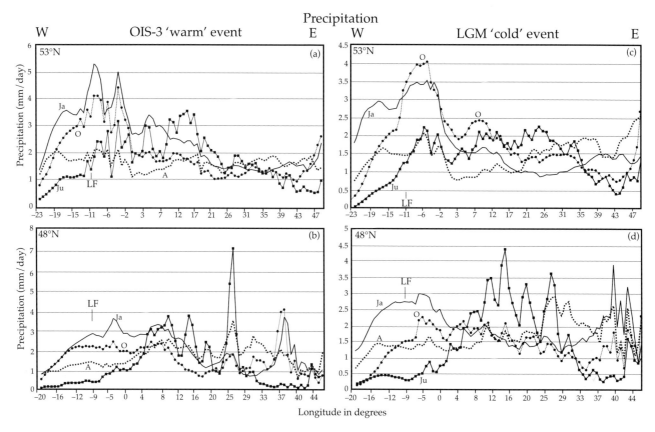

Figure 5.13. *West-to-east transects at ~48° and ~53° North latitude of the simulated seasonal mean precipitation (mm/ day) for mid-winter (Ja = January), spring (A = April), mid-summer (Ju = July) and autumn (O = October). Left: OIS-3 'warm' event; right: LGM 'cold' event. Locations on Figure 5.10. LF = landfall. Note scale difference between Stable Warm and LGM graphs. Base line: Longitude in degrees west (negative) and east (positive).*

Precipitation gradients

The precipitation had a much more complex pattern of spatial, seasonal and longer-term changes than temperature (Fig. 5.13). The pattern clearly shows that, although difference maps such as Figure 5.8 are capable of clarifying broad temporal differences in rainfall, where geography is the main point of interest the values of the graphic transects are more informative.

Figure 5.8 showed that during the OIS-3 'warm' phase the winter precipitation equalled or exceeded by 0.5–1.0 mm/day the Modern values over much of western and central Europe. The 53°N transect (Fig. 5.13:a,c) reveals that during the OIS-3 'warm' and LGM 'cold' phases both a substantial winter precipitation was preceded in the British Isles and France by heavy autumnal rains. In central Europe less rain fell (3 mm/day) and the eastern plains were even drier. The same pattern marks the 48°N transect (Fig. 5.13:b,d), but in central Europe and the continental interior beyond the 20°E meridian the rainfall is reduced to 1–3 mm/day on average; that is not very much but it does not qualify as aridity.

The records at 0°E (Fig. 5.14:a,b) show a much more striking difference between the OIS-3 'warm' and LGM 'cold' precipitation simulations. Except for a peak of 14 mm/day where the transect crosses the Pyrenees, the average rainfall was c. 2–3 mm/ day and nowhere exceeded 4 mm/day. In eastern Europe the LGM precipitation was also reduced relative to that of the OIS-3 'warm' phase (25°E transect: Fig. 5.14:c,d), but given values of 1–3 mm/day rising to peaks of up to 5 mm/day in the mountainous Balkans (Fig. 5.10), the term arid would be too excessive here; the winters were dry here but from June through October there was modest rainfall.

A blanket of snow

The snow depths of Figure 5.9, 5.15 and 5.16 are true snow values, not their equivalent in water, but the 60 km spatial resolution does not take account of wind-driven local thickness variations, patchy snow in sheltered areas or sublimation during cold dry

Precipitation

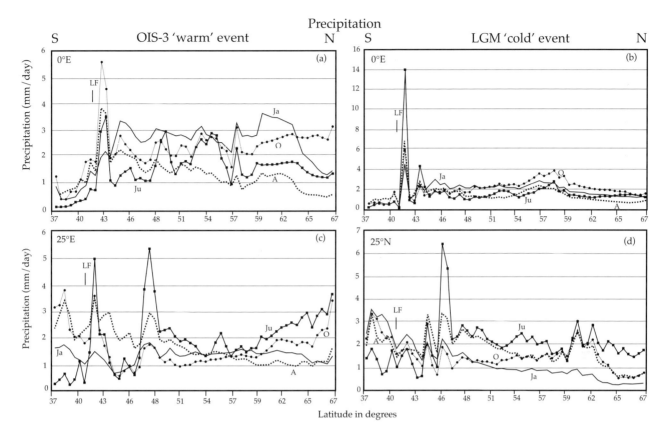

Figure 5.14. *North-to-south transects at ~0° and ~25° East longitude of the simulated seasonal mean precipitation (mm/day) for mid-winter (Ja = January), spring (A = April), mid-summer (Ju = July) and autumn (O = October). Left: OIS-3 'warm' event; right: LGM 'cold' event. Locations on Figure 5.10. LF = landfall. Note scale difference between (a) and (b) graphs. Base line: Latitude in degrees north.*

intervals, all key variables in terms of animal and human use of the winter landscape.

The OIS-3 'warm' simulation (Fig. 5.15) shows a thin (0–15 cm) snow cover at 53°N in the Atlantic maritime zone (0–15 cm); probably it was patchy also in time and space because the winter temperature oscillated narrowly about the freezing point (Fig. 5.11a). Although the average winter precipitation was 3–5 mm/day (Fig. 5.13a), snow was on the ground for only 10–60 days (Fig. 5.9), but the land was not free of snow until late in May. Farther east the snow blanket was much thicker, in part because of heavy snowfall as late as in April, but beyond 35°E the snow cover in that same month had already thinned substantially as an indication of the early spring of the continental region.

Farther south, at 48°N (Fig. 5.15b), the coastal zone and most of the rest of France eastward to the 4°E meridian was free of snow most of the time (Fig. 5.15b), although some snow fell on up to 120 days each year. This mild wet winter climate in the Stable Warm Phase (Fig. 5.13:b) was due to low precipita-

tion combined with winter and spring temperatures that were much of the time above freezing (Fig. 5.11b). Farther inland, as far as and even beyond 30°E, snow covered the ground for at least half of the year (Fig. 5.9) and melted late in April or early May. The Carpathians, crossed by the transect at about 25°E, are marked by a sharp snowfall peak (Fig. 5.15:b) and in the mountains the depth of snow fallen in spring exceeded even the winter pack. East of the mountains, however, the winter and spring snow cover thinned rapidly eastward and the main melting came before April.

During the LGM the 53°N transect crossed the southernmost lobes of the British and Fennoscandian ice sheets (Fig. 5.9); they stand out in the graphs by a snow cover of at least 500 cm (Fig. 5.15:c). In contrast, the open country around the ice lobes had hardly any snow at all, partly because of a winter precipitation of little more than 1 mm/day and partly because the sublimation at temperatures below −15°C (Fig. 5.11:d) under clear polar skies was strong.

Snow depth

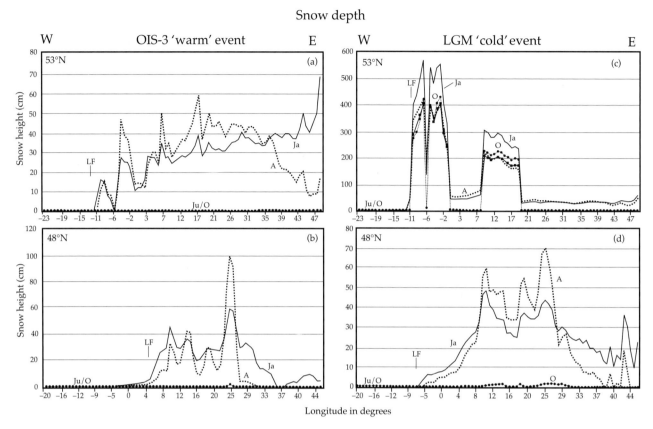

Figure 5.15. *West-to-east transects at ~48° and ~53° North latitude of the simulated actual snow depth (cm) for mid-winter (Ja = January), spring (A = April), mid-summer (Ju = July) and autumn (O = October). Left: OIS-3 'warm' event; right: LGM 'cold' event. Locations on Figure 5.10. LF = landfall. Note scale differences between all four graphs. Base line: Longitude in degrees west (negative) and east (positive).*

At the same time, conditions at 48°N in the Atlantic maritime zone (Fig. 5.16:b,d) were not unlike those of the Stable Warm Phase, although with even less snow because of the very low precipitation and relatively warm winter temperatures of less than –5°C.

The south-to-north transects at 0°E and 25°E (Fig. 5.16) also show the contrast between full maritime and full continental climate conditions during the Stable Warm and LGM Phases. The Stable Warm record on the 0°E transect is almost free of snow except for the crossing of the Pyrenees at *c.* 42°N. A snow cover of no more than 20 cm appears where the transect crosses the east side of the British Isles and the spring snowfall dominates over the winter (Fig. 5.16:a,b). The transect then passes into the North Sea but touches land at two points (Fig. 5.10) which show that the snow cover thickened a good deal towards the north, again with a large spring component.

The transect along the 25°E meridian shows a substantial snow blanket in the Stable Warm Phase (Fig. 5.16:b), again with a large contribution of spring snow that thickened the blanket northward. Since this is a region of minor precipitation, the accumulation may be due to nearly complete preservation under very cold conditions during the winter and spring. During the LGM, snow depths of no more than 20 cm in the southern part of the region highlight its seasonal aridity. Only near the margin of the Scandinavian ice-sheet at *c.* 55°N the snow accumulation on the ice was as much as 4 m per year.

Acknowledgements

This chapter and its predecessors in *Quaternary Research* (Alfano *et al.* 2003; Barron & Pollard 2002; Pollard & Barron 2003) rests on several input and validation data bases compiled by panels of the Stage 3 Project: (a) Ice Sheets, Isostasy and Shorelines (Kurt Lambeck, Neil Arnold, John T. Andrews and Tjeerd van Andel); (b) Modelling (Eric Barron, Brian Huntley, David Pollard and Paul Valdes); (c) Sea-Surface Temperatures (Mark Chapman, Michael Sarnthein, Uwe Pflaumann and Nick Shackleton); (d) Vegetation (Judy Allen, Jacques-Louis de

Snow depth

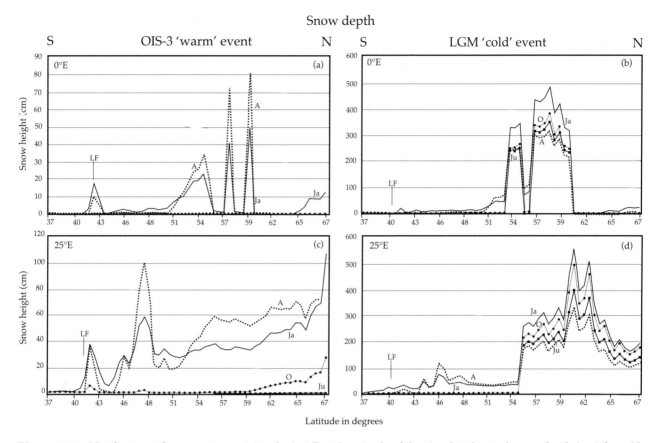

Figure 5.16. *North-to-south transects at ~0° and ~25° East longitude of the simulated actual snow depth (cm) for mid-winter (Ja = January), spring (A = April), mid-summer (Ju = July) and autumn (O = October). Left: OIS-3 'warm' event; right: LGM 'cold' event. Locations on Figure 5.10. LF = landfall. Note scale differences between all four graphs. Base line: Latitude in degrees north.*

Beaulieu, Eberhard Grüger, Brian Huntley, Chronis Tzedakis and William Watts) and (e) Geological Proxies (Pierre Antoine, Ko van Huissteden and Jamie Woodward), as well as the many European palynologists who provided pollen spectra. Piers Gollop extracted for us numerical data from the simulations that turned out to be critical for the second half of this Chapter. We are most grateful to them all and to the many other Project members and staff from participating institutions who contributed observations and data and participated in discussions and debates during Stage 3 meetings. Eric Barron and David Pollard were directly supported by the National Science Foundation through grant ATM98-09239; generous support of the computational capability was through the PMESH Project (Partnership in Modelling Earth System History – ATM00-00545). Sharon Capon did all the black and white drawings.

Notes

1. Calendric dates (ice core, TL, OSL, ESR and U-Th dates) and calibrated ^{14}C dates (van Andel *et al.* 2003a) are used throughout and the phrase 'xxxx ka BP' refers to calibrated ^{14}C and other calendrical dates (calibration by CalPal.v1998: see van Andel *et al.* 2003a). The raw versions of ^{14}C dates are available from the Stage 3 Project's chrono-archaeological data base at its website[3]) or in references cited there.

2. The term p*hase*, as in *climate phase*, is used for temporal units of climate change in contrast to *climate zone* which is used in its geographic sense.

3. The Stage 3 Web site can be consulted at: http://www.esc.esc.cam.ac.uk/oistage3/Details/Home page/html.

References

Aiello, L.C. & P. Wheeler, 2003. Neanderthal thermoregulation and the glacial climate, in *Neanderthals and Modern Humans in the European Landscape during the Last Glaciation*, Chapter 9, eds. T.H. van Andel & W. Davies. (McDonald Institute Monographs.) Cambridge: McDonald Institute for Archaeological Research, 147–66.

Alfano, M.J., E.J. Barron, D. Pollard, B. Huntley & J. Allen,

2003. Comparison of climate model results with European vegetation and permafrost during Oxygen Isotope Stage Three. *Quaternary Research* 59, 97–107.

Allen, J.M.R., W.A. Watts & B. Huntley, 2000. Weichselian palynostratigraphy, palaeovegetation and palaeoenvironment: the record from Lago Grande di Monticchio. *Quaternary International* 73/74, 91–110.

Andersen, B.G. & J. Mangerud, 1989. The last interglacial-glacial cycle in Fennoscandia. *Quaternary International* 3–4, 21–9.

Arnold, N., T.H. van Andel & V. Valen, 2002. Extent, dynamics and isostatic compensation of the Scandinavian ice-sheet during Oxygen Isotope Stage 3 (60,000–30,000 cal years BP). *Quaternary Research* 57, 38–48.

Barnola, J.M., D. Raynaud, Y.S. Korotkevich & C. Lorius, 1987. Vostok ice core provides 160,000-year record of atmospheric CO_2. *Science* 329, 408–14.

Barron, E.J. & D. Pollard, 2002. High-resolution climate simulations of Oxygen Isotope Stage 3 in Europe. *Quaternary Research* 58, 296–309.

Berger, A. & M.F. Loutre, 1991. Insolation values for the climate of the last 10 million years. *Quaternary Science Reviews* 10, 297–316.

Chappell, J.A. & N.J. Shackleton, 1986. Oxygen isotopes and sea-level. *Nature* 324, 137–40.

Chappell, J.A. Omura, T. Esat, M. McCulloch, J. Pandolfi, Y. Ota & B. Pillans, 1996. Reconciliation of late Quaternary sea-levels derived from coral terraces at Huon Peninsula with deep-sea oxygen isotope records. *Earth and Planetary Science Letters* 141, 227–36.

CLIMAP (Climate: Long-Range Investigation, Mapping and Prediction Project) Members, 1981. Seasonal reconstructions of the Earth's surface at the last glacial maximum. Boulder (CO): *The Geological Society of America*, 18 sheets.

Dansgaard, W., S.J. Johnsen, H.B. Clausen, D. Dahl-Jensen, N.S. Gundestrup, C.U. Hammer, C.S. Hvidberg, J.P. Steffensen, H. Sveinbjörnsdottir, J. Jouzel & G. Bond, 1993. Evidence for general instability of past climate from a 250-kyr ice-core record. *Nature* 364, 218–20.

Davies, W. & P. Gollop, 2003. The human presence in Europe during the Last Glacial Period II: climate tolerance and climate preferences of mid- and late glacial hominids, in *Neanderthals and Modern Humans in the European Landscape during the Last Glaciation*, Chapter 8, eds. T.H. van Andel & W. Davies. (McDonald Institute Monographs.) Cambridge: McDonald Institute for Archaeological Research, 131–46.

Dickinson, R.E., R.M. Errico, F. Giorgi & G.T. Bates, 1989. A regional climate model for the Western United States. *Climate Change* 15, 383–422.

Donner, J., 1995. *The Quaternary of Scandinavia*. Cambridge: Cambridge University Press.

Giorgi, F., M.R. Marinnucci & G.T. Bates, 1993a. Development of a second generation regional climate model (RegCM2), part I: Boundary-layer and radiative transfer processes. *Monthly Weather Review* 121, 2794–813.

Giorgi, F., M.R. Marinucci & G.T. Bates, 1993b. Development of a second generation regional climate model (RegCM2), part II: Convective processes and assimilation of lateral boundary conditions. *Monthly Weather Review* 121, 2814–32.

Grootes, P.M., M.J. Stuiver, J.W. White, S. Johnsen & J. Jouzel, 1993. Comparison of oxygen isotope records from the GISP2 and GRIP Greenland ice cores. *Nature* 366, 552–4.

Guthrie, D. & T. van Kolfschoten, 1999. Neither warm and moist, nor cold and arid: the ecology of the Mid Upper Palaeolithic, in *Hunters of the Golden Age*, eds. W. Roebroeks, M. Mussi, J. Svoboda & K. Fennema. Leiden: *Analecta Prehistoric Leidensia*, 1–20.

Haxeltine, A. & I.C. Prentice, 1996. BIOME3: an equilibrium terrestrial biosphere model based on ecophysical constraints, resource availability and competition among plant functional types. *Global Biogeochemical Cycles* 10, 693–709.

Huntley, B. & J.R. Allen, 2003. Glacial environments III: palaeo-vegetation patterns in Late Glacial Europe, in *Neanderthals and Modern Humans in the European Landscape during the Last Glaciation*, Chapter 6, eds. T.H. van Andel & W. Davies. (McDonald Institute Monographs.) Cambridge: McDonald Institute for Archaeological Research, 79–102.

Huntley, B., P.C. Tzedakis, J.-L. de Beaulieu, D. Pollard, M. Alfano & J. Allen, 2003. Comparison between Biome 3.5 models and pollen data. *Quaternary Research*, 195–212.

Johnsen, S.J., D. Dahl-Jensen, N. Gundestrup, J.P. Steffensen, H.B. Clausen, H. Miller, V. Masson-Delmotte, A.E. Sveinbjörnsdottir & J. White, 2001. Oxygen isotope and paleotemperature records from six Greenland ice-core stations, Camp Century, Dye-3, GRIP, GISP2, Renland and North GRIP. *Journal of Quaternary Science* 16, 299–308.

Kageyama, M. & P.V. Valdes, 2000. Impact of the North American ice-sheet orography on the Last Glacial Maximum eddies and snowfall. *Geophysical Research Letters* 27, 1515–18.

Kageyama, M., P.J. Valdes, G. Ramstein, C. Hewitt & U. Wyputta, 1999. Northern Hemisphere storm tracks in present day and Last Glacial Maximum climate simulations: a comparison of the European PMIP models. *Journal of Climate* 12, 742–60.

Lambeck, K., 1995a. Late Pleistocene and Holocene sea-level change in Greece and southern Turkey: a separation of eustatic, isostatic and tectonic contributions. *Geophysical Journal International* 106, 1022–44.

Lambeck, K., 1995b. Late Devensian and Holocene shorelines of the British Isles and North Sea from models of glacio-hydro-isostatic rebound. *Journal of the Geological Society of London* 152, 437–48.

Lambeck, K., 1995c. Constraints on the Late Weichselian ice sheet over the Barents Sea from observations of raised shorelines. *Quaternary Science Reviews* 14, 1–16.

Lambeck, K., 1996. Sea-level change and shore-line evolution: a general framework of modelling and its application to Aegean Greece since Palaeolithic time. *Antiquity* 70, 588–611.

Lambeck, K. & J. Chappell, 2001. Sea level change through the last glacial cycle. *Science* 292, 679–86.

Mangerud, J., R. Løvlie, S. Gulliksen, A.-K. Hufthammer, E. Larsen & V. Valen, 2003. Paleomagnetic correlations between Scandinavian ice-sheet fluctuations and Greenland Dansgaard/Oeschger events, 45,000–25,000 yrs BP. *Quaternary Research* 59, 213–22.

Markova, A.K., A.N. Simakova, A.Y. Puzachenko & L.M. Kitaev, 2002. Environments of the Russian Plain during the Middle Valdai Briansk Interstade (33,000–24,000 yr BP) indicated by fossil mammals and plants. *Quaternary Research* 57, 391–400.

Meese, D.A., A.J. Gow, R.B. Alley, G.A. Zielinsky, P.M. Grootes, M. Ram, K.C. Taylor, P.A. Mayewski & J.F. Bolzan, 1997. The Greenland Ice Sheet Project 2 depth-age scale: methods and results. *Journal of Geophysical Research* 102, 26,411–23.

Musil, R., 2003. The Middle and Upper Palaeolithic game suite in central and southeastern Europe, in *Neanderthals and Modern Humans in the European Landscape during the Last Glaciation*, Chapter 10, eds. T.H. van Andel & W. Davies. (McDonald Institute Monographs.) Cambridge: McDonald Institute for Archaeological Research, 167–90.

Olsen, L., V. Mejdahl & S.F. Selvik, 1996. Middle and late Pleistocene stratigraphy, chronology and glacial history in Finnmarken, North Norway. *Norges Geologisk Undersøgelse Bulletin* 429, 111pp.

Olsen, L., K. van der Borg, B. Bergstrøm, H. Svelan, S.-E. Lauritzen & G. Hansen, 2001a. AMS radiocarbon dating of glacigenic sediments with low organic content — an important tool for reconstructing the history of glacial variations in Norway. *Norsk Geologisk Tidsskrift* 81, 59–92.

Olsen, L., H. Svelan & B. Bergstrøm, 2001b. Rapid adjustments of the western part of the Scandinavian ice-sheet during the mid- and Late Weichselian — a new model. *Norsk Geologisk Tidsskrift* 81, 93–118.

Peltier, W.R., 1994. Ice-age paleotopography. *Science* 265, 195–201.

Pollard, D. & E.J. Barron, 2003. Causes of model-data discrepancies in European climate during Oxygen Isotope Stage 3 with insights from the last glacial maximum. *Quaternary Research* 59, 108–13.

Pollard, D. & S.L. Thompson, 1994. Sea-ice dynamics and CO2 sensitivity in a global climate model. *Atmospheres-Oceans* 32, 449–67.

Pollard, D. & S.L. Thompson, 1995. Use of a land-surface-transfer scheme (LSX) in a global climate model: the response to doubling stomatal resistance. *Global and Planetary Change* 10, 129–61.

Porter, S.C., 1989. Some geological applications of average Quaternary glacial conditions. *Quaternary Research* 32, 245–61.

Prentice, I.C., W. Cramer, S.P. Harrison, R. Leemans, R.A. Monserud & A.M. Solomon, 1992. A global biome model based on plant physiology and dominance, soil properties and climate. *Journal of Biogeography* 19, 117–34.

Shackleton, J.C., T.H. van Andel & C.N. Runnels, 1984. Coastal paleogeography of the central and western Mediterranean during the last 125,000 years and its archaeological implications. *Journal of Field Archaeology* 11, 307–14.

Shinn, R.A. & E.J. Barron, 1989. Climate sensitivity to continental ice sheet size and configuration. *Journal of Climate* 2, 1517–37.

Stewart, J.R., T. van Kolfschoten, A. Markova & R. Musil, 2003a. The mammalian faunas of Europe during Oxygen Isotope Stage Three, in *Neanderthals and Modern Humans in the European Landscape during the Last Glaciation*, Chapter 7, eds. T.H. van Andel & W. Davies. (McDonald Institute Monographs.) Cambridge: McDonald Institute for Archaeological Research, 103–30.

Stewart, J.R., T. van Kolfschoten, A. Markova & R. Musil, 2003b. Neanderthals as part of the broader Late Pleistocene megafaunal extinctions?, in *Neanderthals and Modern Humans in the European Landscape during the Last Glaciation*, Chapter 12, eds. T.H. van Andel & W. Davies. (McDonald Institute Monographs.) Cambridge: McDonald Institute for Archaeological Research, 221–32.

Stringer, C., H. Pälike, T. van Andel, B. Huntley, P. Valdes & J.R.M. Allen, 2003. Climatic stress and the extinction of the Neanderthals, in *Neanderthals and Modern Humans in the European Landscape during the Last Glaciation*, Chapter 13, eds. T.H. van Andel & W. Davies. (McDonald Institute Monographs.) Cambridge: McDonald Institute for Archaeological Research, 233–40.

Stuiver, M. & P. Grootes, 2000. GISP2 oxygen isotope ratios. *Quaternary Research* 53, 277–84.

Svelan, O.L., H. Bergstrøm, B. Selvik, S.F. Lauritzen, S.-E. Stokland & K. Grøsfjeld, 2001. Methods and stratigraphies used to reconstruct Mid- and Late Weichselian palaeoenvironmental and palaeoclimatic changes in Norway. *Norges geologiske undersøkelse Bulletin* 438, 21–46.

Thompson, S.L. & D. Pollard, 1995a. A global climate model (GENESIS) with a land-surface-transfer scheme (LSX), part 1: Present-day climate. *Journal of Climate* 8, 732–61.

Thompson, S.L. & D. Pollard, 1995b. A global climate model (GENESIS) with a land-surface-transfer scheme (LSX), part 2: CO2 sensitivity. *Journal of Climate* 8, 1104–21.

Thompson, S.L. & D. Pollard, 1997. Greenland and Antarctic mass balances for present and doubled atmospheric CO2 from the GENESIS version 2 global climate model. *Journal of Climate* 10, 871–900.

van Andel, T.H., 2003. Humans in an Ice Age — the Stage 3 Project: overture or finale?, in *Neanderthals and Modern Humans in the European Landscape during the Last Glaciation*, Epilogue, eds. T.H. van Andel & W. Davies. (McDonald Institute Monographs.) Cambridge: McDonald Institute for Archaeological Research, 257–65.

van Andel, T.H. & J.C. Shackleton, 1982. Late Paleolithic and Mesolithic coastlines of Greece and the Aegean.

Journal of Field Archeology 9, 445–54.

van Andel, T.H. & P.C. Tzedakis, 1996. Palaeolithic landscapes of Europe and Environs, 150,000–25,000 years ago. *Quaternary Science Reviews* 15, 481–500.

van Andel, T.H. & P.C. Tzedakis, 1998. Priority and opportunity: reconstructing the European Middle Palaeolithic climate and landscape, in *Science and Archaeology: an Agenda for the Future*, ed. J.S. Bailey.

London: English Heritage, 37–46.

van Andel, T.H., W. Davies, B. Weninger & O. Jöris, 2003a. Archaeological dates as proxies for the spatial and temporal human presence in Europe: a discourse on the method, in *Neanderthals and Modern Humans in the European Landscape during the Last Glaciation*, Chapter 3, eds. T.H. van Andel & W. Davies. (McDonald Institute Monographs.) Cambridge: McDonald Institute for Archaeological Research, 21–30.

van Andel, T.H., W. Davies & B. Weninger, 2003b. The human presence in Europe during the last glacial period I: human migrations and the changing climate, in *Neanderthals and Modern Humans in the European Landscape during the Last Glaciation*, Chapter 4, eds. T.H. van Andel & W. Davies. (McDonald Institute Monographs.) Cambridge: McDonald Institute for Archaeological Research, 31–56.

van Huissteden, J., D. Pollard & J. Vandenberghe, 2003. Paleotemperature reconstructions of the European permafrost zone during Oxygen Isotope Stage 3 compared with climate model results. *Journal of Quaternary Research* 18, 453–64.

Appendix 5.1. *Climatic variables ('fields') relevant for human–climate responses and interactions derived from meso-scale simulations and selected from the 78 of the ReGCM2 set. Also available from the Stage 3 Project website at: http://www.esc.cam.ac.uk/oistage3/Details/Homepage/htm.*

Climate end members of the last glaciation
Modern (MODERN)
Stable Warm Phase (OIS-3 'warm')
Early Cold Phase (OIS-3 'cold')
Late Glacial Maximal Phase (LGM)

Seasons
Winter (December, January, February, DJF)
Spring (March, April, May, MAM)
Summer (June, July, August, JJA)
Autumn (September, October, November, SON)

Levels above the surface
Are given in sigma co-ordinates.
Example: for air pressure or surface pressure ? = .995 is approximately 995 Mb (millibars).

RCM output	GCM output
.210 =	= .251
.510 =	= .501
.815 =	= .866

Codes for selected simulations ('fields')

First part of file name	Field	First part of file name	Field
grid	land-ocean-ice mask	*Wind*	
topog	topography (m above sea-level)	wind	lowest-level (~40 m) wind speed (m/sec)
		wbin6	fraction of time lowest-level wind speed >6 m/s
Temperature		u210	wind vectors, sigma (pressure/surface pressure) = 0.210
chill	wind chill (°)		
t	lowest-level (~40 m) air temp. (°C)	u510	wind vectors, sigma (pressure/surface pressure) = 0.510
dt	Modern minus palaeo. lowest level air temp. (°C)		
tmax	maximum monthly lowest-level air temp. (°C)	u815	wind vectors, sigma (pressure/surface pressure) = 0.815
tmin	minimum monthly lowest-level air temp. (°C)		
trange	tmax minus tmin (°C)	*Cloudiness*	
trans24	diurnal range of lowest level air temp. (°C)	cloud	cloudy fraction of whole sky (0–1)
tsoi	soil temperature, upper one metre.		
ts2	air temperature at 2 m (°C)		
Precipitation, hydrology		*Vegetation Biome3.5 output*	
relhum	relative humidity (0–1)	aetopet	actual/potential annual evapotranspiration ratio
pme	precipitation minus evaporation (mm/day)	biome types	Plant Functional Types
precip	precipitation (mm/day)	gdd0	annual growing-days above 0°C
dprecip	Modern minus palaeo precipitation (mm/day)	gdd5	annual growing-days above 5°C
evap	evapotranspiration (mm/day)	lai	seasonal maximum Leaf Area Index
eop	evapotranspiration/precipitation ratio	NPP	Net Primary Productivity, gC/m2/yr
pbin3	fraction of time with precipitation >3 mm/day	npppft	NPP of selected Plant Functional Types, gC/m2/yr
runoff	surface runoff (mm/day)		
snowd	number of days/year with snow cover (1–365)		
snowh	snow depth, actual (cm), not the liquid equivalent		
wsoi	soil moisture relative to pore space (0–1 ma), whole column (~3 m)		

Chapter 6

Glacial Environments III: Palaeo-vegetation Patterns in Last Glacial Europe

Brian Huntley & Judy R.M. Allen

Why should we reconstruct the glacial plant cover?

The attempts described in Chapter 5 (Barron *et al.* 2003) to simulate the climates of the extreme states of oxygen isotope stage 3 (OIS-3) (Fig. 6.1), exemplified by a long warm event at *c.* 45–42 ka BP and a cold event at *c.* 30 ka BP, and of the Last Glacial Maximum (LGM – OIS-2), can not in themselves constitute a useful description of the palaeoenvironments of Europe that were the habitat of Neanderthals and Anatomically Modern Humans.

The environment in which these humans lived was shaped by the complex interactions that exist between climate and biota. Various climates encourage specific types of vegetation that in turn affect key terrain characteristics such as surface roughness and reflectivity; these in turn influence regional and local precipitation and evaporation, as well as seasonal climate contrasts. As climate changes affect the vegetation, so the vegetation affects the kinds and abundance of the fauna.

Was it the direct effects of the glacial climate, or the influence of climate upon the vegetation, and hence upon the availability of food resources, that shaped human presence in glacial Europe? Although plant foods probably provided only a minor portion of the mid-glacial human diet, they did sustain, directly or indirectly, the mammalian fauna that constituted the larger part of the human food supply (Balter *et al.* 2001). Thus, the reconstruction of the plant cover is an important step in our understanding of the palaeoenvironment and how it changed, and its potential impacts on human behaviour and the human chance of survival.

OIS-3 plant cover: the state of knowledge before the Stage 3 Project

In 1995, in preparation for the Stage 3 Project, van Andel & Tzedakis (1996; 1998) used existing data to construct a qualitative and subjective overview of the palaeoenvironmental history of glacial Europe and North Africa. This included climate and plant cover, and also attempted a sketch of European and Mediterranean vegetation for typical warm and cold Dansgaard/Oeschger events (Fig. 6.1). Assuming a relatively small Fennoscandian ice cap, they suggested that during much of the dominantly warm period between about 60,000 and about 40,000 years ago the vegetation was comparable to that during the Hengelo warm/moist event (*c.* 40 ka BP). Figure 6.2 illustrates the vegetation pattern that they inferred. Sparse vegetation, similar to modern high-Arctic tundra or polar desert, covered the ice-free portions of Fennoscandia, whereas shrub tundra with

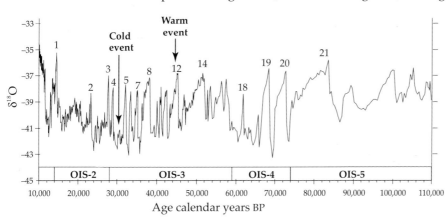

Figure 6.1. *The GISP2 ice-core δ¹⁸O record. Selected Dansgaard-Oeschger warm events are labelled.*

79

Figure 6.2. *The vegetation of Europe during OIS-3. Sketch map of typical warm type Dansgaard/Oeschger event, c. 41–38 ka BP (modified after van Andel & Tzedakis (1998): Fig. 7; dates in calendar years). The Fennoscandian ice cap was modified from Andersen & Mangerud (1989); coastline was based on –70 m isobath (modern isobaths from Lambeck 1995). Baltic Sea coast line is sketched.*

juniper, dwarf birch and willow extended from Russia across northern Germany and the Low Countries (Behre 1989), with scattered spruce parkland in the eastern Baltic region (Liivrand 1991). A very open parkland of pine, spruce and birch, although lacking other deciduous trees, was present in France and across the Alpine Foreland (de Beaulieu & Reille 1992a,b; 1984; Grüger 1989; Reille & de Beaulieu 1990). Open deciduous woodlands of oak, hazel, elm, lime and beech were extensive throughout southern Europe, with mixed deciduous and evergreen open woodlands of oak, pine and juniper in the furthest south, extending from southern Spain (Carrión *et al.* 1999; Pons & Reille 1988) to southern Greece (Tzedakis 1999).

More recent studies, however, have revealed evidence of repeated and often rapid fluctuations in the character of the vegetation of this interval. In Italy, at Lago Grande di Monticchio, two major

warm/moist events, perhaps equivalent to the Glinde and Hengelo warm events (Watts *et al.* 2000), generated a tree sequence that began with birch and oak, was followed by elm, ash and hornbeam, and closed with beech and fir, resembling a compressed variant of the interglacial succession in northern Europe (de Beaulieu & Reille 1992a). Shorter warm events were marked by a quasi-synchronous expansion of hazel, oak, beech, lime and elm (Follieri *et al.* 1998; Leroy *et al.* 1996). In Catalunya, in Spain, warm events brought a mixed parkland of pine and deciduous oak (Perez-Obiol & Julia 1994), while in Greece a similar deciduous woodland marked the northern regions (Tzedakis 1994; Wijmstra 1969).

After about 40 ka BP a sequence of increasingly frequent cold and dry events meant that dominant conditions were more similar to those during the ice advance of late OIS-4 (van Andel & Tzedakis 1996, fig. 13), although with a smaller Fennoscandian ice cap. In the maritime west of Europe a shrub-tundra landscape prevailed, whereas in the continental east the steppe vegetation was dominated by grasses requiring at least 300 mm yr^{-1} of precipitation (Rossignol-Strick 1995), a landscape quite unlike that of the drier late OIS-4 when wormwoods and chenopods dominated. Very low relative abundances of tree pollen, only of coniferous taxa, imply at most a discontinuous scatter of trees. This landscape, quite different from the tundra and polar deserts of the present high Arctic latitudes, extended southward to the Pyrenean, Alpine and Carpathian mountain ranges and to the northern Mediterranean region.

No-analogue biomes

At regional scales, biotic assemblages can usefully be described using the concept of biomes. Biomes are the geographically most extensive biotic communities it is convenient to recognize; they are characterized in terms of the ecology, physiognomy and phenology of the dominant taxa in their vegetation (Allaby 1994, 50). At the present time the dominant biome over much of central Europe is temperate deciduous forest, with evergreen taiga/montane forest to the north and in mountainous regions, and warm mixed forest and temperate sclerophyll woodland to the south (Ozenda 1994).

When we attempt to reconstruct the biomes of Europe during OIS-3, it is important to remember that, even if temperatures in much of Europe at that time were as low as those of the Arctic today, the insolation intensity was markedly higher because of the lower latitude. Furthermore, throughout OIS-3 the summer sun rose higher in the sky in northern mid-latitudes, giving even higher summer insolation than today (Berger 1978). As a result, the contrasts between temperatures within a few centimetres of the ground surface and those that meteorologists conventionally measure 2 m above the ground would have been greater than in Arctic regions today (Andersen 1993). This contrast differentially affects trees and other taller-growing plants, as opposed to low-growing plants and surface-living invertebrates, the latter effectively experiencing a longer 'growing season' than the former to a greater extent than is the case in the Arctic today. It must, therefore, not be assumed that glacial environmental conditions in mid-latitude Europe were similar to those of northern Canada, Lapland and Siberia today.

In addition, the glacial atmosphere had a markedly lower concentration of carbon dioxide than the post-glacial atmosphere; concentrations ranged between c. 180 and c. 220 ppmv during OIS-3, in contrast to their typical post-glacial value of c. 280 ppmv (Petit et al. 1999). Higher plants utilizing the more widespread photosynthetic pathway (C3), as opposed to the modified pathway (C4) used by some tropical grasses and succulents, suffer a reduction in water-use efficiency at lower carbon dioxide concentrations; that is to say, at any given level of moisture availability, they experience greater effective water limitation because they must keep their stomata open longer to gain the carbon that they require, and as a result lose more water through transpiration. Furthermore, this once again applies especially to those plants whose leaves are located in the relatively well-mixed air 2 m or more above the ground in an open canopy, as opposed to those growing in the few centimetres above the ground where carbon dioxide concentrations often may even be locally elevated as a result of soil respiration. The c. 20 per cent lower carbon dioxide concentration during OIS-3 is likely to have contributed to a greater limitation of trees to mesic environments than is seen in contemporary landscapes.

Given that the combinations of environmental conditions during OIS-3 were unlike any today, it is to be expected that many mid-glacial biomes have

no modern analogues. Evidence to support this can be found in the OIS-3 assemblages of fossils of several groups. For example, Coope (1977) inferred, from a rich, diverse temperate insect fauna found in association with a flora totally lacking in trees at Upton Warren (UK), that uniformitarianism could not be straightforwardly called upon for help in understanding the major biomes of Northwest Europe during the late Pleistocene. Mammalian palaeontologists also have argued against the idea that the mid-latitude glacial vegetation cover of much of Europe was a tundra analogous to that found in the Arctic today. On the contrary, the no-analogue biome variously referred to as steppe-tundra or 'mammoth steppe' must of necessity have been endowed with considerable productivity in order to sustain the rich megafauna of the glacial period (Guthrie 1990; Lister 1987; Lister & Sher 1995). The megafauna included not only such large extinct herbivores as the woolly mammoth (*Mammuthus primigenius*), the woolly rhinoceros (*Coelodonta antiquitatis*) and the giant deer (*Megaloceros giganteus*), but also many species that survived into the Holocene.

A related and striking characteristic of the glacial landscape was its many unusual species associations. Communities without modern analogues had combinations of species found today in high northern latitudes or at high altitudes with others that today inhabit dry continental areas, as well as some normally associated with more temperate climates. An example is the mid-glacial fauna of western Aquitaine (Delpech 1993) which included reindeer (*Rangifer tarandus*), chamois (*Rupicapra rupicapra*), steppe ass (*Equus hydruntinus*) and wild boar (*Sus scrofa*).

Simulating plant cover for OIS-3

Figure 6.2 presents a useful initial picture of the landscapes that might have been the environmental backdrop when the Neanderthals flourished, encountered modern humans and were eventually replaced by them. The data that supported these reconstructions, however, have patchy geographic coverage and are derived mainly from locations in western maritime Europe, with an almost complete absence of data for the continental climate of eastern and south-eastern Europe. The wide spacing of data points also leads to uncertainties of hundreds of kilometres in the positions of biome boundaries, as well as to an absence of potentially important details. A more important shortcoming is the qualitative and subjective nature of the inferences with respect to the character of the vegetation and climate. Furthermore, the intuitive reliance on modern analogues that may be inappropriate leads to a degree of vagueness in the definition of landscape units. Finally, the inferences may also be flawed as a result of an incomplete understanding of the interactions between climate, vegetation and atmospheric composition.

If we wish to specify the environmental conditions under which the human genus was reduced to a single species, and to assess the role that changes of climate and vegetation might or might not have played, biome reconstructions or simulations that make the minimum of uniformitarian assumptions are required. To achieve this we sought to characterize the vegetation contrasts between typical warm and cold phases of the Dansgaard/Oeschger oscillations (hereafter referred to as warm type (D/O) and cold type (D/O) events: Fig. 6.1) using two complementary approaches:

1. inferring palaeovegetation objectively and systematically from palynological data;
2. modelling palaeovegetation using Biome 3.5 (J. Kaplan pers. comm.; see also Haxeltine & Prentice 1996; Prentice *et al.* 1992) and output from a Regional Climate Model (RegCM2), the latter linked to an atmospheric General Circulation Model (GENESIS version 2.0a: see Barron *et al.* 2003).

By comparing the *inferred palaeovegetation* record derived from palynological data with the *modelled vegetation* patterns based upon the simulated climate, we hoped to identify those aspects of palaeoclimate that may have been primarily responsible for the differences in vegetation between cold and warm phases of D/O oscillations during OIS-3 (for details see Alfano *et al.* 2003; Huntley *et al.* 2003).

Before making any reconstructions or simulations, we had first to decide which time periods to focus upon within OIS-3. The total amount of incoming solar radiation and its seasonal distribution varied little between 45 and 30 ka BP (Berger 1978); selecting events within this interval, therefore, allowed the same orbital parameters to be used for both simulations. We selected the relatively long warm event centred upon *c.* 45 ka BP in the GISP2 ice core (Meese *et al.* 1997) to represent the typical warm type D/O event. The cold event centred upon *c.* 30 ka BP in the GISP2 time scale was selected to represent the typical cold type D/O event. The latter cold event occurs after the last abundance peak of temperate taxa in continuous terrestrial pollen records from central and southern Europe, but precedes the onset of the LGM. In terms of mean annual temperatures inferred from the Greenland ice-core $\delta^{18}O$ records, however, it is very similar to the LGM (Fig. 6.1).

Palaeovegetation inferences from palynological data
Both macro- and microfossils of terrestrial plants, preserved in many deposits, provide direct evidence of the nature of the vegetation of the past. Plant macrofossils (e.g. seeds, leaves, wood) typically provide greater taxonomic resolution, although the records are often difficult to quantify and strongly biased towards taxa growing close to the site of deposition. Macrofossil remains from sites of human habitation are likely to be particularly severely biased towards taxa that were selectively collected and used by the human population, or else were favoured by human disturbance. A good example of the potential for such biases is the relative abundances of different taxa in charcoal assemblages from archaeological sites; these are more likely to reflect selection of favoured species than to portray accurately the relative abundance of woody taxa in the surrounding landscape. Furthermore, although the presence of some woody taxa during the last glacial has been documented by charcoal studies, such as Willis *et al.* (2000) for central and southeastern Europe, the presence of charcoal on archaeological sites does not necessarily imply extensive woodland cover; the trees need not have been abundant, nor even present, in the local vegetation. In practice, systematic studies of plant macrofossil remains from OIS-3 deposits are still very few; such data, therefore, were not used as the basis for our reconstructions.

Plant microfossils, mainly pollen grains and spores, provide a less biased and more regional record of the composition of past vegetation. Plant microfossil records also are more widely available and better quantified; studies representing some or all of OIS-3 are available from some tens of localities throughout Europe. Accordingly, we have compiled plant microfossil data as the basis for inferring the European OIS-3 vegetation.

The primary problem encountered in compiling the palaeovegetation data set arose from uncertainties with respect to the chronologies at the localities from which data were made available. The early part of OIS-3, before *c.* 42 ka BP, lies beyond the limit of conventional radiocarbon (^{14}C) dating; even for the later part of OIS-3, the reliability of ^{14}C dates is questionable and the statistical uncertainties of the ^{14}C age determinations are often comparable in magnitude to the millennial time-scale of many of the D/O oscillations. Few European pollen records spanning this interval have been dated by other techniques; the only one available to us was that from Lago Grande di Monticchio in southern Italy (Allen & Huntley 2000; Allen *et al.* 2000).

In order to select appropriate samples from

which to compile pollen data for the chosen warm type and cold type D/O events, continuous records were matched to one another and to the GISP2 $\delta^{18}O$ record on the basis of the sequence of fluctuations they recorded, utilizing in addition any age constraints available. In the case of discontinuous records, however, we were forced to rely upon ^{14}C dates to identify appropriate samples from which to compile data. For the warm event we utilized data from interstadial sites with finite ^{14}C ages between 45 and 30 ka BP. No data were available from such sites for the cold interval, however, because in northern Europe, from where the discontinuous records mostly came, cold events are represented principally by periglacial deposits lacking palaeovegetation evidence. Figure 6.3 shows the locations of the 30 sites from which data were compiled (for details of the samples used see Huntley et al. 2003, table 2).

The character of past vegetation can be inferred from palynological data in several ways, of which the subjective and qualitative approach of van Andel & Tzedakis (1996; 1998) is one example. For the Stage 3 Project we preferred to adopt a quantitative and systematic approach, because

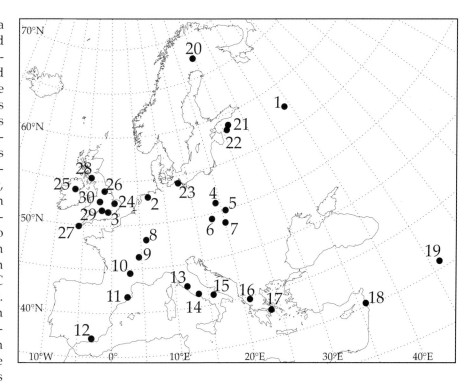

Figure 6.3. *Locations of sites from which palynological data were compiled. Numbers indicate sites as follows: 1) Nero (Aleshinskaya & Gunova 1976); 2) Hengelo (Ran et al. 1990); 3) Isleworth + Ismaili Centre (Coope et al. 1997; Kerney et al. 1983); 4) Wola Grzymalina (Krzyszkowski et al. 1993); 5) Sadowie (Srodon 1987); 6) Piersiec (Niedzialkowska & Szczepanek 1994); 7) Dobra (Srodon 1968); 8) La Grande Pile (de Beaulieu & Reille 1992b); 9) Les Echets (de Beaulieu & Reille 1984); 10) Lac du Bouchet (Reille & de Beaulieu 1990); 11) Banyoles (Perez-Obiol & Julia 1994); 12) Padul (Pons & Reille 1988); 13) Lagaccione (Follieri et al. 1998); 14) Valle di Castiglione (Follieri et al. 1988); 15) Lago Grande di Monticchio (Allen et al. 2000); 16) Ioannina (P.C. Tzedakis unpublished); 17) Kopaïs (Tzedakis 1999); 18) Ghab (Niklewski & van Zeist 1970); 19) Zeribar (van Zeist & Bottema 1977); 20) Sokli (Helmens et al. 2000); 21) Töravere (Liivrand 1990); 22) Valguta (Liivrand 1991); 23) Sassnitz-Stubbenkammer (K. Erd pers. comm. 1999); 24) Beetley (Phillips 1976; West 1991); 25) Derryvree (Colhoun et al. 1972); 26) Oxbow (Gaunt et al. 1970); 27) Porth Seal, Isles of Silly (Scourse 1991); 28) Sourlie (Jardine et al. 1988); 29) Stanton Harcourt (Seddon 1985); 30) Upton Warren (Coope 1961). (For details of samples see Huntley et al. 2003, table 2.)*

this would facilitate objective comparisons of the inferred and simulated vegetation patterns. In making palaeovegetation inferences, it was also important to infer vegetation attributes that were available from the output of the model used to simulate the palaeovegetation. The principal model outputs are, firstly, estimates of net primary productivity (NPP: net mass of carbon accumulated per square metre per year (g C m^{-2} yr^{-1})) for a limited set of plant functional types (pfts) and, secondly, biome assignments based upon these NPP simulations and a set of rules defining the interactions between pfts. A pft

is a broad class of plants defined by stature, leaf form, phenology and climatic thresholds (Woodward & Cramer 1996), e.g. temperate summergreen trees — including oak, maple, ash and other similar trees. It should be noted that, in characterizing pfts, the terms summergreen and raingreen are used to discriminate between deciduous taxa that lose their leaves as an adaptation to surviving winter cold as opposed to a season of drought.

A method for reconstructing biomes from pollen data was first developed by Prentice *et al.* (1996) and subsequently modified to utilize all of the

Table 6.1. *Assignment of pollen taxa to plant functional types (pfts).*

Pft code	Pollen taxa
cgs	*Calligonum*-type (calligonum), *Hippophaë* (sea buckthorn), *Oxyria*-type (mountain sorrel), Polygonaceae undiff. (dock and knotweed family), *Polygonum persicaria* (redshank), *Polygonum bistorta*-type (bistort), *Rumex* (dock).
wgs	*Armeria* (thrift), Boraginaceae (borage family), Cruciferae (Brassicaceae) (cress family), Crassulaceae (stonecrop family), *Echium*-type (bugloss), *Euphorbia* (spurge), *Hypericum* (St John's wort), Labiatae (Lamiaceae) (mint family), *Lathyrus/Vicia*-type (vetch), Leguminosae (Fabaceae) (pea family), *Mentha*-type (mint), Malvaceae (mallow family), *Papaver* (poppy), Scrophulariaceae (figwort family), Solanaceae (nightshade family), Thymelaceae cf. *Daphne* (daphne family), 'Steppics'.
sf	*Aconitum* (monkshood), cf. *Ajuga* (bugle), cf. *Allium* (onion), *Asphodelus* (asphodel), *Astragalus/Oxytropis* (milkvetch), Campanulaceae (bellflower family), Caryophyllaceae (pink family), *Centaurea* (knapweed), Compositae (Asteraceae) (daisy family), Compositae subfam. Tubuliflorae (Asteroideae) (aster sub-family), Compositae subfam. Liguliflorae (Cichoriodeae) (chicory sub-family), *Chamaenerion* (rosebay willowherb), *Cirsium*-type (thistle), Dipsacaceae (teasel family), *Epilobium* (willowherb), *Filipendula* (meadowsweet), *Fumana* (fumana), *Galium*-type (bedstraw), *Geranium* (cranes-bill), *Geum* (avens), *Helianthemum* (rockrose), Liliaceae (lily family), *Linum* (flax), *Matricaria*-type (chamomile), *Onobrychis*-type (sainfoin), *Plantago* undiff. (plantain), *Pleurospermum* (pleurospermum), *Potentilla* (cinquefoil), Ranunculaceae (buttercup family), Resedaceae (mignonette family), Rosaceae undiff. (rose family), Rubiaceae (bedstraw family), *Sanguisorba* (burnet), *Sanguisorba* cf. *S. officinalis* (great burnet), *Scabiosa* (scabious), *Thalictrum* (meadow rue), Umbelliferae (Apiaceae) (carrot family), *Urtica* (nettle), *Valeriana* (valerian), *Veronica* (speedwell), 'Steppics'.
wdf	*Ephedra fragilis*-type (joint pine).
df	*Ephedra* undiff. (joint pine).
sf/df	*Artemisia* (wormwood), Chenopodiaceae (chenopod family).
g	Gramineae (grass family), Cerealia (cereals).
h	Ericales (heath family), *Calluna* (ling), *Empetrum* (crowberry).
s	Cyperaceae (sedge family).
aa	*Betula nana*-type (dwarf birch), *Dryas octopetala* (mountain avens), Gentianaceae (gentian family), *Pedicularis* (lousewort), Primulaceae (primrose family), Saxifragaceae (saxifrage family), *Saxifraga* cf. *S. stellaris* (starry saxifrage), *Salix* (willow).
ab	*Rubus chamaemorus* (cloudberry).
bec	*Picea* (spruce).
bec/cbc	*Pinus* subgen. *Haploxylon* (white pine).
bs	*Betula* (birch), *Larix* (larch).
ctc$_1$	*Cedrus* (cedar), *Taxus* (yew).
bec/ctc	*Abies* (fir).
ec	*Juniperus*-type (juniper), *Pinus* subgen. *Diploxylon* (yellow pine).
bts	Cornaceae (dogwood family), *Lonicera* (honeysuckle), *Sambucus* (elder), *Sorbus* (mountain ash), *Viburnum* (guelder rose).
bs/ts	*Alnus* (alder), *Populus* (poplar), *Salix* (willow).
ts	*Acer* (maple), *Euonymus* (spindle tree), *Fraxinus excelsior*-type (ash), *Quercus robur*-type (oak), *Viscum* (mistletoe).
ts$_1$	*Bruckenthalia* (bruckenthalia), *Carpinus betulus* (hornbeam), *Corylus* (hazel), *Fagus* (beech), *Frangula* (alder buckthorn), *Tilia* (lime), *Ulmus* (elm).
ts$_2$	Cannabaceae (hemp family), *Castanea* (sweet chestnut), *Celtis* (southern nettle tree), *Fraxinus ornus* (manna ash), *Juglans* (walnut), *Ostrya/Carpinus orientalis*-type (hop and eastern hornbeam), *Platanus* (plane), *Pterocarya* (wingnut), Rhamnaceae (buckthorn family), *Vitis* (vine), *Zelkova* (Caucasian elm).
wte	*Quercus ilex*-type (holm oak, evergreen oaks).
wte$_1$	*Buxus* (box), *Hedera* (ivy), *Ilex* (holly).
wte$_2$	*Olea* (olive), *Phillyrea* (phillyrea), *Pistacia* (turpentine tree), *Cistus* (sunrose).

Key to pft codes:
cgs - cool herb/shrub; wgs - warm herb/shrub; sf - steppe forb/shrub; wdf: warm desert forb/shrub; df - desert fob/shrub; sf/df - steppe/desert forb/shrub; g - grass; h - heath; s - sedge; aa - arctic-alpine dwarf shrub/herb; ab - arctic/boreal herb; bec - boreal evergreen coniferous tree; bec/cbc - boreal evergreen/cold boreal coniferous tree; bs - boreal summergreen tree; ctc$_1$ - cool-temperate gymnosperm tree; bec/ctc - boreal evergreen/cool-temperate coniferous tree; ec - eurythermic coniferous tree/shrub; bts - boreal/temperate summergreen shrub; bs/ts - boreal/temperate summergreen tree/shrub; ts - temperate summergreen tree/shrub/liane; ts$_1$ - cool-temperate summergreen tree/shrub; ts$_2$ - warm-temperate summergreen tree/shrub/liane; wte - warm-temperate broad-leaved evergreen tree/shrub; wte$_1$ warm-temperate broad-leaved evergreen tree/shrub/liane; wte$_2$ - warm-temperate sclerophyll tree/shrub.

information in the pollen data by Tarasov *et al.* (1998) and Allen *et al.* (2000). The first step is to assign each pollen taxon to a pft (Table 6.1). For each pollen spectrum a score is calculated for each pft using the relative abundances of the component pollen taxa. To reconstruct biomes these pft scores are summed in groups relating to each biome (Table 6.2). The biome assigned to the spectrum is that with the highest total score, except in cases where two or more biomes have equal total scores; in such cases the biome assigned is that characterized by the fewest pfts.

This 'biomization' process has one major drawback when applied to pollen assemblages representing combinations of environmental conditions very different from those of today: the only biomes that may be assigned are those that are recognized in the present global vegetation. As a result, no allowance is made for past biomes without a modern analogue that may have existed under past no-analogue conditions. In addition, the biome classification step discards potentially important information about the scores for the pfts that are present. This information may provide insights into the nature of the palaeovegetation, and possibly also into the causes of mismatches between the inferred and simulated palaeovegetation. To overcome these problems we have based our comparisons not principally upon the biomes inferred or simulated, but upon the aggregated scores for nine groups of pfts recognized for the pollen data (Table 6.3) that are equivalent to nine of the limited set of pfts used in the model (see below and Barron *et al.* 2003). The three additional pfts for which Biome 3.5

provides estimates of NPP either are irrelevant in the present context (tropical broadleaf trees and tropical grass), or else represent a type of plant for which we have no more than very rare pollen data (cushion-forb — included in the Arctic–Alpine dwarf shrub/herb pft).

In order to assess this approach, and thus give a clearer meaning to the assessment of the OIS-3 results, we applied the same method to an extensive set of surface (modern) pollen samples from Europe, comparing the results both with the known contemporary vegetation of Europe and with the output from a model simulation for modern climate.

Table 6.2. *Assignment of plant functional types (pfts) to biomes.*

Biome	PFTs
deciduous taiga/montane forest	(h), bec/cbc, bs, ec, bs/ts, ab
evergreen taiga/montane forest	(h), bec, bec/cbc, bs, bec/ctc, ec, bs/ts, ab
cold mixed forest	(h), bs, ctc_1, bec/ctc, ec, bs/ts, (ts_1), ab
temperate conifer forest	(h), bec, bec/cbc, bs, bec/ctc, ec, (bts), bs/ts, (ts_1)
temperate deciduous forest & temperate woodland*	(h), bs, ctc_1, bec/ctc, ec, bts, bs/ts, ts, ts_1, (ts_2), (wte_1)
cool mixed forest	(h), bec, bec/cbc, bs, bec/ctc, ec, bts, bs/ts, ts, ts_1
warm mixed forest	(h), ec, (bts), bs/ts, ts, ts_1, ts_2, wte, (wte_1)
temperate sclerophyll woodland/scrub	g, ec, wte, wte_2
temperate grassland	cgs, sf, sf/df, g
steppe tundra/warm steppe	wgs, sf, sf/df, g
desert: shrubland and steppe	df, sf/df
hot desert	wdf, df, sf/df
shrub tundra/dwarf-shrub tundra/ prostrate shrub tundra/cushion-forb-lichen–moss tundra	g, h, s, aa, ab

Notes
[1] pfts in parentheses are restricted to part of the biome.
[2] *Spectra with >70 per cent pollen of woody taxa are classified as temperate deciduous forest. Those with <70 per cent pollen of woody taxa are re-classified as wooded steppe (Allen *et al.* 2000).

Table 6.3. *Composition of aggregated plant functional types (pfts).*

Aggregated pft	Component pollen-based pfts
(a) Cold herbaceous	cgs, sf, s, aa, ab
(b) Boreal summergreen tree	bs, bs/ts
(c) Boreal evergreen tree	bec, bec/cbc
(d) Tundra shrub	h
(e) Cool coniferous tree	ctc_1, bec/ctc, ec
(f) Temperate summergreen tree	ts, ts_1, ts_2, bts
(g) Temperate grass	g
(h) Temperate broadleaved evergreen tree	wte, wte_1, wte_2
(i) Woody desert plants and steppe forbs	wgs, wdf, df, sf/df

Palaeovegetation simulations based upon modelled palaeoclimate

The model simulations used for comparison with inferred palaeovegetation derive from the third set of model experiments described by Barron & Pollard (2002) and Pollard & Barron (2003); sensitivity tests of the results of these model experiments have been presented by Alfano *et al.* (2003). In these experiments the boundary conditions used for the GENESIS 2a global climate and vegetation simulations were: 30 ka BP orbit; 200 ppmv CO_2; ice sheets and shorelines for 42 ka BP ('warm') and 30 ka BP ('cold') and sea-surface temperatures (SSTs) prescribed on the basis of available data (Barron & Pollard 2002). Vegetation was modelled in GENESIS 2a using BIOME 3 (Haxeltine & Prentice 1996). For the regional simulations using RegCM2, boundary conditions were as above, but included also the vegetation simulated by GENESIS 2a/BIOME 3 in the global simulation and a 'small' configuration of the Scandinavian ice sheet (Arnold *et al.* 2002). The European scale regional vegetation simulations were made using BIOME 3.5 and the climate simulated by ReGCM2; they are portrayed in terms of NPP for nine pfts as well as the biomes assigned.

Unfortunately, throughout the series of modelling experiments, the boundary conditions used for simulations of the 30 ka BP cold event stubbornly refused to generate plausibly cold conditions. This most probably resulted from limitations in the sea-surface temperature data available for this interval, as well as from poorly-defined limits of sea-ice cover both in terms of seasonal duration and of spatial distribution. Sensitivity tests emphasized the importance of assumptions about the seasonal cycle of temperatures, a relatively long 'winter' and short 'summer' duration being required to approximate plausibly cold conditions (Alfano *et al.* 2003). The pollen data thus probably provide the only reliable evidence of vegetation during the 30 ka BP cold type D/O event.

In addition to the OIS-3 simulations, a RegCM2/BIOME 3.5 simulation of the present-day climate and vegetation of Europe was also performed to enable comparison with the pollen-derived pft scores and inferred biomes for the present. For this modern simulation the boundary conditions were: modern orbit; 345 ppmv CO_2; modern ice and shorelines; and prescribed modern SSTs.

Data–model comparisons

The vegetation inferences from the pollen data and the vegetation simulations from the model are presented in a series of maps. Figures 6.4 and 6.5 present the results for the nine pfts for the present day. The panels in Figure 6.4 show the aggregated scores from the pollen data, whereas Figure 6.5 shows the simulated NPP values. The equivalent results for the warm type D/O event and for the cold type D/O event are illustrated by two further pairs of figures (Figs. 6.6 & 6.7 (warm); Figs. 6.9 & 6.10 (cold)). Figure 6.11 shows the inferred and simulated biomes.

Modern vegetation

For the modern vegetation, comparison of the spatial patterns of aggregated scores and simulated NPP values for the nine pfts reveals some degree of general agreement but also some discrepancies. For the tree pfts (Figs. 6.4 & 6.5: panels b, c, e, f & h) the overall patterns are generally similar; for example, temperate broadleaved evergreen trees are found mainly in the southern parts of the region, whereas temperate summergreen trees occur mainly between 40°N and 55°N. The discrepancies relate principally to the spatial patterns for the cold herbaceous (Figs. 6.4 & 6.5: panel a) and tundra shrub (Figs. 6.4 & 6.5: panel d) pfts; these both have wider distributions in the pollen-based inferences than in the model simulations. In the case of the cold herbaceous pft this most likely relates to the inclusion of sedges (Cyperaceae) in the pollen-based pft. Although sedges are widespread and often abundant in Arctic and montane vegetation, they also occur locally in wetlands throughout the region. Similarly, because their pollen cannot be, or is not routinely, identified to a finer taxonomic level, the tundra shrub pollen-based pft includes all of the Ericales, i.e. not only the Boreal and Arctic dwarf-shrub species, but also the tall heaths of southwestern and Mediterranean Europe such as the tree heath (*Erica arborea*); inclusion of the latter group likely accounts for the discrepancy in this case.

Comparison of the inferred and simulated biomes (Figs. 6.11a & b) reveals somewhat greater discrepancies. In eastern and central Europe, where the model simulates a temperate deciduous forest, the biome inferred from the pollen data is cool mixed forest. In southern Iberia, temperate coniferous forest is simulated by the model, whereas temperate sclerophyll woodland is the predominant biome inferred from the pollen data. In both cases, published maps of modern vegetation suggest that the pollen-based inference is more realistic (see e.g. Ozenda 1994; Ozenda & Borel 2000). The problem with the model simulation for southern Iberia is related to a feature specific to BIOME 3.5 (the use of absolute minimum temperature as a constraint upon some pfts); published vegetation simulations based upon other versions of BIOME exhibit better agreement with the actual southern European vegetation (e.g. Haxeltine & Prentice 1996, BIOME 3; Kaplan 2001, BIOME 4). Despite these problems, there is broad-scale agreement between the pollen-based and model results with respect to the general position and orientation of the boundaries of the principal biomes.

Although the discrepancies outlined above must be remembered and taken into account when examining the results for the OIS-3 warm and cold D/O events, they are not in themselves of sufficient magnitude to rule out either the pollen-based or the model approach as a basis for gaining useful insights into the environmental conditions at these times.

Warm type D/O event

Although the results for the modern vegetation suggest that data–model comparisons for climatic events in OIS-3 should focus upon spatial patterns in the pft scores and NPP values, the inferred and modelled biomes are also presented because they have some additional value.

The pollen-based pft scores for the warm type D/O event indicate a much more restricted presence, as well as reduced overall abundance, of tree pfts (Figs. 6.6b, c, e, f & h) than at the present day. Temperate summergreen trees (Fig. 6.6f) and cool coniferous trees (Fig. 6.6e) display high scores mainly in southern Europe (cf. Figs. 6.4f & e). Boreal summergreen and boreal evergreen trees (Figs. 6.6b & c) have low scores in northern mainland Europe, except in areas east of the Fennoscandian ice sheet where high scores are achieved by boreal summergreen trees. The most striking features of the palaeovegetation data for the warm type D/O event are the high scores for the cold herbaceous pft (Fig. 6.6a), coupled to relatively high scores at many sites for the temperate grass (Fig. 6.6g) and woody desert plants and steppe forbs (Fig. 6.6i) pfts.

For the warm type D/O event simulated NPP values (Fig. 6.7) do not exceed 600 g C m^{-2} yr^{-1} for any pft. This contrasts with the modern simulation, in which NPP values for six of the pfts exceed 700 g C m^{-2} yr^{-1} resulting in total NPP values comparable to other published data (e.g. Fischer 1997). This simulated lower productivity of the vegetation during the warm type D/O event is paralleled by the simulation of a markedly lower leaf area index (LAI) than under modern conditions across most of Europe. Together, these results imply a generally more open or sparse cover of vegetation during the warm type D/O event than today.

There are some striking contrasts between the inferred and simulated spatial distributions of individual pfts: For example, the cold herbaceous pft is simulated only in regions of northern Europe close to the Fennoscandian ice sheet and in the Alps (Fig. 6.7a), whereas the pollen data indicate a much more widespread occurrence (Fig. 6.6a). There also are marked discrepancies between the simulated NPP values and the pollen-based scores for some pfts with respect to the differences between the warm type D/O event and the present day. The temperate grass (Figs. 6.5g & 6.7g) and woody desert plants and steppe forbs (Figs. 6.5i & 6.7i) pfts both have generally lower simulated NPP values but higher pollen-based scores than today, especially in northern Europe. The most striking of such discrepancies, however, relates to the principal tree pfts that have simulated NPP values >200 g C m^{-2} yr^{-1} over large areas where their pollen-based scores are very low or even zero; as a result, trees dominate the simulated vegetation over large areas where the pollen data show non-tree pfts to have been dominant.

These discrepancies are further highlighted when inferred and modelled biomes are compared (Figs. 6.11c & d). The widespread, substantial NPP simulated for boreal evergreen trees results in the evergreen taiga/montane forest biome dominating most of Europe in the model results. Temperate deciduous forest and temperate woodland biomes are also simulated in scattered areas across a central latitudinal belt, the latter being more common further east. Tundra biomes, however, are simulated only in limited areas of the north, close to the Fennoscandian ice sheet, with temperate grassland simulated as the predominant biome for southern Europe. In contrast, inference from the pollen data indicates that non-forest biomes, principally dwarf-shrub tundra, steppe tundra/warm steppe and temperate grassland, predominated across most of northwestern and central Europe. In northern Europe, tundra with deciduous taiga/montane forest was present to the east of the Fennoscandian ice sheet, with evergreen taiga present only further still to the east. At sites in west central Europe the cool mixed forest biome occurred along with temperate grassland, becoming more dominant further south, whereas east of the Mediterranean treeless steppe tundra/warm steppe is inferred.

Cold type D/O event

For the cold type D/O events the pollen-based scores for tree pfts (Figs. 6.9b, c, e, f & h) are generally lower than during warm type D/O events (see Fig. 6.6). The one exception is the site in the northeast where the boreal summergreen tree pft still achieves

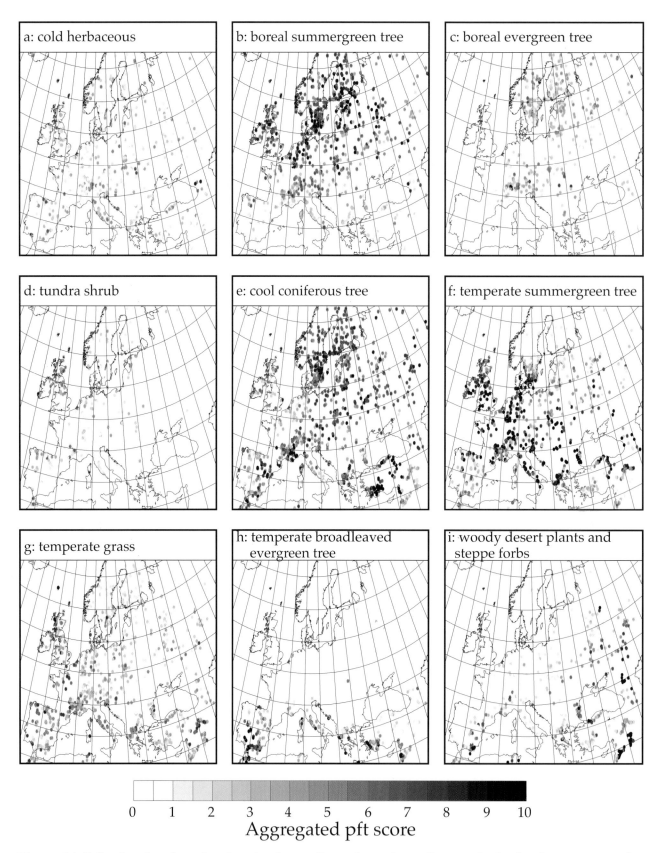

Figure 6.4. *Pollen-based modern plant functional types. Scores for modern pollen samples for the nine aggregates of plant functional types listed in Table 6.2.*

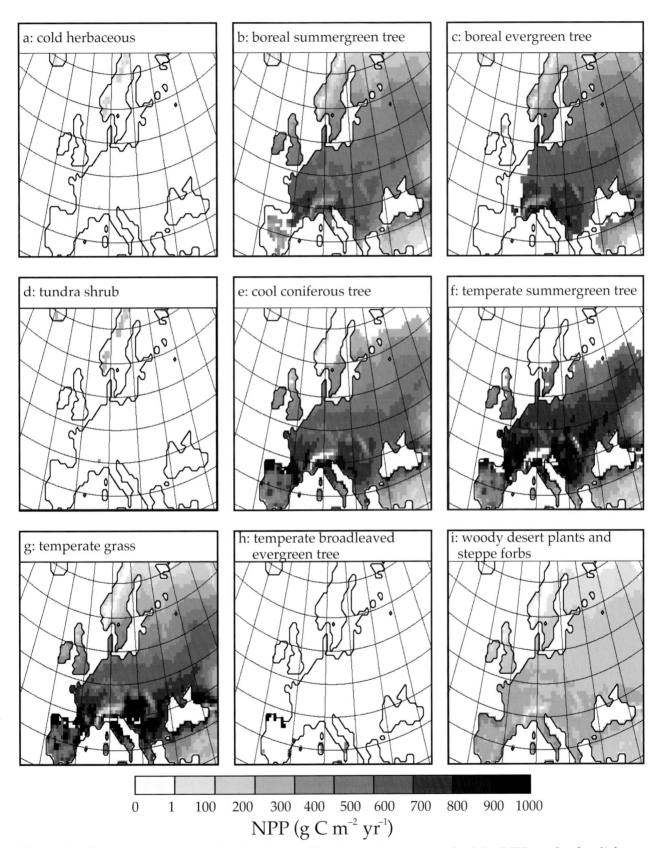

Figure 6.5. *Simulated modern plant functional types. Simulated net primary productivity (NPP – g C m⁻² yr⁻¹) for nine aggregated plant functional types. Values from BIOME 3.5 linked to the simulated modern climate.*

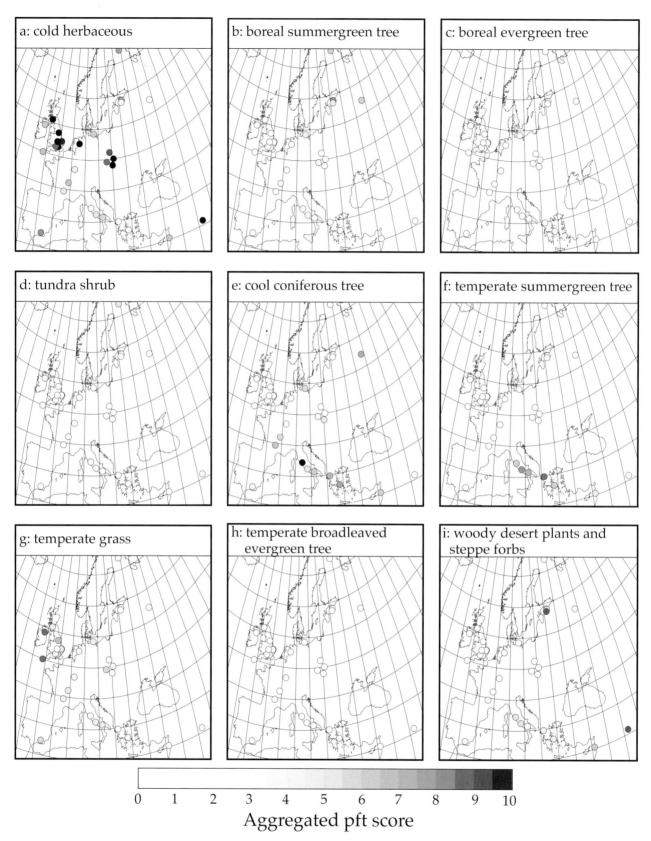

Figure 6.6. *Pollen-based plant functional types for Stage 3 warm type D/O event. Pft scores for the relevant pollen samples are plotted for aggregates of plant functional types.*

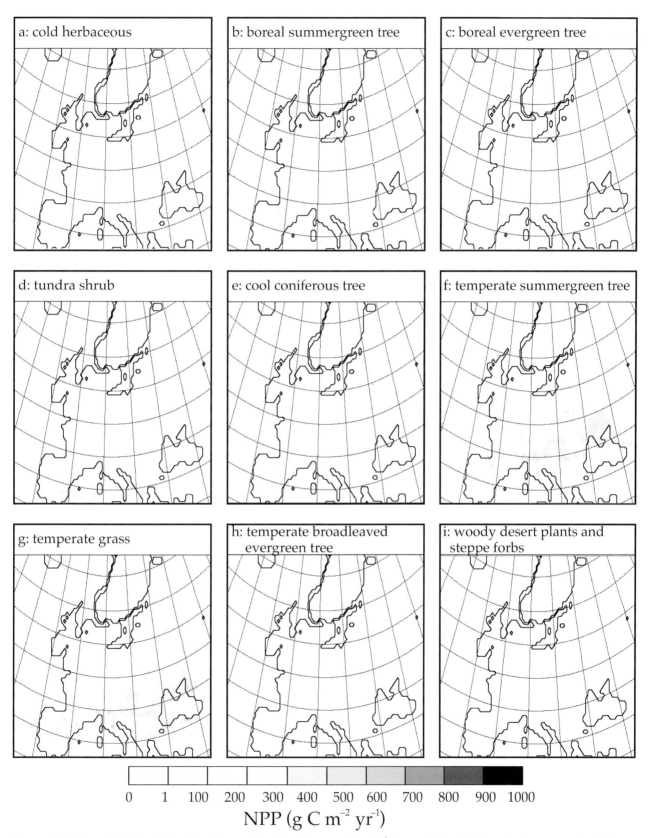

Figure 6.7. *Simulated plant functional types for Stage 3 warm type D/O event. Simulated net primary productivity (NPP – g C m^{-2} yr^{-1}) for nine aggregated plant functional types. Values are derived from BIOME 3.5 applied to simulated Stage 3 warm type climate.*

a relatively high score (Nero Lake; Fig. 6.3: site 1). Cool coniferous trees also are still widespread (Fig. 6.9e), whereas temperate summergreen trees are more restricted to the south than in the warm type D/O event (Fig. 6.9f). High scores are achieved by the cold herbaceous, temperate grass, and woody desert plants and steppe forbs pfts (Fig. 6.9a, g & i).

Although the pollen data thus indicate substantial contrasts between the warm and cold type D/O events, simulated NPP values for the cold type D/O event (Fig. 6.10) are generally very similar to those for the warm type D/O event (Fig. 6.7). This is an inevitable consequence of the failure of the general circulation model (GENESIS 2a), given the cold type D/O event boundary conditions applied, to simulate conditions substantially colder than those simulated for the warm type D/O event boundary conditions (see Barron *et al.* 2003).

As would be expected, given the pollen-based scores for the various pfts, non-forest biomes are inferred from the pollen data for cold type D/O events, with the sole exception of the single site in the northeast where the evergreen taiga/montane forest biome is inferred (Fig. 6.11e). The modelled biomes (Fig. 6.11f), however, are very similar to those simulated for the warm type D/O event.

What have we achieved?

Three notable features of the inferred and modelled palaeovegetation arise from the comparisons presented above:

1. The vegetation patterns inferred from the pollen data for the warm and cold type D/O events differ markedly from present-day patterns; they do not represent simple latitudinal shifts of present vegetation belts.
2. Whereas the vegetation patterns inferred for the warm and cold type D/O events differ substantially from each other, the simulated vegetation patterns are remarkably similar to each other.
3. The simulated Stage 3 vegetation has a much lower NPP, and LAI, than the simulated modern vegetation. The vegetation simulated for Stage 3 is thus structurally different from modern vegetation; labelling it as equivalent to a modern biome hence is potentially misleading.

The vegetation patterns inferred from the pollen data for warm and cold type D/O events within Stage 3 are consistent with the climate variations over Greenland and the North Atlantic as inferred from ice and ocean sediment core data (Bond *et al.* 1993; Meese *et al.* 1997). The OIS-3 cold events were characterized by temperatures markedly lower than those of the

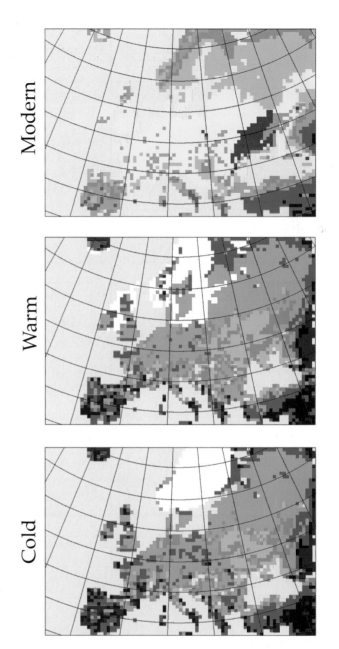

Figure 6.8. *Simulated leaf area index. Simulated maximum leaf area index ($m^2 m^{-2}$) for modern, warm and cold type D/O events.*

warm events which, in turn, were somewhat less warm than interglacial conditions (i.e. OIS-5e or OIS-1, see Chapter 2, van Andel 2003, Fig. 2.3).

At most sites used in this study, neither the warm nor the cold D/O event pft spectra have extensive modern analogues; in particular, the abundance and widespread occurrence of the cold herbaceous pft implies a prevalence of no-analogue vegetation. The inferred palaeovegetation also reveals evidence of steep climatic gradients in Europe during Stage 3. The occurrence of boreal summergreen trees and, to a lesser extent, boreal evergreen trees east of the Fennoscandian ice sheet, especially during the warm events, is particularly interesting. If there was, as commonly assumed, an extensive Fennoscandian ice sheet, the presence of these boreal trees may reflect regional warming resulting from katabatic air flow off the ice sheet under the influence of prevalently westerly air flow. However, if, as others have argued (Arnold *et al.* 2002), such an extensive ice sheet did not exist during Stage 3, then the presence of boreal trees in northeast Europe is more difficult to explain. The limited area in the central and eastern Mediterranean where the temperate summergreen tree pft occurred during warm type D/O events, and the similar but even more restricted area of occurrence of this pft during cold events, also suggests a longitudinal temperature and/or moisture availability gradient related to continentality. Such a gradient is not seen as a marked feature of the present climate of the western central Mediterranean.

The major discrepancy between the inferred and simulated vegetation patterns for the warm type D/O events is the substantial NPP simulated for several tree pfts throughout much of Europe in regions where pollen data show that trees were generally absent, or at best formed a cover that was sparse and discontinuous (see van Andel 2003, Figs. 2.5 & 2.6; Allen *et al.* 2000). Why were these pfts — boreal summergreen, boreal evergreen, cool coniferous and temperate summergreen trees — mainly absent from these regions? Although sufficiently low absolute minimum temperatures can exclude boreal evergreen, cool coniferous and temperate summergreen trees, boreal summergreen trees have no lower limit for absolute cold and thus cannot be excluded by this alone (Larcher 2003; Prentice *et al.* 1992). In common with the other tree pfts, however, boreal summergreen trees are apparently excluded by sufficiently low values for the accumulated warmth of the growing season, as well as by low moisture availability (Prentice *et al.* 1992). Thus, either very low accumulated warmth of the growing season or reduced effective moisture availability could exclude all of these tree pfts. Given the comments above relating to the impact of the reduced atmospheric CO_2 concentration during OIS-3, and the relatively high insolation during that period, it is perhaps most plausible to hypothesize that the no more than sparse presence of trees throughout most of northern Eu-

rope at that time reflects primarily effective moisture limitation. Furthermore, the thermally most favourable sites for trees at that time (i.e. south- to southwest-facing slopes) would also have been the most moisture deficient.

An alternative hypothesis, embedded in the conventional definition of an interstadial, as opposed to an interglacial, proposes that the climate during OIS-3 warm type D/O events was suitable to support tree pfts at sufficient densities to form forest biomes, as simulated by the model, but that these pfts were absent or sparse in occurrence across most of Europe because they had failed to migrate from the refugia where they had survived the climatic extremes of OIS-4 and of the cold stadial events during OIS-3. Given the long duration of most of the Stage 3 warm events, especially between about 60 and 40 ka BP (Fig. 6.1), however, migrational lag of sufficient magnitude is unlikely. In the early Holocene, European trees, including temperate summergreen trees, moved out of their areas of glacial distribution at rates of 0.2–2 km yr^{-1} (Huntley 1988; 1991), with some taxa advancing their range boundaries by more than 2000 km in the first millennium after 10,000 ^{14}C yr BP. There is no evidential basis to suggest that the same tree taxa did not move equally rapidly during OIS-3 warm type D/O events.

Nonetheless, in addition to limitations of dispersal distance and time to maturity that influence migration rates, OIS-3 tree populations may also have been subject to 'population lag' if their populations during cold events were very sparse. Effective migration demands a population source capable of generating a 'propagule pressure' adequate to ensure that long-distance dispersal of even a very small fraction of the propagules suffices to maintain migration (Collingham *et al.* 1996). Modelling studies (Huntley *et al.* 1996) have shown that if trees survived cold events only as minute populations, the spatial distribution of these units plays a critical role in determining the magnitude of the lag in response to new, more favourable conditions. In southern Europe, however, there is evidence showing very rapid response of the vegetation to changing conditions (Allen *et al.* 1999; Nimmergut *et al.* 1999). This implies that, while during cold events only small, widely scattered groups of individuals survived (Huntley *et al.* 1996), these enabled larger populations to build up rapidly in response to more favourable conditions. This would have generated, within a matter of centuries, a source population sufficient to sustain rapid large-scale migration. Given that the duration of the warm events of early Stage 3 was of the order of several millennia (Fig. 6.1), this should in princi-

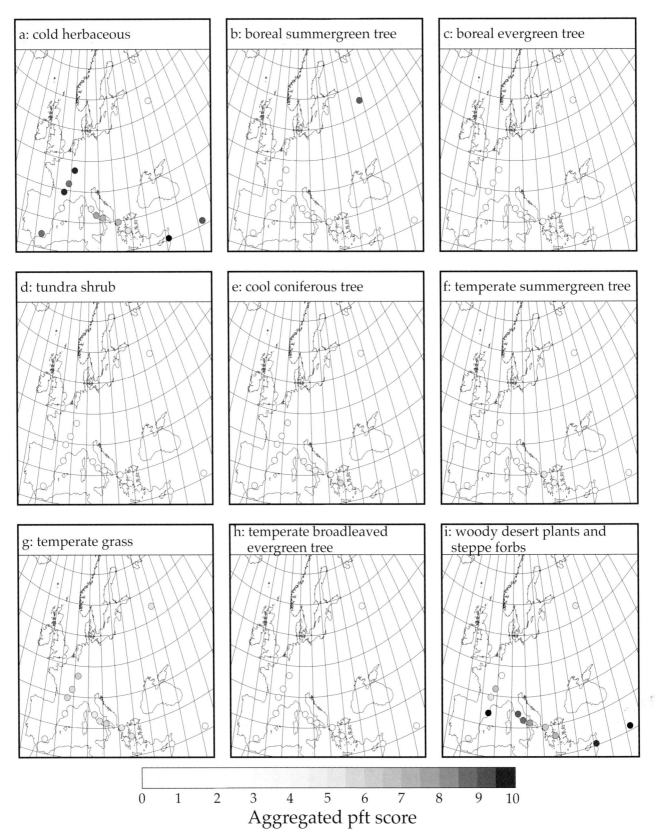

Figure 6.9. *Pollen-based plant functional types for Stage 3 cold type D/O event. Mean pft scores for the relevant pollen samples from each site are plotted for nine aggregates of plant functional types.*

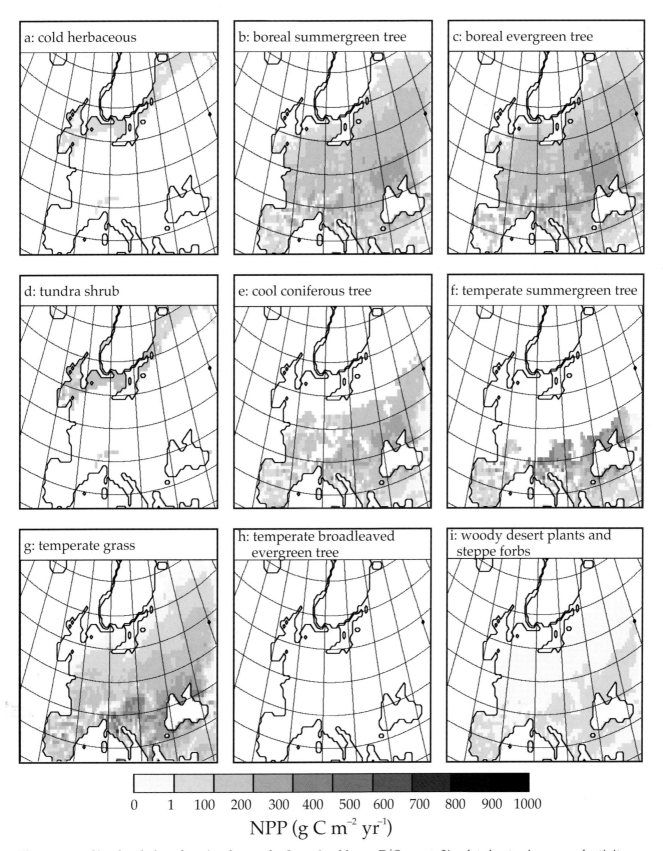

Figure 6.10. *Simulated plant functional types for Stage 3 cold type D/O event. Simulated net primary productivity (NPP – g C m⁻² yr⁻¹) for nine aggregated plant functional types. Values are derived from BIOME 3.5 applied to simulated Stage 3 cold type climate.*

Inferred biomes   Simulated biomes

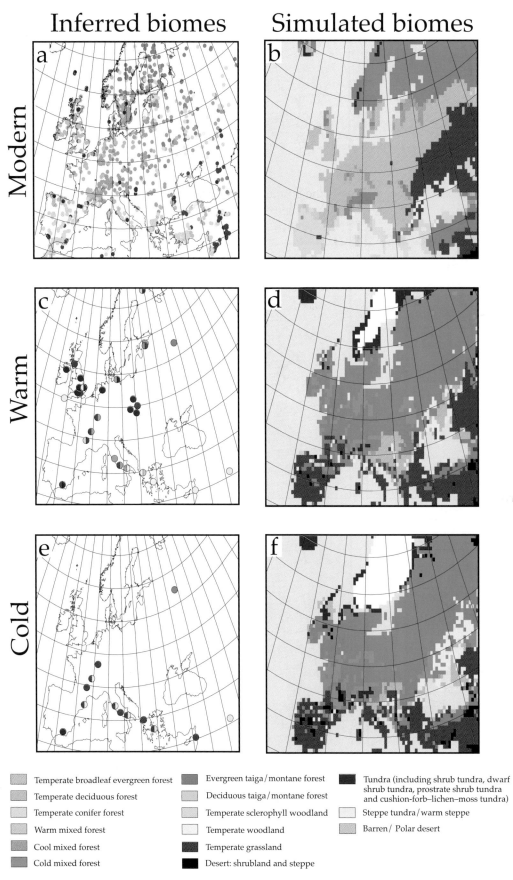

Temperate broadleaf evergreen forest

Temperate deciduous forest

Temperate conifer forest

Warm mixed forest

Cool mixed forest

Cold mixed forest

Evergreen taiga/montane forest

Deciduous taiga/montane forest

Temperate sclerophyll woodland

Temperate woodland

Temperate grassland

Desert: shrubland and steppe

Tundra (including shrub tundra, dwarf shrub tundra, prostrate shrub tundra and cushion-forb–lichen–moss tundra)

Steppe tundra/warm steppe

Barren/Polar desert

ple have allowed tree taxa, including temperate summergreen trees, to extend their ranges through-out Europe, had the climate been suitable. Although the intermittent occurrence between 60–37 ka BP of up 20–30 per cent tree pollen in some records from localities in northern Europe attests to the occurrence either of scattered woodland stands in the landscape, or of vegetation of a parkland or 'savannah-like' structure (van Andel & Tzedakis 1996; 1998), the trees represented are predominantly members of the cool coniferous and boreal summergreen pfts (Fig. 6.6e & b), with only sparse occurrences of the boreal evergreen pft (Fig. 6.6c). In contrast, pollen of members of the temperate summergreen pft is hardly present north of the Alps, even during warm type D/O events (Fig. 6.6f), strongly indicating their exclusion by unsuitable climatic conditions. Furthermore, other studies of last glacial climate and vegetation in northern Europe (e.g. Bos et al. 2001; Caspers & Freund 2001; Coope 1977; 2002; Peyron et al. 1998) have failed to find plant macrofossil evidence even of thinly scattered woodland in that region during OIS-3 and OIS-2. Given the differential sensitivity to thermal conditions, but not to moisture availability, of the temperate summergreen pft, as opposed to the other tree pfts that were present in northern Europe, conditions too cold for their occurrence, as opposed to too dry, must be inferred. This inference appears to gain support from the evidence for widespread long-lasting permafrost conditions in northern mainland Europe during Stage 3 (van Huissteden et al. 2003) that also conflicts with the simulated climatic conditions, although some uncertainty remains with respect to the dating of the permafrost evidence.

It is probable that the simulated Stage 3 warm-type D/O event climate differs from the actual climate of such events with respect to either accumulated

warmth or availability of moisture. Additionally, the similarity between the simulated climate of the warm and cold type D/O events suggests that some critical aspect(s) of the system, most likely those relevant to the cold event, may have been specified and/or simulated incorrectly. Sensitivity tests (Alfano et al. 2003) show that a combination of a 2°C reduction in the annual maximum and a 6°C reduction in the annual minimum temperature, together with a much extended winter period, is able to render cold event simulations consistent with the palaeoenvironmental data from northern Europe. However, application of these 'corrections' to areas south of the Pyrenean–Alpine ranges leads to increased mismatches between inferred and simulated vegetation in these areas.

Simulated Stage 3 NPP and LAI values are consistently much lower than those at the present day. This may in part reflect the simulated direct effects upon the vegetation of the lowered atmospheric CO_2 concentration during Stage 3 (200 ppmv as opposed to the 345 ppmv used in the modern simulations). As discussed above, the lowered CO_2 concentration may also partly account for the apparent moisture limitations indicated by the inferred vegetation (Jolly & Haxeltine 1997); prevalence of grassland and other non-forest biomes over broad areas may not solely reflect reduced moisture availability, but also reduced water-use efficiency of the vegetation (Cowling & Sykes 1999). Another important aspect of the climate simulations that may contribute to discrepancies between the simulated and inferred vegetation is the land-surface feedback from the simulated extensive cover of dark evergreen forests in areas where the pollen data indicate that open herbaceous vegetation or, during warm events, no more than sparse coniferous parkland or patchy woodland, was predominant. The contrasting seasonal albedo of taiga as opposed to tundra has been shown to have a large regional effect upon simulated Holocene temperatures (Foley et al. 1994).

The underlying cause of the failure by the GENESIS 2a/ReGCM2 modelling approach to simulate significantly colder climatic conditions in northern Europe for the cold type D/O event than for the warm type D/O event remains a matter of conjecture. Erroneous prescribed sea-surface temperatures and sea-ice limits used to force the model may account for this, and further studies along this line are underway using IMAGES cores (M. Sarnthein pers. comm.). However, for present purposes the wisest approach seems to be to disregard the cold type D/O event simulations. In some cases (e.g. Stewart et al. 2003; Davies & Gollop 2003) it is possible to replace the cold type D/O event simulations by a

Figure 6.11. *(On left) Inferred and simulated biomes. Biomes inferred from pollen analytical data (panels a, c & e) and simulated using BIOME 3.5 applied to the simulated climates (panels b, d & f). Panels (a) and (b) show the results for the present day, panels (c) and (d) for the Stage 3 warm type D/O event and panels (e) and (f) for the Stage 3 cold type D/O event. Panel (a) shows the biomes inferred from pollen surface samples, each sample being assigned to one biome. In panels (c) and (e), however, the biomisation procedure was applied to the series of relevant samples from each site. Where one biome prevailed in the results it alone is shown for the site; where two biomes were almost equally frequently inferred, however, a split symbol is used to indicate both inferred biomes.*

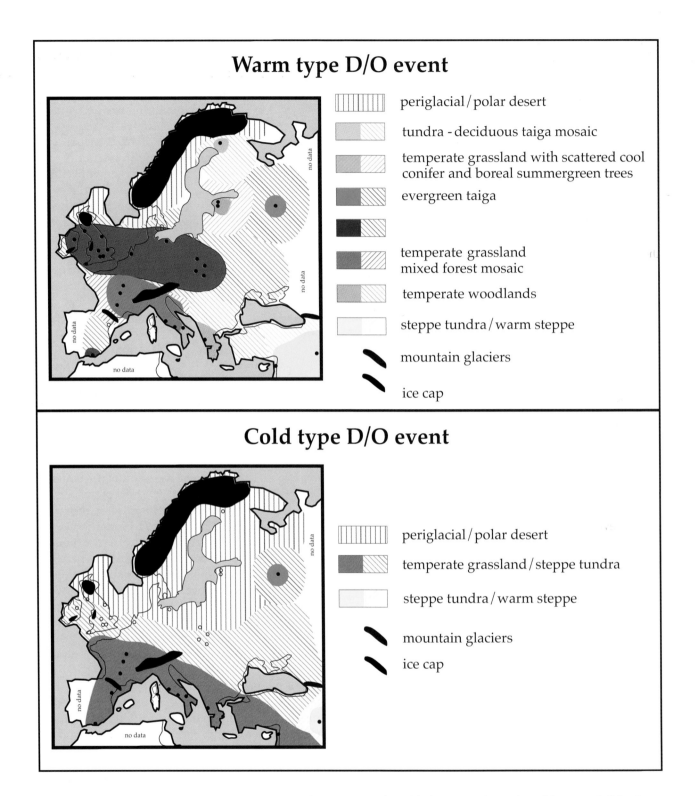

Figure 6.12. *Maps of the vegetation reconstructed for a warm and a cold climatic end member of Stage 3. Solid colours represent areas of justifiable extrapolation from data points; hatched regions represent areas of less certainty.*

simulation of Last Glacial Maximum (i.e. *c.* 21 ka BP) climate (Barron & Pollard 2002). This can be justified on the basis that, according to the ice-core record, temperatures in Greenland during cold type D/O events from about 33 ka BP onward were similar to or even somewhat colder than those of the Last Glacial Maximum (Fig. 6.1). However, no vegetation simulations were run for the Last Glacial Maximum simulated climate, and nor were pollen data compiled for that interval. The pollen-based pft scores thus provide the only sound basis for any discussion of the vegetation of the cold type D/O events, as well as for the warm type D/O events.

What might the plant cover of glacial cold and warm events have looked like?

It is evident from the above that we are still some way from achieving either a reliable set of simulations or a reasonably spatially and temporally comprehensive set of observations relating to variations in the climate and vegetation of Europe during OIS-3. In the interim, however, even a somewhat speculative set of vegetation maps representing the principal climatic end members of the last glacial may serve to summarize the current state of our understanding of the plant cover and to enable more precise questions to be posed regarding the human use of natural resources.

In Figure 6.12 we present generalized maps of the probable vegetation cover of Europe during warm and cold type D/O events, drawn based upon the pft scores and biomes inferred from the available pollen data. Note, however, that we regard it as unlikely that any two warm or cold type D/O events had identical vegetation. Just as the climate was not simply oscillating between two states, the vegetation development in response to the unique climate of each event, and to the many interrelated driving variables and feedbacks, would have been different, resulting in subtle but potentially important distinctions between the vegetation patterns of successive warm or cold type D/O events.

As these maps portray, closed canopy forests were extremely limited in extent in Europe even during warm type D/O events. Forests of a boreal character, albeit dominated by pine rather than spruce, were limited in occurrence to the northeast of Europe, whereas elsewhere closed forest stands formed at most a minor element of a landscape mosaic dominated by herbaceous vegetation. Either such mosaics with patchy woodland, or parkland/savannah-like vegetation with scattered individual trees, dominated across much of Europe during the warm type D/O events. The herbaceous matrix of these landscapes was apparently comprised of a no-analogue mixture of taxa characteristic of steppe, tundra and temperate grasslands. Given the high insolation, the relatively high NPP simulated by the model may accurately reflect highly productive no-analogue herbaceous vegetation. Such vegetation, with scattered trees or woodland stands, might plau-

sibly have provided the primary production necessary to support the large grazing and browsing herbivores whose remains abound in sediments from the warm type D/O events. In turn, the productive ecosystems of these warm events may have favoured human populations. In contrast, however, the cold type D/O events were characterized by a reduction in the abundance of trees throughout Europe, and by the predominance of herbaceous vegetation with relatively high abundance of steppic taxa indicative probably of lower moisture availability and hence of reduced productivity compared to the herbaceous vegetation characteristic of the warm type D/O events. The conditions during these cold events would have likely led to reductions in the range and population sizes of the large herbivores, and hence of the associated carnivores and of humans.

Acknowledgements

We are indebted to Tjeerd van Andel for devising the Stage 3 Project and maintaining its momentum and for his assistance in making this chapter more accessible to the non-specialist reader. We are especially grateful to our fellow members of the Stage 3 Project 'Vegetation Panel': J.-L. de Beaulieu, E. Grüger, P.C. Tzedakis and W.A. Watts. We thank the many European palynologists who supported our work and provided pollen spectra. These include Z. Balwierz, S.J.P. Bohncke, S. Bottema, K. Erd, P.L. Gibbard, K.F. Helmens, D. Kryszkowski, E. Liivrand, D. Magri, R. Perez-Obiol and G. van Reenen. Data for Lake Nero were obtained from the European Pollen Data Base. K. Mamakowa obtained the Polish data for us through the Polish Pollen Database. BH holds a Royal Society – Wolfson Foundation 'Research Merit Award'.

References

Aleshinskaya, Z.K. & V.S. Gunova, 1976. History of Nero as reflection on the surrounding landscape dynamics, in *Problemy Paleohidrologii*, eds. G.P. Kalinin & R.V. Kligo. Moscow: Nauka, 214–22.

Alfano, M.J., E.J. Barron, D. Pollard, B. Huntley & J.R.M. Allen, 2003. Comparison of climate model results with European vegetation and permafrost during oxygen isotope stage three. *Quaternary Research* 59, 97–107.

Allaby, M., 1994. *The Concise Oxford Dictionary of Ecology*. Oxford: Oxford University Press.

Allen, J.R.M. & B. Huntley, 2000. Weichselian palynological records from southern Europe: correlation and chronology. *Quaternary International* 73/74, 111–26.

Allen, J.R.M., U. Brandt, A. Brauer, H.-W. Hubberten, B. Huntley, J. Keller, M. Kraml, A. Mackensen, J.

Mingram, J.F.W. Negendank, N.R. Nowaczyk, H. Oberhänsli, W.A. Watts, S. Wulf & B. Zolitschka, 1999. Rapid environmental changes in southern Europe during the last glacial period. *Nature* 400, 740–43.

Allen, J.R.M., W.A. Watts & B. Huntley, 2000. Weichselian palynostratigraphy, palaeovegetation and palaeo-environment: the record from Lago Grande di Monticchio, southern Italy. *Quaternary International* 73/74, 91–110.

Andersen, B.G. & J. Mangerud, 1989. The last interglacial-glacial cycle in Fennoscandia. *Quaternary International* 3/4, 21–9.

Andersen, J., 1993. Beetle remains as indicators of the climate in the Quaternary. *Journal of Biogeography* 20, 557–62.

Arnold, N.S., T.H. van Andel & V. Valen, 2002. Extent and dynamics of the Scandinavian ice sheet during Oxygen Isotope Stage 3 (65,000–25,000 yr BP). *Quaternary Research* 57, 38–48.

Balter, V., A. Person, N. Labourdette, D. Drucker, M. Renard & B. Vandermeersch, 2001. Were Neandertalians essentially carnivores? Sr and Ba preliminary results of the mammalian palaeobiocoenosis of Saint-Cesaire. *Comptes Rendus de l'Academie des Sciences Serie II Fascicule A- Sciences de la Terre et des Planetes* 332 (1), 59–65.

Barron, E. & D. Pollard, 2002. High-resolution climate simulations of Oxygen Isotope Stage 3 in Europe. *Quaternary Research* 58, 296–309.

Barron, E., T.H. van Andel & D. Pollard, 2003. Glacial environments II: reconstructing the climate of Europe in the Last Glaciation, in *Neanderthals and Modern Humans in the European Landscape during the Last Glaciation*, Chapter 5, eds. T.H. van Andel & W. Davies. (McDonald Institute Monographs.) Cambridge: McDonald Institute for Archaeological Research, 57–78.

Behre, K.-E., 1989. Biostratigraphy of the last glacial period. *Quaternary Science Reviews* 8, 25–44.

Berger, A., 1978. Long-term variations of caloric insolation resulting from the Earth's orbital elements. *Quaternary Research* 9, 139–67.

Bond, G., W. Broeker, S. Johnsen, J. McManus, L. Labeyrie & G. Bonani, 1993. Correlations between climate records from North Atlantic sediments and Greenland ice. *Nature* 365, 143–7.

Bos, J.A.A., S.J.P. Bohncke, C. Kasse & J. Vandenberghe, 2001. Vegetation and climate during the Weichselian early glacial and pleniglacial in the Niederlausitz, eastern Germany: macrofossil and pollen evidence. *Journal of Quaternary Science* 16, 269–90.

Carrión, J.S., M. Munuera, C. Navarro, F. Burjachs, M. Dupré & M.J. Walker, 1999. The palaeoecological potential of pollen records in caves: The case of Mediterranean Spain. *Quaternary Science Reviews* 18, 1061–75.

Caspers, G. & H. Freund, 2001. Vegetation and climate in the Early- and Pleni-Weichselian in northern central Europe. *Journal of Quaternary Science* 16, 31–48.

Colhoun, E.A., J.H. Dickson, A.M. McCabe & F.W. Shotton, 1972. A Middle Midlandian freshwater series at Derryvree, Maguiresbridge, County Fermanagh, Northern Ireland. *Proceedings of the Royal Society of London, B* 180, 273–92.

Collingham, Y.C., M.O. Hill & B. Huntley, 1996. The migration of sessile organisms: a simulation model with measurable parameters. *Journal of Vegetation Science* 7, 831–46.

Coope, G.R., 1961. A Late Pleistocene flora and fauna from Upton Warren, Worcestershire. *Philosophical Transactions of the Royal Society of London B* 244, 379–421.

Coope, G.R., 1977. Fossil coleopteran assemblages as sensitive indicators of climatic change during the Devensian (Last) Cold Stage. *Philosophical Transactions of the Royal Society, series B* 280, 313–40.

Coope, G.R., 2002. Changes in the thermal climate in North-western Europe during marine oxygen isotope stage 3, estimated from fossil Insect assemblages. *Quaternary Research* 57, 401–8.

Coope, G.R., P.L. Gibbard, A.R. Hall, R.C. Preece, J.E. Robinson & A.J. Sutcliffe, 1997. Climate and environmental reconstructions based on fossil assemblages from Middle Devensian (Weichselian) deposits of the River Thames at South Kensington, Central London, UK. *Quaternary Science Reviews* 16, 1163–95.

Cowling, S.A. & M.T. Sykes, 1999. Physiological significance of low atmospheric G02 for plant–climate interactions. *Quaternary Research* 52, 237–42.

Davies, W. & P. Gollop, 2003. The human presence in Europe during the Last Glacial Period II: climate tolerance and climate preferences of mid- and late glacial hominids, in *Neanderthals and Modern Humans in the European Landscape during the Last Glaciation*, Chapter 8, eds. T.H. van Andel & W. Davies. (McDonald Institute Monographs.) Cambridge: McDonald Institute for Archaeological Research, 131–46.

de Beaulieu, J.-L. & M. Reille, 1984. A long Upper Pleistocene pollen record from Les Echets, near Lyon, France. *Boreas* 13, 111–32.

de Beaulieu, J.-L. & M. Reille, 1992a. Long Pleistocene pollen sequences from the Velay Plateau (Massif Central, France), 1. Ribains Maar. *Vegetation History and Archaeobotany* 1, 223–42.

de Beaulieu, J.-L. & M. Reille, 1992b. The last climatic cycle at La Grande Pile (Vosges, France): a new pollen profile. *Quaternary Science Reviews* 11, 431–8.

Delpech, F., 1993. The fauna of the Early Upper Palaeolithic: biostratigraphy of large mammals and current problems in chronology, in *Before Lascaux: the Complex Record of the Early Upper Palaeolithic*, eds. H. Knecht, A. Pike-Tay & R. White. Ann Arbor (MI): CRC Press, 71–84.

Fischer, A., 1997. Seasonal features of global net primary productivity models for the terrestrial biosphere, in *Past and Future Rapid Environmental Changes: the Spatial and Evolutionary Responses of Terrestrial Biota*, eds. B. Huntley, W. Cramer, A.V. Morgan, H.C. Prentice

& J.R.M. Allen. (NATO ASI Series I: Global Environmental Change.) Berlin: Springer-Verlag, 469–83.

Foley, J.A., J.E. Kutzbach, M.T. Coe & S. Levis, 1994. Feedbacks between climate and boreal forests during the Holocene epoch. *Nature* 371, 52–4.

Follieri, M., D. Magri & L. Sadori, 1988. 250,000-year pollen record from Valle di Castiglione (Roma). *Pollen et Spores* 30, 329–56.

Follieri, M., M. Giardini, D. Magri & L. Sadori, 1998. Palynostratigraphy of the last glacial period in the volcanic region of central Italy. *Quaternary International* 47/48, 3–20.

Gaunt, G.D., G.R. Coope & J.W. Franks, 1970. Quaternary deposits at Oxbow opencast coal site in the Aire Valley, Yorkshire. *Proceedings of the Yorkshire Geological Society* 38, 175–200.

Grüger, E., 1989. Palynostratigraphy of the last interglacial/glacial cycle in Germany. *Quaternary International* 3/4, 69–79.

Guthrie, R.D., 1990. *Frozen Fauna of the Mammoth Steppe; the Story of Blue Babe.* Chicago (IL): University of Chicago Press.

Haxeltine, A. & I.C. Prentice, 1996. BIOME3: An equilibrium terrestrial biosphere model based on ecophysiological constraints, resource availability and competition among plant functional types. *Global Biogeochemical Cycles* 10, 693–710.

Helmens, K.F., M.E. Räsänen, P.W. Johansson, H. Jungnerd & K. Korjonen, 2000. The Last Interglacial-Glacial cycle in NE Fennoscandia: a nearly continuous record from Sokli (Finnish Lapland). *Quaternary Science Reviews* 19, 1605–23.

Huntley, B., 1988. Glacial and Holocene vegetation history: Europe, in *Vegetation History*, eds. B. Huntley & T. Webb III. Dordrecht: Kluwer Academic Publishers, 341–83.

Huntley, B., 1991. How plants respond to climate change: migration rates, individualism and the consequences for plant communities. *Annals of Botany* 67, 15–22.

Huntley, B., Y.C. Collingham & M.O. Hill, 1996. Tree species' responses to environmental changes during the last glacial: Evaluating alternative hypotheses using a spatially-explicit migration model. *Il Quaternario* 9, 617–26.

Huntley, B., M.J. Alfano, J.R.M. Allen, D. Pollard, P.C. Tzedakis, J.-L. de Beaulieu, E. Grüger & W.A. Watts, 2003. European vegetation during Marine Oxygen Isotope Stage 3. *Quaternary Research* 59, 195–212.

Jardine, W.G., J.H. Dickson, P.D.W. Haughton, D.D. Harkness, D.Q. Bowen & G.A. Sykes, 1988. A late Middle Devensian interstadial site at Sourlie, near Irvine, Strathclyde. *Scottish Journal of Geology* 24, 288–95.

Jolly, D. & A. Haxeltine, 1997. Effect of low glacial atmospheric CO_2 on tropical African montane vegetation. *Science* 276, 786–8.

Kaplan, J.O., 2001. Geophysical Applications of Vegetation Modelling. Unpublished PhD thesis, Lund University.

Kerney, M.P., P.L. Gibbard, A.R. Hall & J.E. Robinson, 1983. Middle Devensian river deposits beneath the 'Upper Floodplain' of the River Thames at Isleworth, west London. *Proceedings of the Geologists' Association* 93, 385–93.

Krzyszkowski, D., Z. Balwierz & W. Pyszynski, 1993. Aspects of Weichselian Middle Pleniglacial stratigraphy and vegetation in central Poland. *Geologie en Mijnbouw* 72, 131–42.

Lambeck, K., 1995. Late Devensian and Holocene shorelines of the British Isles and North Sea from models of glacio-hydro-isostatic rebound. *Journal of the Geological Society London* 151, 437–48.

Larcher, W., 2003. *Physiological Plant Ecology: Ecophysiology and Stress Physiology of Functional Groups.* Berlin: Springer.

Leroy, S.A.G., S. Giralt, P. Francus & G. Seret, 1996. The High sensitivity of the palynological record in the Vico Maar Lacustrine Sequence (Latium, Italy) highlights the climatic gradient through Italy for the last 90 ka. *Quaternary Science Reviews* 15, 189–202.

Liivrand, E., 1990. Type section of the lower and middle-Valdaian interstadial deposits at Töravere in southeast Estonia. *Proceedings Estonian Academy of Sciences* 39, 12–17.

Liivrand, E., 1991. *Biostratigraphy of the Pleistocene Deposits in Estonia and Correlations in the Baltic Region.* (Department of Quaternary Research Report 19.) Stockholm: Stockholm University.

Lister, A.M., 1987. Giant deer and giant red deer from Kent's Cavern and the status of *Strongyloceros spelaeus* Owen. *Transactions of the Torquay Natural History Society* 91, 189–98.

Lister, A.M. & A.V. Sher, 1995. Ice cores and mammoth extinction. *Nature* 378, 23–4.

Meese, D.A., A.J. Gow, R.B. Alley, G.A. Zielinski, P.M. Grootes, M. Ram, K.C. Taulor, P.A. Mayewski & J.F. Bolzan, 1997. The Greenland Ice Sheet Project 2 depth-age scale: methods and results. *Journal of Geophysical Research* 102, 26,411–23.

Niedzialkowska, E. & K. Szczepanek, 1994. Utwory pylowe Vistulianskiego stozka Wisly w kotlinie Oswiecimskiej. *Studia Geomorphologia Carpatho-Balcanica* 27/28, 29–43.

Niklewski, J. & W. van Zeist, 1970. A late Quaternary pollen diagram from northwestern Syria. *Acta Botanica Neerlandica* 19, 737–54.

Nimmergut, A.P., J.R.M. Allen, V.J. Jones, B. Huntley & R.W. Battarbee, 1999. Submillennial environmental fluctuations during marine Oxygen Isotope Stage 2: a comparative analysis of diatom and pollen evidence from Lago Grande di Monticchio, South Italy. *Journal of Quaternary Science* 14, 111–23.

Ozenda, P., 1994. Vegetation du Continent Européen. *Lausanne: Delachaux et Niestlé.*

Ozenda, P. & J.-L. Borel, 2000. An ecological map of Europe: why and how? *Life Sciences* 323, 983–94.

Perez-Obiol, R. & R. Julia, 1994. Climatic changes on the Iberian Peninsula recorded in a 30,000-yr pollen record from lake Banyoles. *Quaternary Research* 41, 91–8.

Petit, J.R., J. Jouzel, D. Raynaud, N.I. Barkov, J.-M. Barnola, I. Basile, M. Bender, J. Chappellaz, M. Davis, G.

Delaygue, M. Delmotte, V.M. Kotlyakov, M. Legrand, V.Y. Lipenkov, C. Lorius, L. Pépin, C. Ritz, E. Saltzman & M. Stievenard, 1999. Climate and atmospheric history of the past 420,000 years from the Vostok ice core, Antarctica. *Nature* 399, 429–36.

Peyron, O., J. Guiot, R. Cheddadi, P. Tarasov, M. Reille, J.-L. de Beaulieu, S. Bottema & V. Andrieu, 1998. Climatic reconstruction in Europe for 18,000 yr BP from pollen data. *Quaternary Research* 49, 183–96.

Phillips, L.M., 1976. Pleistocene vegetational history and geology in Norfolk. *Philosophical Transactions of the Royal Society of London, B* 275, 215–86.

Pollard, D. & E.J. Barron, 2003. Causes of model–data discrepancies in European climate during oxygen isotope stage 3 with insights for the Last Glacial Maximum. *Quaternary Research* 59, 108–13.

Pons, A. & M. Reille, 1988. The Holocene and upper Pleistocene record from Padul (Granada, Spain): a new study. *Palaeogeography, Palaeoclimatology, Palaeoecology* 35, 145–214.

Prentice, I.C., W. Cramer, S.P. Harrison, R. Leemans, R.A. Monserud & A.M. Solomon, 1992. A global biome model based on plant physiology and dominance, soil properties and climate. *Journal of Biogeography* 19, 117–34.

Prentice, I.C., J. Guiot, B. Huntley, D. Jolly & R. Cheddadi, 1996. Recontructing biomes from palaeoecological data: a general method and its application to european pollen data at 0 and 6 ka. *Climate Dynamics* 12, 185–94.

Ran, E.T.H., S.J.P. Bohncke, K.J. van Huissteden & J. Vandenberge, 1990. Evidence of episodic permafrost conditions during the Weichselian Middle Pleniglacial in the Hengelo Basin (The Netherlands). *Geologie en Mijnbouw* 69, 207–18.

Reille, M. & J.L. de Beaulieu, 1990. Pollen analysis of a long upper Pleistocene continental sequence in a Velay maar (Massif Central, France). *Palaeogeography, Palaeoclimatology, Palaeoecology* 80, 35–48.

Rossignol-Strick, M., 1995. Sea–land correlation of pollen records in the eastern Mediterranean for the glacial-interglacial transition: biostratigraphy versus radiometric time-scale. *Quaternary Science Reviews* 14, 893–915.

Scourse, J.D., 1991. Late Pleistocene stratigraphy and palaeobotany of the Isles of Scilly. *Philosophical Transactions of the Royal Society of London, B* 334, 404–48.

Seddon, M., 1985. Evidence of sustained regional permafrost during deposition of fossiliferous Late Pleistocene river sediments at Stanton Harcourt (Oxfordshire, England). *Proceedings of the Geologists' Association* 96, 53–71.

Srodon, A., 1968. On the vegetation of the Paudorf interstadial (last glaciation) in the western Carpathians. *Acta Palaeobotanica* 9, 3–29.

Srodon, A., 1987. A communication on the locality of the periglacial flora at Sadowie in the Miechow uppland (Vistulian, Southern Poland). *Acta Palaeobotanica* 21, 71–5.

Stewart, J.R., T. van Kolfschoten, A. Markova & R. Musil, 2003. The mammalian faunas of Europe during Oxygen Isotope Stage Three, in *Neanderthals and Modern Humans in the European Landscape during the Last Glaciation*, Chapter 7, eds. T.H. van Andel & W. Davies. (McDonald Institute Monographs.) Cambridge: McDonald Institute for Archaeological Research, 103–30.

Tarasov, P.E., R. Cheddadi, J. Guiot, S. Bottema, O. Peyron, J. Belmonte, V. Ruiz-Sanchez, F. Saadi & S. Brewer, 1998. A method to determine warm and cool steppe biomes from pollen data; application to the Mediterranean and Kazakhstan regions. *Journal of Quaternary Science* 13, 335–44.

Tzedakis, P.C., 1994. Vegetation change through glacial-interglacial cycles: a long pollen sequence perspective. *Philosophical Transanctions of the Royal Society, London Series B* 345, 403–32.

Tzedakis, P.C., 1999. The last climate cycle at Kopais, central Greece. *Journal of the Geological Society, London* 156, 425–34.

van Andel, T.H. 2003. Glacial environments I: the Weichselian climate in Europe between the end of the last interglacial and the Last Glacial Maximum, in *Neanderthals and Modern Humans in the European Landscape during the Last Glaciation*, Chapter 2, eds. T.H. van Andel & W. Davies. (McDonald Institute Monographs.) Cambridge: McDonald Institute for Archaeological Research, 9–20.

van Andel, T.H. & P.C. Tzedakis, 1996. Palaeolithic landscapes of Europe and environs, 150,000–20,000 years ago: an overview. *Quaternary Science Reviews* 15, 481–500.

van Andel, T.H. & P.C. Tzedakis, 1998. Priority and opportunity: reconstructing the European Middle Palaeolithic climate and landscape, in *Science in Archaeology, an Agenda for the Future*, ed. J. Bayley. London: English Heritage, 37–45.

van Huissteden, K., J. Vandenberghe & D. Pollard, 2003. Palaeotemperature reconstructions in the European permafrost zone during marine oxygen isotope Stage 3 compared with climate model results. *Journal of Quaternary Science* 18, 453–64.

van Zeist, W. & S. Bottema, 1977. Palynological investigations in Western Iran. *Palaeohistoria* 19, 19–86.

Watts, W.A., J.R.M. Allen & B. Huntley, 2000. Palaeoecology of three interstadial events during oxygen-isotope Stages 3 and 4: a lacustrine record from Lago Grande di Monticchio, southern Italy. *Palaeogeography, Palaeoclimatology, Palaeoecology* 155, 83–93.

West, R.G., 1991. *Pleistocene Palaeoecology of Central Norfolk.* Cambridge: Cambridge University Press.

Wijmstra, T.A., 1969. Palynology of the first 30 metres a 120 m deep section in northern Greece (Macedonia). *Acta Botanica Neerlandica* 25, 297–312.

Willis, K.J., E. Rudner & P. Sümegi, 2000. The full-glacial forests of central and southeastern Europe. *Quaternary Research* 53, 203–13.

Woodward, F.I. & W. Cramer, 1996. Plant functional types and climatic changes: Introduction. *Journal of Vegetation Science* 7, 306–8.

Chapter 7

The Mammalian Faunas of Europe during Oxygen Isotope Stage Three

John R. Stewart, Thijs van Kolfschoten, Anastasia Markova & Rudolf Musil

As part of the Stage Three Project, an endeavour to better understand the environment for the period of time during which the human species *Homo neanderthalensis* became extinct (van Andel 2002), an investigation into the contemporary mammals was initiated. To this end a data base was constructed of western and central European mammalian fossils dated between 60 and 20 ka BP, the period that contains the Oxygen Isotope Stage 3 (OIS-3) and the beginning of OIS-2 between 24 and 20 ka BP. The rationale behind creating the data base was that the character of the environment of OIS-3, and indeed the late Pleistocene as a whole, remains a subject of debate (Guthrie 1990; Davies *et al.* 2000; Musil 1999 *versus* Colinvaux 1986; Ritchie 1984; Ritchie & Cwynar 1982).

Overview papers on the palaeoecology of the Late Pleistocene of northern latitudes and particularly those dealing with the mammalian fossil record have generally treated it without further subdividing the time period. Most literature also tends to be limited in its geographical extent such as Cantabria (Altuna & Mariezkurrena 1988), Southwest France (Delpech 1975), Moravia (e.g. Musil 1955; 1959a,b; 1994a,b; 1997; and references in Chapter 10, Musil 2003), the former Soviet Union (Markova *et al.* 1995) and Russian Plain (Markova *et al.* 2002) and Britain and Ireland (Stuart 1977; 1982; Stuart & van Wijngaarden-Bakker 1985; Currant & Jacobi 1997; 2001). The mammalian assemblages of Europe (OIS-2; 24–12 ka BP) have been collated in the palaeogeographical atlas/monographs of Markova (1982) and for all of the Northern Hemisphere by Baryshnikov & Markova (1992). No literature is known to the authors, however, which specifically treats the mammalian fauna of Europe during OIS-3 with the exception of Musil (1986). In that paper European mammalian faunas are considered through the whole of the Last Glacial. Previous data bases that can be compared with the Stage Three Project data base are those whose coverage includes the end of the Late Pleistocene and earlier Holocene of the United States of America (Faunmap Working Group 1994; 1996) and the Late Pleistocene of the former Soviet Union, including the Mikulino (Eemian) Interglacial faunas, as well as those of the Last Glacial (Markova *et al.* 1995).

Literature detailing the vegetation of Europe during OIS-3 is almost solely based on studies of pollen from single sites (see Huntley *et al.* 2003). In that paper the pollen data was transformed into plant functional types and then into Biomes. The concensus view seems to be that the mid and northern latitudes of Europe are dominated by non-arboreal open vegetational habitats (Huntley *et al.* 2003). Challenges to this have been made from both charcoal studies, particularly in Moravia and Hungary (Lityńska-Zajac 1995; Musil 1999; Willis *et al.* 2000) and inferences drawn from mammalian fossils (Musil 2001; Stewart & Lister 2001) although these other proxies are generally either dismissed or not addressed.

The Stage Three Project had as one goal the collection of various types of existing data documenting the archaeology and pollen sequences of Europe during OIS-3 as well as the fossil mammals. In addition, a regional scale climate simulation model was run for Europe giving as output maps for various parameters during the LGM (21k) and an arbitrary warm phase (42k) of OIS-3 (Barron & Pollard 2002; Pollard & Barron 2003). As well as output for these hypothesized past conditions observed values were available for today. A Biome 3 model was coupled to the regional climate model and used to predict biomes that have been compared to biomes reconstructed from pollen data (Huntley *et al.* 2003).

The intention was that these sources of information, the existing data and model outputs, be used to perform various comparison exercises. Modern mammalian distribution data are also available (Mitchell-Jones *et al.* 1999) which can be compared to the climate for today. This then allows a comparison to be made between the climate associated with current animal populations and those hypothesized from the models to have accompanied the past populations. A similar comparison can then be made between the Biome 3 model outputs associated with mammals and the vegetation forming the habitats of those same species at the present. A further comparison can be made between the OIS-3 mammalian fossil data and the vegetational data collected by the panel of palynologists (Huntley *et al.* 2003).

Therefore, the present paper describes the mammalian fossil data collected, constituting the Stage Three Project data base and compares it with the climate and the vegetation simulated by the OIS-3 regional climate model which was coupled to a Biome 3 model and inferred (by pollen) for OIS-3. The latter comparisons should not be seen as either tests of the mammalian distributional data, the pollen data or of the climate and vegetation model simulations. They are instead ways of better understanding each set of information as it was anticipated that each would have its own set of problems and biases.

Materials and methods

The Stage Three Project Mammalian Faunal Data Base
A data base was constructed from absolutely dated mammalian faunas from sites in western and central Eastern Europe from between 20 and 50 ka BP (approximately Oxygen Isotope Stage Three). The data base (called the Stage Three Mammalian Faunal Data Base[1]) is made up of existing data from literature published before March 2001.

The data base contains most of the available radiometrically dated mammalian faunal sites for Oxygen Isotope Stage 3 from Europe, but eastern European mammalian data are partially represented in this paper. More detailed materials and reconstructions can be found in Markova *et al.* (1995) and Markova *et al.* (2002). Europe is here defined as the western Palaearctic bordered by the Atlantic to the West, the Mediterranean to the South, the Arctic Ocean to the North and the 40° longitude line to the East. (This is not the traditional definition of the land area of Europe as the Eastern border is usually taken to be the Urals at *c.* 60° longitude. However, since the Stage Three Project includes, as one of its main

aims, a comparison with the outputs from a regional climate model, the area covered by that model defines the area considered for the present purposes.)

In all 468 dated faunas from 294 sites are included with 1912 radiometric dates dating 119 mammalian taxa in total. Most sites are dated by radiocarbon whether conventional or by the Accelerator Mass Spectrometry technique. Also included are thermoluminescence dates (TL), optically-stimulated luminescence dates (OSL), electron spin resonance dates (ESR) and uranium-series dates when available. Archaeological and palaeontological sites are presented in the Stage Three Project Archaeological Data Base.[2]

Each line of the data base is an absolute date and attached to each date are all the attributes of the layer at each site from which the date derives. Attributes include whether the site is a cave or open-air sites, the sediment type making up the layer, the archaeological industry found in the layer if any, as well as the fauna. Longitudes and latitudes were recorded for each site as digital degrees that would allow mapping using a Geographical Information System (GIS) program.

In the data base the mammals have been arbitrarily divided into large and small mammals. Small mammals include all rodents, insectivores and lagomorphs while large mammals comprise carnivores, ungulates and proboscideans. Bats (Chiroptera) were excluded from the data base because they are rarely identified to a specific taxonomic level as were marine mammals because the primary goal of the project was to investigate the terrestrial environment and again they are relatively rare. Human fossils were included as part of the mammalian faunal field.

The data was collected by first making a systematic search for absolute dates, most of which were radiocarbon dates from the two main journals publishing date lists: *Radiocarbon* and *Archaeometry*. Both finite and infinite dates were included in the data base although the latter were not used in the analysis that follows. The dates are published with their standard errors, the latitude and longitude in decimel degrees, the horizon and the material dated, as well as a section where notes from the site excavator are recorded. The latter may include references that were subsequently followed up. If not, a literature search would often lead to access to publications regarding site characteristics including whether there was any fauna identified from the dated layers. The search for published faunas was necessarily less systematic because the locations where sites are published can often be ill-defined. In such cases a broad approach including corresponding with mammalian faunal specialists from various parts of Europe was necessary.

It soon became apparent from the data base was that many modern European species listed in *The Atlas of European Mammals* (Mitchell-Jones *et al.* 1999) were not present in the fossil record. Species not described from Oxygen Isotope Stage 3 include mostly small mammals whose modern geographical ranges are limited in extent and which are usually congeneric with more common species today. The reasons for the absence of these species in the written fossil record is either that they had not been considered relevant due to the position of a site compared to the modern range of that taxon. An example of this is the complete absence of taxa such as *Sorex alpinus*, *Microtus tatricus* and *M. multiplex*, among many others, from the OIS-3 fossil record of Europe. This is no doubt due to the fact that some such species have been recently accepted as distinct taxa on the basis of characters such as chromosome counts, that their distributions are limited to small areas and either skeletal comparative material is not available or that there are no discernable skeletal differences between them and one or more of their congeners. Another problem with identification was that it quickly became apparent that different 'schools' of palaeontology were using different nomenclature. A good example is the use of *Panthera spelaea* versus *Panthera leo*. In such cases the lead author was forced to take a conservative view, submerging both into *Panthera leo*. This is because 'lumping' was the only option when only half the data base was 'split'. Other examples are where *Equus ferus*, *Equus caballus*, *Equus germanicus* and *Equus* sp. were included in *Equus ferus*. Similarly, *Bos* and *Bison* were lumped as *Bos/Bison*, and no attempt was made to distinguish species in the genera *Capra*, *Martes* and *Mustela* for the purposes of this paper although they are listed as individual species in the data base. Furthermore, only those records that specifically mentioned *Dama dama* rather than *Dama dama/Cervus elaphus* were included. All hyena records were assigned to *Crocuta crocuta* and unspecified bear records were not included. (It should be noted, however, that the different authors of this contribution differed in their approach to this problem.)

Time intervals
The next task was to decide what the temporal resolution of the analysis would be and that involved two primary considerations. The two factors concern the practical potential of making meaningful comparisons between the absolutely dated mammalian faunas and the global climate signature. The first involved the precision of the dates in relation to the amplitude of the climatic cycles during OIS-3 while the second simply involved the problems of turning the ^{14}C dates into calendrical years. Global temperatures from 60 ka BP to 20 ka BP have been elucidated recently through such initiatives as the GRIP and GISP ice-core projects in Greenland (e.g. Dansgaard *et al.* 1993). This has led to a greater detail than was previously available from the ocean records (e.g. Shackleton & Opdyke 1973) and while there is a net decrease in global temperature during this episode there appears to have been a larger number of temperature oscillations than previously believed. It has become apparent, however, that despite detailed knowledge of the global climate throughout this episode and the availability of absolute dates associated with the faunas in the data base, correlation of the terrestrial sequence with specific oscillations would be marred by the large standard error of those dates. Furthermore, problems exist with the radiocarbon record due to variations in the production of ^{14}C in the atmosphere through time (Jöris & Weninger 1996). Therefore, the interval represented by the data base was divided into three temporal zones designed to allow comparisons between dated faunas. Table 7.1 outlines these three zones that encompass an early, middle and late phase of which the early and late phases can certainly be compared climatically as they represent relatively warmer and colder conditions respectively

Mapping
The first stage of the analysis was to map individual taxa dated to three time zones (Table 7.1). The mapping program Atlasnow! was used to plot the maps. Most taxa were mapped in this way and examples of the maps can be seen in Figures 7.2–7.4. Many taxa were so rare thoughout Europe as to make interpretation of any geographical change through time unreliable.

Ecological classification of taxa
The taxa listed in the data base were divided into 10 categories based on ecological and historical biogeographical criteria. Their extant/extinct status was initially used and then the taxa in each group were subdivided on the basis of geographical and/or ecological associations that the taxa may have today or had in the past.

Provinciality analysis
This analysis was performed with two aims in mind.

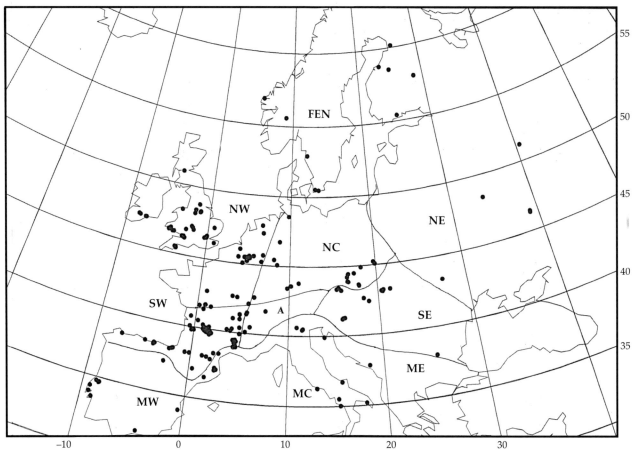

Figure 7.1. *Map of absolutely dated sites with large mammal taxa in Europe during OIS-3 of the Stage Three Project Mammalian Data Base showing the regional provinces of Europe (after Gamble 1986). FEN = Fennoscandia, NW = North West, NC = North Central, NE = Northeast, SW = Southwest, A = Alpine, SE = Southeast, MW = Mediterranean West, MC = Mediterranean Central, ME = Mediterranean East.*

Table 7.1. *Oxygen Isotope Stage Three time zones.*

Stage Three Project foci	Stage 3 time zones	Calendric ka BP	¹⁴C ka [BP]	Ice-core interstadials	Peat stratigraphy
–	Late	28–20	< 27	Cold: 4, 3, 2	Towards cold maximum
30 ka	Intermediary	28–37	27–35	Colder: 5, 6, 7, 8	'Denekamp'
42 ka	Early	60–37	> 35	Warm: 9, 10, 11, 12	'Hengelo'

Note: There is a problem with the corresponding climates of each time zone because the cold event for Stage Three Project purposes falls in the block of time that also includes the Denekamp interstadial. Therefore we are dealing with climates relative to other time zones i.e. the Intermediary time zone will be colder on average than the early one and warmer than the cold one.

The first being to ascertain the ecological character and faunal composition of the different geographical regions of Europe and to consider whether they constitute biogeographical faunal provinces. The second aim was to give an idea of the taxa available for exploitation by the various Palaeolithic human populations of OIS-3.

The regions used as subdivisions of Europe for this analysis are those first described by Gamble (1986) and subsequently by Davies (2001) (see Fig. 7.1). They were based on the three latitudinal zonal provinces (North, South and Mediterranean) that are subdivided into nine longitudinal regions; Northwest, North Central, Northeast, Southwest, Alpine, Southeast, Mediterranean West, Mediterranean Central and Mediterranean East. An additional region, Fennoscandia, has been added for the purposes of this study. The mammalian faunas present in each

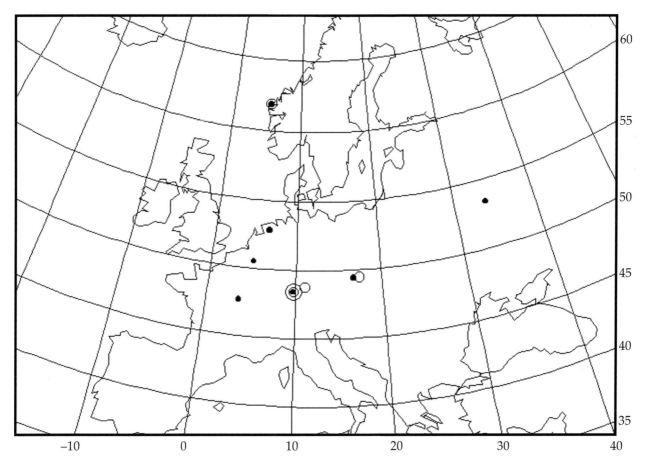

Figure 7.2a. *Geographical distribution of* Lemmus *sp. in Europe during the three time zones of OIS-3 described in Table 7.1. Small solid circle = Early time zone; medium circle = Intermediate time zone; large circle = Late time zone.*

region through the three time zones of OIS-3 (and the start of OIS-2) defined in Table 7.1 were then described.

Climate/mammal comparison

To compare the geographic distribution of mammal species for our three time zones (Figs. 7.2a, 7.3a & 7.4a) with the relevant climate models we have used a procedure developed in Chapter 8 (Davies & Gollop 2003, 134–5; Fig. 8.4) that counts the number of findspots of each species between each pair of isotherms (or isopleths for other variables such as precipitation) on a (palaeo)climate map. Histograms are then plotted of numbers of findspots against the spectrum of climatic variables, in this case winter and summer mean air temperatures (Figs. 7.2b, 7.3b & 7.4b) and precipitation (Figs. 7.2c, 7.3c & 7.4c) using 'warm' (42 ka) and 'cold' (21 ka) palaeoclimatic and modern simulations for the three OIS-3 intervals. The 'modern' data permitted us to see whether the taxa were associated with the same climatic variables in the past as today.

Only a few mammalian taxa in the data base were useful as their modern distributions must be relatively undisturbed by modern factors absent in the past such as human interference. This excludes most larger mammals. The other criterion was that the mammal should have a relatively limited distribution likely to be related to climate. Taxa such as the pygmy shrew *Sorex minutus* are of little use as they cover almost the whole of Europe today (Mitchell-Jones *et al.* 1999). The taxa chosen were the following:

Lemmus sp. (Norway lemming *Lemmus lemmus* and Siberian lemming *L. sibiricus*): The taxon has a reasonable fossil record in terms of numbers of occurrences during OIS-3 (Fig. 7.2a) and is unlikely to be mistaken for other species of microtine rodents although it may be difficult to separate the two species of *Lemmus* osteologically. *L. lemmus* has a present-day distribution that is restricted to Scandinavia in alpine and subarctic tundra habitats and sometimes in birch and pine woodland (Macdonald & Barrett

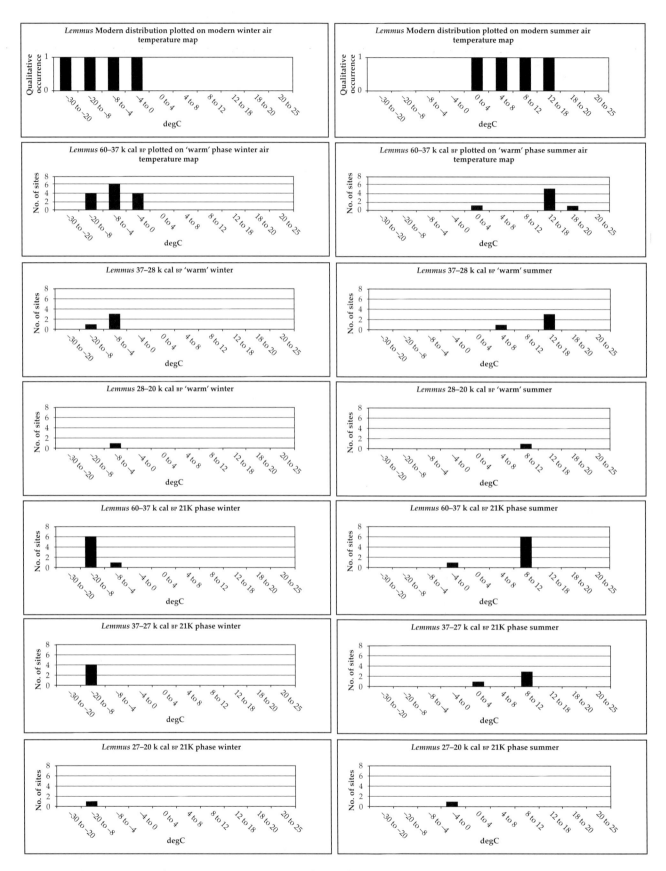

Figure 7.2b. *Comparison between air temperature isotherms (winter and summer) for modern and OIS-3* Lemmus *sp.*

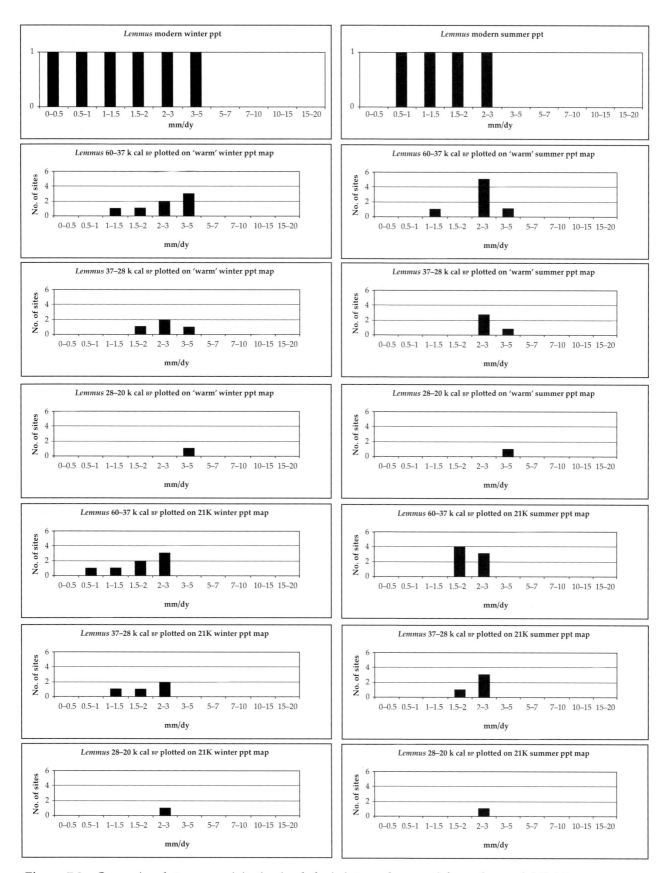

Figure 7.2c. *Comparison between precipitation isopleths (winter and summer) for modern and OIS-3* Lemmus *sp.*

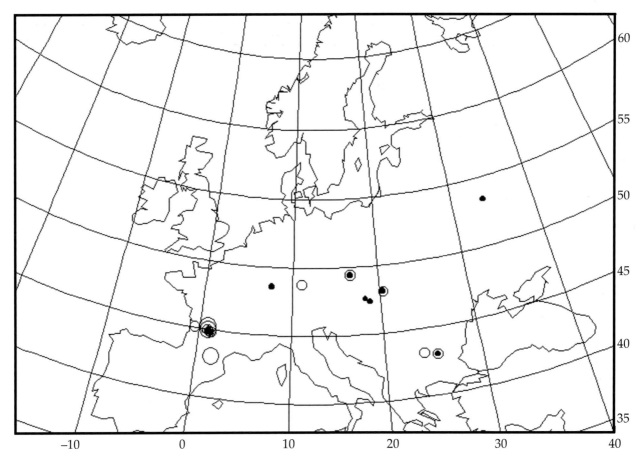

Figure 7.3a. *Geographical distribution of* Spermophilus *sp. in Europe during the three time zones of OIS-3 described in Table 7.1. Small solid circle = Early time zone; medium circle = Intermediate time zone; large circle = Late time zone.*

Table 7.2. *European OIS-3 ecological provinciality table.*

Mammalian ecological categories by region
(excluding the Neanderthal, extant upland taxa and *E. hydruntinus*)

Fennoscandia
Extant cold taxa
Extinct cold taxa

Northwest	*North Central*	*Northeast*
Extant ubiquitous taxa	Extant ubiquitous	Extant ubiquitous
Extant cold taxa	Extant cold taxa	Extant cold taxa
Extant continental taxa	Extant continental taxa	Extant continental taxa
Extant warm taxa	Extant warm taxa	Extant warm taxa
Extinct cold taxa	Extinct cold taxa	Extinct cold taxa

Southwest	*Alpine*	*Southeast*
Extant ubiquitous	Extant ubiquitous	Extant ubiquitous
Extant cold taxa	Extant cold taxa	Extant cold taxa
Extant continental taxa	Extant warm taxa	Extant continental taxa
Extinct cold taxa	Extinct cold taxa	Extant warm taxa
Extinct 'interglacial survivors'		

Mediterranean West	*Mediterranean Central*	*Mediterranean East*
Extant ubiquitous	Extant ubiquitous	Extant ubiquitous
Extant warm taxa	Extant warm taxa	Extant warm taxa
Southern endemic taxa	Southern endemic taxa	
Extinct cold taxa	Extinct 'interglacial survivors'	
Extinct 'interglacial survivors'		

1993; Mitchell-Jones *et al.* 1999). *L. sibiricus* has a siberian tundra zone distribution. This taxon can therefore be used to investigate the extent of the colder temperatures present at southern latitudes in Europe during OIS-3.

Alopex lagopus (Arctic fox): The taxon has a better fossil record in OIS-3 than *L. lemmus* although it is possible that it is sometimes mistaken for the red fox *Vulpes vulpes*. This is apparent in those records in the Stage Three Project Mammalian data base where *Alopex* and *Vulpes* are not distinguished[1] and for the purposes of this analysis those

110

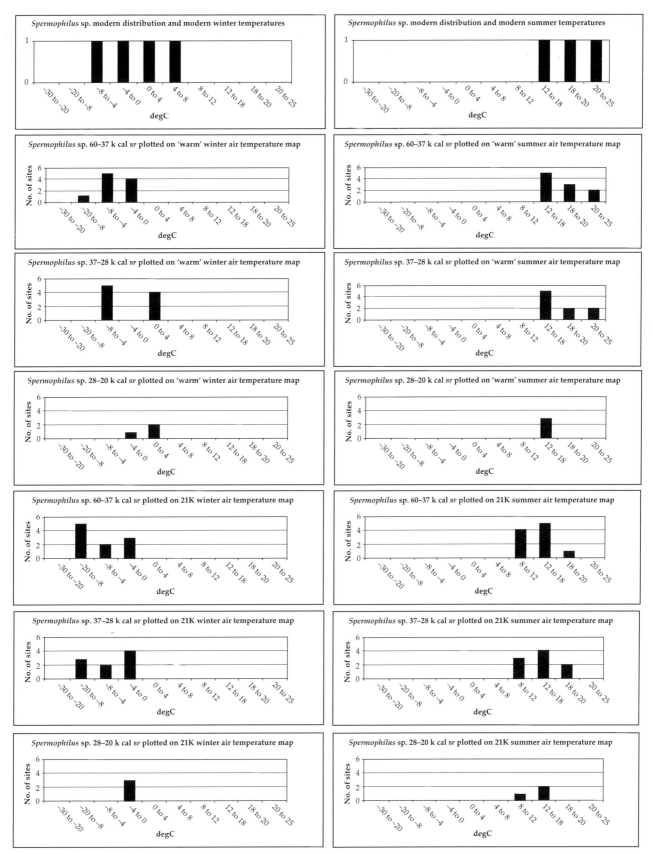

Figure 7.3b. *Comparison between air temperature isotherms (winter and summer) for modern and OIS-3* Spermophilus *sp.*

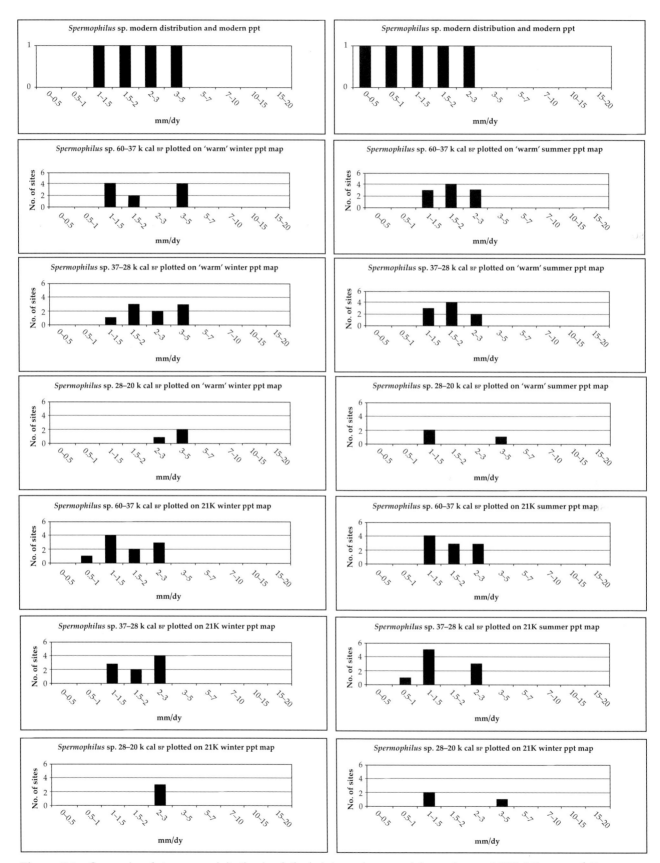

Figure 7.3c. *Comparison between precipitation isopleths (winter and summer) for modern and OIS-3* Spermophilus *sp.*

unspecific records were omitted. The species has a present-day distribution in Europe that is restricted to Scandinavia and the tundra and forest tundra zones of Eastern Europe (Flint *et al.* 1970) in alpine and arctic tundra habitats above the treeline, although it is found in open woodland during the winter (Macdonald & Barrett 1993; Mitchell-Jones *et al.* 1999). This taxon can also be used to investigate the extent of the colder temperatures present at southern latitudes in Europe and can therefore complement *Lemmus* sp.

Spermophilus sp. (Sousliks): There are two species of souslik in Europe today; the European souslik *Spermophilus* (*Colobotis*) *citellus* and the spotted souslik *S.* (*Colobotis*) *suslicus*. Both species live on short grass steppe on open and well-drained soil. Today they are both endemic to Europe the European souslik living in the Czech Republic, Austria, Hungary, Slovakia, western Romania and Yugoslavia while the spotted souslik lives in southern part of Russia, the Ukraine and Moldova (Macdonald & Barrett 1993; Mitchell-Jones *et al.* 1999). The species have not been separated as they are similar in distribution and there is a possibility that they are not correctly distinguished in the fossil record which is extensive during OIS-3. The taxon represents an eastern, relatively continental steppe species and can therefore be used to investigate the extent of such habitats and the climates associated with them during OIS-3. In the area just east of that covered by this paper there are also the great souslik *Spermophilus* (*Colobotis*) *major*, the yellow souslik *Spermophilus* (*Colobotis*) *fulvus* and the small souslik *Spermophilus* (*Colobotis*) *pygmaeus* (Flint *et al.* 1970; Panteleev *et al.* 1990; Gromov & Erbaeva 1995).

Apodemus sylvaticus (Wood mouse): The wood mouse is widespread in Europe today from the Iberian peninsula to westernmost Russia and as far north as southern Norway and Sweden. The taxon is very variable in the habitats that it lives in today which is dependent on the geographical area. In the west it is very diversely associated while in the east it is a woodland edge species and in the more extreme parts of Eastern Europe it is a steppeland species (Mitchell-Jones *et al.* 1999). There are other species in the genus *Apodemus* in Europe today which could possibly be mistaken for *A. sylvaticus*. They are the yellow-necked mouse *A. flavicolis*, the rock mouse *A. mystacinus*, the striped mouse *A. agrarius* and the Alpine mouse *A. alpicola*. None of these species, however, has a distribution that exceeds the distribution

of the wood mouse and so they do not cause difficulties when their fossils are used to characterize more southern climes in Europe.

Talpa sp. (Moles): The genus *Talpa* in Europe today contains five species including the common mole *Talpa europaea*, the blind mole *T. caeca*, the Iberian mole *T. occidentalis*, the Roman mole *T. romana* and the Balkan mole *T. stankovici*. The fossil records are either described as *T. europaea* or not specifically designated; therefore the genus is lumped in the data base. As a whole the genus has a similar distribution in Europe to *A. sylvaticus* and the taxon is used as a complementary one to investigate relatively mild climatic conditions in Europe during OIS-3.

Vegetation/mammal comparison
The other comparison that can be made relates to the vegetation of OIS-3 and involves both the Biome 3 model output and data derived from pollen sequences (Huntley *et al.* 2003). An ongoing debate concerning OIS-3 and the late Pleistocene in general concerns the nature of the vegetation in the northern latitudes of the Palaearctic and Nearctic (e.g. Guthrie 1990 *versus* Colinvaux 1986; Ritchie 1984; Ritchie & Cwynar 1982). The distribution of members of all ecological groups, defined on the basis of their present distribution and knowledge about their modern behaviour, diets and habitat preferences as well as their extinct/extant status, were compared to the vegetation reconstruction based on the Biome 3 model output and the data derived from pollen sequences.

Results

Mapping and ecological classification of taxa
A number of categories have been defined. The definitions were based partly on modern understanding of their ecologies and partly on differences between their geographical distributions through OIS-3 and their modern geographical range.

The categories with examples are listed below:

1 – Extant taxa.
1a – Extant ubiquitous taxa. These are mostly large taxa and particularly include the carnivores whose geographical ranges during OIS-3 are fairly ubiquitous. Examples are the wolf *Canis lupus*, the brown bear *Ursus arctos*, the red fox *Vulpes vulpes*. Also included are herbivores such as the red deer *Cervus elaphus*. Finally, it would seem that Anatomically Modern Humans (*Homo sapiens*) are in this category.

The spotted hyaena *Crocuta crocuta*, the lion *Panthera leo* and the leopard *Panthera pardus* were also distributed widely during OIS-3.

1b – Extant cold-tolerant taxa. Include here are large and small taxa whose modern geographical ranges are in the boreal zone. They include reindeer *Rangifer tarandus*, arctic fox *Alopex lagopus*, wolverine *Gulo gulo*, musk ox *Ovibos moschatus*, collared lemming *Dicrostonyx torquatus* and Norway and Siberian lemming *Lemmus lemmus* and *L. sibiricus*. They include tundra species as well as ones primarily associated with northern coniferous forests or taiga.

1c – Extant continental or steppe-adapted taxa. These include the saiga antelope *Saiga tatarica*, the ground squirrels *Spermophilus* sp., hamsters *Cricetus* sp., *Cricetulus* sp. and *Mesocricetus* sp., the mole rat *Spalax* sp., the steppe lemming *Lagurus lagurus*, and possibly the marmot *Marmota* sp. (although they also have an Alpine distribution in Europe today).

1d – Extant temperate taxa. These include small mammals often associated with woodlands such as wood mouse *Apodemus sylvaticus* (Fig. 7.4a) or the bank vole *Clethrionomys glareolus*. The wild boar *Sus scrofa* and the roe deer *Capreolus capreolus* may at times be indicative of such habitats although the wild boar can also be found in more open areas, particularly when associated with wetlands and the roe deer occurs today also in steppe regions (Mitchell-Jones *et al.* 1999).

1e – Extant taxa found in upland or montane areas today. These include the chamoix *Rupicapra rupicapra* and the ibex *Capra ibex* and to a lesser extent the marmot *Marmota marmota*.

1f – Extant southern European peninsular endemics. They include the voles *Microtus (Terricola) savii* from the Istrian peninsula, and *Microtus (Terricola) duodecimcostatus* and *M. (Terricola) brecciensis/cabrerae* from the Iberian Peninsula, and the Iberian lynx *Lynx pardina*.

2 – Extinct cold taxa.
2a – Extinct cold taxa. They include a few large taxa that are thought to be cold adapted based on their temporal distributions, on their faunal associations and on anatomical features. Included here are the mammoth *Mammuthus primigenius* and the woolly rhino *Coelodonta antiquitatis* and probably the giant deer *Megaloceros giganteus* and cave bear *Ursus spelaeus*. The latter two species are found in the last

interglacial and so are clearly not as cold-adapted as the mammoth and woolly rhino. It should be noted, however, that the woolly rhino may be questioned as extreme cold-adapted as no OIS-3 dated fossils are known from Scandinavia for this species.

2b – 'Interglacial survivors'. These are solely large mammals like the straight-tusked elephant *Elephas antiquus*, the extinct rhinos *Stephanorhinus hemitoechus* and *S. kirchbergensis*. They all possibly became extinct towards the end of OIS-3 and certainly before 20 ka although the process of extinction was time transgressive (Stuart 1991). The fallow deer should also perhaps be included here as they disappeared from much of Europe at the end of OIS-3. A final species which when mapped seems to fit in this category is the Neanderthal *Homo neanderthalensis*.

2c – *Equus (Asinus) hydruntinus*. This extinct ass is presumed to be a steppic and semi-desert species (Batyrov & Kuzmina 1991; Vereshagin & Baryshnikov 1980). Unlike the other continental species, however, *E. hydruntinus* became extinct, although its demise was in the early Holocene. In one regard the distribution change of *E. hydruntinus* through OIS-3 is not unlike Neanderthals as they too move south while losing territory to the North.

There are also taxa in the data base that are problematic in as much as they may not be reliable stratigraphically due to their habit of burrowing. The rabbit *Oryctolagus cuniculus*, badger *Meles meles* and fox *Vulpes vulpes* are perhaps the most important culprits although the habit of burying their dead also potentially puts both human species *H. sapiens* and *H. neanderthalensis* into this category.

These categories and the mapping of taxa through time suggest that while diversity exists across Europe during OIS-3 very little change occurred in the area occupied by individual taxa. Exceptions to this rule are the 'interglacial survivors' that became extinct, including the Neanderthal and the ass *E. hydruntinus*. However, further work on the data base has found that the numbers of records of many mammals decreased towards the LGM during OIS-3 (Stewart in press, and Chapter 12 this volume, Stewart *et al.* 2003).

Provinciality analysis
The mammalian faunas present in each region though the three time zones of OIS-3 defined in Table 7.1 are described in turn using the Stage Three Project mammalian data base.[1]

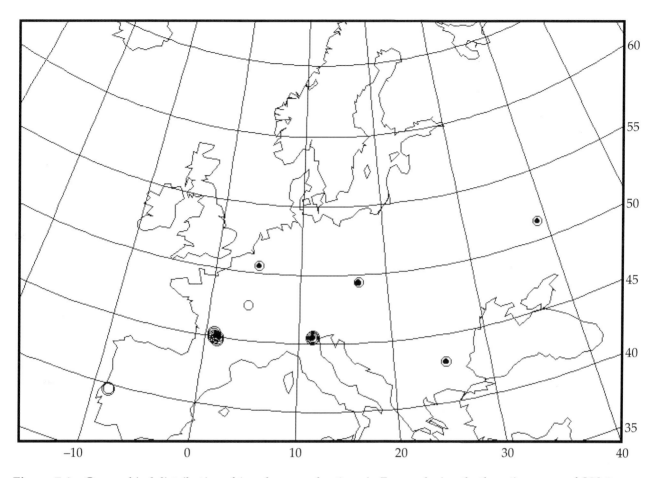

Figure 7.4a. *Geographical distribution of* Apodemus sylvaticus *in Europe during the three time zones of OIS-3 described in Table 7.2. Small solid circle = Early time zone; medium circle = Intermediate time zone; large circle = Late time zone.*

Fennoscandia: This region has a surprising number of dated sites although all but two are isolated finds of mammoths (Ukkonen *et al.* 1999). The two exceptions, Hamnsundhelleren Cave and Skjonghelleren Cave, both on the Norwegian coast at about 61.8° latitude, are of great significance as they yielded interesting faunas. At Hamnsundhelleren Cave arctic fox, reindeer, harp seal, bearded seal and ringed seal were found (Valen *et al.* 1996) while at Skjonghelleren Cave arctic fox, reindeer, ringed seal, otter and Norway lemming were found (Larsen *et al.* 1987).

This signifies that although these northern regions might be thought to be quite barren there must have been enough vegetation to sustain the mammoth, one of the two largest herbivores alive during OIS-3. In addition the kinds of boreal species one would expect to find at these latitudes today such as reindeer, arctic fox and Norway lemming were also present and presumably represent tundra vegetation. The mammoth, however, implies a more lush

vegetation in this region than is found there today with a significant grass and/or shrub component to the vegetation. The otter and the various seals tell us that the sea off the coast of Norway could not have been permanently frozen. This is all in agreement with the more recent view of the smaller extent of the Scandinavian ice sheet during OIS-3 (Arnold *et al.* 2002).

Northwest: This area has two concentrations of sites; SW Britain (including Devon, South Wales and the Mendips) and the Belgian Ardennes. Both are limestone areas with corresponding concentrations of caves and rock-shelter sites. Other localities including caves and open-air sites are scattered through the rest of the region.

There are examples of most categories mammals in the Northwest region including extinct cold (e.g. mammoth, woolly rhino), extant cold (e.g. reindeer, collared lemming) and extant ubiquitous taxa (e.g. red deer, wolf) which are present in both main

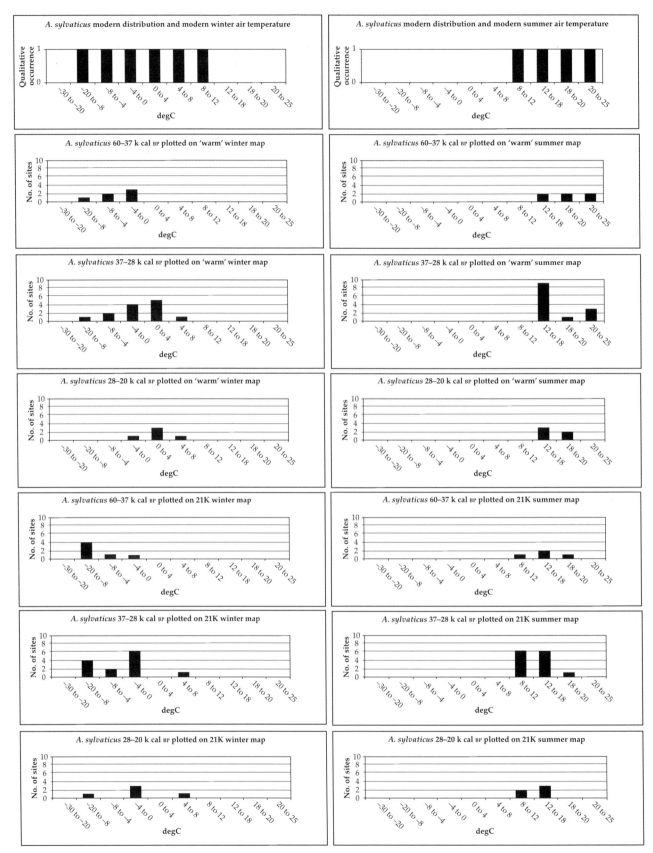

Figure 7.4b. *Comparison between air temperature isotherms (winter and summer) for modern and OIS-3* Apodemus sylvaticus.

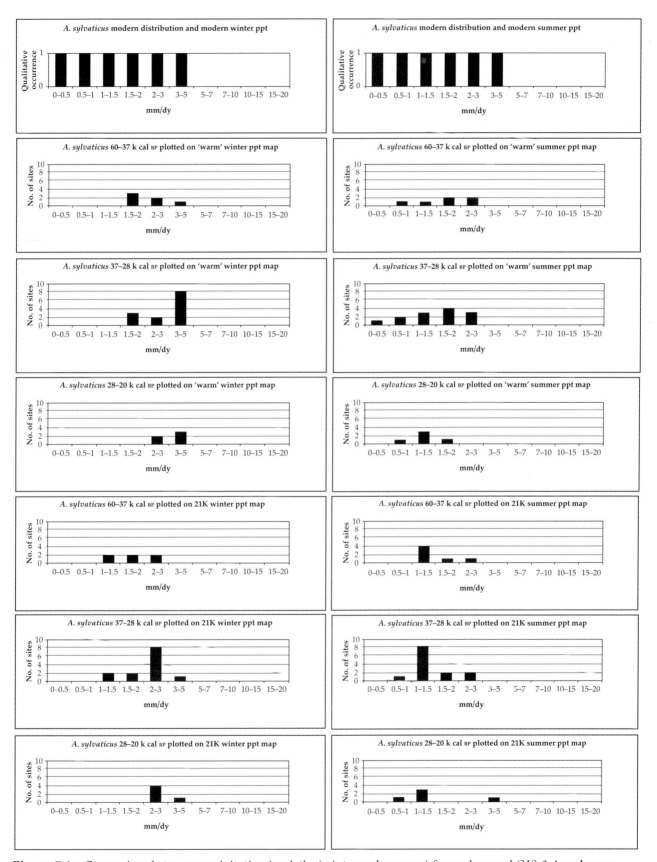

Figure 7.4c. *Comparison between precipitation isopleths (winter and summer) for modern and OIS-3* Apodemus sylvaticus.

concentrations of sites. The montane taxa (e.g. ibex, chamoix and marmot), possibly the temperate taxa (e.g. hedgehog, mole and woodmouse) and certainly the boar and roe deer appear to be absent from Britain although they are present in the Ardennes. This discrepancy may be caused by the difference in topographies between the British and Belgian areas that yield fossil mammals. The Ardennes, where both montane and temperate species are found, have steep rocky terrain on the sides of valleys which may afford a habitat for montane species and sheltered woodlands on the valley floors for the temperate woodland taxa like wood mice. The presence of temperate woodland taxa was recently described as a significantly different perspective from the generally held view that thermophilous refugia during the Late Pleistocene were exclusively much further south (Stewart & Lister 2001 *versus* Hewitt 1996; 1999; 2000). Eastern steppic taxa like *Equus hydruntinus* and the saiga are absent from Britain but present in the Ardennes at one site each, saiga at Trou du Renard and *Equus hydruntinus* at Sclayn Cave. The interglacial survivors are absent from the area (except for the Neanderthals who decrease in number of sites in this area though time).

North Central: Most of the sites in this area are concentrated along its edges and in particular along its southern edge. This is the concentration of sites in Moravia. The lack of sites to the north of the region is in part due to the relatively low topography of the more northern parts of Germany and Poland that make up the central and northern parts of the region. Chapter 10 of the present volume by Musil (2003) deals solely with the area of Moravia and of southeast Europe and we refer readers there for further details if required.

Many of the larger taxa in the data base are present in this region including extinct cold, extant cold, extant ubiquitous and extant montane mammals. Interglacial survivors are absent, as are the southern peninsular endemics. The temperate taxa like wood mice, wild boar and roe deer may not be well represented indicating that deciduous woodlands may not have existed in profusion.

The giant deer is less well represented at sites on mainland Europe than in Britain and is only present at a few sites in the North Central region. The smaller mustelids like the weasel and stoat may be at few sites compared with other areas, although this could be due to recovery bias as they are small and hence may not be recovered without fine-mesh sieving. The taxon most noticeably abundant compared to

other areas of Europe is the wolverine *Gulo gulo*. This taxon is a specialist mustelid found in low population densities in the areas they frequent today. They are associated today with coniferous forests which may be the reason for their being well represented in Moravia. It has been suggested by Musil (1994a), however, that these occurrences along with those of other carnivorous mammals such as wolves on archaeological sites like Pavlov I may be due to their use for pelts by Upper Palaeolithic peoples.

Northeast: The Northeast region is relatively poor in absolutely dated sites making it more difficult to make reliable comments on the fauna there during OIS-3. Montane taxa are absent which is probably because the region constitutes the Russian plain which is flat providing no habitat for the chamoix or ibex. The mammoth, woolly rhino, red deer, reindeer, auroch/bison, saiga and musk ox are the large herbivores present. The interglacial survivors and the peninsular endemics are absent as expected due to the remoteness from the southern areas of Europe that they persisted in. The extant temperate taxa are probably absent due to the lack of any sheltered topography and possibly because of extreme cold winter air temperatures.

Southwest: This is the region with the largest number of dated sites as it includes the Dordogne, which has long received concentrated attention from archaeologists, and to a lesser extent the northern Rhone, Cantabria and the Pyrenees. Consequently any assessment of the mammalian fauna of the area will be much more reliable than elsewhere in Europe.

The extant ubiquitous taxa, the extant cold taxa, the extinct cold taxa, the extant montane taxa and the extant temperate taxa are mostly present. The wolverine and Norway lemming are absent and seem to be confined to the north east of Europe during OIS-3. The extant and extinct cold taxa seem to creep into the northwest and northeast of Iberia and as such this is the southernmost area that these species are found. The fact that some of the species such as arctic fox, reindeer and mammoth may simply be there due to the transport of artefacts made from their bones cannot be discounted completely. However, because this pattern is also true of the giant deer, the cave bear and the woolly rhino (a record from the Pyrenees) suggests that the pattern may be a true representation. This part of Iberia is distinct from the rest of the peninsula even today in terms of its ecological affinities which may be the explanation. The Cantabro-Pyrenean Mountains cause this part of Spain to have a higher precipitation than the more southern parts and hence make it more similar to the more northern parts western Europe than the area to the south. It is likely that this phenomenon is at least partly responsible for the concept of the Ebro frontier (Zilhão 1993; 2000) to explain human population patterns of the later Palaeolithic. The proximity of the river Ebro to the southern boundary of the more northern vegetation zone is more likely to be the cause of the phenomenon than the river itself. A number of modern animals have distributions with boundaries at this latitude in Iberia (Mitchell-Jones *et al*. 1999).

The presence in this region of interglacial survivors such as the straight-tusked elephant, the steppe rhino and Merck's rhino (together with Mediterranean taxa like the rabbit) underscore the mixed nature of this region forming the distribution overlap area between the northern and Mediterranean faunal provinces (Davies *et al*. in press).

Extant temperate taxa are well represented and species such as the garden dormouse, the wood mouse and the mole are all recorded (Fig. 7.4a). The more continental species such saiga and souslik are also present although not as visibly as in more northern reaches of western Europe. Finally, extant montane taxa such as the chamoix and ibex and many extant ubiquitous taxa like red deer, lion and wolf are well represented.

Alpine: The Alpine region is the area of Europe yielding the fewest absolutely dated OIS-3 mammalian fossils. Little of interest can therefore be concluded from such an analysis. The only site that appears to be in this area is Schnurenloch in Switzerland dated to between 16,514±747 and 33,294±834 ka BP which yielded red deer, chamoix, ibex, musk ox, cave bear, arctic and red fox, wolverine, dhole, wolf, wild cat, leopard, marmot, brown hare and undetermined rodents (Jequier 1975). This list implies that a number of habitats were never-the-less available in areas of the Alpine region such as tundra for the musk ox and/or grassland plains for the marmot and the brown hare, wooded areas for the wild cat, the wolverine and the leopard as well as to the craggy montane landscapes indicated by the chamoix and ibex.

Southeast: This region includes the Hungarian basin and the Carpathian ring and is divided from the Mediterranean East region by the Dinaric Alps and follows a line between the Morava and Axios rivers and then strikes for Istambul via Sofia. The sites in this region are all to the west of the region in Hun-gary and Croatia and include Istálloskö Cave, Szeleta Cave, Vindija and Krapina.

The region being in the overlap zone between the Mediterranean and the Northern zones is characterized by elements of each. Porcupine is present together with mammoth, arctic fox, reindeer, cave bear, elk and wolverine. Temperate taxa are well represented including woodmouse, fat dormouse and mole with steppic elements like souslik, grey hamster and mole rat. Montane taxa are also found as is the merck's rhino. Ubiquitous taxa are also found in profusion.

Mediterranean West: The Mediterranean West region constitutes the major part of Iberia south of the latitude of the Cantabrian Mountains (*c*. 42° Latitude).

The fauna of this region includes extant ubiquitous taxa (red deer, horse, lion, hyaena, wolf, red fox, brown bear), temperate taxa (roe deer, wild boar, wood mouse), montane taxa (chamoix, ibex), interglacial survivors (Merck's rhino, steppe rhino, straight-tusked elephant) and peninsular endemics (Iberian lynx and the rodents *Microtus* (*Terricola*) *duodecimcostatus* and *M.* (*Terricola*) *brecciensis/cabrerae*). Missing are the extant and extinct cold taxa and the continental taxa. There is no evidence for the extension of the northern steppe-tundra into the interior of the region during OIS-3 as described by Finlayson *et al*. (2000) although this may be the result of a lack of dated sites in the Meseta as most sites in Iberia are situated within a relative short distance from the coast.

The extinct hamster *Allocricetus bursae* is recorded at Cova Negra (Perez Ripoll 1977) and Caldeirao Cave[1] although there is a suggestion that this taxon is in fact conspecific with *Phodopus sungorus* (Stuart 1991; Nadachowski 1982).

Mediterranean Central: This region encompasses the Italian peninsula as well the mainland area south of the Alps extending a little way either side of the peninsula.

The faunal elements included in the area differ between the penninsula itself and the more northern part of the region. Cold taxa such as the reindeer, the musk ox, the mammoth and woolly rhino are totally absent, while the giant deer, the elk, the arctic fox, the wolverine are present only in the north of the region and are absent from the Istrian peninsula itself. The exception is the cave bear that was identified in the Grotta del Fossellone (Blanc 1953). Extant ubiquitous, extant montane and extant temperate taxa are all well represented in the faunas.

A peninsular endemic exists in the form of the Savi's pine vole *Pitymys savii* and the region acts as a refugium for taxa such the straight-tusked elephant and the fallow deer.

Mediterranean East: This region encompasses the Dinaric coast and Greece and all the area below the Dinaric Alps. There are only two sites in this region, Bacho Kiro in Bulgaria and Crvena Stijena in Montenegro.

The list of taxa found as fossils in this region consists of extant ubiquitous species, extant montane species, the two more warm tolerant extinct cold taxa (*U. spelaeus* and *M. giganteus*), an interglacial survivor (*S. hemitoechus*), extant continental (common, grey and Romanian hamsters, steppe lemming and mole rat), extant temperate taxa (wild boar and wood mouse) and even an extant cold taxon (arctic fox). This list is only based on two sites but is comparable to the other two Mediterranean peninsular areas, certainly if the more northern parts of the other peninsulae are included. The cold taxa, both extinct and extant, are found at Bacho Kiro which has the most reliably complete fauna including rodents with continental and temperate affinities but none of the arctic ones.

Climate/mammal comparison
Lemmus sp.: The Norway lemming and Siberian lemming are northern Palaearctic taxa today living in alpine and tundra habitats. In the past the genus lived in Europe from Norway down to 47° latitude (Fig. 7.2a). The temperatures that characterize their present distribution are from –30 – –20 to –4–0°C in the winter (December, January and February) and 0–4 to 12–18°C in the summer (June, July and August) (Fig. 7.2b). These temperature ranges were compared to those which may have existed in the areas of *Lemmus* sp. distributions during OIS-3. Maps of *Lemmus* sp. distribution were compared to the two possible climatic conditions (the 21k and the 'warm' simulations) modeled for OIS-3 during the Stage Three Project (Barron & Pollard 2002; Pollard & Barron 2003). The results of these comparisons can be seen in Figure 7.2b. The winter temperatures with which *Lemmus* sp. coincides either from the 'cold' 21k or the 'warm' simulations both fall within the winter temperature range experienced by the genus today. The distinction between the 'warm' and the 21k winter simulations in relation to *Lemmus* sp. is that the modal value for the 21k simulation is lower (–20 to –8° C) than for the 'warm' simulation (–8 to –4°C) but their ranges overlap completely. For the summer temperatures the situation is similar and most sites fall within the range of the summer temperatures defining the modern distribution of the species. A single site (Skjonghelleren Cave) causes outlying summer temperatures to be suggested for *Lemmus* during the Early and Middle time zones for OIS-3 that are colder than that predicted using the uniformitarian principle and the modern summer air temperatures in the geographical range of *Lemmus* sp. (Fig. 7.2b). This site is in Scandinavia on the Norwegian coast close to the OIS-3 ice-sheet margin (Arnold *et al.* 2002) and may suggest that the cline in temperature adjacent to the ice was steeper than predicted by the model. Similarly, a single site falls on a 'warm' simulation summer temperature range that is too warm according to modern values.

The comparison of the modern distribution of *Lemmus* sp. with the modern precipitations rates (mm/day) yielded a range from 0–0.5 to 3–5 mm/day for the winter months and 0.5–1 to 2–3 mm/day for the summer months (Fig. 7.2c). The comparison between the fossil distributions in the three time zones compared with 'warm' simulations for winter produced slightly higher rates of precipitation than with the 21k 'cold' simulation (Fig. 7.2c). The results for the comparison with the summer simulation values showed the 'warm' output to have a site that falls on precipitation levels above the results that might be expect from the modern precipitation and *Lemmus* sp. distribution comparison while the 21k simulation was more similar to modern *Lemmus* associations. The 21k simulation for summer is more in line with that predicted by uniformitarianism than the 'warm' simulation (Fig. 7.2c).

Therefore the distribution of *Lemmus* sp. in Europe during OIS-3 when compared with the possible values simulated by the climate model coincided with similar air temperatures and precipitation levels than they are associated with today.

Alopex lagopus: The arctic fox is another northern taxon living in Scandinavia today but extended from Scandinavia down to northern Spain during OIS-3. The air temperature values with which the arctic fox is associated today are between –30 and –20 to –4 to 0°C for the winter and –8 to –4 to 12–18 for the summer. The values produced for all time zone comparisons with the 21k winter and summer simulations include many sites that exceed the modern range of *A. lagopus* temperatures today. The 'warm' simulation while it produces figures that overlap with the modern temperature range have many sites falling on values above the modern range.

120

Table 7.3. *Comparison between observed modern climate small mammal associations with simulated OIS-3 climate small mammal fossil associations.*

	'Warm'		21k	
	Air temperature	Precipitation	Air temperature	Precipitation
Lemmus sp.	≈ winter	≈ winter	≈ winter	≈ winter
	1 site too warm in summer	1 site too wet in summer	1 site too cold in summer	≈ summer
Alopex sp.	A few sites too warm in winter	≈ winter	1 site too warm in winter	≈ winter
	A few sites too warm in summer	Skewed too wet in summer	A few sites too warm in summer	A few sites too wet in summer
Spermophilus sp.	1 site too cold in winter	≈ winter	A few sites too cold in winter	1 site too dry in winter
	≈ summer	1 site too wet in summer	A few sites too cold in summer	1 site too wet in summer
Apodemus sylvaticus	≈ winter	≈ winter	≈ winter	≈ winter
	≈ summer	≈ summer	≈ summer	≈ summer
Talpa sp.	≈ winter	≈ winter	≈ winter	≈ winter
	≈ summer	1 site too dry in summer	≈ summer	1 site too dry in summer

The precipitation values with which todays *A. lagopus* is found are 0.5–1 to 2–3 mm/day during the winter and 0–0.5 to 2–3 in the summer. The OIS-3 comparisons on the whole lead to the conclusion that the precipitation simulated for OIS-3 is too wet in summer although the winter simulations equate better with the precipitation values associated with *A. lagopus* today.

Therefore the arctic fox climate/mammal distribution analysis seems to indicate that the temperatures modelled in winter and summer by both the 'warm' and 21k climate simulations are too warm and that the precitation predicted may be accurate in winter but too wet in summer (Table 7.3).

Spermophilus sp.: The modern distribution of both European *Spermophilus* species, the spotted souslik *S. suslicus* and the European souslik *S. citelus*, are in the southeast of the area. In OIS-3 they extended as far west as the Dordogne in southwest France. The comparison of their distribution with air temperature values today indicates that they occur in winter temperatures of between –8 and 4 to 4 and 8°C and summer temperatures of 12–18°C to 20–25°C. The OIS-3 'warm' and 21k simulations of winter air temperature suggest that the 'warm' simulation compares best with the temperatures that the sousliks live under today. The 21k winter simulation appears to produce too cold a range of figures with many souslik sites falling in the –20 to –8°C temperature

range. The summer simulations are similar to each other although even the 21k simulation produces air temperatures that are below the range of modern *Spermophilus* in Europe today while the 'warm' simulation is in agreement with modern values.

The precipitation rates associated with *Spermophilus* in Europe today are as follows: 1–1.5 to 3–5 mm/day in winter and 0–0.5 to 2–3 mm/day in summer. The 'warm' simulation for precipitation at *Spermophilus* sites is in line with the modern range. The 21k simulation for *Spermophilus* is mostly in line with the values that the taxon is associated with today although one site with values falls below the modern range. The range of summer precipitation values with which *Spermophilus* is associated today is mostly in line with both the 'warm' and 21k outputs except for one site that appears too cold for the species during the latest time zone.

The results of the *Spermophilus* comparisons show that both the 'warm' and 21k simulations of both winter and summer temperatures are a little too cold although the 'warm' simulation produces temperatures more comparable with modern souslik distributions than does the 21k simulation. The 'warm' simulation for winter and summer produces values for *Spermophilus* sites that only slightly differ from precipitation values they are associated with today (Table 7.3).

Apodemus sylvaticus: The wood mouse *Apodemus*

sylvaticus lives in most of Europe up to southern Scandinavia and in much of the same area in OIS-3 (Fig. 7.4a). The air temperature values with which the wood mouse is associated today in winter are −20 to −8 to 8–12°C and in summer are 8–12 to 20–25°C. The summer and winter air temperatures that fossil *Apodemus sylvaticus* coincides with in the 'warm' and 21k simulations are consistent with those of today (Fig. 7.4b). There is complete overlap between modern and simulated temperature values (Fig. 7.4b).

The precipitation values with which the wood mouse falls today (0–0.5 to 3–5 mm/day in winter and 0–0.5 to 3–5 mm/day in summer) are wide and hence there is little surprise that the values with which the fossils are associated when compared to both the 'warm' and 21k simulation outputs are within the modern range (Fig. 7.4c).

The climatic values simulated for air temperature and precipitation for both the 'warm' and 21k simulations completely agree with the temperatures needed by *A. sylvaticus* according to the modern range with which they are associated (Table 7.3).

Talpa sp.: The genus *Talpa* is found today over much of Europe up to 62° latitude in Finland. At present they are associated with winter air temperatures of −20 to −8 to 8–12°C and 8–12 to 20–25°C in summer.

The 'warm' and 21k air temperature simulations for winter and summer when compared to the mole distribution in relation to the winter and summer temperatures that they enjoy today suggest that temperatures are accurate.

The precipitation values with which *Talpa* are associated at present are 0.5–1 to 3–5 mm/day in winter and 0.5–1 to 3–5 mm/day in summer. The OIS-3 fossil occurrences of *Talpa* when compared to simulated precipitation values for winter overlap completely giving no discrepancy whether it be the 'warm' or the 21k simulation. The summer simulations on the other hand produce one site that appears too dry for the species.

Again, as with *A. sylvaticus* the other relatively temperate taxon, the genus *Talpa* shows the 21k simulation to be in agreement with modern values with which they are associated both for summer and winter. The 'warm' and 21k simulation suggest that one site is associated with precipitation values that are too low for *Talpa* in summer in comparison with modern values.

Vegetation/mammal comparison

The other comparison that can be made is that where fossil mammalian distribution data are compared with vegetation model output and pollen data both divided into warm and cold interval biomes. The results of this comparison exercise will be discussed by region.

Fennoscandia: One site, Sokli, is the only source of pollen data considered by Huntley *et al.* (2003) for the warm interval; they infer that dwarf-shrub tundra/tundra and cool mixed forest were present in northern Finland at that time. The modelled vegetation is not very different and predicts shrub tundra, dwarf-shrub tundra/tundra and ice. The difference between the warm and cold intervals was that the ice was more extensive in the cold interval.

The mammalian data partially agree with the pollen evidence as tundra with taiga would support the reindeer, arctic fox and Norway lemming found on the coast of Norway, although these sites are at a considerable distance from the site providing pollen data in northern Finland (Helmens *et al.* 2000). The mammoth sites are more difficult to reconcile with the tundra and taiga, a greater grass or herb component would presumably have existed to feed these megaherbivores. The vegetation simulated by the model is similar to the pollen data and would thus represent the same problems for the mammoth while adequately supporting the reindeer, Norway lemming and arctic fox.

Northwest: A number of sites (10) were used by Huntley *et al.* (2003) to characterize the region that includes the British Isles. The biomes represented in the warm interval were mostly dwarf-shrub tundra/tundra with occurrences of steppe tundra/warm steppe and temperate grasslands and cool mixed forest. In the cold interval there were no sites in the more northern parts of the region and only La Grand Pile is present in the south with temperate grasslands. The model simulation on the other hand shows a dominance of evergreen taiga/montane forest. Patches of cold mixed forest are also present and increase in the warm interval; and shrub tundra in the north and even barren polar desert in parts of Scotland and Ireland which increase in the cold interval.

The extant cold mammals are consistent with the tundra and cool mixed forest inferred by the pollen data. The extinct cold mammals and the general richness of the fauna (including sparse steppe elements) would have been sustained by the steppe tundra/warm steppe and temperate grassland also indicated by the palynology. The temperate woodland mammalian taxa may possibly have relied on the cool mixed forest or else are anomalous. The simulated vegetation may have accommodated the extant cold taxa (taiga, tundra and cool mixed forest). However, the cold extinct megafaunal elements together with many of the other grazers and browsers, animals relying on grasslands and more temperate woodlands like red deer, roe deer and horse, may have had difficulties existing on the vegetation modelled.

North Central: Only a small number of sites represents the warm interval in the North central area of Europe in Huntley *et al.* (2003) and none in the cold interval. The biomes represented by the pollen sites for the warm interval are cool mixed forest, temperate grassland and dwarf-shrub tundra/tundra. The model predicts evergreen taiga/montane forest in both the warm and cold intervals with increasing cool mixed forest in the warm interval.

As with the Northwest region the extant cold animals, like the reindeer and Norway and collared lemmings, would have been accommodated by the tundra vegetation. However, many other taxa may have had problems subsisting on such vegetation although the temperate grasslands indicated may have been sufficient to feed the extinct cold animals (mammoth and woolly rhino) and the steppe elements (saiga, steppe lemming and common hamster) while the cool mixed forest may have sufficed for the woodland mammals (wild boar, the bank vole, the wood mouse, the wolverine and the marten) although this is not certain. The evergreen taiga of the model simulation is only consistent with a limited number of the mammalian taxa present like the wolverine, elk and marten, if it were a pine marten. The more sparse cool mixed forest of the model may have provided the necessary habitat for the other forest animals although the mammals that rely on extensive grasslands like many of the extinct cold taxa and the extant steppic and ubiquitous herbivores are difficult to reconcile with the vegetation simulated.

Northeast: The only pollen sites referred by Huntley *et al.* (2003) suggest that evergreen taiga/montane forest, cool mixed forest and temperate grasslands are present in the Northeast region of Europe. The dominant biome simulated by the model is again evergreen taiga/montane forest and increasing amounts of cool mixed forest in the warm interval compared to the cold interval. To the east of the region temperate woodlands, temperate deciduous forest and temperate grasslands are present in the simulations, with the grasslands decreasing at the expense of woodlands towards the cold interval.

The pollen-inferred vegetation would provide adequate habitats for most of the mammals found from OIS-3 sites in the Northeast region. The evergreen taiga/montane forest are consistent with the wolverine and the elk while the temperate grassland agree with the megafaunal elements and presumably supported the steppic elements and even the reindeer. The few mammals that required more in the way of deciduous tree cover, like the wood mouse, may have been accommodated by the cool mixed forest. The modelled vegetation is more problematic as the evergreen taiga/montane forest with scattered cool mixed forest would not easily support the large grazers seen in the fossil record like the mammoth, woolly rhino, horse or even the saiga and steppic rodents like the common hamster, bobak marmot, souslik and steppe lemming.

Southwest: Two pollen sites in the Southwest region suggest the presence of temperate grassland, cool mixed forest and temperate sclerophyll woodland in the warm interval and temperate grassland and steppe tundra/warm steppe in the cold interval (Huntley *et al.* 2003). The model on other hand shows evergreen taiga/montane forest, temperate woodland and temperate grassland in the warm interval and evergreen taiga/montane forest, desert: shrubland and steppe and temperate grassland in the cold interval.

The pollen data described by Huntley *et al.* (2003) for the Southwest area provide a wide range of habitats across the warm and cold episodes which would agree broadly with the mammals like temperate grassland and steppe tundra/warm steppe for steppic and other grazing herbivores, cool mixed forest and temperate sclerophyll woodland for the animals requiring a range of woodland types. The difference between the vegetation types inferred from the pollen for the cold and warm intervals is not matched with any obvious qualitative difference in the mammalian assemblages. There is no obvious reason provided by either the pollen or simulated vegetation to account for the more southern taxa such as the rabbit and the *Stephanorhinus* rhinos. These, together with the cold elements, emphasize that this is a region of overlap between the northern and Mediterranean regions. It is just possible that the desert: shrubland and steppe are the requisite biome to account for the southern elements.

Alpine: No pollen sites are used to characterize this region in Huntley *et al.* (2003). The model simulations show shrub tundra in the higher parts of the Alps with a surrounding of evergreen taiga/montane forest and cold mixed forest. The shrub tundra increases at the expence of other biomes during the cold interval.

The mammalian assemblage at Schnurenloch would seem to be adequately provided for by the biome model vegetation. The evergreen taiga/montane forest would have been particularly favoured by the wolverine, the cool mixed forest would suit the red deer and the shrub tundra would may have attracted the musk ox. There must have been more in the way of grassland for the brown hare unless it is a misidentified mountain hare for which more sparse vegetation would be adequate. The other taxa are either montane specialists or carnivores that would probably have been accommodated by a variety of the habitats indicated by the models.

Southeast: Again no pollen sites were referred for this region by Huntley *et al.* (2003). The model simulation has temperate grassland, temperate deciduous forest, temperate woodland and evergreen taiga/montane forest with an increase of temperate grassland at the expense of deciduous forest from the warm to the cold interval.

The modelled temperate grassland may have sustained grazers such as the mammoth, horse and giant deer while the temperate deciduous forest and temperate woodland would have provided habitat for the wild boar and roe deer and woodland rodents such the fat dormouse, woodmouse and bank vole. The evergreen taiga/montane forest is not inconsistent with the wolverine, marten, lynx and elk although they could persist in other woodlands. The more steppic rodents like the mole rat, steppe lemming and the hamsters as well as the saiga are difficult to adequately accomodate with the vegetation biomes simulated.

Mediterranean West: The Iberian Peninsula has pollen indicative of temperate grassland and dwarf-shrub tundra/tundra in the warm interval and only dwarf-shrub tundra/tundra in the cold interval. The model simulates a great deal of variation with a dominance of temperate grassland and varying proportions of desert: shrubland (a term used in Huntley *et al.* 2003) and steppe, evergreen taiga/montane forest and temperate woodland. The difference between the warm and cold phases is not great and is mainly exemplified by a decrease in temperate grassland and a corresponding increase in the other biome types.

The pollen-based biomes are indicative of dwarf-shrub tundra/tundra which is not in agreement with the mammalian data apart from in the northernmost part of the Peninsula where reindeer is found at sites like La Riera. The grassland, desert: shrubland and steppe are consistent with the many herbivores like the horse, steppe rhino, *Equus hydruntinus* and the bovids although there is no evidence in the biomes of the deciduous or coniferous woodlands that would be needed to support the straight-tusked elephant, Merck's rhino, wild cat, lynx, roe deer, garden dormouse and woodmouse. The simulated vegetation does, however, have a woodland component, both deciduous and coniferous, as well as the grasslands to support all the mammals.

Mediterranean Central: The pollen data used to infer biomes for the Istrian peninsula show cool mixed forest, temperate woodlands and temperate grassland for the warm episode and temperate grassland and steppe tundra, warm steppe in the cold episode. The model simulated mostly warm grasslands for the cold episode with a little temperate woodland, desert: shrubland and steppe and temperate deciduous forest. The warm interval model predicted mostly the same components but with less temperate woodland and more temperate deciduous forest and no desert: shrubland and steppe.

The OIS-3 mammal fossils from Italy are possibly consistent with the pollen data. Open and forest taxa can be accommodated by the range of vegetation inferred. There is no pollen data for the most northern part of this region where a few northern species like the giant deer, elk and arctic fox are present. Therefore any latitudinal variation in biomes cannot be confirmed. The model simulates much the same although without the steppe tundra/warm steppe of the pollen data and the addition of desert: shrubland and steppe and temperate deciduous forest. It is not clear how different this truly is and whether it would better provide habitats for the mammals or not.

Mediterranean East: The two pollen sites in the Southeast region show temperate woodland and temperate deciduous forest in the warm interval and temperate grasslands and steppe tundra/warm steppe in the cold interval. The model has a dominance of temperate grassland and temperate deciduous forest with some evergreen taiga/montane forest increasing in the cold interval, temperate woodland and desert: shrubland and steppe.

The mammals in this area include some large herbivores that may have lived in temperate grasslands or steppe tundra/warm steppe as well as some that requires forest vegetation. However, the grasslands are inferred for the cold episode and the woodland for the warm episode while the mammalian evidence indicates that there was a mixture of both biomes at both times.

Discussion and conclusion

Mapping and ecological classification of taxa and provinciality analysis
The analysis of the mammalian taxa present by province during the three time intervals of OIS-3 demonstrates that there was a distribution cline running north–south and another less marked cline running west–east across Europe. The north–south cline can broadly be divided along the lines of the three latitudinal provinces defined by Gamble (1986) with Northern, Southern and Mediterranean provinces. These provinces were defined as a Northern faunal province where Mediterranean taxa were absent, a Mediterranean province where northern taxa were absent and a southern province where northern and Mediterranean taxa could potentially be present together (Davies *et al.* in press). The southern province therefore forms an area of overlap and only taxa such as the few peninsular endemics are restricted to the Mediterranean province.

The west–east breakdown of Europe is less pronounced although there are taxa that are restricted to the east such as the mole rat (*Spalax*) and there are no taxa that are truly restricted to the west. The west–east cline is an oceanic/continentality gradient.

Climate model simulation/mammalian distribution comparison
Climate–mammal comparisons were achieved where smaller mammal distributions today were compared with observed modern winter and summer air temperatures and precipitations to gain an insight into the values that coincide with, and possibly control, their geographical ranges. The same was done for the distribution of fossils of those taxa that were compared to the summer and winter air temperatures and precipitation values for 'warm' and cold 21k climate model simulations of OIS-3. The results of this exercise indicate that it depends on the taxon and its present geographic range as to which of the simulations seem to conform with the taxon's air temperature and precipitation tolerance (Table 7.3). The taxa living in the north of Europe today, *Lemmus* sp. and *Alopex lagopus*, indicate that the winter and summer air temperatures simulated by both the 'warm' and 21k models were often too warm when compared with the temperatures that these species live under today. This was particularly true of the arctic fox. One locality, near the Scandinavian ice sheet, at which *Lemmus* sp. was found suggests temperatures during winter during OIS-3 (in both the 'warm' and 21k simulation) that are too cold for the taxon implying that the temperature gradient away from the ice sheet was steeper than modelled. The precipitation values of both simulations on the other hand appear to be too wet for either *A. lagopus* or *Lemmus* sp.

The past distribution of the souslik *Spermophilus* sp., an eastern taxon, suggests that air temperatures modelled by both simulations are too cold, whether for summer or winter. The precipitation values are mostly comparable with those in the modern *Spermophilus* range although individual sites cause outlying associated values.

The distributions of *Apodemus* and *Talpa* during OIS-3, both of which are associated with the mixed desciduous woodlands of the lower and mid latitudes of Europe today, suggest that the summer air temperatures predicted by the 21k simulation and the 'warm' simulations are in accordance with the temperatures they are associated with today. Precipitation values are mostly in accordance with modern ranges although *Talpa* indicates that summers may be too dry in the simulations for one site.

The lack of agreement between the different taxa regarding which of the values predicted by the model that are inconsistent requires an explanation. This phenomenon has been recognized previously as that of the 'non-analogue community' (e.g. Faunmap Working Group 1994; 1996; Stafford *et al.* 1999). While these mixtures have previously been recognized they have not been considered using observed and simulated climatic values for temperature or precipitation. Table 7.3 summarizes the degree to which the mammals above are inconsistent.

There are a number of ways to explain the discrepancy between the different divergences between taxonomic 'tolerances' inferred from modern geographic ranges and simulated climates for the fossil localities. They are:
1. The modern geographical ranges of the taxa, and hence the climatic ranges they live under today do not extend to the maximum possible. Human and other factors may be restricting their present geographical and climatic ranges giving a falsely restricted view of their 'tolerances'.
2. The fossils were misidentified and hence give a false negative indication of the climates simulated.
3. The dates of the fossils may be misleading. Time

averaging may be falsely creating regional assemblages that are in fact mixtures of animals living at different times under different climatic regimes during OIS-3 and beyond.

4. The taxa in the past may not be the same genotypic populations as those we see today. Either they may have evolved into the populations we see today or the populations of the past may have become extinct and hence not contributed to the gene pools and climatic tolerances of the species today.

5. The regional climate model does not encompass a consideration of fine-scale topography. Topography may produce a wider range of climates in any one area of Europe than simulated by the climate model because smaller-scale microclimatic differences are invisible to the model. This phenomenon was discussed recently by Stewart & Lister (2001), who showed that northern refugia existed that may explain the colder than expected temperatures for the existence of taxa such as *Talpa* and *Apodemus*. These would be particularly important to the smaller mammals being used in the comparative study performed here.

6. The climatic parameters, such as temperature and precipitation, are not as important to mammals, as snow cover, type of vegetation, and co-occurences with other mammal species (Markova 1992 and Musil 2003, Chapter 10 this volume).

7. Finally the regional climate model simulations may simply be wrong at all scales. This possibility was raised by Huntley *et al.* (2003) when comparing BIOME model simulated vegetation with inferred palaeovegetational data from pollen studies.

The seven possible explanations for the discrepancies caused by the different mammal–climate comparisons should be considered in turn. The first factor, that the modern distributions have been restricted by human landscape alterations, is difficult to address adequately as detailed knowledge of pre-agricultural distributions is not available for small mammal taxa. However, the use of smaller mammals was designed to minimize this problem as smaller areas of suitable habitat can support the taxa. The possibility of misidentification was negated by knowledge of the practical problems of identification of each taxon. Hence most were not dealt with beyond the genus. The possibility that time averaging is causing apparent mixtures of taxa which in fact belong to distinct episodes (interstadials and stadials) is difficult to negate but the consistency of the non-analogue combinations may be proof that they are genuine. In North America and Russia the

coeval nature of non-analogue taxonomic elements of faunas has been demonstrated using direct dates on bones (Stafford *et al.* 1999). The fourth possible problem is likely to be a factor as it would seem unlikely for instance that the populations of arctic foxes and lemming in southern Europe during OIS-3 contributed to the modern gene pools of those taxa in Scandinavia today. The topographic averaging effect is also likely to be a partial explanation as the model does not address topographic variation at a fine local scale. The fact that other factors may be guiding the presence/absence of mammalian taxa such as snow cover or the presence/absence of other mammal taxa cannot be ignored and such criteria have already been given as influencing factors in the distribution of certain taxa (Markova 1992). The model may, however, be the most important problem as its output was meant to be tested by the various proxy-data collected during the Stage Three Project. The comparative exercise above would therefore cause one to question the validity of the climate model output comparisons with mammalian taxa without any reference to modern climate associations. Such exercises are seductive for extinct species as their climatic tolerances are not known although based on the comparisons achieved above it would seem premature to do so.

Vegetation/mammal comparison
A number of issues arises from the comparison between the OIS-3 mammalian distributions with both the pollen data and the Biome 3 simulated vegetation. They are:

The mammoth steppe versus tundra: Arguments have existed for some time regarding the nature of the late Pleistocene vegetation of the higher latitudes of the Palaearctic and Nearctic (Guthrie 1990 *versus* Colinvaux 1986; Ritchie 1984; Ritchie & Cwynar 1982). The polarized views are characterized by the insistence amongst some palynologists that the dominant biome present in these regions was a relatively barren tundra while mammalian palaeontologists have argued for a richer flora to sustain the mega-herbivores of the time. Some of the taxa present in northern and mid latitudes of Europe of OIS-3 such as the reindeer and other extant cold taxa like the musk ox and the Norway/Siberian lemmings are consistent with tundra. However, the diversity and number of grazers of the time are not easily reconciled with tundra of the type present in arctic regions today. The same is true of the taxa which today would be associated with continental steppic regions of the interior of the Palaearctic like the saiga and the sousliks.

Presence of woodland in northern regions: The debate regarding the presence of woodland refugia in northern areas has existed in the main between palynologists who reject the possibility and its advocates, mostly vertebrate palaeontologists, those studying charcoal and latterly molecular biogeographers (for review see: Stewart & Lister 2001). Huntley *et al.* (2003) have recently suggested that the question can not be answered using charcoal from sites as the human populations may have selectively collected wood species. They also say that charcoal cannot be used to 'infer extensive woodlands'. The fact that mammalian fossils in the same deposits suggesting the presence of trees is more difficult to argue away. It is also true that pollen can be more readily quantified but this does not address the qualitative presence of charcoal. Directly dated taxonomically determined charcoal must indicate the presence of tree species in unknown numbers at that time at a relative proximity of the site where it was found. The northern areas of Europe with charcoal that would presumably be disputed are the Belgian Ardennes and Moravia. The fact remains that both areas contain temperate mammalian species associated with absolutely dates and Moravia even has absolutely dated deciduous trees (Musil 1999).

Davis (2000) recently explored the problem originally asked by Hesselman regarding the source area of pollen, i.e. How could pollen from a few trees in an unusual local habitat be distinguished from distant forests? This problem is at the heart of any reconstruction of the vegetation of the Late Pleistocene (including that of OIS-3).

Southern refugia: The provinciality analysis demonstrated that for the period between 60 and 20 thousand calendrical years BP there is no evidence that the southern European peninsulas such as Iberia, Istria and the southern Balkans formed exclusive refugia. None of the animals that might be expected to retreat into southern refugia as the cold advanced as described by Hewitt (1996; 1999; 2000) are seen to do so during OIS-3. In fact they are all found in more northern areas. The conclusions from recent genetic work that these peninsular areas formed areas for endemism rather than refugia (Bilton *et al.* 1998) is thus confirmed as the only animals exclusively in these peninsulae are the few peninsular endemics such as voles *Microtus* (*Terricola*) *savii* from the Istrian peninsula, and *Microtus* (*Terricola*) *duodecimcostatus* and *M.* (*Terricola*) *brecciensi* / *cabrerae* from the Iberian peninsula and the Iberian lynx *Lynx pardina*.

Acknowledgements

J.S. would like to acknowledge the Leverhulme Trust grant that funded his contribution to this paper. Tjeerd van Andel is thanked for the providing the fertile environment of the Stage Three Project which gave rise to this contribution. William Davies is thanked for helping collect the dates as is the McDonald Institute which partially funded him. Bernie Weninger and Olaf Jöris are thanked for providing their Calpal program which calibrated the ^{14}C dates and Bernie Weninger for advising on the time slices used and illustrated in Table 7. 1. The following are thanked for numerous helpful discussions: T. van Andel, W. Davies, P. Gollop, C. Gamble, K. Tzedakis, A. Lister, A. Stuart, S. Parfitt, M. Collard, J.-L. de Beaulieu, J. Allen and K. van Huisteden. Finally, Alexis Willett is thanked for numerous helpful discussions regarding the management and handling of the data base.

Notes

1. Stewart, J.R., M. van Kolfschoten, A. Markova & R. Musil, 2001. *Stage Three Project Mammalian Data Base.* http://www.esc.esc.cam.ac.uk/oistage3/Details/Homepage.html.
2. Davies, S.W.G., 2001. *Stage Three Project Archaeological Data Base.* http://www.esc.esc.cam.ac.uk/oistage3/Details/Homepage.html.

References

Altuna, J. & K. Mariezkurrena, 1988. Les macromammifières du Paléolithique moyen et supérieur ancien dans la région cantabrique. *Archaeozoologia* 12, 179–96.

Arnold, N., T.H. van Andel & V. Valen, 2002. Extent and dynamics of the Scandinavian ice sheet during Oxygen Isotope Stage 3 (65–25 ka BP). *Quaternary Research* 57, 38–48.

Barron, E.J. & D. Pollard, 2002. High-resolution climate simulations of Oxygen Isotope Stage 3 in Europe. *Quaternary Research* 58, 296–309.

Baryshnikov, G.F. & A.K. Markova, 1992. Main mammal assemblages between 24,000 and 12,000 yr BP, in *Atlas of Paleoclimates and Paleoenvironments of the Northern Hemisphere (Late Pleistocene–Holocene)*, eds. B. Frenzel, M. Pechi & A. Velichko. Budapest, 127–9.

Batyrov, B.H. & I.E. Kuzmina, 1991. The Pleistocene ass *Equus hydruntinus* regalia in Eurasia. Paleontological studies of the fauna of the SSSR. *Trudy Zoologicheskogo Instituta RAN, St-Peterburg* 238, 121–38. [In Russian.]

Bilton, D.T., P.M. Mirol, S. Mascheretti, K. Fredga, J. Zima & J.B. Searle, 1998. Mediterranean Europe as an area of endemism for small mammals rather than a source for northwards postglacial colonisation. *Proceedings of the Royal Society London B* 265, 1219–26.

Blanc, A.C., 1953. *Excursion au Mont Circe.* (IV Congres International Roma - Pisa 1953.) INQUA.

Colinvaux, P.A., 1986. Plain thinking on Bering Land Bridge vegetation and mammoth populations. *Quarterly Review of Archaeology* 7, 8–9.

Currant, A.P. & R. Jacobi, 1997. Vertebrate faunas of the British Late Pleistocene and chronology of human settlement. *Quaternary Newsletter* 82, 1–8.

Currant, A.P. & R. Jacobi, 2001. A formal mammalian biostratigraphy of the Late Pleistocene of Britain. *Quaternary Science Reviews* 20, 1707–16.

Dansgaard, W., S.J. Johnsen, H.B. Clausen, D. Dahl-Jensen, N.S. Gundestrup, C.U. Hammer, C.S. Hvidberg, J.P. Steffensen, A.E. Sveinbjörnsdottir, J. Jouzel & G. Bond, 1993. Evidence for general instability of past climate from a 250-kyr ice-core. *Nature* 364, 218–20.

Davies, S.W.G., 2001. A very model of a modern human industry: new perspectives on the origin and spread of the Aurignacian in Europe. *Proceedings of the Prehistoric Society* 67, 195–217.

Davies, W. & P. Gollop, 2003. The human presence in Europe during the Last Glacial Period II: climate tolerance and climate preferences of mid- and late glacial hominids, in *Neanderthals and Modern Humans in the European Landscape during the Last Glaciation*, Chapter 8, eds. T.H. van Andel & W. Davies. (McDonald Institute Monographs.) Cambridge: McDonald Institute for Archaeological Research, 131–46.

Davies, S.W.G, J.R. Stewart & T.H. van Andel, 2000. Neanderthal landscapes: a promise, in *Neanderthals on the Edge,* eds. C.B. Stringer, R.N.E. Barton & J.C. Finlayson. Oxford: Oxbow Books, 1–8.

Davies, S.W.G., J.R. Stewart & T.H. van Andel, in press. *An Overview of the Results of the Stage Three Project: Hominids in the Landscape. CALPE 2001.* Chicago (IL): Chicago University Press.

Davis, M.B., 2000. Palynology after Y2K: understanding the source area of pollen in sediments. *Annual Review of Earth and Planetary Science* 28, 1–18.

Delpech, F., 1975. Les Faunes du Paleolithique Superieur dans le Sud-Ouest de la France. Unpublished thesis, Universite de Bordeaux I.

Faunmap Working Group, 1994. *Faunmap: a Data Base Documenting Late Pleistocene Distributions of Mammal Species in the United States.* Springfield (IL): Illinois State Museum.

Faunmap Working Group, 1996. Spatial response of mammals to Late Quaternary environmental fluctuations. *Science* 272, 1601–6.

Finlayson, C., R.N.E. Barton, F. Giles Pacheco, G. Finlayson,

D.A. Fa, A.P. Currant & C.B. Stringer, 2000. Human occupation of Gibraltar during Oxygen Isotope Stages 2 and 3 and a comment on the late survival of Neanderthals in Southern Iberian Peninsula, in *Paleolitico da Peninsula Iberica.* Actas do 3.°Congresso de Arqueologia Peninsular, vol 2. Porto, ADECAP 2000, 277–86.

Flint, V.E., U.D. Chugunov & V.M. Smirin, 1970. *Mammals of the Soviet Union.* Moscow: Mysl Press. [In Russian.]

Gamble, C., 1986. *The Palaeolithic Settlement of Europe.* Cambridge: Cambridge University Press.

Gromov, I.M. & M.A. Erbaeva, 1995. The mammals of Russia and adjacent territories, in *Lagomorphs and Rodents,* vol. 167, eds. A.A. Aristov & G.I. Baranova. St Petersburg: Zoological Institute RAS, 1–520. [In Russian.]

Guthrie, R.D., 1990. *Frozen Fauna of the Mammoth Steppe: the Story of Blue Babe.* Chicago (IL): University of Chicago Press.

Helmens, K.F., M.E. Räsänen, P.W. Johansson, H. Jungner, & K. Korjonen, 2000. The last interglacial–glacial cycle in NE Fennoscandia: a nearly continuous record from Sokli (Finnish Lapland). *Quaternary Science Reviews* 19, 1605–24.

Hewitt, G., 1996. Some genetic consequences of ice ages, and their role in divergence and speciation. *Biological Journal of the Linnean Society* 58, 247–76.

Hewitt, G., 1999. Post-glacial re-colonisation of European biota. *Biological Journal of the Linnean Society* 68, 87–112.

Hewitt, G., 2000. The genetic legacy of the Quaternary ice ages. *Nature* 405, 907–13.

Huntley, B., M. J. Alfano, J. Allen, D. Pollard, P.C. Tzedakis, J.-L. de Beaulieu, E. Grüger, & W. Watts, 2003. European vegetation during Marine Oxygen Isotope Stage 3. *Quaternary Research* 59, 195–212.

Jequier, J.-P., 1975. *Le Mousterien Alpin: revision critique.* Yverdon: Institut d'Archeologie Yverdonnoise.

Jöris, O. & B. Weninger, 1996. Calendric age-conversion of glacial radiocarbon data at the transition from the Middle to Upper Palaeolithic in Europe. *Bulletin de la Sociétée Préhisorique Luxembourgoise* 18, 43–55.

Larsen, E., S. Gulliksen, S.-E. Lauritzen, R. Lie, R. Lövlie, & J. Mangerud, 1987. Cave stratigraphy in western Norway; multiple Weichselian glaciations and interstadial vertebrate fauna. *Boreas* 16, 267–92.

Lityńska-Zajac, M., 1995. Anthracological analysis, in *Complex of Upper Palaeolithic Sites near Moravany.* Krakow: Jagellonian University Press, 74–9.

Macdonald, D. & P. Barrett, 1993. *Mammals of Britain and Europe.* London: Collins.

Markova, A.K., 1982. Fauna of mammals of the Late Valdai time, in *Paleogeography of Europe during the Last 100,000 Years: Atlas-monography,* eds. I.P Gerasimov & A.A. Velichko. Moscow: Nauka Press, 51–62. [In Russian.]

Markova, A.K., 1992. Influence of paleoclimatic changes in the Middle and Late Pleistocene on the composition of small mammal faunas: data from Eastern

Europe — Mammalian migration and dispersal events in the European Quaternary. *Courier Forschung-Institut Senckenberg* 153, 93–100.

Markova, A.K., N.G. Smirnov, A.V. Kozharinov, N.E. Kazantseva, A.N. Simakova, & L.M. Kitaev, 1995. Late Pleistocene distribution and diversity of mammals in northern Eurasia. *Paleontologia I Evolucio* 28–29, 5–143.

Markova, A.K., A.N. Simakova, A.Y. Puzachenko & L.M. Kitaev, 2002. Environments of the Russian Plain during the Middle Valdai Briansk Interstade (33,000–24,000 yr BP) indicated by fossil mammals and plants. *Quaternary Research* 57, 391–400.

Mitchell-Jones, A.J., G. Armori, W. Bogdanowicz, B. Kryštufek, P.J.H. Reijnders, F. Spitzenberger, M. Stubbe, J.B.M. Thissen, V. Vohralīik, & J. Zima, 1999. *The Atlas of European Mammals.* London: The Academic Press.

Musil R., 1955. Osteologický materiál z paleolitického sídliště v Pavlově [Das osteologische Material aus der paläolith. Siedlungssstätte in Pollau]. *Acta Academiae Scientiarum Čechoslovenicae*, basis brunensis 27, spis 318, seš. 6, 279–320.

Musil, R., 1959a. Osteologický materiál z paleolitického sídliště v Pavlově. Část II. (Das osteologische Material aus der paläolithischen Siedlungsstätte in Pavlov. II. Teil). *Anthropozoikum* 8 (1958), 83–106.

Musil, R., 1959b. Poznámky k paleontologickému materiálu z Dolních Věstonic [Bemerkungen zum paläontologischen Material aus Dolní Věstonice (Unterwisternitz]. *Anthropozoikum* 8 (1958), 73–82.

Musil, R., 1986. Paleobiogeography of terrestrial communities in Europe during the Last Glacial. *Acta Musei Nationalis Pragae* XLI B, 1–2, 1–83.

Musil, R., 1994a. Hunting game of the culture layer of Pavlov, in *Pavlov I: Excavations 1952–53*, eds. B. Klima *et al.* (The Dolní Věstonice Studies 2.) Liège: Études et Recherches Archéologiques de l'Université de Liège (ERAUL), 169–96.

Musil, R., 1994b. The knowledge of the Pleistocene: an assumption for the differentiation of natural regularities and human interventions. *GeoLines* 1, 25–6.

Musil, R., 1997. Hunting game analysis. *The Dolní Věstonice Studies, Pavlov - Northwest* 4, 443–68.

Musil, R., 1999. Životní prostředí v posledním glaciálu [The environment in the Last Glacial on the territory of Moravia]. *Acta Musei Moraviae, Scientia Geologica* 84, 161–86.

Musil, R., 2001. Natural environment. *Anthropologie* 38, 327–31.

Musil, R., 2003. The Middle and Upper Palaeolithic game suite in central and southeastern Europe, in *Neanderthals and Modern Humans in the European Landscape during the Last Glaciation*, Chapter 10, eds. T.H. van Andel & W. Davies. (McDonald Institute Monographs.) Cambridge: McDonald Institute for Archaeological Research, 167–90.

Nadachowski, A., 1982. *Late Quaternary Rodents of Poland, with Special Reference to Morphotype Dentition Analysis of Voles.* Warsaw / Krakow: Polish Academy of Sciences.

Panteleev, P.A., A.N. Terekhina & A.A. Varshavski, 1990. *Ecogeographic Variations of the Rodent.* Moscow: Nauka Press. [In Russian.]

Perez Ripoll, M., 1977. *Los Mamiferos del Yacimiento Musteriense de Cova Negra (Jativa, Valencia).* (Servicio de investigcion Prehistorica, sere de trabajos Varios 3.) Valencia: Servicio de investigcion Prehistorica.

Pollard, D. & E.J. Barron, 2003. Causes of model-data discrepancies in European climate during Oxygen Isotope Stage 3 with insights from the Last Glacial Maximum. *Quaternary Research* 59, 108–13.

Ritchie, J.C., 1984. *Past and Present Vegetation of Far Northwest of Canada.* Toronto: University of Toronto Press.

Ritchie, J.C. & L.C. Cwynar, 1982. The Late Quaternary vegetation of the northern Yukon, in *Paleoecology of Beringia*, eds. D.M. Hopkins *et al.* New York (NY): Academic Press, 113–26.

Shackleton, N.J. & N.D. Opdyke, 1973. Oxygen isotope and palaeomagnetic stratigraphy of equatorial Pacific core V28–238: oxygen isotope temperatures and ice volumes on a 10^5 and 10^6 year scale. *Quaternary Research* 3, 39–55.

Stafford, T.M., H.A. Semken, R.W. Graham, W.F. Klipel, A. Markova, N. Smirnov & J. Southon, 1999. First accelerator mass spectrometry ^{14}C dates documenting contemporaneity of nonanalogue species in late Pleistocene mammal communities. *Geology* 27, 903–6.

Stewart, J.R., in press. The fate of the Neanderthals: a special case or simply part of the broader Late Pleistocene megafaunal extinctions? *Proceedings of the XIVth Congres of the l'UISPP, Liege.*

Stewart, J.R. & A.M. Lister, 2001. Cryptic northen refugia and the origins of modern biota. *Trends in Ecology and Evolution* 16, 608–13.

Stewart, J.R., T. van Kolfschoten, A. Markova & R. Musil, 2003. Neanderthals as part of the broader Late Pleistocene megafaunal extinctions?, in *Neanderthals and Modern Humans in the European Landscape during the Last Glaciation*, Chapter 12, eds. T.H. van Andel & W. Davies. (McDonald Institute Monographs.) Cambridge: McDonald Institute for Archaeological Research, 221–32.

Stuart, A.J., 1977. The vertebrates of the last cold stage in Britain and Ireland. *Philosophical Transactions of the Royal Society of London B* 280, 295–312.

Stuart, A.J., 1982. *Pleistocene Vertebrates in the British Isles.* London: Longman.

Stuart, A.J., 1991. Mammalian extinction in the Late Pleistocene of northern Eurasia and North America. *Biological Review* 66, 453–562.

Stuart, A.J. & L.H. van Wijngaarden-Bakker, 1985. Quaternary vertebrates, in *The Quaternary of Ireland*, eds. K.J. Edwards & W.P. Warren. London: Academic Press.

Ukkonen, P., J.P. Lunkka, H. Junger & J. Donner, 1999. New radiocarbon dates on Finnish mammoths indicate large ice-free area in Fennoscandia during the Middle Weichselian. *Journal of Quaternary Science* 14, 711–14.

Valen, V., E. Larsen, J. Mangerud & A.K. Hufthammer, 1996. Sedimentology and stratigraphy in the cave Hamnsundhelleren, western Norway. *Journal of Quaternary Science* 11, 185–201.

van Andel, T., 2002. The climate and landscape of the Middle Part of the Weichselian Glaciation in Europe: the Stage 3 Project. *Quaternary Research* 57, 2–8.

Vereshagin, N.K. & G.F. Baryshnikov, 1980. Paleoecology of the late Mammoth complex at Arctic zone of Eurasia. *Bulletin Moskovskogo Obshchestva Ispytatelei Prirody, biol. sec.* 85/2, 5–19. [In Russian.]

Willis, K.J., E. Rudner, & P. Sümegi, 2000. The full-glacial forests of central and southeastern Europe. *Quaternary Research* 53, 203–13.

Zilhão, J., 1993. Le passage du Paléolithique Moyen au Paléolithique Supérieur dans le Portugal, in *El Origen del Hombre Moderno en el Suroeste de Europa*, ed. V. Carbrera. Madrid: Universidad Nacional de Educación a Distancia, 127–45.

Zilhão, J., 2000. The Ebro frontier: a model for the late extinction of Iberian Neanderthals, in *Neanderthals on the Edge*, eds. C.B. Stringer, R.N.E. Barton & J.C. Finlayson. Oxford: Oxbow Books, 111–21.

Chapter 8

The Human Presence in Europe during the Last Glacial Period II: Climate Tolerance and Climate Preferences of Mid- and Late Glacial Hominids

William Davies & Piers Gollop

Climate and the human colonization of glacial Europe

Aims and resources of this study

During the Weichselian glaciation, Anatomically Modern Humans (AMH) entered a Europe which had until then been inhabited solely by Neanderthals, whose demise 10,000 years later left them as the sole hominid species. Explanations for this major event in human history range from genocide by modern humans to a lack of cognitive ability among the Neanderthals (Golding 1955; Mellars *et al.* 1999, 349; etc.), but the potential role of Weichselian climate changes has never been thoroughly examined. Was it a matter of differences in physiological adaptation between the two human species? Or was the role of climate changes indirect, acting through an intermediary such as the limits of the adaptability of faunal and floral food resources (Chapter 6: Huntley & Allen 2003; Huntley *et al.* 2003; Chapter 7: Stewart *et al.* 2003; Chapter 10: Musil 2003)? In Chapter 4 (van Andel *et al.* 2003b) we have examined the changing distribution of Neanderthal and modern humans across Europe, in the setting of a changing climate that, from relatively mild some 60,000 years ago, gradually deteriorated towards the Last Glacial Maximum.

In Chapter 4 (van Andel *et al.* 2003b) we raised the question of whether the wanderings of the Neanderthals and modern humans across Europe were responses to those large-scale regional climate changes. Below we further examine that question using simulations of the coldest and warmest end members of the mid-glacial climate (Chapter 5: Barron *et al.* 2003; Barron & Pollard 2002; Pollard & Barron 2003).

The study draws on two main sources: 1) an analysis of human distributions in time and space, based on archaeological dates obtained from the Stage 3 project's chrono-archaeological data base and used as proxies for human presence (Chapter 4: van Andel *et al.* 2003b); and 2) high-resolution simulations of the main states (end members) of the glacial climate in Europe between 70 and 20 ka BP.[1] We use the Greenland GISP2 ice-core record, divided into temporal climate phases (Fig. 8.1; Table 8.1), as the standard to compare human dispersal patterns with the conditions of the 'Stable Warm', 'Transitional', 'Early Cold' and 'Last Glacial Maximum Cold' Phases.[2]

The neologism 'isozone' is used extensively throughout this paper. We use it to mean the surface area covered between two adjoining isopleth values on the mapped Pennsylvania State University simulations. Each different colour on an output map is taken as a separate isozone, e.g. the yellow (between 12°C and 18°C) isozone in Figure 8.2.

Climate changes of the last glaciation

About 115 ka ago, the penultimate interglaciation ended with a long, slow cooling that led to the Weichselian Glaciation, but no direct evidence exists for a sizeable Scandinavian ice sheet until about 65 ka BP,[3] during OIS-4. That first ice sheet lasted a mere few thousand years, and may not even have reached the southern and western shores of the Baltic (Donner 1995, 66–7). At the start of OIS-3, about 59 ka BP, sudden substantial warming initiated a series of long relatively warm events interrupted only by a few brief cold spells (Fig. 8.1). The early warm period lasted until about 43 ka BP, when the climate began to deteriorate. Cold events now came in closer sets, and the last warm event, shorter and cooler than its predecessors, ended 37 ka BP. By 30 ka BP, the lowest temperatures equalled or even exceeded those of the Last Glacial Maximum (OIS-2). The latter, however, started 5000 years later, when a major growth of the

131

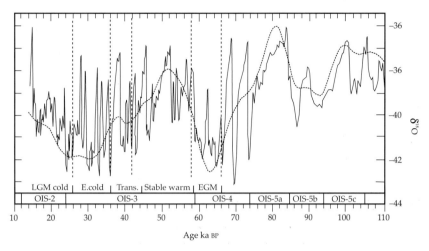

Figure 8.1. *Second- (labelled) and third-order (Dansgaard/Oeschger) climate changes for the period 110–10 ka BP from the GISP2 ice core, Greenland (Meese et al. 1997; Johnsen et al. 2001; Stuiver & Grootes 2000). The broken heavy line depicts the general trend of climate changes for the interval (from Chapter 4: van Andel et al. 2003b, Fig. 4.1). Climate Phases: Last Glacial Maximum Cold Phase (LGM Cold); Early Cold Phase (E. Cold); Transitional Phase (Trans.); Stable Warm Phase (Stable warm); Early Glacial Maximum (EGM).*

Fennoscandian ice sheet began. For details of the climate history itself, see Chapter 2 (van Andel 2003).

Simulating the late Quaternary climate of Europe
Even supercomputers lack the capacity to display at high resolution the millennial-scale climate changes typical of OIS-3. Therefore, the Stage 3 models were limited to 'snapshots' intended as typical for the end members of the glacial climate range. An extensive set of modern climate data was used to evaluate the modelling programme itself and to provide a comparison with the rigours of the climates of the last glacial period (Chapter 5: Barron & Pollard 2002; Pollard & Barron 2003). The results give us reasonable confidence in the use of high-resolution, nested global and regional circulation models.

When evaluating hypotheses that changing mid- and late glacial human settlement patterns in Europe were partly or mainly responses to climatic forcing, contrasts between ranges of climate variables are often more informative than absolute values. Among the about 50 variables of the Stage 3 simulation set (Chapter 5: Barron *et al.* 2003, Appendix) that may be directly or indirectly relevant to the human case, we believe the following to have had the largest impact: summer and winter monthly mean temperatures, day/night temperature contrasts, wind-chill, precipitation, including thickness of snow-cover and annual snow days, as well as the seasonal ranges of those variables.

The European climate today
The European climate is dominated by two major gradients, a strong N–S gradient from arctic Scandinavia to the subtropical Mediterranean (Fig. 8.2: top), and an equally strong transition from the maritime Atlantic coasts to the continental conditions in the east. The N–S gradient is disrupted by the high W–E mountain barrier of the Pyrenees, Alps and Carpathians, which separates the temperate European zone from the hot, summer-dry Mediterranean zone. Farther east, beyond the 25th meridian, the subarctic climate zone passes more gradually into the subtropical Mediterranean.

The transition from the maritime Atlantic region to continental eastern Europe has temperature contrasts as stark as those between the Mediterranean and Arctic regions; note, for instance the 15°C northward shift of the summer-warm zone east of the 15°E meridian (see also Chapter 5: Barron *et al.* 2003, Figs. 5.11–5.16). The eastward increasing continentality transforms the mild, moist winters and cool summers of the west coast into the harsh winters and hot summers of Russia. Striking regional contrasts in the amounts of precipitation add further complexity to the European climate (Chapter 5: Barron *et al.* 2003).

The climate of the last glaciation
The climate pattern of the Stage 3 Stable Warm Phase, about 59–42 ka BP (Fig. 8.2: centre), resembles that of the present, except for being 4°–6°C cooler than the modern state at each latitude. Mean summer temperatures above 18°C occurred only in southern Spain and the western Mediterranean, but in the east, well beyond 20°E, they reached as far north as the 50°–55°N parallel in Scandinavia (only in western Russia did they extend below the 50°N one).

During the Last Glacial Maximum, conditions were more extreme (Fig. 8.2: bottom). Summer days above 18°C existed only in the southern Mediterranean and the extreme southeast, south of 45°N. In western Europe, the summer rarely reached 10°C, while in winter only the Atlantic shores of southern France, Iberia and the Mediterranean coasts had 'mild' average temperatures of 4°–6°C. Elsewhere,

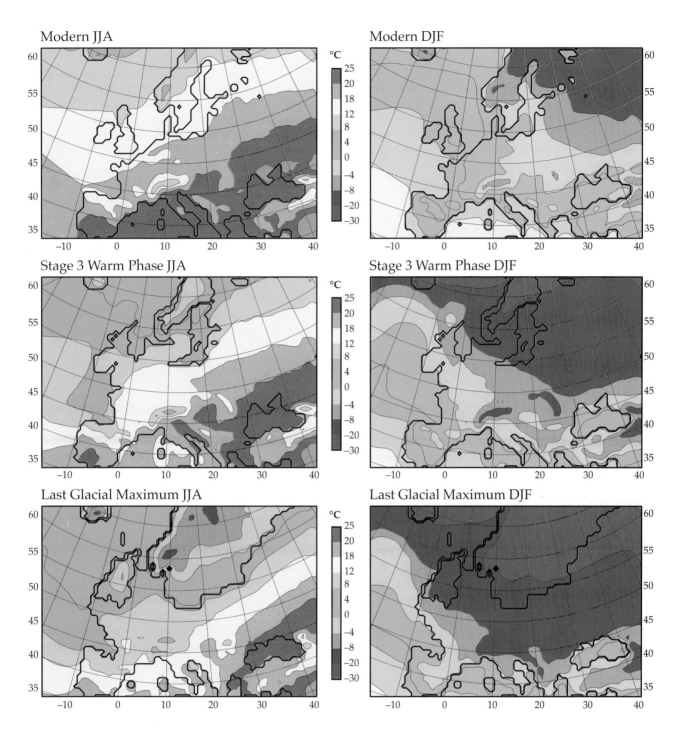

Figure 8.2. *Simulated patterns of late Quaternary mean temperatures in Europe for summer (JJA: left) and winter (DJF: right). Central isotherm scale is in °C. Top: Modern; heavy black line indicates today's coastline. Middle: Warm event typical for the interval 60–40 ka BP; heavy black line shows the mid-glacial coastline at −70 m with the shores of the Baltic and Black Seas and a small Scandinavian ice sheet. Bottom: Last Glacial Maximum (LGM) c. 25– 16 ka BP; heavy black lines are the sea-coasts at lowest sea-level (−120 m), the shores of the Black Sea and the margins of the Fennoscandian ice-sheet.*

mean winter temperatures were well below –8°C to –20°C near the ice sheet, and –30°C on it. Because the complex relief of Europe imprints a strong orographic control on precipitation, regional patterns are complex. The low temperature of oceanic surface waters reduced precipitation in glacial times, and in many areas created arid conditions. The Mediterranean, for example, was winter-dry, instead of winter-wet as it is today.

Questions and procedures

Questions raised in Chapter 4

Matching the patterns of the Neanderthal and modern human (AMH) presence in Europe (Chapter 4: van Andel *et al.* 2003b) against simulations for end members of the middle and late glacial climate change (Chapter 5: Barron *et al.* 2003) allows us to address a series of questions raised at the end of Chapter 4 (van Andel *et al.* 2003b).

A. Neanderthal and AMH adaptations to the glacial climate: similar or dissimilar?

- Question A-1. Is there a relationship between Neanderthal settlement and migration patterns and climate changes?
- Question A-2. Is there a relationship between the immigration and settlement choices of early modern humans and climate changes?
- Question A-3. Are the responses to climate changes of Neanderthals and early modern humans similar?

B. Impact of the glacial climate on Neanderthals and AMH: direct or indirect?

- Question B-1. Are observed similarities between climate changes and human settlement and migration due to a direct impact of climate on humans?
- Question B-2. Or did the changing climate control the resources on which humans depended for survival?

C. Possible causes of Neanderthal extinction and AMH survival

- Question C-1. Did the Neanderthals become extinct because of their inability to adjust to an arctic mode of living, such as seasonal migrations to hunt migratory arctic fauna?
- Question C-2. Did some of the early modern humans (those who migrated south with the coming cold: Fig. 4.6b) suffer from the same inability, and therefore also become extinct?

To address the A and B sets of questions we have available the spatio-temporal distribution of three main pan-European Palaeolithic technocomplexes between 60 and 20 ka BP (Chapter 4: van Andel *et al.* 2003b), the Mousterian serving as proxy for the Neanderthals and the Aurignacian and Gravettian as proxies for AMH. Because the temporal resolution of our archaeological data base is limited to about ±1000 to ±1500 calendar years (Chapter 3: van Andel *et al.* 2003a), we must assess hominid activities in relation to broad climatic phases rather than to millennial scale D/O events (Fig. 8.1). Using a regional perspective scaled to the 60 × 60 km pixels of the output grid of the climate simulations (Chapter 5: Barron *et al.* 2003), we can identify and compare the climatic tolerance ranges and preferences of the Neanderthals and AMH.

- 'Tolerance' refers to the *ranges* of climate parameters thought to be important for human behaviour which were apparently accepted by Stage 3 hominids, such as temperature, precipitation, wind-chill, snow-depths and annual days of snow-cover.
- Conversely, 'preference' implies the apparently *active* selection of a particular combination of climate conditions by hominids, as demonstrated by the relative frequencies of sites within given climate isozones.

Some ambiguity is attached to the preference term, because a set of sites indicating strong preference for extremely cold conditions, as for instance is the case for the Gravettian, may merely represent the aestival occupation of people preferring the benefits of the local summer, but who lived elsewhere in winter. Since our data do not include much information on seasonal or year-round occupation, a rigorous treatment of this issue is not yet possible.

Humans today find ways to withstand large climatic extremes as long as the resources (raw materials, food) are worth exploiting, and we see no reason why Stage 3 humans were any different. Therefore, resources rather than climate may have directly determined the presence of humans in certain regions, while areas elsewhere with similar climates but dissimilar resources had no hominid presence.

Spatio-temporal human site patterns and selected climate variables

The Stage 3 Project has attempted a series of end member simulations of the climate of the last glacial period. The '42 ka warm', '30 ka cold' and '21 ka cold' simulations (Barron & Pollard 2002; Pollard & Barron 2003) were designed to depict *typical* warm and *typical* cold climate events for the time between 60 ka BP and *c.* 20 ka BP (van Andel 2002). For that period, they represent the end members of the climatic range, and we shall apply them here singly and in combination to represent the different climate phases. Characterizations of the latter can then be used to evaluate human site spatio-temporal distributions in a climatic context (Table 8.1; Fig. 8.1).

Unfortunately, no satisfactory simulations of the '30 ka cold' end member were obtained, because, notwithstanding repeated adjustments, the simulations stubbornly continued to be too warm to be acceptable. This left us with one warm and one cold climate end member, and a gap for the 'Transitional climate phase' between 37 and 27 ka BP. Having no choice, we accepted the '21 ka' simulation as representative for the entire cold interval from about 37 to 27 ka BP. Fortunately, the long-term temperature trend of the Greenland ice-core record displays little change between 37 and 27 ka BP (Fig. 8.1; heavy dashed line). Thus we dealt with the inevitable by plotting the frequency diagrams for the Transitional phase (44–37 ka BP) separately against both cold and warm end members, to get a sense of the possible climatic extremes for this period.

The Stage 3 climate simulations are displayed on maps of 83 longitudinal by 56 latitudinal grid points spaced 60 km apart, which cover a mapped area of 16,732,800 km². During the Stable Warm Phase (59–44 ka BP), 6,552,000 km² (39 per cent) of that area was covered by sea and 2 per cent by ice (417,000 km²), leaving 9,763,200 km² suited in principle for human occupation. During the Last Glacial Maximum Cold Phase, the Scandinavian ice sheets expanded at the expense of both sea (5,554,000 km²) and land (7,948,800 km²) to occupy *c.* 10 per cent of the mapped area (3,240,000 km²). Such figures lend substance to the image of a huge land surface available to a relatively small hominid population during warm and cold conditions.

The dated archaeological assemblages selected from the Project data base (Chapters 3 & 4: van Andel *et al.* 2003a,b) have been divided into four temporal units, corresponding approximately to the climate phases of Table 8.1 and Figure 8.1, and plotted on transparent acetate sheets which matched the scale and projection (Lambert conformal) of the climate simulation maps. To relate the spatio-temporal patterns of archaeological sites to the selected climatic parameters, in this case mean monthly summer and winter temperatures, wind-chill, and snow days/

Table 8.1. *Second order climate phases of the Weichselian Glaciation between OIS-5a and OIS-1, based on the GISP2 Greenland ice core and used to place middle and late Palaeolithic human history in a climatic context.*

SPECMAP	Climate phase	Age ka BP
OIS-5a	Early Glacial Warm Phase	>74
OIS-4	Transitional Phase	74–66
OIS-4	First Glacial Maximum	66–60
OIS-3	Stable Warm Phase	59–44
OIS-3	Transitional Phase	44–37
OIS-3	Early Cold Phase	37–27
OIS-2	Last Glacial Maximum	27–16

Table 8.2. *Numerical scales for the base lines of Figures 8.5–8.10. Top: monthly means.*

Mean winter T(°C)		Wind-chill T(°C)		Mean summer T(°C)	
a.	–20 to –8	h.	–34 to –29	n.	–4 to 0
b.	–8 to –4	i.	–29 to –23	o.	0 to –4
c.	–4 to 0	j.	–23 to –18	p.	4–8
d.	0–4	k.	–18 to –12	q.	8–12
e.	4–8	l.	–12 to –7	r.	12–16
f.	8–12	m.	–7 to 16	s.	16–20
g.	12–16			t.	20–24

Snow conditions

Cover (days/year)		Depth (cm)	
a.	<10	l.	0–1
b.	10–30	m.	1–5
c.	30–60	n.	5–10
d.	60–90	o.	10–20
e.	90–120	p.	20–50
f.	120–150	q.	50–70
g.	150–180	r.	70–100
h.	180–210	s.	100–150
i.	210–240	t.	150–200
j.	240–270		
k.	270–300		

year and depth (Fig. 8.3), all site maps were overlain on climate simulation maps (Fig. 8.4). For each technocomplex and each climate phase, all sites within each pair of isopleths (*isozones*) were counted and the site counts in each isozone plotted as frequency histograms (Figs. 8.5–8.10). These histograms form the basis for the analysis discussed below.

Hominids and the glacial climate

The histograms provide clear measures of the interaction between humans and some of the climate variation. Tolerance for climate variables is measured as the spread of occupation sites relative to the full scale of the variables (Figs. 8.5–8.10), and expressed as the scale distance between the most distant outlying sites. Preference is defined by the scale width of

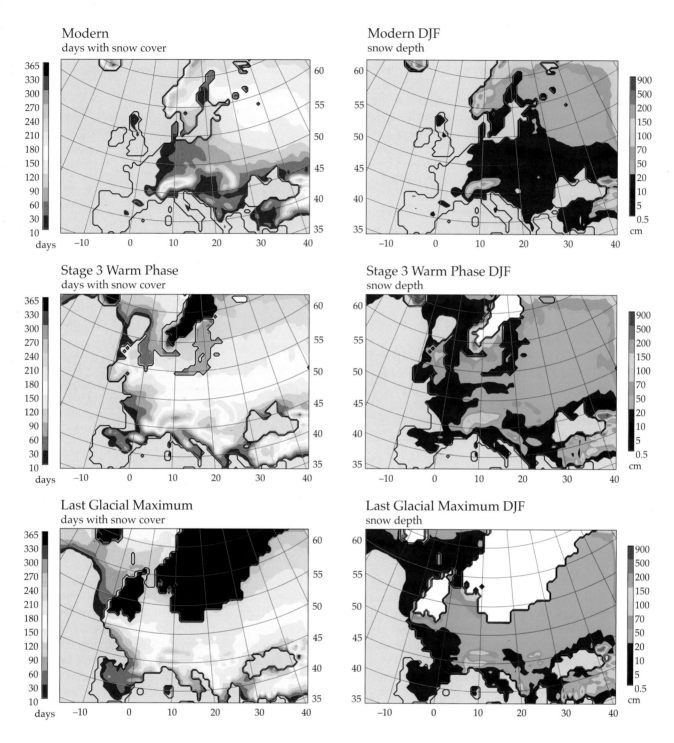

Figure 8.3. *Simulated patterns of late Quaternary days with snow cover (left and snow depth (right) in Europe for winter (DJF). Left scale is in days/year. Right scale is in centimetres. Top: Modern; black line indicates today's coastline. Middle: Warm event typical for the interval 60–40 ka BP; heavy black line shows the mid-glacial coastline at –70 m with the shores of the Baltic and Black seas and a small Fennoscandian ice sheet. Bottom: Last Glacial Maximum c. 25–17 ka BP; black line shows the sea-coasts at lowest sea-level (–120 m), the shores of the Black Sea and the margins of the large Fennoscandian ice sheet.*

any clusters of sites. For details of the base scales in Figures 8.5 to 8.10, see Table 8.2.

Human tolerance of a range of climates

The earliest Neanderthal sites we are concerned with here date to the later OIS-4 period from 70 to 60 ka BP. They are too small in number (12) to be of much value, but it is worth noting that even at that early date the responses to all five climate variables show a strong resemblance to the patterns of the later time-slices (Table 8.3). Curiously, the tolerance for wind-chill factors throughout the entire time from 60 to 37 to 27 ka BP is strikingly greater than that for the mean air temperatures.

Temperature and snow tolerances of the modern humans, early and late, are quite like those of the Neanderthals, showing little variation over the entire studied interval. All three groups display a remarkable tolerance for snow-cover, lasting up to seven, or a little more than seven, months per year.

Points of note

- The temperature, wind-chill and snow tolerance ranges of all three technocomplexes are very similar (Table 8.3) and consistently cover the entire or almost entire range of the climate isozones in winter and summer (Figs. 8.5–8.10).
- This supports the hypothesis that the Neanderthals and the incoming modern humans were not limited by any inherent inability of coping with even severe cold conditions.

Human preferences with regard to temperature

NOTE: For regional maps of the simulated summer and winter temperatures, snow days/year and mean snow depth, see Figures 8.2 and 8.3.

The Neanderthals: Mousterian technocomplex

The few sites dating from 60 ka BP and before have winter temperatures from –8°C to 12°C, with a marked preference for 0°C or values just below and wind-chill values warmer than –23°C. In summer, temperatures between 16°C and 25°C were preferred.

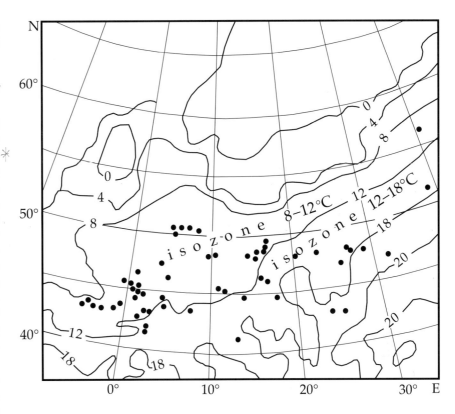

Figure 8.4. *Counting sites by 'isozones' bound by pairs of isopleths.*

This pattern persists during the Stable Warm Phase (Fig. 8.5: 59–44 ka BP), when the majority of the sites were located in the –8°C to 8°C isozones; in winter, temperatures below –8°C were shunned and those from –4°C to 4°C strongly preferred. Wind preferences are between –18°C and 7°C. Most sites are located in regions with mean summer temperatures ranging from 12°C to 25°C. In terms of temperature, site-isozone patterns before and after 60 ka BP are very similar, with a consistent, understandable preference for the warmer parts of Europe through the year. The preferred parts, limited before 60 ka to the Mediterranean, extended after that time to vast areas north of the Pyrenean-Alpine-Carpathian mountain barrier.

During the Transitional (44–37 ka BP) and Early Cold (37–27 ka BP) Phases, Mousterian sites declined in numbers from a peak of about thirty in the late Stable Warm Phase at 44 ka BP, to half a dozen at 30 ka BP (Chapter 4: van Andel *et al.* 2003b). Interestingly, and perhaps counter-intuitively, there appears for the first time a strong Mousterian presence in the –20°C to –8°C isozones during cold events in the Transitional Phase, but winter temperatures near zero are preferred in both cold and warm simulations. Throughout OIS-3, summer temperatures lay be-

Table 8.3. *Tolerance ranges for some principal parameters of the glacial climate of Neanderthals and anatomically modern humans. Based on archaeological site maps plotted on simulations of temperature, wind-chill and snow for cold and warm climate end members (Table 8.1). See Figures 8.5 to 8.10 for the ranges.*

Interval (ka)	Variable	Mousterian	Aurignacian	Gravettian
			Temperature	
60–44	winter	–8 to 12		
44–37(w)	T°C	–8 to 12	–8 to 8	
44–37(c)		–20 to 8	–20 to 8	
37–27		–20 to 12	–20 to 12	–20 to 8
37–20			–20 to 8	–20 to 8
60–44	windchill	–23 to 16		
44–37(w)	T°C	–23 to 16	–34 to 16	
44–37(c)		–23 to 16	–29 to 16	
37–27		–34 to 16	–29 to 16	–34 to 16
27–20			–29 to 16	–34 to 16
60–44	summer	8–25		
44–37(w)	T°C	8–25	8–25	
44–37(c)		4–25	4–20	
37–37		4–25	4–25	4–20
27–20			4–25	0–20
		Snow		
60–44	depth	0–50		
44–37(w)		0–50	0–50	
44–37(c)		0–70	0–60	
37–27		0–90	0–70	0–70
27–20			0–50	0–150
60–44	days/year	<10–180		
44–37(w)		<10–180	<10–180	
44–37(c)		<10–210	<10–240	
37–27		<10–290	<10–210	<10–210
27–20			<10–210	<10–210

Note: (w) and **(c)** denote processed with 42 ka warm and 21 ka cold event simulations, respectively.

across Europe at this time shows increasing isolation in western Europe (Chapter 4: van Andel *et al.* 2003b, Figs. 4.4c–d). This anomaly can be explained by the presence of scattered, mainly Croatian, southern German and Romanian dated sites in the sample. Wind-chill isozones from –23°C to –7°C appear to have been acceptable, but even during the final run-up to the glacial maximum, the preferred summer mean temperatures remained, as before, in the range from 8°C to about 16°C or 20°C.

Points of note
- In terms of temperature, site-isozone patterns both before and after 60 ka BP are similar, consistently preferring the warmer parts of Europe throughout the year. Limited before 60 ka BP to the Mediterranean, the preferred territory expanded across large parts of Europe during the Stable Warm Phase.
- The strong preference for the mildest winter temperatures implies that, in the main, areas were chosen that permitted permanent occupation.
- During the Transitional and Early Cold Phases, Mousterian sites show for the first time a strong presence in the –20°C to –8°C isozones when compared against the Cold Phase simulations. The emphasis upon near-zero winter temperatures in both cold and warm events of the Transitional Phase shifts strongly towards ones closer to –8°C, raising for the first time the question of permanent occupation of areas with such severe conditions.
- In summer, preferred mean temperatures remained, as before, in the range from 8°C to about 16°C or 20°C.

Modern humans: Aurignacian and Gravettian technocomplexes
Anatomically Modern Humans, represented by Aurignacian (Churchill & Smith 2000) and certain other Early Upper Palaeolithic technocomplexes, entered Europe in OIS-3 between 47 and 44 ka BP, just when, give or take a few millennia, the Stable Warm Phase began to make way for the Transitional Phase. 21 ka cold phase simulations are appropriate for most of the time from 44 ka BP onward.

The Aurignacian temporal record, in number of sites/unit time, shows a steady population increase until *c.* 31 ka BP, followed by a steep decline (Chapter 4: van Andel *et al.* 2003b, Fig. 4.5) until dwindling to insignificance early during the Last Glacial Maximum Cold Phase. The Gravettian technocomplex appears in the Early Cold Phase when

tween 8°C and 20°C (Fig. 8.5), and isozones colder than 8°C were empty, but interestingly, site numbers also declined towards the warmest isozones. Compared to the wind-chill levels of –18°C to 7°C preferred in the Stable Warm Phase and warm parts of the Transitional Phase, much colder wind-chill values appear to have become acceptable in cold events of the 44–37 ka BP time-slice (Fig. 8.5), although mid-level wind-chill values were more popular in terms of site numbers. Of course, areas with winter temperatures not modified by the wind may have been the norm for occupation.

Finally, during the Early Cold Phase (37–27 ka BP), Neanderthals' site patterns distinctly preferred the colder winter isozones between –20°C and –4°C over warmer ones: another apparently counter-intuitive behaviour, especially since their distribution

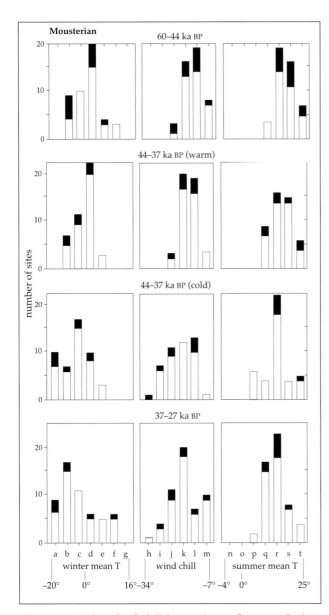

Figure 8.5. *Neanderthal (Mousterian technocomplex) mean temperature preferences for Stable warm, Transitional, and Early cold climate phases. White: sum of cave and abri sites. Black: open-air sites. For complete temperature calibration, labelled 'a, b, c, etc.' on the base scale of each column, see Table 8.2. For temperature tolerance, see Table 8.3.*

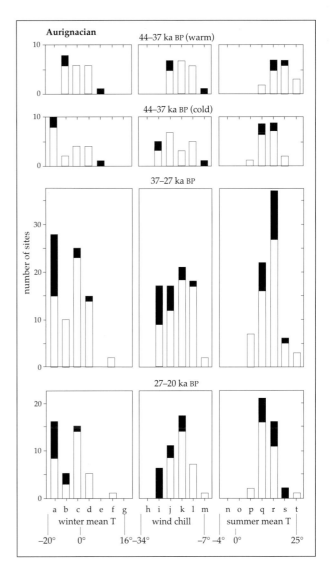

Figure 8.6. *Early modern human immigrants (Aurignacian and other Early Upper Palaeolithic techno-complexes) mean temperature preferences for the Stable warm, Transitional and Early cold climate phases. White: sum of cave and abri sites. Black: open-air sites. For the complete temperature calibrations labelled 'a, b, c, etc.' on the base scales of each column, see Table 8.2. For temperature tolerance, see Table 8.3.*

the Aurignacian begins to fade, and its temporal record rises steeply between 32 and 27 ka BP (Chapter 4: van Andel *et al.* 2003b, Fig. 4.5). In the absence of any evidence to the contrary, this was probably due to an *in situ* 'acculturation' to arctic means of survival under high-glacial conditions (Chapter 4: van Andel *et al.* 2003b), rather than being the result of the arrival and vigorous independent growth of a

'new Gravettian' population as has been suggested by human genetic studies (e.g. Richards *et al.* 2000; Semino *et al.* 2000).

Between 44 and 37 ka BP, the Aurignacian winter temperature preferences (Fig. 8.6) range from −8°C to 8°C during warm phases, and from −20°C to 8°C in cold ones, a pattern similar to the Mousterian, and including the same high site numbers towards the lower end of the winter temperature range when matched against cold simulations. Winter wind-chill preferences also differ little from those of the Mousterian, with many sites falling within the –12°C to –7°C wind-chill isozone (Fig. 8.6). Summer temperature preferences are also much like those of the Mousterian (Fig. 8.6).

The Early Cold Phase (37–27 ka BP) saw the final decline and extinction of the Mousterian, but brought the acme of the Aurignacian era and the onset and major rise of the Gravettian. The distribution of Aurignacian sites across the winter and summer temperature and wind-chill ranges is again quite similar to that of the Mousterian, but the slight bimodality also visible in the Mousterian lower temperature range is much enhanced throughout the interval from 37 to 20 ka BP (Fig. 8.6). While there are many sites within the range of –4°C to 4°C, a separate concentration exists between –20°C to –8°C, the coldest isozone. AMH sites (whether Aurignacian or Gravettian) appear to be less common in the –8°C to –4°C isozone, which conversely appears to be preferred by contemporary Mousterian populations. Preferred summer temperatures range throughout from 8°C to 20°C with highest site numbers concentrated in the 8°–12°C isozone. Much the same pattern is displayed during the Last Glacial Maximum Cold Phase (Fig. 8.6).

The Gravettian pattern of climate preferences, covering only the Early Cold and Last Glacial Maximum Cold Phases, much resembles that of the Aurignacian, except for the much higher concentration of sites in the coldest two isozones (Fig. 8.7). However, only a weak bimodality can be seen for the Early Cold Phase, and the Last Glacial Maximum shows a clear peak in the coldest isozone.

Aurignacian and Gravettian wind-chill preferences in the Early Cold Phase (Figs. 8.6 & 8.7) differ little from those of the warm phases, except that the cold end extends down to –34°C where the distribution is bimodal, with half to two-thirds in the coldest isozone. The Gravettian sites tolerate a slightly wider range of wind-chill values than the Aurignacian, but for both preference and tolerance are nearly identical, and the variation across the range of values between 29°C and –7°C is small.

The picture is essentially the same for summer temperatures in the Early Cold Phase. The Aurignacian site pattern (Fig. 8.6) is similar to that of the Transitional Phase, with almost all sites in the 8°–12°C and 12°–16°C isozones and a few as high as 20°C. The picture is essentially the same for the early Gravettian, with its preference peak lying entirely between 8°C and 16°C.

The Last Glacial Maximum (20–27 ka BP) marks the end of the Aurignacian and the acme of the Gravettian. While the Aurignacian winter temperature distribution remains strongly bimodal, the Gravettian shows a steep rise in numbers of sites towards the cold end. It is also slightly more common above 4°C than the Aurignacian. As regards wind-chill, Gravettian sites are more numerous than before, especially in isozones between –18°C and –7°C, although the tolerance ranges of both technocomplexes are unchanged from those seen in the Early Cold Phase. Summer patterns are as in the previous phase, with both technocomplexes preferring 8°C to 16°C with a few sites in warmer isozones. The Gravettian also has a few sites with summer temperatures just above 0°C.

Dated Mousterian open-air sites are uncommon, but in both the Aurignacian and Gravettian such sites appear in increasing numbers with time, ultimately making up more than half the number in the Gravettian. They are strikingly concentrated in isozones with the lowest tolerated winter temperature and wind-chill values. This peculiar preference for open-air occupation in areas that were very cold in winter raises the question of whether we are looking at seasonally-abandoned winter locations of people preferring those hunting grounds in the northern summer. This issue needs to be pursued in future research.

Points of note
- During the Transitional Phase, Aurignacian winter temperature preferences range from –8°C to 8°C when matched against the warm phase simulation, and preferred wind-chill values are above –23°C. Overall, the 44–37 ka BP pattern is very similar to that of the Neanderthals.
- Matched against cold simulations, Aurignacian sites show a clear concentration in the coldest (tolerated) winter isozones. This preference is, as foreshadowed already in the Mousterian, distinctly bipolar. The Gravettian has the same site patterning but, although also concentrated in the coldest region, is not bipolar. Instead, it increases steeply towards the coldest temperatures.
- Although in the Early Cold Phase Aurignacian and, to some degree even the Mousterian, sites shifted to colder winter isozones, it is only in the Gravettian that the vast majority of the sites occupy regions with winter temperatures well below 0°C.
- Taking the Aurignacian and Gravettian open-air sites separately, they also are concentrated in the

isozones with lowest winter temperatures and largest wind-chills. This odd preference raises again the question whether we are looking at winter temperatures in areas that were preferred hunting sites in the summer.

Hominids and snow

Snow-depth probably affected hominids mainly indirectly through its impact upon prey, inhibiting large mammal grazing and mobility if it became excessive (Guthrie 1990, 251), e.g. >60 cm for reindeer (Pruitt 1970, in Gamble 1986, 108). However, the simulated values do not account for wind drift, and the number of snowy days/year can vary locally depending upon site orientation and degree of shelter. As in the case of preference for very cold winters, snow preferences may merely indicate conditions in winter at temporarily deserted, seasonally occupied sites. These *caveats* aside, variations in snow-cover do yield a useful insight in the environments across which hominids migrated over time.

Neanderthals: Mousterian technocomplex
The pre-60 ka BP Mousterian exhibits a distinct preference for areas with less than 5 cm of snow, and areas with snow-depths of 50 cm or more were not occupied at all. Areas with <60 days of annual snow-cover were preferred, and only a few sites occur in areas with up to 179 snowy days.

The Mousterian distribution in the Stable Warm Phase (59–44 ka BP) is similar to the previous one; most sites are located in areas with a mean snow-depth of <5 cm and <29 days of snow-cover per year (Fig. 8.8). The distribution of sites across both variables is bimodal. The same pattern marks the warm intervals of the Transitional Phase (44–37 ka BP). During the intervening cold events, the tolerance increases; most sites then occur in areas with up to 10 cm of snow and some in areas with up to 70 cm, exceeding the previous maximum of 50 cm. In terms of their occurrence in snow-cover (days per year) isozones, the site distribution of the Transitional Phase does not differ much from that of earlier periods (Fig. 8.8). Especially when we scale site frequencies to the areas of their respective isozones, we can see that the preference for areas with <30 days' cover continues, but an increased presence in the 30–59 day isozone may indicate climate change.

The increase in the Mousterian tolerance of deep snow continues in the Early Cold Phase (37–27 ka BP), when sites are found even in the 70–99.99 cm isozone (Fig. 8.8), but a similar increase in the <0.09

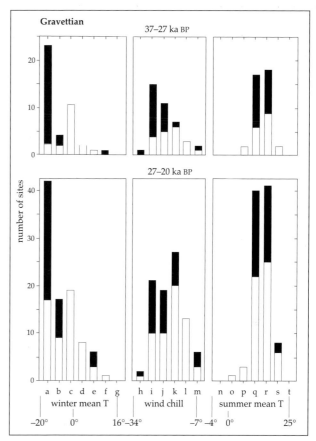

Figure 8.7. *Later modern human (Gravettian and other Upper Palaeolithic technocomplexes) mean temperature preferences for the Stable warm, Transitional, and Early cold climate phases. White: sum of cave and abri sites. Black: open-air sites. For the complete temperature calibration, labelled 'a, b, c, etc.' on the base scales of each column, see Table 8.2. For temperature tolerance, see Table 8.3.*

cm snow-depth isozone suggests a final retreat to, or preference for, areas with little snowfall. The latter pattern is even more pronounced if we scale site frequencies to their respective isozone areas.

Modern humans: Aurignacian and Gravettian
The earliest Aurignacian (44–37 ka BP) tolerates up to 70 cm of snow, with a maximum of 50 cm during warm events (Fig. 8.9). During warm events, as in the contemporary Mousterian, most sites are in areas with less than 20 cm of snow, and areas with <5 cm are preferred: this is especially noticeable if we scale site frequencies to isozone area. Both Mousterian and Aurignacian show bimodal patterns for warm events, with two distinct clusters of Aurignacian sites in snow-cover day isozones. The first is concentrated in areas with fewer than 60 days'

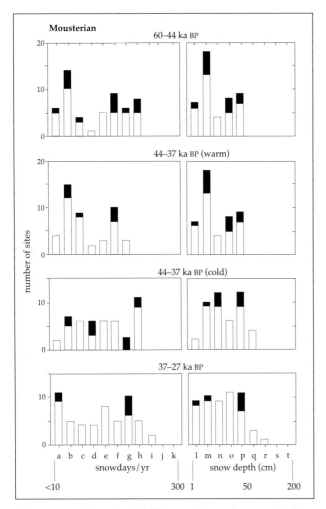

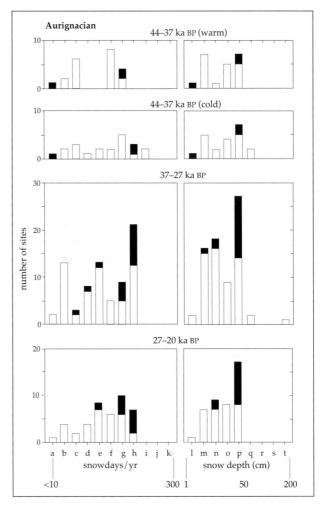

Figure 8.8. *Neanderthal (Mousterian technocomplex) snow condition preferences for Stable warm, Transitional, and Early cold climate phases. Left: days of snow per year; right mean snow depth. White: sum of cave and abri sites. Black: open-air sites. For complete snow calibration, labelled 'a, b, c, etc.' on the base scales of each column, see Table 8.2. For tolerances with regard to snow, see Table 8.3.*

Figure 8.9. *Early modern human immigrants (Aurignacian and other Early Upper Palaeolithic technocomplexes) snow condition preferences for the Stable warm, Transitional and Early cold climate phases. Left: days of snow per year. Right: mean snow depth. White: sum of cave and abri sites. Black: open-air sites. For a complete temperature calibration labelling on the base scales of each column, see Table 8.2. For tolerance with regard to snow, see Table 8.3.*

cover (preferred by AMH, if one chooses to believe the frequencies after scaling against isozone area), while the second is in areas (notably the Ardennes and the Danube) with between 120 and 180 days of snow per year. The pattern during cold events is similar, but more evenly-spread, with regard to snow-cover days (Fig. 8.9). Aurignacian snow-cover tolerances appear to reach up to 240 days per year when matched against the cold simulations, seeming to exceed both the tolerances seen in the contemporary Mousterian and the Aurignacian patterns during the warm events. Nevertheless, Aurignacian sites in the Transitional Phase are generally to be found in

isozones with <60 days' snow-cover, if frequencies are scaled to isozone area.

During part of the cold phases of 37–20 ka BP, the Aurignacian and Gravettian co-existed. In the Early Cold Phase (37–27 ka BP), about half of the Aurignacian sites (some open-air) occur in areas with less than 10 cm snow-depth (Fig. 8.9) — preferentially so if frequencies are scaled against isozone area — with one outlying site in the 150–200 cm isozone (Uphill Quarry, cave 8). The apparent preference for the 20–50 cm snow isozone, however, is partly due to its large area (as is the case in the Last

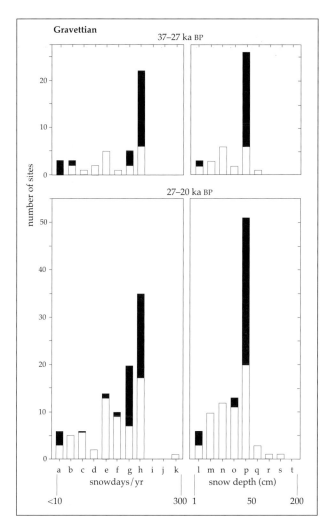

Figure 8.10. *Later modern human (Gravettian and other Upper Palaeolithic technocomplexes) snow condition preferences for the Stable warm, Transitional and Early cold climate phases. Left: days of snow per year. Right: mean snow depth. White: sum of cave and abri sites. Black: open-air sites. For a complete temperature calibration, labelled 'a, b, c, etc.' on the base scales of each column, see Table 8.2. For tolerance with regard to snow, see Table 8.3.*

Glacial Maximum Cold Phase also). Overall, the pattern is similar to that of the cold events of the Transitional Phase. With regard to days of snow-cover, many Aurignacian sites occur in areas with <30 snow days/year, but sizeable numbers are found also in the 90–119 and 180–209 day isozones (Fig. 8.9); the latter is a very extensive isozone.

The Gravettian preference for the 20–50 cm snow-depth isozone exhibited during the Early Cold Phase for its majority of open-air sites seems stronger than that of the Aurignacian (Fig. 8.10), but when

adjusted for isozone area the real preference of both is for areas with <10 cm snow-depth. The maximal Gravettian tolerance does not exceed 70 cm, casting doubt upon the validity of the contemporary Aurignacian outlier in the 150–200 cm isozone (cf. Figs. 8.9 & 8.10): it is a solitary dated bone point, with no other associated artefacts (Jacobi & Pettitt 2000). Gravettian sites are more consistently present in areas with snow-cover for up to 209 days per year than the contemporary Aurignacian, a difference that persists even when scaled against isozone size.

In the Last Glacial Maximum Cold Phase, the snow-depth tolerances of the Gravettian increase (Fig. 8.10) while those of the Aurignacian decrease commensurately, due primarily to the geographical expansion of the former at the expense of the latter. While the snow-depth preferences of the Aurignacian do not really differ from those it displays in the Early Cold Phase, its maximal snow-depth tolerance was reduced from 70 cm to 50 cm (Fig. 8.9).

In contrast, average snow-depth tolerances of the Gravettian extended to areas with up to 150 cm, although most sites are still found within isozones of 50 cm or less (Fig. 8.10). Unlike the patterning from the previous climatic phase, the Gravettian sites are also more strongly present in the areas with no (<0.9 cm) snowfall (if one scales the site frequencies to isozone area), implying a niche expansion that may perhaps be linked to the decline of the Aurignacian behaviour and responses. Gravettian snow-depth preferences generally lie below 20 cm; as noted earlier, the apparent 'preference' for the 20–50 cm isozone can be attributed to its extensive coverage.

Points of note
- Neanderthal snow-cover preferences are remarkably consistent throughout most of OIS-3 (Stable Warm and Transitional Phases), generally remaining in areas with <5 cm snow-depth and <60 days' snow-cover per year, with tolerances hardly exceeding 50 cm depth and 210 days' cover per year.
- The Aurignacian (AMH) initially shows similar bimodal patterning in response to snow conditions as the contemporary Neanderthals (during the Transitional Phase), with similar tolerance and preference ranges.
- From 37 ka BP onwards, both the Aurignacian and Gravettian start to show increasingly strong (unimodal) preferences for snow-depths of up to 20 cm. Depth tolerances for the Gravettian reach 150 cm during the Last Glacial Maximum Cold Phase, and AMH seem to be able to tolerate <210 days' snow-cover per year. Such preferences are

no doubt linked to the stronger cold temperature preferences noted previously for AMH (especially the Gravettian) during the winters of cold events, although the Gravettian also consolidates its presence in areas with little snow (maritime western Europe) at 27–20 ka BP.

- Neanderthal patterns in relation to snow conditions seem to be less clear during cold events than in warm ones. Only towards the end of their existence do Neanderthal tolerances (not their preferences) appear to increase, perhaps owing to climatic deterioration, to 100 cm snow-depth and 240 days' snow-cover.

Concluding remarks

The crucial question of this paper is: what effect did changing climate have upon the settlement patterns of Middle and Upper Palaeolithic hominids in Europe during OIS-3? More detailed 'landscape' approaches to this problem, discussing the importance of resources with regard to human preferences and tolerances for what might often appear to be hostile climates, merit separate consideration in another paper. While 'tolerance' is used for the passive acceptance by hominids of surrounding physical conditions, 'preference' is more difficult to assess: how can we be sure people are actively selecting the colder areas, for example, if we do not have the seasonality evidence to suggest that they overwintered at the dated site[s]? People may have left the area during the winter, or alternatively they may have withstood inclement weather because they stood to gain much in resources.

In order to answer the seven questions listed earlier in this paper, we must be certain how much of the human spatio-temporal distributions we have described can be ascribed directly to climatic conditions. Only having quantified the effects of the latter can we assess what role resources, mobility and social networks all played. Questions A-1 to A-3 and B-1 can mostly be answered in this paper, while the remainder will have to wait for another occasion to be addressed fully.

Regarding question A-1, we can say that there is some relationship between Neanderthal settlement/migration patterns and climate changes. Neanderthals seem to prefer warmer climatic conditions (whether compared against 'warm' or 'cold' simulations) than AMH, although it must be said that early Aurignacian spatial patterning is very similar (cf. later Aurignacian and especially Gravettian site distributions: see question A-2). As climate deteriorated during OIS-3, the putative lesser mobilities

of Neanderthals (inferred from their preferences for mild winter conditions) must have been challenged by declining winter temperatures, even though their *summer* temperature preferences remain virtually unchanged. Therefore, did Neanderthals change their mobility and/or behaviour as the climate changed?

Likewise, the answer to question A-2 would acknowledge a strong climatic component in the immigration and settlement choices. Early AMH, as defined by the early Aurignacian, appear to show similar climatic preferences and tolerances to the contemporary Neanderthals, but as time progresses, the two species begin to diverge, no doubt partly affected by the spatial expansion of AMH (later Aurignacian and Gravettian) at the expense of Neanderthals. The later Aurignacian and Gravettian sites proceed to show an increased climatic tolerance and a more diverse set of climatic preferences than Neanderthals ever appear to have done. This conclusion would gainsay the claim that Neanderthals are adapted to — and thus prefer — extreme cold conditions (Trinkaus 1981; Holliday 1997).

The answer to question A-3 (differences between Neanderthals and AMH in their respective climatic responses) is inextricably linked to that of A-2. While Neanderthals and AMH scarcely differ in the broad ranges of their climatic *tolerances*, they show increasingly diverse preferences in terms of temperatures and snow-cover after the climatic downturn from 37 ka BP. The Gravettian in particular seems to expand both into milder and colder winter isozones during the Last Glacial Maximum Cold Phase, and this pattern (certainly at the colder end of the spectrum) cannot simply be attributed to dispersal into land (once) occupied by Aurignacian and Mousterian groups, for we have evidence that they expanded into effectively 'empty' landscapes also, e.g. much of the Russian Plain.

The answers given to questions A-1 to A-3 depend on an affirmative answer to question B-1: that climatic factors *did* have a direct effect on human settlement and migration patterns. If we conclude that they did not, then we have to believe that the answer to question B-2 is 'yes': that climatic changes were felt indirectly by human groups. One line of climatic evidence which could be seen as having had indirect effects is snow-cover: did it affect humans through its effects on the grazing of preferred prey species? Thick snow cover could conceivably have affected humans' efforts to provision themselves with fresh lithic raw material nodules had they decided to stay in the area, but in many areas they could probably still have obtained usable materials from

144

the beds of watercourses or been more parsimonious in their re-use of lithic materials already at hand.

Question C-1 demands a greater consideration of resources and individual site attributes than can be afforded in this chapter. Here, we might just say that Neanderthals do not appear to show responses to deteriorating climatic conditions of similar flexibility as seen in the later Aurignacian and Gravettian groups. The suspicion that their responses were restricted by the substantial AMH presence from 37 ka BP must be qualified by their marked thermophilic preferences during OIS-4.

What we call the Gravettian arose at around 32 ka BP, in response to deteriorating climate, and its territorial 'strongholds' appear to be the middle or even higher latitudes of Europe and the Russian Plain; its presence in Mediterranean zones (apart from the Dordogne) is scattered and perfunctory. The Aurignacian shares many of these characteristics, although they are less well-developed, and the earliest Aurignacian has more in common with the contemporary Neanderthals than with the later manifestations of the same technocomplex or with the Gravettian. It is therefore entirely feasible that some of the early AMH (represented by early Aurignacian sites) succumbed to climatic downturns, as postulated by question C-2. The scattered earliest Aurignacian assemblages are frequently small and impoverished in their tool-type diversity, and it has been claimed that they represent ephemeral, economically-unspecialized and highly mobile AMH groups (Davies 2001). Such groups might well have been more susceptible to climatic fluctuations.

As noted earlier, an eastwards-increasing continentality can be discerned across Europe. The focusing of archaeological attention on several climatically-differentiated European regions, known to have extensive and consistent OIS-3 records, would help us to understand how humans responded to climatic and environmental change at the finer scale. Such regions might include the Dordogne (Mediterranean climate), the Ardennes (temperate Maritime climate) and the Middle Danube basin (continental climate). The exploitation of resources (and associated seasonality data), mobility and exchange systems (*sensu* Gamble 1982), site complexity and diversity, and settlement patterning across the landscape can all be explored more profitably in such regions. However, the challenge is what to make of the apparent lacunae in between these rich areas: does an absence of evidence for human presence in regions with similar climatic and environmental conditions to ones already known to have been inhabited actually reflect Palaeolithic reality, or a sampling bias?

Acknowledgements

We would like to thank Prof. Tjeerd van Andel for all his input and support during the writing of this paper. Dr Steven Kuhn, Prof. Clive Gamble and Dr John Stewart must also be thanked for their very valuable suggestions and assistance. Thanks are owed to David Pollard for Figures 8.2 and 8.3. Any mistakes which remain are of course our responsibility.

Notes

1. The archaeological data base and the climate and biome simulations for the last glaciation can both be found on the Stage 3 Project website: http://www.esc.esc.cam.ac.uk/oistage3/Details/Homepage.html.

2. The term climate *zone* will be used in the geographic sense, whilst units of climate change will be labelled *phases*.

3. Because the Greenland ice-core climate record is calibrated in calendrical years, calendar dates (Chapter 2: van Andel 2003) are used throughout this Chapter. The phrase 'xxxx ka BP' refers to calibrated ^{14}C dates (calibration by CalPal.v1998: Jöris & Weninger 2000; see also Chapter 3: van Andel *et al.* 2003a) and other calendrical dates such as TL, OSL, U/Th series and ESR. The raw versions of the ^{14}C dates can be found in the Stage 3 Project's chrono-archaeological data base at its website or in references cited there (see note 1 for the url).

References

Barron, E. & D. Pollard, 2002. High-resolution climate simulations of Oxygen Isotope Stage 3 in Europe. *Quaternary Research* 58, 296–309.

Barron, E., T.H. van Andel & D. Pollard, 2003. Glacial environments II: reconstructing the climate of Europe in the Last Glaciation, in *Neanderthals and Modern Humans in the European Landscape during the Last Glaciation*, Chapter 5, eds. T.H. van Andel & W. Davies. (McDonald Institute Monographs.) Cambridge: McDonald Institute for Archaeological Research, 57–78.

Churchill, S.E. & F.H. Smith, 2000. Makers of the Early Aurignacian of Europe. *Yearbook of Physical Anthropology* 43, 61–115.

Davies, W., 2001. A very model of a modern human industry: new perspectives on the origins and spread of the Aurignacian in Europe. *Proceedings of the Prehistoric Society* 67, 195–217.

Donner, J., 1995. *The Quaternary History of Scandinavia.* Cambridge: Cambridge University Press.

Gamble, C., 1982. Interaction and alliance in Palaeolithic society. *Man* 17, 92–107.

Gamble, C., 1986. *The Palaeolithic Settlement of Europe.* Cambridge: Cambridge University Press.

Golding, W., 1955. *The Inheritors*. London: Faber & Faber.

Guthrie, R.D., 1990. *Frozen Fauna of the Mammoth Steppe: the Story of Blue Babe*. London: University of Chicago Press.

Holliday, T.W., 1997. Postcranial evidence of cold adaptation in European Neandertals. *American Journal of Physical Anthropology* 104, 245–58.

Huntley, B. & J.R.M. Allen, 2003. Glacial environments III: vegetation patterns, productivity and non-analogue biomes, in *Neanderthals and Modern Humans in the European Landscape during the Last Glaciation*, Chapter 6, eds. T.H. van Andel & W. Davies. (McDonald Institute Monographs.) Cambridge: McDonald Institute for Archaeological Research, 79–102.

Huntley, B., M.J. Alfano, J.R.M. Allen, D. Pollard, P.C. Tzedakis, J.-L. de Beaulieu, E. Grüger & W.A. Watts, 2003. European vegetation during Marine Oxygen Isotope Stage-3. *Quaternary Research* 59, 195–212.

Jacobi, R.M. & P.B. Pettitt, 2000. An Aurignacian point from Uphill Quarry (Somerset) and the earliest settlement of Britain by *Homo sapiens sapiens*. *Antiquity* 74, 513–18.

Johnsen, S.J., D. Dahl-Jensen, N. Gundestrup, J.P. Steffensen, H.B. Clausen, H. Miller, V. Masson-Delmotte, A.E. Sveinbjörnsdottir & J. White, 2001. Oxygen isotope and paleotemperature records from six Greenland ice-core stations, Camp Century, Dye-3, GRIP, GISP2, Renland and North GRIP. *Journal of Quaternary Science* 16, 299–308.

Jöris, O. & B. Weninger, 1999. Possibilities of calendric conversion of radiocarbon data for the glacial periods, in *Actes du 3ème Congrès International 'Archéologie et 14C', Lyon, 1998*, eds. J. Evin, J.P. Daugas & J.F. Salles. *Revue d'Archéometrie*, Supplément 1999 et *Société Préhistorique de France Mémoire* 26, 87–92.

Jöris, O. & B. Weninger, 2000. Radiocarbon calibration and the absolute chronology of the Late Glacial, in *L'Europe Centrale et Septentrionale au Tardiglaciaire*, Table Ronde de Nemours 13–16 mai 1997, eds. B. Valentin, P. Bodu & M. Christensen. Mémoires du Musée de Préhistoire de l'Ile de France, 19–54.

Meese, D.A., A.J. Gow, R.B. Alley, G.A. Zielinski, P.M. Grootes, M. Ram, K.C. Taylor, Mayeweski & J.F. Bolzan, 1997. The Greenland Ice Sheet Project 2 depth-age scale: methods and results. *Journal of Geophysical Research* 102, 26,411–23.

Mellars, P.A., M. Otte & L.G. Straus, 1999. The Neanderthal problem continued (with reply from J. Zilhão & F. d'Errico). *Current Anthropology* 40, 341–64.

Musil, R., 2003. The Middle and Upper Palaeolithic game suite in central and southeastern Europe, in *Neanderthals and Modern Humans in the European Landscape during the Last Glaciation*, Chapter 10, eds. T.H. van Andel & W. Davies. (McDonald Institute Monographs.) Cambridge: McDonald Institute for Archaeological Research, 167–90.

Pollard, D. & E. Barron, 2003. Causes of model-data discrepancies in European climate during Oxygen Isotope Stage 3 with insights from the Last Glacial Maximum. *Quaternary Research* 59, 108–13.

Pruitt, W.O., 1970. Some ecological aspects of snow, in *Ecology of the Subarctic Regions: Proceedings of the Helsinki Symposium (1966)*. Paris: CNRS, 83–99.

Richards, M., V. Macaulay, E. Hickey, E. Vega, B. Sykes, V. Guida, C. Rengo, D. Sellitto, F. Cruciani, T. Kivisild, R. Villems, M. Thomas, S. Rychkov, O. Rychkov, Y. Rychkov, M. Golge, D. Dimitrov, E. Hill, D. Bradley, V. Romano, F. Cali, G. Vona, A. Demaine, S. Papiha, C. Triantaphyllidis, G. Stefanescu, J. Hatina, M. Belledi, A. Di Rienzo, A. Novelletto, A. Oppenheim, S. Nørby, N. Al-Zaheri, S. Santachiara-Benerecetti, R. Scozari, A. Torroni & H.-J. Bandelt, 2000. Tracing European founder lineages in the Near Eastern mtDNA pool. *American Journal of Human Genetics* 67, 1251–76.

Semino, O., G. Passarino, P.J. Oefner, A.A. Lin, S. Arbuzowa, L.E. Beckman, G. de Benedictis, P. Francalacci, A. Kouvatsi, S. Limborska, M. Marcikiae, A. Mika, B. Mika, D. Primorac, A.S. Santachiara-Benerecetti, L.L. Cavalli-Sforza & P.A. Underhill, 2000. The genetic legacy of Paleolithic *Homo sapiens sapiens* in extant Europeans: a Y-chromosome perspective. *Science* 290, 1155–9.

Stewart, J.R., T. van Kolfschoten, A. Markova & R. Musil, 2003. The mammalian faunas of Europe during Oxygen Isotope Stage Three, in *Neanderthals and Modern Humans in the European Landscape during the Last Glaciation*, Chapter 7, eds. T.H. van Andel & W. Davies. (McDonald Institute Monographs.) Cambridge: McDonald Institute for Archaeological Research, 103–30.

Stuiver, M. & P. Grootes, 2000. GISP2 oxygen isotope ratios. *Quaternary Research* 53, 277–84.

Trinkaus, E., 1981. Neanderthal limb proportions and cold adaptation, in *Aspects of Human Evolution*, ed. C.B. Stringer. London: Taylor & Francis, 187–224.

van Andel, T.H., 2002. Reconstructing climate and landscape of the last mid-pleniglacial in Europe: the Stage 3 Project. *Quaternary Research* 57, 2–8.

van Andel, T.H., 2003. Glacial environments I: the Weichselian climate in Europe between the end of the OIS-5 interglacial and the Last Glacial Maximum, in *Neanderthals and Modern Humans in the European Landscape during the Last Glaciation*, Chapter 2, eds. T.H. van Andel & W. Davies. (McDonald Institute Monographs.) Cambridge: McDonald Institute for Archaeological Research, 9–20.

van Andel, T.H., W. Davies, B. Weninger & O. Jöris, 2003a. Archaeological dates as proxies for the spatial and temporal human presence in Europe: a discourse on the method, in *Neanderthals and Modern Humans in the European Landscape during the Last Glaciation*, Chapter 3, eds. T.H. van Andel & W. Davies. (McDonald Institute Monographs.) Cambridge: McDonald Institute for Archaeological Research, 21–30.

van Andel, T.H., W. Davies & B. Weninger, 2003b. The human presence in Europe during the Last Glacial period I: human migrations and the changing climate, in *Neanderthals and Modern Humans in the European Landscape during the Last Glaciation*, Chapter 4, eds. T.H. van Andel & W. Davies. (McDonald Institute Monographs.) Cambridge: McDonald Institute for Archaeological Research, 31–56.

Chapter 9

Neanderthal Thermoregulation and the Glacial Climate

Leslie C. Aiello & Peter Wheeler

Neanderthals are often described as having an arctic (Ruff *et al.* 1993) or hyperarctic (Holliday 1997a) body form that arose as a consequence of long-term climatic selection under glacial conditions in Europe and European Russia during the later part of the Middle Pleistocene and early Late Pleistocene (Holliday 1997a,b; Ruff 1994; Trinkaus 1981; 1983; Trinkaus *et al.* 1991). For example, Steegmann and colleagues (Steegmann *et al.* 2002) estimate that Neanderthals lived under cold glacial conditions for a total of 155 ka since they emerged as a recognizable species about 250 ka BP. This long period of cold corresponds to the penultimate glaciation between 250–180 ka plus the most recent glacial advance between 115–30 ka.

Neanderthals are described as being cold adapted because their body form corresponds to, and sometimes exceeds, that found in modern cold-adapted peoples and is consistent with theoretical expectations derived from the application of Bergmann's Rule (1847) and Allen's Rule (1877) to human body form. These 'rules' state that within species those homeothermic animals living in colder climates (nearer the poles) will have a higher body mass (Bergmann's Rule) and shorter extremities (Allen's Rule). In comparison to modern and fossil people occupying warmer climate conditions, Neanderthals have large rib-cages and barrel-shaped chest (Franciscus & Churchill 2002), very wide trunks (Ruff 1993; 1994; Trinkaus & Ruff 1989) and short extremities (Holliday 1997a,b).

In a series of papers Ruff (Ruff 1991; 1993; 1994; Ruff & Walker 1993) has developed the cylinder model of body form, demonstrating that alterations in trunk width, and not changes in stature, determine the ratio between surface area and body volume. The wider the body the smaller is the surface area to volume ratio and the greater is the efficiency in retaining heat and keeping (relatively) warm. Ruff has demonstrated that Neanderthals had very wide bi-iliac breadths, and hence wide trunk widths, be-

ing most similar to some modern Eskimos (Ruff 1994).

Neanderthals also have relatively short legs, and relatively short distal extremities giving them low brachial (radius length/humerus length) and crural (tibia length/femur length) indices (Coon 1962; Trinkaus 1981; Holliday 1997a,b; Churchill 1998). Trinkaus (1981) shows that the brachial index is higher in the assumed warmer-adapted Middle Eastern Neanderthals than in colder adapted European Neanderthals. There is a similar ecogeographical patterning for femur length relative to femoral head size (Ruff 1994). Both of these observations are consistent with the predictions of Allen's Rule.

Other aspects of Neanderthal morphology have also been suggested to reflect cold adaptation. These include the long Neanderthal head (Steegmann & Platner 1968) large noses, large paranasal sinuses and big brains (Churchill 1998). For the purposes of the current discussion of thermoregulation, it is the body form, and particularly the low relative surface area that is examined. Our major questions in relation to Neanderthal life under glacial conditions are:

1. In terms of both energetic cost and survival, what quantitative advantage did the Neanderthal cold adapted body form provide at low environmental temperatures?
2. What were the actual environmental conditions that prevailed in Europe and European Russia at the time of Neanderthal occupation?

Neanderthal temperature tolerance

From basic physical principles and estimates of basal metabolic rate (BMR), skin surface area, thermal conductance (C), and the maximum sustainable elevation in resting metabolic rate (RMR), it is possible to produce general estimates of both the relative energetic costs of thermoregulation for Neanderthals and modern humans at low environmental temperatures and the minimum temperature these hominins could have tolerated, given access to sufficient dietary resources.

The lower limit of the thermoneutral zone within which a mammal can regulate its core temperature solely by controlling its thermal conductance is known as the lower critical temperature (T_{lc}). As the ambient temperature (T_a) falls below this level, homeostasis can only be maintained by progressively increasing internal heat production, and incurring the additional energetic costs associated with this increase in heat production. Reducing the thermal conductance of a mammal will lower both T_{lc} itself and the rate of heat loss at all ambient temperatures below T_{lc}. If it is assumed that Neanderthals had a typical human body temperature of 37°C, then their lower critical temperature (T_{lc}) is given by:

$$T_{lc} = 37°C - (BMR / Total\ conductance)$$

where BMR is Basal Metabolic Rate (W) and total conductance is mean thermal conductance (C) multiplied by total body surface area (m^2).

The maximum metabolic rate that can be sustained indefinitely by a modern human is approximately 3 times BMR (Burton & Edholm 1955). If it is assumed that the maximum sustainable metabolic rate of Neanderthals was a similar ratio to BMR then the minimum ambient temperature (T_{min}) at which they could have maintained normal body temperature is given by:

$$T_{min} = 37°C - (3 \times BMR / Total\ conductance)$$

However, as discussed below it is possible that adaptive or acclimatory factors could have enabled Neanderthals to physiologically sustain higher maximum levels of internal heat production than those of modern humans (Steegman et al. 2002).

In the initial analyses Neanderthal responses to low ambient temperatures were predicted on the basis of relatively conservative assumptions about the similarity of their physiology to modern humans. Basal metabolic rate was calculated using the allometric equation relating this parameter to body mass in eutherian mammals (Kleiber 1961):

$$BMR\ (W) = 3.4 \times mass\ (kg)^{0.75}$$

The values predicted by this equation show good agreement with those of living humans and other primates (Aiello & Wheeler 1995).

Thermal conductance (C) below the lower critical temperature (T_{lc}), which is the mean rate of heat loss from each square metre of vasoconstricted skin

for every degree Celsius gradient between core and ambient temperatures, was taken as a standard value of 5W.m^{-2}.°C^{-1}, calculated from representative modern human data (Mount 1979). Lower values of this parameter would correspondingly reduce the absolute estimates of lower critical temperatures and minimum sustainable temperatures, but the proportionate changes in the predicted values for all three hominin physiques examined would be similar.

In subsequent analyses these parameters were modified to explore both the possible effect of strategies that would have increased internal heat production or reduced heat loss. These include the influence of climatic and dietary factors that may have resulted in minimum rates of metabolic rate being elevated above those predicted by the Kleiber equation, and the potential contributions to reducing thermal conductance made by biological insulation, such as increased muscle mass, subcutaneous fat and hair cover, as well as the cultural protection provided by clothing, fire and shelter.

Body surface areas (Table 9.1) were calculated from estimates of body mass and stature using the equation of Dubois & Dubois (1916). For comparison we have also modelled early African *Homo erectus* based on the Nariokotome youth (Ruff & Walker 1993; Ruff 1994; Ruff et al. 1997) and early anatomically modern *Homo sapiens* based on the mean of four Upper Palaeolithic skeletons (Skhul 4, Předmost 3, Předmost 9, Grotto des Enfants 4) (Ruff 1994). The physique of early African *Homo erectus* is widely believed to represent a tall, linear body form that is adapted to hot, arid conditions (Ruff & Walker 1993; Wheeler 1993; Ruff 1994). The early Anatomically Modern Humans represent an overall body shape that is within the range of recent Sub-Saharan Africans and at the extreme of recent European variation (Holliday 1997b; see also Pearson 2000a).

Based on the inferred body mass and stature data presented in Table 9.1 and the assumptions outlined above, the three hominin body forms show very little difference in the ambient temperature at which thermoregulatory thermogenesis must be initiated or in the minimum sustainable ambient temperature at which a heat production equivalent to 3 times BMR is reached (Table 9.2; Fig. 9.1). Neanderthals with a BMR predicted by the Kleiber equation would need to initiate additional heat production at 27.3°C (81.2°F)[1], *Homo erectus* at 28.5°C (83.3°F) and *Homo sapiens* at 28.2°C (82.7°F). The minimum sustainable ambient temperature for Neanderthals would be 8.0°C (46.5°F), for *Homo erectus* 11.6°C (52.8°F) and for *Homo sapiens* 10.5°C (50.9°F).

This analysis demonstrates that Neanderthals probably only had a small advantage of 1°C over *Homo sapiens* and *Homo erectus* in the temperature at which they would have incurred additional energetic costs associated with thermoregulation and of *c.* 2.5–3.5°C in their minimum sustainable ambient temperature. These differences are extremely small in comparison with those that might be expected to provide a significant advantage for life under the glacial conditions in European Late Pleistocene.

This conclusion may be surprising in the context of the literature emphasizing the potential cold-adapted features of Neanderthal morphology, including the energetic advantages of foreshortened limbs and low relative surface area. Although the Neanderthal physique was modified in a direction consistent with increased body heat retention the actual quantitative advantages resulting from this morphology alone are remarkably small. This contrasts with the apparent benefits conferred by the distinctive linear physique of early *Homo erectus* and many living humans inhabiting open tropical habitats in coping with the opposite thermal problem of high environment heat loads. In these conditions this body morphology has been shown to reduce both energy and water stress by reducing the gain of heat from the environment and facilitating its dissipation (Wheeler 1993; Ruff 1991; 1993; 1994). However, it should be recognized that the Neanderthal physique could have associated thermoregulatory advantages other than energy conservation alone. The relatively short limbs would have helped in maintaining the temperature of the extremities closer to that of the core. In extremely cold conditions this would have helped prevent thermal damage to the distal portions of the limbs and may have been particularly important in maintaining the normal manipulative ability and sensitivity of the fingers, which is progressively impaired as their temperature falls.

The above analysis assumes that the Kleiber equation correctly predicts BMR for Neanderthals

Table 9.1. *Fossil hominin data.*

	Body mass (kg)	Stature (cm)	BMR[1]	Body surface area[2]
Neanderthal[3]	80.8	167	91.630	1.898
Homo sapiens[4]	70.0	177	82.282	1.862
Homo erectus[5]	68.0	185	80.512	1.900

[1] BMR = $3.4 \times$ mass $(kg)^{0.75}$ (Kleiber 1961).
[2] Body surface area $(m^2) = 0.00718 \times$ mass $(kg)^{0.425} \times$ stature $(cm)^{0.725}$ (Dubois & Dubois 1916).
[3] mean data for male Neanderthals (Ruff *et al.* 1997).
[4] mean data for 4 male early Anatomically Modern *Homo sapiens* (Předmost 3, Předmost 9, Skhul 4, Grotte des Enfants 4) (Ruff 1994).
[5] KNM-WT 15000 (adult) (Ruff 1994).

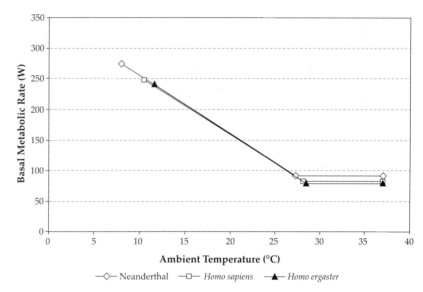

Figure 9.1. *The critical points and the minimum sustainable ambient temperatures for Neanderthals,* Homo sapiens *and* Homo erectus. *(Data from Table 9.1. See text for the equations used to predict the temperatures.)*

(and other hominins). There is reason to believe that this might not be a correct assumption. Among modern people there is a strong negative correlation between temperature and BMR (Leonard *et al.* 2002) with arctic-adapted peoples having significantly higher BMRs than those adapted to the tropics. The elevated BMR appears to derive from two sources, a diet based on a high intake of animal fat and protein and the effects of temperature and day length on thyroid function (Leonard *et al.* 2002). Leonard and colleagues demonstrate that arctic populations eating an indigenous high fat and protein diet had BMRs that were up to 25 per cent higher than expected while the BMRs of arctic peoples eating a more mixed diet were elevated by about 10 per cent.

Richards *et al.* (2000) have suggested on the basis of stable isotope analysis that Neanderthals

Table 9.2. *Lower critical and minimum sustainable ambient temperatures.*

	C[1]	Lower critical T[2]		Minimum sustainable T[3]	
		°C	°F	°C	°F
Neanderthal					
Kleiber BMR	9.489	27.3	81.2	8.0	46.5
Elevated BMR[4]	9.489	25.9	78.6	3.7	38.6
Homo sapiens	9.312	28.2	82.7	10.5	50.9
Homo erectus	9.498	28.5	83.3	11.6	52.8

[1] C (total conductance) = typical human conductance (5 W.m^{-2}.°C^{-1}) multiplied by surface area (m^2) from Table 9.1. W = watts.
[2] Critical Temperature (°C) = 37°C – (BMR/total conductance).
[3] Minimum sustainable Ambient Temperature °C = 37°C – ((3 × BMR)/total conductance).
[4] BMR elevated by 15 per cent to account for climate and dietary induced increases. See text for explanation.

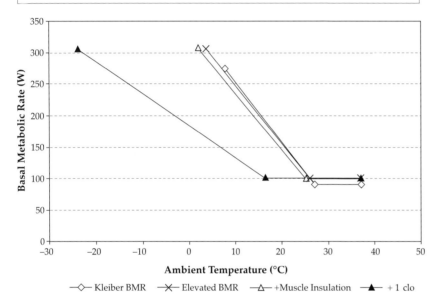

Figure 9.2. *The lower critical temperature and minimum sustainable ambient temperatures for a Neanderthal with a Kleiber-estimated BMR, with an elevated BMR, with an elevated BMR and a 5 per cent reduction in heat loss as the result of additional muscle insulation, and with an elevated BMR, a 5 per cent reduction in heat loss as the result of addition muscle insulation plus 1 clo insulation. (Data and equations as in Figure 9.1.)*

had a diet high in animal-based resources and that their diet most probably contrasted with that of Upper Palaeolithic modern humans who ate a more mixed diet (Richards *et al.* 2001). If Neanderthal BMR is increased by a conservative 15 per cent to reflect a higher climate and dietary-induced level, and it is assumed that their maximum sustainable metabolic rate was also three times their elevated BMRs, Neanderthals would initiate thermoregulatory thermogenesis at 25.9°C (78.6°F) rather than at 27.3°C (81.2°F) and their minimum sustainable ambient temperature would be 3.7°C (38.6°F) rather than 8.0°C

(46.5°F) (Table 9.2; Fig. 9.2).

These analyses assume no form of insulation such as increased muscle mass, subcutaneous fat, hair or clothing. All of these factors can act singly or together to buffer the effects of cold weather. But not all are equally likely to have been employed by the Neanderthals. It is known that Neanderthals had increased muscle mass in relation to modern humans (Trinkaus & Ruff 1989; Pearson 2000a,b). Glickman-Weiss *et al.* (1993) demonstrated that increased muscle mass can result in a reduction in heat loss of up to 5 per cent in cold water when subjects differing in body mass by up to 20 kg are matched for amount of subcutaneous fat. If this same degree of insulation were to be assumed for Neanderthals in relation to modern humans, Neanderthals with a Kleiber-predicted BMR would initiate thermoregulatory thermogenesis at 26.8°C (80.3°F) and their minimum sustainable ambient temperature would be 6.5°C (43.7°F). Neanderthals with an elevated BMR would initiate thermoregulatory thermogenesis at 25.3°C (77.6°F) and their minimum sustainable ambient temperature would be 1.9°C (35.5°F) (Table 9.3; Fig. 9.2).

This degree of muscle mass insulation is considerably below the level that would be provided by 1 clo, or the amount of insulation that would be necessary to maintain a resting sitting human whose metabolism is 50 kcal.m^{-2}.hr indefinitely comfortable in an environment of 21°C (70°F). A clo is roughly equivalent to the insulation provided by a western business suit and would reduce heat loss by about forty-four per cent from the assumed typical human conductance of 5W.m^{-2}.°C^{-1} to 2.8169 W.m^{-2}.°C^{-1}. A clo of insulation would reduce the lower critical temperature of a Neanderthal with an elevated BMR to 17.3°C (63.1°F) and the minimum sustainable ambient temperature to –22.1°C (–7.8°F). If this was combined with the insulation afforded by the increased Neanderthal muscle mass, the critical temperature

Table 9.3. *Insulated hominins.*

	C^1	Total conductance	Kleiber BMR[3] Lower critical temperature °C	°F	Minimum sustainable temperature °C	°F	Elevated BMR[4] Lower critical temperature °C	°F	Minimum sustainable temperature °C	°F
Neanderthal										
Muscle[4]	4.750	9.014	26.8	80.3	6.5	43.7	25.3	77.6	1.9	35.5
1 clo[5]	2.817	5.346	19.9	67.7	−14.4	6.0	17.3	63.1	−22.1	−7.8
Muscle + 1 clo	2.736	5.192	19.4	66.8	−15.9	3.3	16.7	62.1	−23.9	−11.0
Homo sapiens										
1 clo	2.817	5.246	21.3	70.4	−10.1	13.9	19.0	66.1	−17.1	1.2

[1] C = conductance in $W.m^{-2}.°C^{-1}$.
[2] Total conductance = (C × surface area in m^2).
[3] Kleiber BMR = BMR predicted from the Kleiber (1961) equation (see Table 9.1).
[4] BMR elevated by 15 per cent to account for climate and dietary induced increases. See Table 9.2 and text for explanation.
[5] C is reduced by 5 per cent to account for Neanderthal muscularity.

would be further marginally lowered to 16.7°C (62.1°F) and the minimum sustainable ambient temperature to –23.9°C (-11.0°F) (Table 9.3; Fig. 9.2).

One clo of insulation is approximately equivalent to that provided by either 3.2 cm of subcutaneous fat or 3.9 cm of relatively sparse body hair covering the entire body surface. These estimates are based on the insulation provided by 1 cm of fat being $0.049°C.M^{-2}.W^{-1}$ (Blurton & Edholm 1955) and that provided by 1 cm of body hair, similar in thermal properties to that of a modern pig, being $0.040°C.M^{-2}.W^{-1}$ (Blaxter 1989). Arctic adapted people living an indigenous life style are generally very lean and Neanderthals may have been similar. Furthermore the 3.2 cm of fat necessary to produce 1 clo of insulation would weigh in excess of 52 kg, an amount that would leave an 80 kg Neanderthal very little mass for muscle, bone and other tissue! How hirsute the Neanderthals might have been is unknown.

What is clear from this modelling is that Neanderthals would have had to have adopted some form of insulation approaching 1 clo if they were to survive away from shelter and fire at temperatures below freezing for any length of time. If Neanderthals were not hair-covered this could have been achieved by the use of roughly tailored skins, by the use of plant material or even felt. The numerous stone scrapers characteristic of the Mousterian could equally well have been used to scrape hair from the outer side of animal skins as to prepare their inner surfaces as commonly assumed (Bar-Yosef pers. comm. 2002). How sophisticated Neanderthal insulation may have been is dependent on the central question of how cold it actually was in the areas inhabited by

Neanderthals in glacial Europe?

The Stage 3 temperature environment

In the preceding section, our analyses of hominin thermoregulation were based on changes in ambient temperature alone and are therefore based on the assumption that any additional losses associated with significant airflow over the body are negligible. One relatively simple method to incorporate the influence of this important variable is by the use of wind chill temperature (WC) to provide a more relevant estimate of the environmental temperature actually experienced by a human. Wind chill is based on the rate of heat loss from exposed skin resulting from the combined effects of cold and wind speed and is calculated from the formula:

$$WC = 35.74 + 0.6215.T_a - 35.75.V^{0.16} + 0.4275T.V^{.016}$$

WC is wind chill temperature (°F), T_a is ambient air temperature, and V is wind speed (mph).

Wind chill is one of the outputs of the Stage 3 modelling which has been described in detail elsewhere in this volume (Chapter 2: van Andel 2003; Chapter 5: Barron *et al.* 2003). Wind chill temperatures for 457 Mousterian, Aurignacian and Gravettian sites spanning the period between 70–22 ka BP recorded in the Stage 3 archaeological data base (Chapter 4: van Andel *et al.* 2003b) are given in Appendices 9.1–9.3.

For the purposes of the present analyses, we have employed the time ranges and archaeological affiliation for these sites that are also used by van Andel *et al.* (Chapter 4: van Andel *et al.* 2003b). We

Table 9.4. *Median wind chill and differences between seasons and periods.*

| | Degrees Fahrenheit | | | Degrees Celsius | | |
| | | Temp Differences | | | Temp Differences | |
	Median WC[1]	Modern	Warm period	Median WC	Modern	Warm period
Summer						
Modern	61.1			16.2		
Warm period	56.0	5.1		13.3	2.9	
Glacial Maximum	48.5	12.7	7.6	9.1	7.0	4.2
Winter						
Modern	23.2			−4.9		
Warm period	8.8	14.4		−12.9	8.0	
Glacial Maximum	0.7	22.4	8.1	−17.4	12.5	4.5

[1] The grand median of wind chill estimates for all of the Mousterian (*n* = 141), Gravettian (*n* = 152) and Aurignacian (*n* = 164) sites in the data base.

Table 9.5. *Second order climate phases of the Weichselian Glaciation between OIS-5a and OIS-1, based on the GISP2 Greenland ice core, together with the chrono-archaeological time slices and the wind chill data sets used to predict temperature.*

SPECMAP	Climate phase[1]	Age[2]	Archaeological time slice	Wind chill data set
OIS-5A	Early Glacial Warm Phase	>74		
OIS-4	Transitional Phase	74–66	70–60	Glacial Max.
OIS-4	First Glacial Maximum	66–59		
OIS-3	Stable Warm Phase	59–45	59–47	
OIS-3	Transitional Phase	45–37	47–43	Warm Period
			42–38	
			37–34	
OIS-3	Early Cold Phase	37–27	33–30	
OIS-2	Last Glacial Maximum	27–16	29–26	Glacial Max.
			25–22	

[1] See Chapters 4 or 5 for a rationale and description.
[2] Age in thousands of years before present (BP).

recognize that there is considerable debate over the dating of some of these sites (e.g. d'Errico & Sánchez-Goñi 2003). In particular, some of the more recent sites may prove to be older than listed here.

Furthermore, even if the dates for the sites accepted by van Andel and colleagues (Chapters 3 & Chapter 4: van Andel *et al.* 2003a,b) are considered to be accurate, there are still considerable difficulties with the resolution of the absolute dates. It is not possible to match sites to specific warm or cold fluctuations (Dansgaard/Oeschger oscillations) that are documented in the GISP2 ice core for the periods considered here.

Because of both of these dating problems we have given six different wind chill (WC) temperatures for each site: Glacial Maximum summer WC, Glacial Maximum winter WC, typical Stage 3 warm period summer WC, typical Stage 3 warm period winter WC, modern summer WC, and modern winter WC. Glacial Maximum wind chill temperatures are used as proxies for the wind chill temperature

during Stage 3 cold periods (Chapter 2: van Andel 2003; Chapter 5: Barron *et al.* 2003). The modern summer and winter wind chill temperatures are provided for comparison.

The median wind chill temperatures for the 457 sites in Appendices 9.1–9.3 show that there is a much greater difference between modern and Stage 3 winter temperatures than there is between modern and Stage 3 summer temperatures (Table 9.4). There is a 12.5°C (22.4°F) difference between modern and Glacial Maximum winters and an 8.0°C (14.4°F) difference between modern and Stage 3 warm period winters. In contrast there is a 7.0°C (12.7°F) difference between modern and a Glacial Maximum summers and a 2.9°C (5.1°F) difference between modern and Stage 3 warm period summers. It is also important to note that Stage 3 warm period summers and winters are considerably chillier than modern summers and winters.

Because of the difficulties with the resolution of the dates for the sites in the chrono-archaeological data base, van Andel and colleagues (Chapter 4: van Andel *et al.* 2003b) have divided the GISP2 ice-core record into longer-duration time slices, or climate sub-units (Fig. 4.1; Table 4.3), that allow comparison with justifiable temporal divisions in the archaeological record. The wind chill data sets that are used in subsequent analyses are given in Table 9.5. The Glacial Maximum wind chill data set is used to estimate temperatures for the earliest archaeological time slice (70–60 ka BP) and for the four most recent periods (22–37 ka BP) that correspond to OIS-4, the early cold phase of OIS-3 and OIS-2. The warm period data set is used to estimate temperatures for OIS-3 between 59–37 ka BP) when warmer conditions prevailed. Although it is recognized that this provides, at best, an approximation of the climatic conditions prevailing at the sites when the hominins were in occupation, the analyses focus primarily on the more recent time slices where it is likely that conditions approximating those estimated by the Glacial Maximum temperature data set prevailed. The implications of these assumptions are discussed in the following section.

The wind chill data sets from Table 9.5 are used

to determine the median wind chill temperatures for each of the major archaeological cultures, Mousterian, Aurignacian and Gravettian, for each of the time slices between 70–22 ka BP. Summer and winter wind-chill temperatures for both the Mousterian and Aurignacian sites during the warm periods of OIS-3 (38–59 ka BP) are, as expected, higher than the wind-chill temperatures for earlier and later periods. During these warm periods median summer wind chill temperatures for the Mousterian sites ranged between 56.3–58.5°F and for the Aurignacian sites between 54.0–54.3°F (Table 9.6). Median winter wind chill temperatures for the Mousterian sites ranged between 8.8–10.2°F and between 7.8–11.3°F for the Aurignacians (Table 9.7).

In the more recent colder time slices representing the early cold phase of OIS-3 and OIS-2 the Neanderthals appear to be occupying sites with consistently warmer winter wind chill temperatures than either the Aurignacians or the Gravettians (Fig. 9.3). Table 9.8 summarizes the median wind chill temperatures, and temperature percentiles, by archaeological culture over the period of the four time slices from 37–22 ka BP. Median summer wind chill temperatures are virtually identical at ~48.0°F over this period. However, median winter temperatures vary from 3.1°F for the Mousterian sites to –4.1°F for the Aurignacian sites and –8.7°F for the Gravettian sites.

It cannot be emphasized more strongly that there is considerable controversy over the precise dates of these sites. However, there is a marked consistency in the median summer and winter wind-chill temperatures for the three archaeological cultures during the period between 37–22 ka BP when cold conditions prevailed (Fig. 9.3; Table 9.8). This lends credence to the conclusion that the Mousterians were se-

Table 9.6. *Summer wind chill (F°) percentiles.*

Time slices[1]		25–22	29–26	33–30	37–34	42–38	47–43	59–47	70–60
Mousterian									
	5	44.4	46.8	30.9	44.4	46.7	43.5	43.5	42.6
	10	44.4	46.8	37.8	44.9	47.9	48.2	50.3	42.7
	25	45.4	46.9	46.7	47.3	53.9	54.0	53.3	46.3
median	50	46.4	47.7	48.0	52.2	56.3	58.5	57.0	50.1
	75	49.2	48.1	52.2	63.3	64.9	64.3	62.1	57.2
	90	.	.	57.3	69.6	69.9	68.7	65.8	61.8
	95	.	.	.	.	70.3	69.8	68.0	.
		$n=6$[2]	$n=4$	$n=18$	$n=12$	$n=25$	$n=33$	$n=32$	$n=11$
Aurignacian									
	5	34.3	20.4	30.3	15.8	39.0	51.2	.	.
	10	34.3	31.8	34.8	33.0	43.9	51.2	.	.
	25	43.8	43.2	42.6	36.9	51.3	52.1	.	.
median	50	48.8	46.8	48.5	47.0	54.0	54.3	.	.
	75	54.3	52.2	53.8	50.1	60.2	60.2	.	.
	90	.	54.4	55.0	54.7	63.8	.	.	.
	95	.	55.5	61.9	63.1	67.1	.	.	.
		$n=7$	$n=33$	$n=48$	$n=36$	$n=32$	$n=8$	$n=0$	$n=0$
Gravettian									
	5	38.3	33.6	34.6	42.6	.	.	.	.
	10	42.0	41.9	41.8	42.6	.	.	.	.
	25	46.5	44.1	43.5	46.0	.	.	.	.
median	50	48.8	47.6	46.7	54.7	.	.	.	.
	75	54.7	52.0	49.6	63.5	.	.	.	.
	90	58.6	54.7	54.8	.	.	.	.	.
	95	61.0	60.6	55.0	.	.	.	.	.
		$n=50$	$n=64$	$n=38$	$n=6$	$n=0$	$n=0$	$n=0$	$n=6$

[1] Time slices ⇨ thousands of years BP.
[2] n = number of sites in the time-slice for which temperature is known.

Table 9.7. *Winter wind chill (F°) percentiles.*

Time slices[1]		25–22	29–26	33–30	37–34	42–38	47–43	59–47	70–60
Mousterian									
	5	–10.6	–0.4	–25.1	–15.2	–9.7	–3.7	–5.7	–7.2
	10	–10.6	–0.4	–16.9	–14.4	–4.1	–0.2	–0.3	–6.0
	25	–5.4	0.4	–0.3	–11.0	2.4	3.0	2.8	0.8
median	50	1.9	3.1	5.2	–1.1	8.8	10.1	10.2	7.7
	75	12.0	26.3	12.1	19.6	13.8	18.4	19.1	24.4
	90	.	.	32.6	22.5	19.5	25.5	29.7	30.9
	95	.	.	.	.	23.4	32.0	32.4	.
		$n=6$[2]	$n=4$	$n=18$	$n=12$	$n=25$	$n=33$	$n=32$	$n=11$
Aurignacian									
	5	–11.4	–30.0	–15.8	–22.8	–6.4	–0.3	.	.
	10	–11.4	–17.5	–15.5	–16.9	–1.4	–0.3	.	.
	25	–10.6	–11.0	–10.5	–10.9	1.2	0.2	.	.
median	50	7.7	–3.6	–1.9	–7.3	7.8	11.3	.	.
	75	1.5	5.8	9.8	9.1	14.0	20.6	.	.
	90	.	16.4	24.6	16.2	20.5	.	.	.
	95	.	28.3	30.4	16.9	28.2	.	.	.
		$n=7$	$n=33$	$n=48$	$n=36$	$n=32$	$n=8$	$n=0$	$n=0$
Gravettian									
	5	–17.0	–16.2	–16.0	–15.5	.	.	.	.
	10	–15.5	–15.5	–15.5	–15.5	.	.	.	.
	25	–12.1	–10.9	–10.2	–15.5	.	.	.	.
median	50	–9.1	–7.1	–7.0	9.7	.	.	.	.
	75	6.5	6.0	9.1	–5.7	.	.	.	.
	90	12.6	11.8	11.3	.	.	.	.	.
	95	20.6	20.3	16.9	.	.	.	.	.

[1] Time slices ⇨ thousands of years BP.
[2] n = number of sites in the time-slice for which temperature is known.

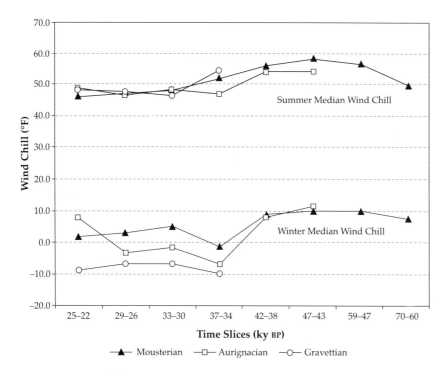

lecting sites with median winter wind-chill temperatures that were considerably warmer than the conditions prevailing at the sites occupied by either the Aurignacians or particularly the Gravettians.

Neanderthals and the Stage 3 temperature environment

The analyses of wind-chill estimates for the Glacial Maximum suggest that the Neanderthal occupants of Mousterian sites could have physiologically survived summer Glacial Maximum summers without insulation other than that estimated for their increased muscle mass (Fig. 9.4). Based on the forgoing modelling (Table 9.3) the minimum sustainable ambient temperature at three times BMR for a Neanderthal with a Kleiber-estimated BMR would be 6.5°C (43.7°F) and for a Neanderthal with

Figure 9.3. *Median summer and winter wind chill temperatures (°F) for eight archaeological time slices spanning the period from 70–22 ka BP. (Data from Tables 9.6 and 9.7.)*

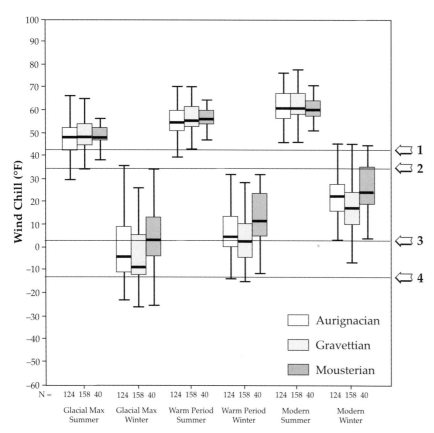

Figure 9.4. *Wind chill temperature (°F) estimates for the Aurignacian, Gravettian and Mousterian sites for the period from 37–22 ka BP. Although the GISP2 ice core suggests that the majority of this period would be experiencing cool temperatures, estimates for Stage 3 warm period temperatures are included along with the estimates for glacial maximum summer and winter temperatures. Estimates for modern day summer and winter temperatures are included for comparison. Lines = minimum sustainable temperatures. Line 1 = (6.5°C, 43.7°F) for a muscled Neanderthal with a Kleiber-estimated BMR. Line 2 = (1.9°C, 35.5°F) for a muscled Neanderthal with an elevated BMR. Line 3 = (–15.9°C, 3.3°F) for a muscled Neanderthal with a Kleiber-estimated BMR and insulation of 1 clo. Line 4 = (–23.9°C, –11.0°F) for a similarly insulated Neanderthal with an elevated BMR. (Minimum sustainable temperatures from Table 9.3.)*

Table 9.8. *Summer and winter wind chill percentiles for the period 37–22 ka BP.*

Percentiles	Summer wind chill (F°)			Winter wind chill (F°)		
	Aurignacian $n = 124$	Gravettian $n = 158$	Mousterian $n = 40$	Aurignacian $n = 124$	Gravettian $n = 158$	Mousterian $n = 40$
5	29.8	41.1	38.9	−19.6	−16.0	−15.9
10	34.5	41.9	44.6	−15.5	−15.5	−12.6
25	42.6	45.1	46.8	−10.6	−11.6	−3.6
median 50	48.0	48.1	48.1	−4.1	−8.7	3.1
75	52.4	54.2	52.5	9.1	5.9	15.5
90	54.7	54.8	64.3	16.5	11.1	31.6
95	57.2	59.9	69.4	25.5	16.	34.1

an elevated BMR 1.9°C (35.5°F). There are forty Mousterian sites in the four time slices representing the period from 22–37 ka BP. Of these only two have summer Glacial Maximum wind chill temperatures colder than 43.7°F (Hyaena Den and Salzofenhöhle) and only Hyaena Den has a temperature colder than 35.5°F (see Appendix 9.1).

The situation is much different for Glacial Maximum winters. The temperatures estimated for all of the sites fall below the minimum sustainable ambient temperature of a Neanderthal without additional insulation (Fig. 9.4). Even if muscled Neanderthals with Kleiber-estimated BMRs were insulated to the level of 1 clo, providing protection to −15.9°C (3.3°F), twenty-three of the forty Mousterian sites would be outside of their temperature limits. The minimum sustainable ambient temperature for a similarly insulated Neanderthal with an elevated BMR is −23.9°C (11.0°F), but there are still six of the forty sites with colder estimated temperatures.

This analysis assumes that all of the forty sites were occupied during the winters of the colder periods of the climate fluctuations that were characteristic of OIS-3. Some of the sites with colder winter temperatures may have been seasonally occupied during the summers. Alternatively, the sites could have been occupied during the increasingly infrequent OIS-3 warm periods. If this were the case and they were occupied in warm period winters, six of the forty sites would still have temperatures lower than the 3.3°F limit of the Kleiber-estimated insulated Neanderthal but all of the sites would be assessable to the insulated Neanderthal with an elevated BMR.

These analyses are designed only to give an indication of the magnitude of temperatures that would have faced Neanderthals during OIS-3 and of Neanderthal temperature tolerance given various assumptions about BMR and insulation. They may prove useful in formulating testable hypotheses in relation to patterns of occupation of some of the sites

that would seem to be outside the range of tolerance during the winters of the colder phases of OIS-3. However it is also clear that some degree of insulation and/or behavioural modification would have been necessary for Neanderthals to have survived as well as they did and for as long as they did during this period.

The winter and summer temperature estimates presented here are twenty-four hour average wind chill temperatures averaged over three-month periods (December, January and February for the winter estimates and June, July and August for the summer estimates) and may overestimate the temperature stress experienced by the Neanderthals. These hominins occupied cave and rockshelter sites, are known to have used simple structures as well as fire and most certainly used behavioural mechanisms such as huddling to keep themselves warm at night or during the periods of coldest temperature. However, these behavioural factors would have to have given them a considerable advantage to negate the necessity of some type of insulation. There is an approximately 46°F difference in minimum sustainable ambient temperature between a muscled Neanderthal with elevated BMR (35.5°F) and one with 1 clo of insulation (−11.0°F) who could have survived in the sites known to have been occupied. There would have to be the same degree of advantage given by behavioural or cultural factors to permit an uninsulated Neanderthal to survive during Glacial Maximum winters.

It should also be noted that physiological capability does not necessarily imply ecological viability. The costs of maintaining internal heat production at the required levels would only have been possible if Neanderthals were able to sustain a correspondingly high level of dietary energy intake.

In Chapter 4 van Andel et al. (2003b) suggest that both the Aurignacians and the Mousterians showed a similar pattern of retreat to warmer western and southern climates in the second half of OIS-3

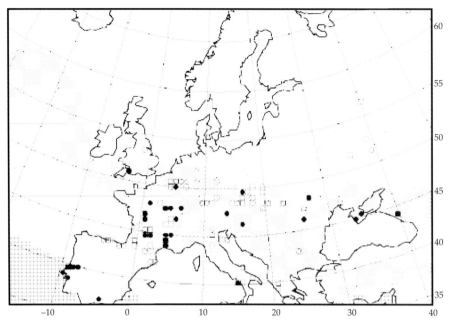

Figure 9.5. *Distribution of Mousterian (black circles), Aurignacial (open diamonds) and Gravettian (open squares) sites dating from the four most recent time slices spanning the period 70–22 ka BP. Hatching represents glacial maximum winter wind chill temperatures warmer than 32 °F. Horizontal lines represent wind chill temperatures between 32 °F and 0 °F. Diagonal lines represent wind chill temperatures between 0 °F and –10 °F. Unshaded areas represent wind chill temperatures below –10 °F. The lower critical temperature and minimum sustainable ambient temperatures for a Neanderthal with a Kleiber-estimated BMR, with an elevated BMR, with an elevated BMR and a 5 per cent reduction in heat loss as the result of additional muscle insulation, and with an elevated BMR, a 5 per cent reduction in heat loss as the result of addition muscle insulation plus 1 clo insulation. (Data and equations as in Fig. 9.1.)*

Conclusion

Although the Neanderthals have frequently been described as have an arctic or hyper arctic body form, the foregoing analyses have demonstrated that they would only have had a modest advantage over Anatomically Modern Humans in their lower critical and minimum sustainable temperatures. This continues to be the case even if allowance is made for the insulating effect of their increased muscle mass and for a dietary related elevated BMR.

When inferred Neanderthal temperature tolerances are set against inferred wind chill temperatures for the sites they were known to have occupied, it is clear that they could not have survived winter temperatures without additional cultural insulation. This applies equally to Stage 3 warm period winters and to Glacial Maximum winter temperatures (here used as a proxy for Stage 3 cold period winters). They would have had little difficulty surviving Stage 3 summers without additional insulation. Stage 3 wind chill estimates also suggest that Mousterian sites were located in areas with warmer Glacial Maximum winter temperatures than where the Aurignacian or Gravettian sites.

If Neanderthals only had a limited cultural capability of insulating themselves, even to the modest level of 1 clo, they would have had difficulty under the increasingly harsh conditions of the latter half of OIS-3. The costs of maintaining internal heat production at the required levels would have only been possible if Neanderthals were able to sustain a correspondingly high level of dietary energy intake. This may have been a significant factor in the disappearance of the Neanderthals, particularly if they were in competition for climate refugia and dietary resources with better-insulated Anatomically Modern Humans.

and that it was the Gravettians who were particularly adapted to the colder environments of Europe. This analysis clearly demonstrates that the Neanderthals were occupying warmer sites than the Gravettians (Fig. 9.3). However, it also suggests that even though the Mousterians and Aurignacians may have had a similar pattern of retreat the Mousterians were tending to select sites with warmer winter temperatures. Figure 9.5 shows the distribution of Mousterian, Aurignacian and Gravettian sites against relevant climate isotherms. It is clear from this that the Gravettians are occupying colder sites of the European Plains (as well as warmer western and southern European sites) as were to a lesser extent the Aurignacians. These are the environments that are largely outside Neanderthal temperature tolerance and this pattern of distribution implies more sophisticated insulation and/or behavioural capability in at least the Gravettians to enable them to cope these their much colder environments.

Acknowledgements

We would like to thank Tjeerd van Andel, the 'father' of the Stage 3 Project, for inviting us to participate and introducing us to the exciting world of palaeoclimate modelling and paleoenvironmental reconstruction. We would also like to thank the other members of the Stage 3 Project for stimulating discussion and criticism over the past few years of project seminars and meetings. We would also like to thank the Leverhulme Trust for making the project possible and members of our respective departments at University College London and Liverpool John Moores University for their input over the years.

Notes

1. Temperatures in both Fahrenheit and Celsius are reported throughout. This is because the wind chill temperature is considered here to provide the most relevant estimate of the environmental temperature actually experienced by humans. Wind chill is conventionally computed and reported in Fahrenheit.

References

Aiello, L.C. & P. Wheeler, 1995. The Expensive-Tissue hypothesis: the brain and the digestive system in human and primate evolution. *Current Anthropology* 36, 199–221.

Allen, J.A., 1877. The influence of physical conditions on the genesis of species. *Radical Review,* 108–40.

Bagmen, C., 1847. Über die Verhältnisse der Wärmeökonomie der Thiere zu ihrer Grösse. *Göttinger Studien* 1, 595–708.

Barron, E., T.H. van Andel & D. Pollard, 2003. Glacial environments II: reconstructing the climate of Europe in the Last Glaciation, in *Neanderthals and Modern Humans in the European Landscape during the Last Glaciation,* Chapter 5, eds. T.H. van Andel & W. Davies. (McDonald Institute Monographs.) Cambridge: McDonald Institute for Archaeological Research, 57–78.

Blaxter, K., 1989. *Energy Metabolism in Animals and Man.* Cambridge: Cambridge University Press.

Burton, A.C. & O.G. Edholm, 1955. *Man in a Cold Environment: Physiological and Pathological Effects of Exposure to Low Temperatures.* London: Edward Arnold.

Churchill, S.E., 1998. Cold adaptation, heterochrony, and Neandertals. *Evolutionary Anthropology* 7, 46–61. ○

Coon, C.S., 1992. *The Origin of Races.* New York (NY): Alfred A. Knopf.

d'Errico, F. & M.F. Sánchez-Goñi, 2003. Neandertal extinction and the millennial scale climatic variability of OIS 3. *Quaternary Science Reviews* 22, 769–88. ○

Dubois, D. & E.F. Dubois, 1916. Clinical calorimetry: a formula to estimate the approximate surface area if height and weight be known. *Archives of Internal Medicine* 17, 863–71.

Franciscus, R.G. & S.E. Churchill, 2002. The costal skeleton of Shanidar 3 and a reappraisal of Neandertal thoracic morphology. *Journal of Human Evolution* 42, 303–56.

Glickman-Weiss, E.L., A.G. Nelson, C.M. Hearon, F.L. Goss, R.J. Robertson & D.A. Cassinelli, 1993. Effects of body morphology and mass on thermal responses to cold water: revisited. *European Journal of Applied Physiology* 66, 299–330.

Holliday, T.W., 1997a. Postcranial evidence of cold adaptation in European Neandertals. *American Journal of Physical Anthropology* 104, 245–58. ○

Holliday, T.W., 1997b. Body proportions in Late Pleistocene Europe and modern human origins. *Journal of Human Evolution* 32, 423–48.

Kleiber, M., 1961. *The Fire of Life.* New York (NY): Wiley.

Leonard, W.R., M.V. Sørensen, V.A. Galloway, G.J. Spencer, M.J. Mosher, L. Osipova & V.A. Spitsyn, 2002. Climatic influences on basal metabolic rates among circumpolar populations. *American Journal of Human Biology* 14, 609–20.

Mount, L.E., 1979. *Adaptation to Thermal Environment.* London: Edward Arnold.

Pearson, O.M., 2000a. Activity, climate, and postcranial robusticity: implications for modern human origins and scenarios of adaptive change. *Current Anthropology* 41, 569–607. ✳

Pearson, O.M., 2000b. Postcranial remains and the origin of modern humans. *Evolutionary Anthropology* 9, 229–47.

Richards, M.P. P.B. Pettitt, E. Trinkaus, F.H. Smith, M. Paunovic & I. Karavanic, 2000. Neanderthal diet at Vindija and Neanderthal predation: the evidence from stable isotopes. *Proceedings of the National Academy of Sciences of the USA* 97, 7663–6.

Richards, M.P., P.B. Pettitt, M.C. Stiner & E. Trinkaus, 2001. Stable isotope evidence for increasing dietary breadth in the European mid-Upper Paleolithic. *Proceedings of the National Academy of Sciences of the USA* 98, 6528–32.

Ruff, C.B., 1991. Climate, body size and body shape in hominid evolution. *Journal of Human Evolution* 21, 81–105.

Ruff, C.B., 1993. Climatic adaptation and hominid evolution: the thermoregulatory imperative. *Evolutionary Anthropology* 2, 53–60.

Ruff, C.B., 1994. Morphological adaptation to climate in modern and fossil hominids. *Yearbook of Physical Anthropology* 37, 65–107.

Ruff, C.B. & A. Walker, 1993. Body size and body shape, in *The Nariokotome Homo Erectus Skeleton,* eds. A. Walker & R.E. Leakey. Cambridge (MA): Harvard University Press, 234–65.

Ruff, C.B., E. Trinkaus & T.W. Holliday, 1997. Body mass and encephalization in Pleistocene *Homo. Nature* 387, 173–6.

Ruff, C.B., E. Trinkaus, A. Walker & C.S. Larsen, 1993. Postcranial robusticity in *Homo* 1. Temporal trends and mechanical interpretation. *American Journal of Physical Anthropology* 91, 21–53.

Steegmann, A.T. & W.S. Platner, 1968. Experimental cold modification of craniofacial morphology. *American Journal of Physical Anthropology* 28, 17–30.

Steegmann, A.T., F.J. Cerny & T.W. Holliday, 2002. Neandertal cold adaptation: physiological and energetic factors. *American Journal of Human Biology* 14, 566–83. ○

Trinkaus, E., 1981. Neanderthal limb proportions and cold adaptation, in *Aspects of Human Evolution*, ed. C.B. Stringer. London: Taylor & Francis, 187–224.

Trinkaus, E., 1983. *The Shanidar Neandertals*. New York (NY): Academic Press.

Trinkaus, E. & C.B. Ruff, 1989. Cross-sectional geometry of Neandertal femoral and tibial diaphyses: implications for locomotion. *American Journal of Physical Anthropology* 78, 315–16.

Trinkaus, E., S.E. Churchill, I. Villemeur, K.G. Riley, J.A. Heller & C.B. Ruff, 1991. Robusticity versus shape - the functional interpretation of Neanderthal appendicular morphology. *Journal of the Anthropological Society of Nippon* 99, 257–78.

van Andel, T.H. 2003. Glacial environments I: the Weichselian climate in Europe between the end of the OIS-5 interglacial and the Last Glacial Maximum, in *Neanderthals and Modern Humans in the European Landscape during the Last Glaciation*, Chapter 2, eds. T.H. van Andel & W. Davies. (McDonald Institute Monographs.) Cambridge: McDonald Institute for Archaeological Research, 9–20.

van Andel, T.H., W. Davies, B. Weninger & O. Jöris, 2003a. Archaeological dates as proxies for the spatial and temporal human presence in Europe: a discourse on the method, in *Neanderthals and Modern Humans in the European Landscape during the Last Glaciation*, Chapter 3, eds. T.H. van Andel & W. Davies. (McDonald Institute Monographs.) Cambridge: McDonald Institute for Archaeological Research, 21–30.

van Andel, T.H., W. Davies & B. Weninger, 2003b. The human presence in Europe during the Last Glacial period I: human migrations and the changing climate, in *Neanderthals and Modern Humans in the European Landscape during the Last Glaciation*, Chapter 4, eds. T.H. van Andel & W. Davies. (McDonald Institute Monographs.) Cambridge: McDonald Institute for Archaeological Research, 31–56.

Wheeler, P., 1993. The influence of stature and body form on hominid energy and water budgets: a comparison of *Australopithecus* and early *Homo* physiques. *Journal of Human Evolution* 24, 13–28.

Appendix 9.1. *Wind-chill temperatures (°F) for Mousterian sites.*

Site no.	Site name	Glacial Maximum Summer	Glacial Maximum Winter	Warm Period Summer	Warm Period Winter	Modern Summer	Modern Winter
25,000–22,000 ka BP	**6 sites**						
11	Salemas [algar]	46.8	34.1	56.3	31.9	57.5	45.0
115	A. Moula [Soyons]	46.1	0.7	50.9	7.0	55.8	17.9
160	Gr. St-Marcel [d'Ardeche] [Bidon]	48.1	3.1	54.7	11.5	59.8	24.6
203	Trou Magrite	44.4	–3.6	51.2	1.1	55.5	18.8
251	Gr. La Cala	52.5	4.6	56.1	25.5	70.8	40.9
304	Kulna Cave	45.7	–10.6	56.0	–0.4	59.7	10.8
29,000–26,000 ka BP	**4 sites**						
84	Les Pecheurs [Casteljau]	48.1	3.1	54.7	11.5	59.8	24.6
11	Salemas [algar]	46.8	34.1	56.3	31.9	57.5	45.0
159	Gr. d'Echenoz-la-Meline [La Baume]	47.3	–0.4	54.6	5.1	58.4	19.6
160	Gr. St-Marcel [d'Ardeche] [Bidon]	48.1	3.1	54.7	11.5	59.8	24.6
33,000–30,000 ka BP	**18 sites**						
5	Caldeirao Cave	52.1	20.3	59.6	26.7	62.4	37.2
11	Salemas [algar]	46.8	34.1	56.3	31.9	57.5	45.0
17	Columbeira, Gruta Nova	46.3	32.5	56.2	28.7	59.8	37.3
42	Zafarraya Cave	56.4	18.9	60.5	22.5	70.0	30.9
76	Grotte du Renne, Arcy-sur-Cure	47.7	2.0	54.6	8.3	58.2	21.9
98	Combe Grenal [Domme]	48.5	9.8	54.6	13.1	61.1	27.4
101	A du Mas Viel [St-Simon]	46.5	5.9	50.5	4.5	58.1	22.7
107	Gr. de La Baume [Gigny sur Suran]	46.8	0.1	53.6	5.3	57.5	19.4
120	A. Brugas [?O/A] [Vallabrix]	50.1	7.7	61.5	19.1	65.0	23.6
138	L'Ermitage [Lussac-les-Chateaux]	48.9	7.1	53.7	10.0	59.4	25.3
160	Gr. St-Marcel [d'Ardeche] [Bidon]	48.1	3.1	54.7	11.5	59.8	24.6
162	A. Sabourin [Dousse]	48.0	5.8	55.6	12.6	59.9	26.1
169	Montagne de Girault [Genay]	47.0	0.7	54.5	7.5	58.2	21.1
188	Hyaena Den	30.9	–16.0	47.5	8.8	51.2	23.4
236	Salzofenhohle	38.6	–25.1	42.9	–11.1	48.5	4.5

Appendix 9.1. *(cont.)*

Site no.	Site name	Glacial Maximum Summer	Glacial Maximum Winter	Warm Period Summer	Warm Period Winter	Modern Summer	Modern Winter
33,000–30,000 ka BP *(cont.)*							
251	Gr. La Cala	52.5	4.6	56.1	25.5	70.8	40.9
326	Ripiceni-Izvor	54.8	−12.0	63.5	−6.1	70.8	9.8
378	Mezmaiskaya	65.1	−1.3	59.4	0.1	62.7	6.9
37,000–34,000 ka BP 12 sites							
3	Foz do Enxarrique	53.8	17.3	60.2	23.1	64.0	33.8
4	Figueira Brava Cave	52.5	23.4	60.8	29.0	63.7	38.1
6	Oliveira Cave [Almonda cave system]	52.1	20.3	59.6	26.7	62.4	37.2
7	Almonda [EVS]	52.1	20.3	59.6	26.7	62.4	37.2
78	Les Cottes [St. Pierre de Maille]	46.9	3.1	54.5	10.8	58.8	24.9
135	Les Rivaux, Loc. 1 [Espaly-St-Marcel]	46.2	1.5	49.5	5.4	55.0	18.6
203	Trou Magrite	44.4	−3.6	51.2	1.1	55.5	18.8
280	Krapina	52.4	−4.6	64.6	5.7	69.0	19.2
331	Gura Cheii-Risnov	48.5	−15.2	54.3	−8.7	60.8	4.9
338	Buran-Kaya III	69.6	−12.6	70.3	13.8	79.0	22.1
340	Kabazi II	66.4	−6.0	69.6	5.7	76.4	16.5
341	Zaskal'naya VI	69.6	−12.6	70.3	13.8	79.0	22.1
42,000–38,000 ka BP 25 sites							
37	L'Arbreda	52.1	25.9	60.2	19.5	65.4	28.1
40	Ermitons Cave	48.0	9.6	48.1	11.6	54.4	21.3
49	Cueva Millan	42.4	8.6	46.4	7.2	56.6	22.6
73	Roche a Pierrot [St-Cesaire]	48.8	11.5	56.3	19.5	59.3	27.3
78	Les Cottes [St Pierre de Maille]	46.9	3.1	54.5	10.8	58.8	24.9
81	La Quina Y-Z [Villebois la Valette]	49.1	9.8	55.9	15.1	60.9	29.2
98	Combe Grenal [Domme]	48.5	9.8	54.6	13.1	61.1	27.4
115	A. Moula [Soyons]	46.1	0.7	50.9	7.0	55.8	17.9
117	Pech de l'Aze II [Carsac]	48.5	9.8	54.6	13.1	61.1	27.4
146	Gr. de la Chenelaz [Hostias]	44.6	−2.1	51.2	3.7	55.4	17.2
153	Gr. Tournal (or Grande Grotte de Bize)	45.8	2.6	55.8	16.5	61.8	27.5
170	Le Moustier	48.8	9.1	53.3	9.2	60.3	26.1
188	Hyaena Den	30.9	−16.0	47.5	8.8	51.2	23.4
230	Sesselfelsgrotte	44.3	−7.0	55.3	2.2	58.7	14.8
297	Tata	51.4	−7.8	63.7	0.7	66.7	15.2
298	Erd	54.5	−5.3	66.2	2.7	68.6	17.0
324	Buzdujeni I Cave	54.8	−12.0	63.5	−6.1	70.8	9.8
333	Cioarei/Borosteni	52.1	−9.2	58.4	−2.7	63.7	10.0
338	Buran-Kaya III	69.6	−12.6	70.3	13.8	79.0	22.1
339	Starosel'e	66.4	−6.0	69.6	5.7	76.4	16.5
340	Kabazi II	66.4	−6.0	69.6	5.7	76.4	16.5
341	Zaskal'naya VI	69.6	−12.6	70.3	13.8	79.0	22.1
353	Betovo	45.2	−20.1	58.7	−11.2	62.7	−0.3
377	Vorontsov Cave	68.4	26.5	67.3	25.0	68.9	28.5
378	Mezmaiskaya	65.1	−1.3	59.4	0.1	62.7	6.9
47,000–43,000 ka BP 33 sites							
6	Oliveira Cave [Almonda cave system]	52.1	20.3	59.6	26.7	62.4	37.2
35	Castillo	53.8	17.2	52.2	36.7	68.6	46.5
36	Abric Romani	44.6	3.2	60.5	21.0	67.2	30.1
37	L'Arbreda	52.1	25.9	60.2	19.5	65.4	28.1
39	Cova Beneito	54.2	15.3	57.1	23.7	67.2	35.1
45	Roca dels Bous	46.4	8.4	59.4	19.1	65.7	29.1
49	Cueva Millan	42.4	8.6	46.4	7.2	56.6	22.6
52	Kurtzia	35.7	16.4	44.0	18.2	58.7	40.5
67	Vanguard Cave	57.2	24.4	65.1	30.0	75.1	40.7
98	Combe Grenal [Domme]	48.5	9.8	54.6	13.1	61.1	27.4
117	Pech de l'Aze II [Carsac]	48.5	9.8	54.6	13.1	61.1	27.4
126	[Gr.] Neron [Soyons]	46.1	0.7	50.9	7.0	55.8	17.9
149	Barbas III [Creysse]	48.6	12.3	56.1	18.6	61.6	30.2
170	Le Moustier	48.8	9.1	53.3	9.2	60.3	26.1
172	Abri du Ranc de l'Arc [Lagorce]	48.1	3.1	54.7	11.5	59.8	24.6
209	Sclayn Cave	43.6	−4.8	52.1	3.2	55.7	19.0
215	Das Geissenklosterle	42.6	-9.7	52.9	-0.1	57.3	17.1

Appendix 9.1. *(cont.)*

Site no.	Site name	Glacial Maximum Summer	Glacial Maximum Winter	Warm Period Summer	Warm Period Winter	Modern Summer	Modern Winter
47,000–43,000 ka BP *(cont.)*							
230	Sesselfelsgrotte	44.3	−7.0	55.3	2.2	58.7	14.8
247	Castelcivita	53.7	5.1	58.5	9.6	67.7	26.7
258	Gr. del Broion	41.5	−12.3	67.1	10.1	72.3	24.0
266	Gr. Guattari	52.2	2.3	42.1	14.8	69.7	39.9
267	Gr. di Sant'Agostino	53.2	0.8	59.9	10.3	67.6	26.7
275	Divje Babe	55.3	−2.8	66.4	5.8	70.5	20.0
285	Crvena Stijena	53.9	−1.0	62.6	11.1	63.1	17.1
298	Erd	54.5	−5.3	66.2	2.7	68.6	17.0
304	Kulna Cave	45.7	−10.6	56.0	−0.4	59.7	10.8
326	Ripiceni-Izvor	54.8	−12.0	63.5	−6.1	70.8	9.8
333	Cioarei/Borosteni	52.1	−9.2	58.4	−2.7	63.7	10.0
337	Samuilitsa	58.1	−11.8	67.7	0.6	72.0	18.7
339	Starosel'e	66.4	−6.0	69.6	5.7	76.4	16.5
341	Zaskal'naya VI	69.6	−12.6	70.3	13.8	79.0	22.1
374	Il'skaya	66.6	−9.8	69.4	9.5	74.9	16.4
378	Mezmaiskaya	65.1	−1.3	59.4	0.1	62.7	6.9
59,000–47,000 ka BP 2 sites							
10	Gruta do Escoural	55.9	22.7	64.2	29.0	68.0	39.9
17	Columbeira, Gruta Nova	46.3	32.5	56.2	28.7	59.8	37.3
35	Castillo	53.8	17.2	52.2	36.7	68.6	46.5
45	Roca dels Bous	46.4	8.4	59.4	19.1	65.7	29.1
47	Cariguela	54.3	16.6	58.1	20.8	67.7	28.9
54	Pena Miel 1	40.3	11.7	44.2	9.7	56.5	26.2
59	Los Moros I [Gabasa]	46.2	8.1	58.8	17.1	66.3	29.5
65	Gorham's Cave	57.2	24.4	65.1	30.0	75.1	40.7
67	Vanguard Cave	57.2	24.4	65.1	30.0	75.1	40.7
100	Regourdou [Montignac]	48.8	9.1	53.3	9.2	60.3	26.1
108	Gr. aux Ours [Gondenans les Moulins]	46.5	−0.7	52.9	2.9	56.7	17.9
116	La Chapelle-aux-Saints	47.4	6.1	49.5	2.6	56.7	21.4
117	Pech de l'Aze II [Carsac]	48.5	9.8	54.6	13.1	61.1	27.4
118	Ioton [Beaucaire]	50.1	7.7	61.5	19.1	65.0	23.6
119	Fonseigner [Bourdeilles]	48.8	9.1	53.3	9.2	60.3	26.1
120	A. Brugas [?O/A] [Vallabrix]	50.1	7.7	61.5	19.1	65.0	23.6
121	La Roquette II [Conquerac]	46.9	5.9	57.7	15.4	62.1	23.4
149	Barbas III [Creysse]	48.6	12.3	56.1	18.6	61.6	30.2
170	Le Moustier	48.8	9.1	53.3	9.2	60.3	26.1
215	Das Geissenklosterle	42.6	−9.7	52.9	−0.1	57.3	17.1
225	Konigsaue	42.5	−9.8	54.0	0.7	57.7	14.5
230	Sesselfelsgrotte	44.3	−7.0	55.3	2.2	58.7	14.8
251	Gr. La Cala	52.5	4.6	56.1	25.5	70.8	40.9
258	Gr. del Broion	41.5	−12.3	67.1	10.1	72.3	24.0
259	Gr. di Gosto	53.3	3.5	60.8	5.6	65.1	23.2
266	Gr. Guattari	52.2	2.3	42.1	14.8	69.7	39.9
267	Gr. di Sant'Agostino	53.2	0.8	59.9	10.3	67.6	26.7
298	Erd	54.5	−5.3	66.2	2.7	68.6	17.0
304	Kulna Cave	45.7	−10.6	56.0	−0.4	59.7	10.8
326	Ripiceni-Izvor	54.8	−12.0	63.5	−6.1	70.8	9.8
339	Starosel'e	66.4	−6.0	69.6	5.7	76.4	16.5
371	Korman' IV	54.5	−12.7	62.4	−5.4	69.3	9.8
70,000–60,000 ka BP 11 sites							
17	Columbeira, Gruta Nova	46.3	32.5	56.2	28.7	59.8	37.3
65	Gorham's Cave	57.2	24.4	65.1	30.0	75.1	40.7
67	Vanguard Cave	57.2	24.4	65.1	30.0	75.1	40.7
98	Combe Grenal [Domme]	48.5	9.8	54.6	13.1	61.1	27.4
117	Pech de l'Aze II [Carsac]	48.5	9.8	54.6	13.1	61.1	27.4
120	A. Brugas [?O/A] [Vallabrix]	50.1	7.7	61.5	19.1	65.0	23.6
125	[Gr.] Aldene [Cesseras]	43.2	2.2	53.3	11.5	58.8	23.9
127	Pie[d] Lombard [cave] [Tour[r]ettes-sur-Loup]	42.6	−1.1	54.1	8.4	59.9	20.0
266	Gr. Guattari	52.2	2.3	42.1	14.8	69.7	39.9
268	Gr. dei Moscerini	53.2	0.8	59.9	10.3	67.6	26.7
336	Temnata Cave	63.0	−7.2	62.2	−4.1	67.4	15.4

Appendix 9.2. *Wind-chill temperatures (°F) for Aurignacian and Early Upper Palaeolithic sites.*

Site no.	Site name	Glacial Maximum Summer	Glacial Maximum Winter	Warm Period Summer	Warm Period Winter	Modern Summer	Modern Winter
25,000–22,000 ka BP	**7 sites**						
53	La Riera	34.3	16.9	37.5	9.7	53.0	27.0
69	Abri Pataud	48.8	9.1	53.3	9.2	60.3	26.1
73	Roche a Pierrot [St-Cesaire]	48.8	11.5	56.3	19.5	59.3	27.3
83	La Salpetriere [Remoulins]	50.1	7.7	61.5	19.1	65.0	23.6
224	Wildscheuer	43.8	−6.5	53.8	2.4	56.7	15.9
321	Climautsy II	57.2	−11.4	64.8	−5.4	73.2	7.4
328	Ceahlau-Cetatoca II	54.3	−10.6	57.7	−7.2	63.0	5.3
29,000–26,000 ka BP	**33 sites**						
2	Pego do Diabo	46.8	34.1	56.3	31.9	57.5	45.0
25	Labeko Koba	34.8	17.2	44.7	13.7	57.0	30.7
36	Abric Romani	44.6	3.2	60.5	21.0	67.2	30.1
37	L'Arbreda	52.1	25.9	60.2	19.5	65.4	28.1
39	Cova Beneito	54.2	15.3	57.1	23.7	67.2	35.1
70	Le Flageolet I [Bezenac]	48.5	9.8	54.6	13.1	61.1	27.4
82	Le Piage [Fajoles]	48.5	9.8	54.6	13.1	61.1	27.4
84	Les Pecheurs [Casteljau]	48.1	3.1	54.7	11.5	59.8	24.6
87	Canecaude I [Villardonel]	42.6	4.2	49.6	10.9	56.2	21.6
154	Le Raysse [Brive-la-Gaillarde]	47.4	6.1	49.5	2.6	56.7	21.4
156	Tuto de Camalhot [St-Jean de Verges]	42.9	5.2	45.7	8.3	52.2	18.2
161	Fontenioux [St Pierre de Maille]	48.0	5.8	55.6	12.6	59.9	26.1
166	Gr. des Fieux [Miers]	46.5	5.9	50.5	4.5	58.1	22.7
174	Kent's Cavern	35.0	−10.3	43.1	4.2	50.5	30.5
177	West Pin Hole	15.8	−22.8	45.6	1.5	51.3	21.1
180	Paviland Cave [Goat's Hole]	29.8	−15.6	43.0	2.8	49.2	28.8
203	Trou Magrite	44.4	−3.6	51.2	1.1	55.5	18.8
205	Gr. de la Princesse [Marche-les-Dames]	43.6	−4.8	52.1	3.2	55.7	19.0
206	Trou du Renard	44.4	−3.6	51.2	1.1	55.5	18.8
207	Trou de l'Abime, Couvin	45.2	−2.2	52.5	4.3	56.6	20.7
212	Gr. du Spy	43.6	−4.8	52.1	3.2	55.7	19.0
213	Gr. du Haleux [Sprimont]	42.7	−6.2	52.1	2.6	55.1	17.6
254	Gr. del Fossellone	52.2	2.3	42.1	14.8	69.7	39.9
278	Sandalja II	54.2	−0.1	70.6	14.6	70.2	34.5
321	Climautsy II	57.2	−11.4	64.8	−5.4	73.2	7.4
327	Bistricioara-Lutarie	54.3	−10.6	57.7	−7.2	63.0	5.3
328	Ceahlau-Dirtu	54.3	−10.6	57.7	−7.2	63.0	5.3
343	Korpach	54.8	−12.0	63.5	−6.1	70.8	9.8
351	Berdyzh	46.3	−18.8	59.5	−13.5	64.0	2.9
367	Korolevo I	51.1	−9.9	62.0	−0.6	66.8	12.3
369	Byzovaya	22.3	−46.8	40.0	−30.6	46.2	−24.7
370	Molochnyi Kamen'	48.9	−12.8	56.6	−5.7	61.8	7.2
371	Korman' IV	54.5	−12.7	62.4	−5.4	69.3	9.8
33,000–30,000 ka BP	**48 sites**						
2	Pego do Diabo	46.8	34.1	56.3	31.9	57.5	45.0
25	Labeko Koba	34.8	17.2	44.7	13.7	57.0	30.7
31	Cueva Morin	34.5	16.2	40.4	14.3	55.5	31.0
33	Mallaetes Cave	39.9	14.4	43.5	17.6	58.9	41.5
36	Abric Romani	44.6	3.2	60.5	21.0	67.2	30.1
37	L'Arbreda	52.1	25.9	60.2	19.5	65.4	28.1
41	Reclau Viver	52.1	25.9	60.2	19.5	65.4	28.1
46	Nerja	54.5	36.1	55.5	36.8	69.8	45.4
65	Gorham's Cave	57.2	24.4	65.1	30.0	75.1	40.7
69	Abri Pataud	48.8	9.1	53.3	9.2	60.3	26.1
71	La Ferrassie	48.8	9.1	53.3	9.2	60.3	26.1
79	Roc de Combe [Nadaillac]	48.5	9.8	54.6	13.1	61.1	27.4
80	Abri Caminade [Caneda]	48.5	9.8	54.6	13.1	61.1	27.4
82	Le Piage [Fajoles]	48.5	9.8	54.6	13.1	61.1	27.4
83	La Salpetriere [Remoulins]	50.1	7.7	61.5	19.1	65.0	23.6
84	Les Pecheurs [Casteljau]	48.1	3.1	54.7	11.5	59.8	24.6

161

Appendix 9.2. (*cont.*)

Site no.	Site name	Glacial Maximum Summer	Glacial Maximum Winter	Warm Period Summer	Warm Period Winter	Modern Summer	Modern Winter
33,000–30,000 ka BP (*cont.*)							
86	Roc de Marcamps [Prignac-et-Marcamps]	48.2	13.0	55.6	20.0	59.7	28.0
91	Brassempouy [Grande Galerie 2]	47.3	13.6	53.7	19.5	61.3	30.5
151	Abri du Facteur	48.8	9.1	53.3	9.2	60.3	26.1
152	La Rochette [St Leon sur Vezere]	48.8	9.1	53.3	9.2	60.3	26.1
174	Kent's Cavern	35.0	−10.3	43.1	4.2	50.5	30.5
177	West Pin Hole	15.8	−22.8	45.6	1.5	51.3	21.1
180	Paviland Cave [Goat's Hole]	29.8	−15.6	43.0	2.8	49.2	28.8
199	Bench Quarry 'Tunnel' cavern	35.0	−10.3	43.1	4.2	50.5	30.5
200	Picken's Hole, Layer 3	30.9	−16.0	47.5	8.8	51.2	23.4
203	Trou Magrite	44.4	−3.6	51.2	1.1	55.5	18.8
204	Trou Walou	42.7	−6.2	52.1	2.6	55.1	17.6
216	Vogelherd Cave	41.9	−10.9	53.7	0.5	57.6	16.5
218	Lommersum	42.5	−6.9	52.6	2.4	55.7	17.1
219	Bockstein-Torle	41.9	−10.9	53.7	0.5	57.6	16.5
222	Paderborn	42.2	−8.5	52.8	0.5	56.1	15.4
223	Kelsterbach	44.1	−5.7	55.3	3.5	58.6	17.5
248	Gr. Paglicci	59.8	10.9	67.1	16.8	72.7	35.4
251	Gr. La Cala	52.5	4.6	56.1	25.5	70.8	40.9
270	Gr. Barbara	52.2	2.3	42.1	14.8	69.7	39.9
276	Vindija Cave	52.4	−4.6	64.6	5.7	69.0	19.2
277	Velika Pecina 2	52.4	−4.6	64.6	5.7	69.0	19.2
278	Sandalja II	54.2	−0.1	70.6	14.6	70.2	34.5
300	Milovice I	47.1	−9.8	57.9	1.4	61.9	12.7
302	Pod Hradem Cave	45.7	−10.6	56.0	−0.4	59.7	10.8
325	Mitoc Malul Galben	54.8	−12.0	63.5	−6.1	70.8	9.8
326	Ripiceni-Izvor	54.8	−12.0	63.5	−6.1	70.8	9.8
327	Bistricioara-Lutarie	54.3	−10.6	57.7	−7.2	63.0	5.3
335	Bacho Kiro	63.6	−7.3	63.6	−2.3	68.7	16.9
344	Siuren I	66.4	−6.0	69.6	5.7	76.4	16.5
354	Kostienki I	54.7	−15.5	61.8	−10.6	68.3	−3.5
356	Kostienki VIII [Tel'manskaya site]	54.7	−15.5	61.8	−10.6	68.3	−3.5
363	Kostienki XVII [Spitsyn site]	54.7	−15.5	61.8	−10.6	68.3	−3.5
37,000–34,000 ka BP 36 sites							
24	Ruso [I]	34.5	16.2	40.4	14.3	55.5	31.0
25	Labeko Koba	34.8	17.2	44.7	13.7	57.0	30.7
30	La Guelga	34.3	16.9	37.5	9.7	53.0	27.0
31	Cueva Morin	34.5	16.2	40.4	14.3	55.5	31.0
51	Cal Coix	48.7	12.8	56.0	17.5	61.6	26.3
69	Abri Pataud	48.8	9.1	53.3	9.2	60.3	26.1
70	Le Flageolet I [Bezenac]	48.5	9.8	54.6	13.1	61.1	27.4
71	La Ferrassie	48.8	9.1	53.3	9.2	60.3	26.1
72	Esquicho-Grapaou	50.1	7.7	61.5	19.1	65.0	23.6
76	Grotte du Renne, Arcy-sur-Cure	47.7	2.0	54.6	8.3	58.2	21.9
78	Les Cottes [St Pierre de Maille]	46.9	3.1	54.5	10.8	58.8	24.9
81	La Quina Y-Z [Villebois la Valette]	49.1	9.8	55.9	15.1	60.9	29.2
91	Brassempouy [Grande Galerie 2]	47.3	13.6	53.7	19.5	61.3	30.5
175	Robin Hood's Cave	15.8	−22.8	45.6	1.5	51.3	21.1
177	West Pin Hole	15.8	−22.8	45.6	1.5	51.3	21.1
180	Paviland Cave [Goat's Hole]	29.8	−15.6	43.0	2.8	49.2	28.8
199	Bench Quarry 'Tunnel' cavern	35.0	−10.3	43.1	4.2	50.5	30.5
214	Trou Al'Wesse	43.6	−4.8	52.1	3.2	55.7	19.0
215	Das Geissenklosterle	42.6	−9.7	52.9	−0.1	57.3	17.1
216	Vogelherd Cave	41.9	−10.9	53.7	0.5	57.6	16.5
217	Hohlenstein-Stadel [IV]	41.9	−10.9	53.7	0.5	57.6	16.5
219	Bockstein-Torle	41.9	−10.9	53.7	0.5	57.6	16.5
221	Hahnofersand	35.2	−19.8	51.8	−1.4	56.7	14.9
224	Wildscheuer	43.8	−6.5	53.8	2.4	56.7	15.9
240	Willendorf II	45.8	−10.2	55.6	−0.3	59.1	10.9
247	Castelcivita	53.7	5.1	58.5	9.6	67.7	26.7
250	Abri Fumane	44.7	−9.5	58.7	3.7	63.5	17.0

Appendix 9.2. *(cont.)*

Site no.	Site name	Glacial Maximum Summer	Glacial Maximum Winter	Warm Period Summer	Warm Period Winter	Modern Summer	Modern Winter
37,000–34,000 ka BP *(cont.)*							
255	Serino	53.7	5.1	58.5	9.6	67.7	26.7
290	Istallosko cave	50.1	−9.7	60.3	−1.4	66.1	13.7
301	Stranska-skala III	47.1	−9.8	57.9	1.4	61.9	12.7
325	Mitoc Malul Galben	54.8	−12.0	63.5	−6.1	70.8	9.8
335	Bacho Kiro	63.6	−7.3	63.6	−2.3	68.7	16.9
336	Temnata Cave	63.0	−7.2	62.2	−4.1	67.4	15.4
354	Kostienki I	54.7	−15.5	61.8	−10.6	68.3	−3.5
359	Kostienki XII [Volkovskaya]	54.7	−15.5	61.8	−10.6	68.3	−3.5
363	Kostienki XVII [Spitsyn site]	54.7	−15.5	61.8	−10.6	68.3	−3.5
42,000–38,000 ka BP 32 sites							
26	Arenillas	34.5	16.2	40.4	14.3	55.5	31.0
28	La Vina	39.5	13.9	36.4	6.9	51.0	23.1
35	Castillo	53.8	17.2	52.2	36.7	68.6	46.5
36	Abric Romani	44.6	3.2	60.5	21.0	67.2	30.1
37	L'Arbreda	52.1	25.9	60.2	19.5	65.4	28.1
38	Mollet Cave	52.1	25.9	60.2	19.5	65.4	28.1
39	Cova Beneito	54.2	15.3	57.1	23.7	67.2	35.1
71	La Ferrassie	48.8	9.1	53.3	9.2	60.3	26.1
79	Roc de Combe [Nadaillac]	48.5	9.8	54.6	13.1	61.1	27.4
80	Abri Caminade [Caneda]	48.5	9.8	54.6	13.1	61.1	27.4
139	Isturitz [Isturits]	44.6	13.2	48.7	16.1	55.4	25.0
141	A. Castanet [Sergeac]	48.8	9.1	53.3	9.2	60.3	26.1
147	A. Combe Sauniere [Sarliac-sur-l'Isle]	48.8	9.1	53.3	9.2	60.3	26.1
174	Kent's Cavern	35.0	−10.3	43.1	4.2	50.5	30.5
175	Robin Hood's Cave	15.8	−22.8	45.6	1.5	51.3	21.1
177	West Pin Hole	15.8	−22.8	45.6	1.5	51.3	21.1
200	Picken's Hole, Layer 3	30.9	−16.0	47.5	8.8	51.2	23.4
203	Trou Magrite	44.4	−3.6	51.2	1.1	55.5	18.8
214	Trou Al'Wesse	43.6	−4.8	52.1	3.2	55.7	19.0
215	Das Geissenklosterle	42.6	−9.7	52.9	−0.1	57.3	17.1
218	Lommersum	42.5	−6.9	52.6	2.4	55.7	17.1
221	Hahnofersand	35.2	−19.8	51.8	−1.4	56.7	14.9
241	Krems-Hundssteig	45.8	−10.2	55.6	−0.3	59.1	10.9
248	Gr. Paglicci	59.8	10.9	67.1	16.8	72.7	35.4
249	Gr. di Paina	41.5	−12.3	67.1	10.1	72.3	24.0
250	Abri Fumane	44.7	−9.5	58.7	3.7	63.5	17.0
252	Riparo Mochi	52.2	19.7	59.1	9.7	63.9	19.7
277	Velika Pecina 2	52.4	−4.6	64.6	5.7	69.0	19.2
291	Pesko cave	50.1	−9.7	60.3	−1.4	66.1	13.7
302	Pod Hradem Cave	45.7	−10.6	56.0	−0.4	59.7	10.8
336	Temnata Cave	63.0	−7.2	62.2	−4.1	67.4	15.4
363	Kostienki XVII [Spitsyn site]	54.7	−15.5	61.8	−10.6	68.3	−3.5
47,000–43,000 ka BP 8 sites							
35	Castillo	53.8	17.2	52.2	36.7	68.6	46.5
36	Abric Romani	44.6	3.2	60.5	21.0	67.2	30.1
37	L'Arbreda	52.1	25.9	60.2	19.5	65.4	28.1
41	Reclau Viver	52.1	25.9	60.2	19.5	65.4	28.1
203	Trou Magrite	44.4	−3.6	51.2	1.1	55.5	18.8
214	Trou Al'Wesse	43.6	−4.8	52.1	3.2	55.7	19.0
215	Das Geissenklosterle	42.6	−9.7	52.9	−0.1	57.3	17.1
240	Willendorf II	45.8	−10.2	55.6	−0.3	59.1	10.9

Appendix 9.3. *Wind-chill temperatures (°F) for Gravettian and Upper Palaeolithic sites.*

Site no.	Site name	Glacial Maximum Summer	Glacial Maximum Winter	Warm Period Summer	Warm Period Winter	Modern Summer	Modern Winter
25,000–22,000 ka BP	**50 sites**						
13	Terra do Manuel	46.3	32.5	56.2	28.7	59.8	37.3
31	Cueva Morin	34.5	16.2	40.4	14.3	55.5	31.0
37	L'Arbreda	52.1	25.9	60.2	19.5	65.4	28.1
44	Roc de la Melca	48.7	12.8	56.0	17.5	61.6	26.3
69	Abri Pataud	48.8	9.1	53.3	9.2	60.3	26.1
70	Le Flageolet I [Bezenac]	48.5	9.8	54.6	13.1	61.1	27.4
76	Grotte du Renne, Arcy-sur-Cure	47.7	2.0	54.6	8.3	58.2	21.9
83	La Salpetriere [Remoulins]	50.1	7.7	61.5	19.1	65.0	23.6
91	Brassempouy [Grande Galerie 2]	47.3	13.6	53.7	19.5	61.3	30.5
99	Laugerie-Haute Est	48.8	9.1	53.3	9.2	60.3	26.1
124	La Pente-des-Brosses [Montigny-sur-Loing]	47.3	2.3	55.6	10.0	59.1	23.6
137	Puy-Jarrige II [Brive-La-Gaillarde]	47.4	6.1	49.5	2.6	56.7	21.4
145	Bouzil [Saint-Thome]	48.1	3.1	54.7	11.5	59.8	24.6
147	A. Combe Sauniere [Sarliac-sur-l'Isle]	48.8	9.1	53.3	9.2	60.3	26.1
166	Gr. des Fieux [Miers]	46.5	5.9	50.5	4.5	58.1	22.7
167	Gr. d'Enlene [Montesquieu-Avantes]	42.9	5.2	45.7	8.3	52.2	18.2
180	Paviland Cave [Goat's Hole]	29.8	−15.6	43.0	2.8	49.2	28.8
212	Gr. du Spy	43.6	−4.8	52.1	3.2	55.7	19.0
219	Bockstein-Torle	41.9	−10.9	53.7	0.5	57.6	16.5
237	Langmannersdorf A	47.3	−9.9	59.2	2.6	62.8	13.6
239	Alberndorf [in der Riedmark]	42.6	−13.0	53.5	−1.8	57.4	10.8
248	Gr. Paglicci	59.8	10.9	67.1	16.8	72.7	35.4
249	Gr. di Paina	41.5	−12.3	67.1	10.1	72.3	24.0
265	Gr. del Romito	59.0	9.7	59.0	11.0	67.8	26.8
278	Sandalja II	54.2	−0.1	70.6	14.6	70.2	34.5
289	Kastritsa	48.1	−1.1	57.7	6.9	63.5	20.1
293	Balla cave	50.1	−9.7	60.3	−1.4	66.1	13.7
308	Dolní Věstonice I	47.1	−9.8	57.9	1.4	61.9	12.7
310	Petrkovice	45.9	−11.2	56.5	−5.6	60.4	10.5
313	Moravany-Zakovska	49.2	−9.1	60.9	2.4	64.6	14.1
314	Moravany-Lopata II	49.2	−9.1	60.9	2.4	64.6	14.1
317	Spadzista St. A	45.9	−12.6	56.1	−7.3	60.1	9.0
318	Krakow	45.9	−12.6	56.1	−7.3	60.1	9.0
320	Oblazowa 1	48.1	−11.0	51.8	−6.8	56.3	5.0
322	Ciuntu Cave	54.8	−12.0	63.5	−6.1	70.8	9.8
322	Ciuntu Cave	54.8	−12.0	63.5	−6.1	70.8	9.8
323	Brinzeni Cave I	54.8	−12.0	63.5	−6.1	70.8	9.8
325	Mitoc Malul Galben	54.8	−12.0	63.5	−6.1	70.8	9.8
327	Bistricioara-Lutarie	54.3	−10.6	57.7	−7.2	63.0	5.3
332	Coto Miculinti	54.8	−12.0	63.5	−6.1	70.8	9.8
336	Temnata Cave	63.0	−7.2	62.2	−4.1	67.4	15.4
346	Avdeevo	51.9	−16.3	61.6	−11.5	67.3	−2.1
349	Leski	62.4	−8.7	69.3	−7.6	78.0	6.6
354	Kostienki I	54.7	−15.5	61.8	−10.6	68.3	−3.5
359	Kostienki XII [Volkovskaya]	54.7	−15.5	61.8	−10.6	68.3	−3.5
361	Kostienki XV [Gorodtsov site]	54.7	−15.5	61.8	−10.6	68.3	−3.5
364	Kostienki XXI [Gmelinskaya]	54.7	−15.5	61.8	−10.6	68.3	−3.5
373	Sungir' [Vladimir]	41.7	−25.7	53.5	−14.7	58.7	−6.3
375	Gagarino	51.3	−17.8	59.8	−11.2	66.1	−3.9
376	Akhchtyr Cave	59.7	3.9	56.9	0.3	60.7	8.4
29,000–26,000 ka BP	**64 sites**						
8	Cabeco de Porto Marinho III	46.3	32.5	56.2	28.7	59.8	37.3
15	Buraca Escura	52.1	20.3	59.6	26.7	62.4	37.2
19	Abrigo do Lagar Velho	52.1	20.3	59.6	26.7	62.4	37.2
46	Nerja	54.5	36.1	55.5	36.8	69.8	45.4
61	Aitzbitarte III	41.8	12.7	46.6	16.2	56.4	29.3
70	Le Flageolet I [Bezenac]	48.5	9.8	54.6	13.1	61.1	27.4
71	La Ferrassie	48.8	9.1	53.3	9.2	60.3	26.1
74	Arcy-sur-Cure [Grande Grotte?]	47.7	2.0	54.6	8.3	58.2	21.9
79	Roc de Combe [Nadaillac]	48.5	9.8	54.6	13.1	61.1	27.4

Appendix 9.3. *(cont.)*

Site no.	Site name	Glacial Maximum Summer	Glacial Maximum Winter	Warm Period Summer	Warm Period Winter	Modern Summer	Modern Winter
29,000–26,000 ka BP *(cont.)*							
82	Le Piage [Fajoles]	48.5	9.8	54.6	13.1	61.1	27.4
83	La Salpetriere [Remoulins]	50.1	7.7	61.5	19.1	65.0	23.6
88	Grotte Chauvet	48.1	3.1	54.7	11.5	59.8	24.6
103	Les Vignes [St-Martin sous Montaigu]	47.5	1.5	55.9	9.2	59.7	21.9
106	Solutre [O/A]	46.8	1.1	55.3	9.3	59.4	21.6
110	Gr de Laraux	48.9	7.1	53.7	10.0	59.4	25.3
145	Bouzil [Saint-Thome]	48.1	3.1	54.7	11.5	59.8	24.6
151	Abri du Facteur	48.8	9.1	53.3	9.2	60.3	26.1
153	Gr. Tournal (or Grande Grotte de Bize)	45.8	2.6	55.8	16.5	61.8	27.5
154	Le Raysse [Brive-la-Gaillarde]	47.4	6.1	49.5	2.6	56.7	21.4
156	Tuto de Camalhot [St-Jean de Verges]	42.9	5.2	45.7	8.3	52.2	18.2
161	Fontenioux [St Pierre de Maille]	48.0	5.8	55.6	12.6	59.9	26.1
168	La Vigne Brun [St-Maurice-sur-Loire]	47.2	2.1	55.3	9.7	59.8	23.4
175	Robin Hood's Cave	15.8	–22.8	45.6	1.5	51.3	21.1
188	Hyaena Den	30.9	–16.0	47.5	8.8	51.2	23.4
204	Trou Walou	42.7	–6.2	52.1	2.6	55.1	17.6
208	Maisieres-Canal	44.2	–3.4	53.1	5.1	56.5	20.9
210	L'Hermitage [Huccorgne]	43.6	–4.8	52.1	3.2	55.7	19.0
215	Das Geissenklosterle	42.6	–9.7	52.9	–0.1	57.3	17.1
216	Vogelherd Cave	41.9	–10.9	53.7	0.5	57.6	16.5
219	Bockstein-Torle	41.9	–10.9	53.7	0.5	57.6	16.5
226	Kniegrotte	42.5	–10.0	53.5	–0.2	56.8	12.6
227	Obere Klause	44.3	–7.0	55.3	2.2	58.7	14.8
231	Hohle[r] Fels	42.6	–9.7	52.9	–0.1	57.3	17.1
232	Magdalenahohle	44.1	–5.1	52.5	2.2	56.0	17.3
238	Horn (Raberstrasse)	45.8	–10.2	55.6	–0.3	59.1	10.9
239	Alberndorf [in der Riedmark]	42.6	–13.0	53.5	–1.8	57.4	10.8
240	Willendorf II	45.8	–10.2	55.6	–0.3	59.1	10.9
244	Langenlois	45.8	–10.2	55.6	–0.3	59.1	10.9
245	Aggsbach	45.8	–10.2	55.6	–0.3	59.1	10.9
248	Gr. Paglicci	59.8	10.9	67.1	16.8	72.7	35.4
258	Gr. del Broion	41.5	–12.3	67.1	10.1	72.3	24.0
271	Bilancino	54.5	2.7	61.0	6.7	67.6	25.3
272	Gr. di Santa Maria di Agnano	60.8	6.6	68.8	19.6	76.5	34.9
286	Asprochaliko	53.5	0.0	60.8	11.2	68.2	23.7
288	Franchthi	43.8	15.3	47.4	20.0	75.0	45.4
293	Balla cave	50.1	–9.7	60.3	–1.4	66.1	13.7
300	Milovice I	47.1	–9.8	57.9	1.4	61.9	12.7
307	Pavlov I	47.1	–9.8	57.9	1.4	61.9	12.7
308	Dolní Věstonice II	47.1	–9.8	57.9	1.4	61.9	12.7
310	Petrkovice	45.9	–11.2	56.5	–5.6	60.4	10.5
312	Nitra-Cerman	49.2	–9.1	60.9	2.4	64.6	14.1
320	Oblazowa 1	48.1	–11.0	51.8	–6.8	56.3	5.0
325	Mitoc Malul Galben	54.8	–12.0	63.5	–6.1	70.8	9.8
336	Temnata Cave	63.0	–7.2	62.2	–4.1	67.4	15.4
342	Molodova I	54.8	–12.0	63.5	–6.1	70.8	9.8
346	Avdeevo	51.9	–16.3	61.6	–11.5	67.3	–2.1
349	Leski	62.4	–8.7	69.3	–7.6	78.0	6.6
357	Kostienki X	54.7	–15.5	61.8	–10.6	68.3	–3.5
358	Kostienki XI [Anosovka site 2]	54.7	–15.5	61.8	–10.6	68.3	–3.5
360	Kostienki XIV [Markina Gora]	54.7	–15.5	61.8	–10.6	68.3	–3.5
362	Kostienki XVI [Uglianka]	54.7	–15.5	61.8	–10.6	68.3	–3.5
364	Kostienki XXI [Gmelinskaya]	54.7	–15.5	61.8	–10.6	68.3	–3.5
368	Korolevo II	51.1	–9.9	62.0	–0.6	66.8	12.3
369	Byzovaya	22.3	–46.8	40.0	–30.6	46.2	–24.7
33,000–30,000 ka BP 38 sites							
58	Alkerdi	41.8	12.7	46.6	16.2	56.4	29.3
60	Amalda Cave	34.8	17.2	44.7	13.7	57.0	30.7
64	Cueto de la Mina	34.3	16.9	37.5	9.7	53.0	27.0
69	Abri Pataud	48.8	9.1	53.3	9.2	60.3	26.1

Appendix 9.3. *(cont.)*

Site no.	Site name	Glacial Maximum Summer	Glacial Maximum Winter	Warm Period Summer	Warm Period Winter	Modern Summer	Modern Winter
33,000–30,000 ka BP *(cont.)*							
70	Le Flageolet I [Bezenac]	48.5	9.8	54.6	13.1	61.1	27.4
71	La Ferrassie	48.8	9.1	53.3	9.2	60.3	26.1
74	Arcy-sur-Cure [Grande Grotte?]	47.7	2.0	54.6	8.3	58.2	21.9
82	Le Piage [Fajoles]	48.5	9.8	54.6	13.1	61.1	27.4
88	Grotte Chauvet	48.1	3.1	54.7	11.5	59.8	24.6
106	Solutre [O/A]	46.8	1.1	55.3	9.3	59.4	21.6
132	Gr. du Castellas [Dourgne]	43.2	3.2	52.8	13.4	59.0	24.3
150	Trou du Rhinoceros [St-Pe-de-Bigorre]	45.1	11.1	48.2	12.9	54.5	23.0
171	Grotte de Courau (Grotte Saucet)	45.1	11.1	48.2	12.9	54.5	23.0
204	Trou Walou	42.7	–6.2	52.1	2.6	55.1	17.6
207	Trou de l'Abime, Couvin	45.2	–2.2	52.5	4.3	56.6	20.7
208	Maisieres-Canal	44.2	–3.4	53.1	5.1	56.5	20.9
210	L'Hermitage [Huccorgne]	43.6	–4.8	52.1	3.2	55.7	19.0
215	Das Geissenklosterle	42.6	–9.7	52.9	–0.1	57.3	17.1
228	Weinberghohlen [Mauern 2]	42.2	–11.5	55.1	2.1	58.8	16.9
231	Hohle[r] Fels	42.6	–9.7	52.9	–0.1	57.3	17.1
240	Willendorf II	45.8	–10.2	55.6	–0.3	59.1	10.9
242	Krems-Wachtberg	45.8	–10.2	55.6	–0.3	59.1	10.9
244	Langenlois	45.8	–10.2	55.6	–0.3	59.1	10.9
245	Aggsbach	45.8	–10.2	55.6	–0.3	59.1	10.9
248	Gr. Paglicci	59.8	10.9	67.1	16.8	72.7	35.4
251	Gr. La Cala	52.5	4.6	56.1	25.5	70.8	40.9
295	Puspokhatvan	52.1	–7.7	63.7	–4.0	68.0	15.7
307	Pavlov I	47.1	–9.8	57.9	1.4	61.9	12.7
308	Dolní Věstonice II	47.1	–9.8	57.9	1.4	61.9	12.7
309	Predmosti	46.7	–10.2	56.8	–0.2	60.6	11.4
311	Nemsova	48.4	–9.8	56.9	–1.0	60.9	11.1
323	Brinzeni Cave I	54.8	–12.0	63.5	–6.1	70.8	9.8
325	Mitoc Malul Galben	54.8	–12.0	63.5	–6.1	70.8	9.8
342	Molodova I	54.8	–12.0	63.5	–6.1	70.8	9.8
354	Kostienki I	54.7	–15.5	61.8	–10.6	68.3	–3.5
360	Kostienki XIV [Markina Gora]	54.7	–15.5	61.8	–10.6	68.3	–3.5
361	Kostienki XV [Gorodtsov site]	54.7	–15.5	61.8	–10.6	68.3	–3.5
373	Sungir' [Vladimir]	41.7	–25.7	53.5	–14.7	58.7	–6.3
37,000–34,000 ka BP 6 sites							
231	Hohle[r] Fels	42.6	–9.7	52.9	–0.1	57.3	17.1
308	Dolni Vestonice I	47.1	–9.8	57.9	1.4	61.9	12.7
336	Temnata Cave	63.0	–7.2	62.2	–4.1	67.4	15.4
355	Kostienki VI	54.7	–15.5	61.8	–10.6	68.3	–3.5
360	Kostienki XIV [Markina Gora]	54.7	–15.5	61.8	–10.6	68.3	–3.5
378	Mezmaiskaya	65.1	–1.3	59.4	0.1	62.7	6.9

Chapter 10

The Middle and Upper Palaeolithic Game Suite in Central and Southeastern Europe

Rudolf Musil

This chapter deals with the game resources available to Palaeolithic cultures during OIS-3 and the early part of OIS-2 in central and southeastern Europe. Across this large region even the present climate varies greatly and so did the glacial palaeoclimate, both in space and through time. This chapter rests on a synthesis of many publications that span a long time and therefore reflect the style and objectives of the time when research at individual sites was carried out. The region has been sub-divided according to their faunal assemblages into six separate provinces: 1) southern Poland and northern Moravia; 2) Central and southern Moravia and northern Austria; 3) southern Slovakia and the Pannonian basin; 4) Slovenia and Croatia; 5) Bulgaria; 6) Romania.

Palaeoenvironments

As regards the faunal assemblages, it is not always possible to compare the present ecological adaptations of individual species with those of past species. The habitats of central Europe during the Weichsel/ Würm Glaciation are not represented anywhere today and species which differ markedly in their ecological requirements today are often found together in non-analogue glacial environments. As the glacial climate changed with time, individual floral and faunal assemblages did not simply shift north or south but often formed disharmonic mixed assemblages that are typical of each period. The author is of the opinion that there was no continuous permafrost nor a deeply frozen soil mantle over most of the study region except at high elevations, that the vegetation was a park landscape with scattered gallery woodland in river valleys, and that the palaeo-environment differed considerably from place to place (Musil 1999; 2000; 2001). Mammalian assemblages and analyses of dated charcoal support this view as also recently acquired gastropod data. Recent papers from the Pannonian basin, for instance, show a mosaic palaeoenvironment with dispersed forest refugia until late Würm time (Sümegi & Krolopp 2000).

Adaptations to a cold climate

Faunistic associations are related directly (herbivores) or indirectly (carnivores) to the vegetation cover which itself reflects the climate of its time and interacts with it as the climate changes over time (Chapter 5: Barron et al. 2003; Chapter 6: Huntley & Allen 2003). Although we only consider hunted game here, its assemblages were influenced by the local environment of the time. It is widely assumed that glacial faunas were so strongly adapted to a cold climate that they could not survive in a warmer one, but recent studies show that many mammalian species are eurythermic and not narrowly adapted to warm or cold climates. Instead, their environmental preferences are influenced strongly by the differences between maritime and continental climates (Chapter 5: Barron et al. 2003).

Faunal deposits in caves and on open-air sites

There is a fundamental difference between faunal remains collected from open-air sites and those found in cave deposits. Cultural layers at open-air sites are stratigraphically constrained by the sterile superposed and underlying layers, and tend to be of limited duration. The fauna contained in human deposits reflects only the game hunted by human beings. In caves the situation is different because cave sedimentation is a long-term semi-continuous process that results in the deposition of natural cave deposits that contain a mix of bones of cave-dwelling animals with those of the prey they introduced into the cave. The sequence is from time to time interrupted by human occupation that was usually infrequent

Figure 10.1. *Location map of the faunal and archaeological sites used in Chapter 10. 1st number is Faunal Province code/2nd number is site number.* **Czech Republic:** *2/1. Kůlna (cave); 2/2. Bohunice (open-air); 2/3. Stránská skálá (open-air); 2/4. Švéduv stůl (cave); 1/5. Čertova dira (cave); 1/6. Šipka (cave); 2/7. Vedrovice (open-air); 2/8. Předmosti (open-air); 2/9. Pavlov (open-air); 2/10. Dolní Věstonice (open-air); 2/11. Bulhary (open-air); 2/12. Jarošov (open-air); 1/13 Petřkovice (open-air); 2/14. Milovice (open-air); 2/15. Pod Hradem (cave); 2/45. Brno, Vídeňská Street (open air); 2/46. Velké Pavlovice (open air).* **Poland:** *1/16. Dzierzyslaw (open-air); 1/17. Spadzista Street, Krakow (open-air); 1/18. Piekary (open-air); 1/19. Oblazowa (cave); 1/20. Mamutowa (cave).* **Slovakia:** *3/21. Čertova Pec (cave); 3/22. Moravany (open-air).* **Austria:** *2/23. Willendorf (open-air); 2/24. Grubgraben (open-air).* **Hungary:** *3/25. Érd (open-air); 3/26. Szeleta (cave); 3/27. Bivak (cave); 3/28. Istállóskö (cave); 3/29. Tata (open-air); 3/30. Bodrogkereztúr (open-air); 3/31. Mende (open-air); 3/32. Ságvár (open-air).* **Croatia:** *4/33. Veternica (cave); 4/34. Vindija (cave); 4/35. Velika Pećina (cave); 4/36. Krapina (cave);* **Slovenia:** *4/37. Šandalja (cave); 4/38. Ciganska jama (cave).* **Bulgaria:** *5/39. Bacho Kiro (cave); 5/40. Temnata (cave).* **Romania:** *6/41. La Adam (cave); 6/42. Gura Cheii-Rasnov (open-air); 6/43. Spurcata (cave); 5/44. Giurgiu-Malul Rosu (open-air).*

and of short duration. During those times the natural bone assemblages are intermingled with the products, including bones of prey, that are the result of human activity. Usually, those two bone categories can not be distinguished easily from each other, and therefore cave samples of animal associations living in the vicinity of the cave are biased by human choice. From the perspective of a palaeo-ecological evaluation this does not present a major problem, but it may affect our conclusions when we try to define the human choice of hunted animals. Therefore, in general cave deposits are better sources of palaeo-environmental information than open-air sites. For this reason cave locations in the catalogue that follows are marked by a ¶ in front of the site name.

Estimating species abundances

In the publications from which the catalogue presented below has been drawn descriptions range from species lists to attempts at quantification by means of labelling individual species with terms such as *dominant, common* and *rare,* but often many species were left *without* estimates of frequency. At several sites, such as Pavlov, Willendorf, Bodrogkerestúr, Ciganska jama, Moravany Lopata and Grubgraben, the mammalian fauna was quantified by means of simple parameters. The minimal number (MNI) of a given species can be determined from the number of identical bone types. Suppose we have 15 distal parts of the *femur dext* and 10 specimens of the distal *femur sin* of a horse species, one concludes that the minimum number of individuals (MNI) was 15 individuals. The frequency of a single species within a bone assemblage can be expressed in terms of all bones as 'total number of bones/MNI'; example *Lepidus timidus* (79/28) = 79 bone fragments divided by MNI. This can be expressed as a percentage, e.g. (No.of bones/No. of bones in %, MNI/MNI in %).

Using appropriate time-scales

Palaeo-ecological studies can be carried out on three temporo-spatial scales (Musil 2001), but palaeo-ecological correlation of data from a wide range of localities requires that all samples derive from the same time-space frame.

1. First-order changes: changes of global extent
An example are marine sediments that display rapid climate changes expressed by sea-surface temperatures derived from oxygen isotope ratios, such as the Dansgaard/Oeschger oscillations (GRIP 1993; Dansgaard *et al.* 1993; Chapter 2: van Andel 2003).

Currently, high-resolution oceanic climatic records are the standard from which warm and cold events on land are inferred.

2. Second-order changes: changes in meridional and latitudinal distribution
Regional variations of insolation, annual temperatures and precipitation and their distribution, elevation, and rock and soil substrates have been used to define climate provinces that are regarded as independent of first order climate changes. The borders of such provinces are usually unstable and keep changing through time in north–south and west–east directions (Chapter 5: Barron *et al.* 2003). If those changes are substantial, they induce extensive changes in fauna and flora and migrations of animal communities.

3. Third-order changes: local changes in the environment
Significant climatic and other environmental differences exist in the Upper Pleistocene, not only between individual provinces but also in more limited areas within those. The territories, which are of a variety of sizes, are not due to zonation by altitude and other such variables; instead, their importance is that they may possess characteristics more favourable than those of their neighbours. When that is the case, they may function as refugia. Refugia are and always have been important for plants and animals both and for Palaeolithic people whose economic basis was primarily hunting, particularly in colder periods.

The chronological data base

The chronology used here rests on radiocarbon dates except for a few dates obtained with TL and ESR methods, and is independent of Palaeolithic cultural divisions. Due to large changes in the rate of production of ^{14}C in the upper atmosphere during the interval 60–20 ka BP (kiloannum Before Present), ^{14}C dates may deviate by up to 3 ka from the calendrical timescale. The commonly used record of climatic changes for the last pleniglacial is derived from Greenland ice cores which use calendar years. Therefore, all ^{14}C dates have been calibrated (Appendix 10.1) using the CalPal.v1998 programme of Weninger and Jöris (Chapter 3: van Andel *et al.* 2003a; Jöris & Weninger 1998; 1999; 2000). Calibrated ^{14}C dates are expressed as xxxx ka BP and their uncalibrated equivalents are written as xxxx uncal yrs b.p. Dates older than *c.* 45 ka BP exist, but they tend to have large standard deviations. This, however, does not make them useless. In this range there are also a few calendrical dates (see Appendix 10.1).

The resolution of the Greenland ice core named

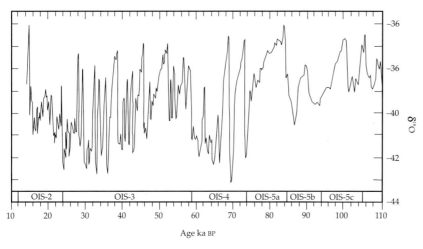

Figure 10.2. *High-resolution δ¹⁸O climate record from the GISP2 ice core (Meese et al. 1997) showing Dansgaard/Oeschger climate events for the interval from the decline of the penultimate interglacial (OIS-5d) to the onset of the Holocene (OIS-1). Higher-order climate phases matched to the confidence limits of our faunal and archaeological data sets are labelled as in Table 10.1A.*

Table 10.1A. *Standard phases of the climatic changes of the Weichselian Glaciation based on the GISP2 Greenland ice core.*

SPECMAP	Climate phase	Age (ka BP)
OIS-4	First Glacial Maximum	66–59
OIS-3	Stable Warm Phase	59–44
OIS-3	Transitional Phase	44–38
OIS-3	Early Cold Phase	38–28
OIS-2	Late Glacial Maximum	28–20

Table 10.1B. *A more detailed division into climatic phases used in some cases with calibrated ¹⁴C with appropriate calibration ranges (SD_cal) ranges.*

Code	Age range	Description
A.	55–47 ka BP	very warm event, most extreme of its time.
	47–45 ka BP	cold event.
	45–40 ka BP	three warm events separated by cold breaks.
	40–38 ka BP	very warm event.
C.	38–32 ka BP	three warm events interrupted by weak cold breaks.
D.	32–28 ka BP	cold event.
E.	24–20 ka BP	cold event.
F.	20–17 ka BP	peak of Late Glacial Maximum.

Table 10.2. *The six regional environmental provinces used in this chapter.*

Faunal Province 1	southern Poland and northern Moravia
Faunal Province 2	central and southern Moravia and northern Austria
Faunal Province 3	southern Slovakia and the Pannonian Basin
Faunal Province 4	Slovenia and Croatia
Faunal Province 5	Bulgaria
Faunal Province 6	Romania

GISP2 (Meese *et al.* 1997) is capable of displaying high-frequency climate changes on a millennial to centennial time-scale (Fig. 10.2) that are called Dansgaard/Oeschger (D/O) events (Dansgaard *et al.* 1993; GRIP 1993). The calibration standard deviations (SD_cal) of our archaeological and faunal dates range from less than ±500 to more than ±2500 years (Appendix 10.1). Matching time-slices based on calibrated faunal and archaeological dates with the D/O oscillations is therefore in most cases not justified. This limits palaeo-environmental studies to climate phases that usually comprise more than one D/O event.

The late glacial climate record

The GISP2 ice-core record has been divided into 'Climate Phases' (Fig. 10.2; Table 10.1A) that are employed in other chapters to compare archaeological time-slices with multi-millennial climate changes (e.g. Chapter 4: van Andel *et al.* 2003b; Chapter 5: Barron *et al.* 2003; Chapter 8: Davies & Gollop 2003). Late in OIS-4 the initial ice-sheet of the Weichsel/Würm glaciation spread over much of Scandinavia, but probably failed to reach the southern shores of the Baltic Sea. This first maximum was short-lived and disappeared around 60 ka BP at the end of OIS-4 (Fig. 10.2). It was succeeded by a series of long, relatively warm events separated by a few brief, not very cold events during the Stable Warm Phase which ended about 44 ka BP. The Stable Warm Phase was followed by a Transitional Phase of many quite cold events separated by a few brief warm ones that ended with a brief warm event between 39 and 37 ka BP, and was followed by an Early Cold Phase consisting of many severely cold events, some even colder than the Late Glacial Maximum (LGM) which began about 25 ka BP with a rapid advance of the main Fennoscandian ice sheet. The deglaciation that rang in the present interglacial began about 16,000 years ago.

In some cases, very good chronological control may allow a more detailed subdivision of the climate (Table 10.1B), but this was limited to cases where the available sets of dates had an SD_cal of better than ±1000 years.

CATALOGUE OF SITES

Notes:

1. All calibrated dates have been rounded to the nearest ka (1000 years); calibrated ^{14}C dates used in the text have been rounded to the nearest 1000 years (ka BP) to avoid suggesting too many meaningful digits. For a full list see Appendix 10.1.
2. Cave sites are indicated with the symbol ¶.
3. Site numbers in () refer to Figure 10.1 and Appendix 10.1.
4. Appendix 10.2 is a list of the scientific and common names of the many mammals discussed in this Chapter. Some have ecological significance.
5. For a list of the Regional Faunal Provinces see Table 10.2.
6. In fauna lists (xx) is the number of bone fragments.

Interval 59–44 ka BP: Stable Warm Phase (interrupted by a colder event around 47 ka BP)

Faunal Province 2: central and southern Moravia and northern Austria

¶ Kůlna (1)

Layer: 7a; **50±5.0 ka** BP (mean of ESR dates Rink *et al.* 1996). **48±3.2 ka** BP (45,660 +2850/–2200 uncal yrs b.p.; ^{14}C date on Neanderthal bone: Mook (1988). Micoquian.

Macrofauna: *Lepus* sp., *Panthera* sp., *Canis lupus*, *Alopex lagopus*, *Vulpes vulpes*, *Gulo gulo*, *Crocuta spelaea*, *Ursus spelaeus*, *Coelodonta antiquitatis*, *Equus* sp., *Equus (Asinus) hydruntinus*, *Mammuthus primigenius* (dominant), *Rangifer tarandus* (dominant), *Alces alces* (rare), *Bovidae* (rare), *?Saiga tatarica*.

Microfauna: *Lemmus lemmus* (number of specimens 222), *Apodemus* sp. (11), *Lagurus lagurus* (693), *Microtus gregalis* (1446), *Microtus arvalis-agrestis* group (52), *Microtus nivalis* (4), *Pitymys subterraneus* (495), *Dicrostonyx torquatus* (204), *Arvicola terrestris* (5), *Ochotona pusilla* (20): Musil (1969; 1988).

Flora (charcoal): *Picea/Larix, Picea excelsa, Pinus* sp., *Acer* cf. *pseudoplatanus*, cf. *Corylus avellana, Euonymus* sp., *Fraxinus excelsior*, indet. deciduous tree: Opravil (1988).

Faunal Province 3: southern Slovakia and the Pannonian Basin

Érd (25)

Layer: e; **51±7.7 ka** BP (44,300 uncal yrs BP). Mousterian.

Macrofauna: *Canis spelaeus* (87), *Vulpes vulpes* (24), *Alopex lagopus* (4), *Ursus arctos priscus* (2), *Ursus spelaeus* (predominant, 13,245), *Martes* sp. (2), *Putorius furo* (1), *Meles meles* (1), *Crocuta spelaea* (281), *Panthera* sp. (1), *Panthera spelaea* (20), *Ochotona spelaea* (3), *Lepus* cf. *timidus* (4), *Mammuthus primigenius* (26), *Equus (Asinus) hydruntinus* (44), *Equus* sp. (717), *Coelodonta antiquitatis* 176), *Sus scrofa* (1), *Rangifer tarandus* (29), *Megaloceros* sp. (20), *Cervus elaphus* (55), *Capra ibex* (17), *Rupicapra* sp. (1), *Bos primigenius* or *Bison priscus* (87).

Microfauna: *Pelobates fuscus* (13), *Bufo* sp. (4), *Lacerta* sp. (1), *Ophidia* indet. (1), *Falco* cf. *tinnunculus* (1), *?Parus* sp. (1), *Sorex araneus* (1), *Plecotus auritus* (1), *Citelllus citelloides* (6), *Spalax leucodon* (3), *?Sicista* cf. *betulina* (1), *Allactaga major* (1), *Myodes (Clethrionomys) rutilus* (1), *Arvicola* sp. (3), *Microtus gregalis* (4), *M. oeconomus* (1), *M. arvalis-agrestis* (8), *Apodemus* sp. (1), (5 layers with rather homogeneous fauna: Jánossy (1986).

Flora: Coniferous trees in all strata, mainly *Pinus sylvestris, Pinus cembra* and *Larix* sp.: Jánossy (1986).

Summary: The long warm and few brief colder climatic events of the 59–44 ka BP interval were marked in Moravia by a cold steppe fauna, but gastropods imply an interstadial climate and a palaeo-ecological interpretation of the flora by Opravil (1988) points to a climate similar to the present one. The conflict can be resolved by assuming that the data may correspond to the interval 47–44 ka BP rather than to the preceding long warmer period.

Interval 44–38 ka BP: Transitional Phase (three warm events with several cold breaks, terminal warm event at 38–37 ka BP)

Faunal Province 2: central and southern Moravia and northern Austria

¶ Kůlna (1)

Layer: 6a; **41±1.0** (38,600 +950/–800 uncal yrs b.p). Micoquian.

Macrofauna: *Aves* indet., *Lepus* sp., *Equus* sp., *Canis lupus*, *Vulpes vulpes*, *Crocuta spelaea*, *Saiga tatarica*, *Ursus spelaeus*, *Mammuthus primigenius* (dominant), *Coelodonta antiquitatis*, *Rangifer tarandus* (dominant), *Bovidae, Saiga tatarica*.

Microfauna: *Dicrostonyx torquatus* (number of specimens 180), *Microtus gregalis* (94), *Microtus arvalis-agrestis* (4), *Microtus nivalis* (8), *Pitymys subterraneus* (38): Musil (1969; 1988; 1990; 1997a).

Flora (charcoal): cf. *Abies, Abies/Picea*, cf. *Picea, Pinus* sp., *Acer* sp., *Acer pseudoplatanus*, indet. conifer: Opravil (1988).

Bohunice (2)

Dates: **45±2.0 ka** BP (42,900 +1700/–1400 uncal yrs b.p.; Bohunician; Mook 1976). **43±2.1 ka** BP (41,400 +1400/–1200 uncal yrs b.p.). **42±2.6 ka** BP (40,173 +1200/–1004 uncal yrs b.p.; Bohunician: Switsur 1976). **38±1.1 ka** BP (36,000±1100 uncal yrs b.p.).

Fauna: *Mammuthus primigenius, Equus* sp. (similar to *Equus taubachensis*): Musil (1976; 1997a).

Flora (charcoal): *Abies* sp. (dominant), cf. *Abies* sp., *Picea* sp. *vel Larix* sp., *Pinus sylvestris*, cf. *Alnus* sp., *Fraxinus* sp.: Opravil (1976).

Flora (pollen): *Betula* sp. (dominant), *Alnus* sp., *Picea* sp., many *Pinus* sp., *Salix* sp., *Corylus* sp., herb pollen much larger than tree pollen: Svoboda & Svobodová (1985), Svobodová (1987).

Stránská skála (3)
Layer: IIIa/4; **43±2.9 ka** BP (41,300 +3100/−2200 uncal yrs b.p.: Svobodová 1987). III/5: **40±1.3 ka** BP (38,200±1100 uncal yrs b.p.). **40±1.9 ka** BP (38,500 + 1700/−1400 uncal yrs b.p.). All Bohunician.
Fauna: *Mammuthus primigenius*, *Bos* sp./*Bison* sp., *Rangifer tarandus*, *Coelodonta antiquitatis*: Musil (1976).
Flora (charcoal): *Pinus* sp., cf. *Pinus*, *Abies* sp., cf. *Abies* sp., *Picea* sp., cf. *Picea*, *Betula* sp., *Salix* sp., cf. *Larix* sp., *Picea* sp./*Larix* sp., *Pinus* cf. *sylvestris*, *Pinus* sp./*Picea* sp., *Sorbus aucuparia*, *Corylus* sp.: Opravil, in Svoboda & Svobodová (1985), *Quercus* sp.: Kyncl (1984).
Flora (pollen): *Pinus* sp., *Abies* sp., *Picea* sp., *Salix* sp., *Corylus* sp., *Alnus* sp., *Betula* sp., *Tilia* sp.; Herb species: *Artemisia*, Asteraceae-Liguliflorae, Asteraceae-Tubiflorae, *Calluna*, *Campanula*, *T. cardamine*, Cyperaceae, Chenopodiaceae, Ericaceae, *Filipendula*, *Geranium*, *Lysimachia*, *Lythrum*, Papaveraceae: *Plantago major-media*, Poaceae, *Polygonum bistorta*, Ranunculaceae, Rosaceae, *Rumex acetosella*, Silenaceae, *Urtica* sp., *Valeriana* sp., Viciaceae, Pteridophyta, Polypodiaceae, *Sphagnum* sp., *Lycopodium clavatum*, Bryophyta: Svoboda & Svobodová (1985); Svobodová (1987).

The faunal assemblage deviates from the climate inferred from the Greenland ice-core record by the presence of thermophile species, but those have been found only in southeastern Europe. The mammal suite itself points to a cold climate, suggesting a cold event in the Transitional Phase rather than one of the warmer ones. In contrast to the fauna the flora suggests a parkland with isolated patches of conifers, interspersed locally with deciduous trees that do not indicate a low annual temperature.

Vedrovice V (7)
Dates: **41±2.6 ka** BP (39,500±1100 uncal yrs b.p.). **40±1.0 ka** BP (37,650±550 b.p.). Both Szeletian.
Flora: *Picea*/*Betula*: 23% herbs: Musil (1997a).

Faunal Province 3: southern Slovakia and the Pannonian Basin

¶ Čertova Pec (21)
Layer: 4; **41±2.8 ka** BP (38,400 +2800/−2100 uncal yrs b.p.); **40±2.8 ka** BP (38,300±2480 uncal yrs b.p.). Both Szeletian.
Fauna: *Lepus* sp., *Marmota marmota*, *Canis lupus*, *Vulpes vulpes*, *Mustela erminea*, *Mustela foina*, *Martes martes*, *Ursus* cf. *taubachensis*, *Ursus spelaeus*, *Equus taubachensis*, *Cervus elaphus*, *Capreolus capreolus*, *Rangifer tarandus*, *Bos* sp., *Bison* sp.: Musil (1998).

Faunal Province 4: Slovenia and Croatia

¶ Veternica (33)
Layer: h,i,j; **45±1.9 ka** BP (43,200±200 uncal yrs b.p.; 306 masl). Mousterian.
Fauna (layers h,i,j): *Ursus spelaeus*, *Ursus arctos*, *Panthera spelaea*, *Panthera pardus*, *Lynx pardina*, *Bos primigenius*, *Canis lupus*, *Sus scrofa*, *Cervus elaphus*, *Megaloceros gigan-*

teus, *Alces alces*, *Dama dama*, *Capreolus capreolus*, *Stephanorhinus kirchbergensis*, *Castor fiber*, *Hystrix* sp., *Cricetus* sp.: Malez (1958; 1961; 1974).

¶ Vindija (34)
Komplex G: layer G3; **>44 ka** BP (>42,000 uncal b.p.; 275 masl). Komplex G, layer G3: **45±4 ka** BP (42,400±4300 uncal yrs b.p.; 58 specimens of *H. neanderthalensis*). Both Mousterian.
Fauna: *Arvicola terrestris*, *Canis lupus*, *Ursus spelaeus*, *Panthera spelaea*, *Equus* cf. *germanicus*, *Sus scrofa*, *Megaloceros giganteus*, *Cervus elaphus* etc.: Malez (1979a,b,c); Malez *et al.* (1984).

Faunal Province 5: Bulgaria

¶ Bacho Kiro (39)
Layers: 11,11a; **>45 ka** BP (>43,000 uncal yrs BP: Hedges, in Bolus & Conard 2001 questionable); **40±1.2 ka** BP (38,500±1100 uncal yrs b.p.); **40±1.6 ka** BP (37,700±1500 uncal yrs b.p.); **37±1.6 ka** BP (34,800±1200 uncal yrs b.p.); **36±1.6 ka** BP (33,800±900 uncal yrs b.p.). All Mousterian or Eastern Balkan Aurignacian?.
Macrofauna (layer 11): Pisces; *Lacerta agilis*, *Lacerta viridis*, *Natrix* cf. *natrix*, *Ursus arctos*, *Ursus spelaeus*, *Canis lupus*, *Vulpes vulpes*, *Alopex lagopus*, *Cuon alpinus*, *Mustela* sp., *Cuon alpinus*, *Mustela erminea*, *Martes martes*, *Crocuta spelaea*, *Panthera pardus*, *Panthera spelaea*, *Dicerorhinus* cf. *hemitoechus*, *Equus* (*Asinus*) *hydruntinus*, *Equus germanicus*, *Cervus elaphus* (6), *Megaloceros giganteus* (4), Cervidae indet. (3), *Capra ibex* (103), *Bos primigenius* or *Bison priscus* (71), *Capra ibex* (103).
Microfauna: *Talpa europaea*, *Sorex araneus*, *Sorex minutus*, *Neomys fodiens*, *Crocidura* sp., *Myotis blythi oxygnathus*, *Citellus citellus*, *Cricetulus migratorius*, *Mesocricetus auratus*, *Cricetus cricetus*, *Clethrionomys glareolus*, *Arvicola terrestris*, *Pitimys subterraneus* (frequently), *Microtus nivalis*, *Microtus arvalis*, *Spalax leucodon*, *Apodemus sylvaticus*, *Sicista subtilis*, *Ochotona* sp.
Aves: *Anas platyrhinchos*, *Aquila chrysaetos*, *Circus aeruginosus*, *Perdix perdix*, *Coturnix coturnix*, *Alectoris graeca*, *Porzana porzana*, *Bubo bubo*, *Anthus campestris*, *Pyrrhocorax graculus*: Kozlowski (1982).

Macrofauna (layer 11a): *Lepus* sp., *Ursus spelaeus*, *Canis lupus*, *Vulpes vulpes*, *Alopex lagopus*, *Cuon alpinus*, *Mustela erminea*, *Putorius putorius*, *Martes martes*, *Crocuta spelaea*, *Panthera pardus*, *Equus germanicus*, *Equus* (*Asinus*) *hydruntinus*, *Cervus elaphus*, *Megaloceros giganteus*, *Capra ibex*.
Aves + Microfauna: *Lacerta agilis*, *Coronella* aff. *austriaca*, *Anas platyrhyncha*, cf. *Delichon urbica*, cf. *Anthus campestris*, *Talpa europaea*, *Sorex araneus*, *Sorex minutus*, *Neomys fodiens*, *Crocidura* sp., *Myotis blythi oxygnathus*, *Myotis dasycneme*, *Vespertilio murinus*, *Barbastella* cf. *schadleri*, *Citellus citellus*, *Cricetulus migratorius*, *Mesocricetus auratus*, *Cricetus cricetus*, *Clethrionomys glareolus*, *Arvicola terrestris*, *Pitymys subterraneus*, *Microtus nivalis*, *Microtus arvalis*, *Spalax leucodon*, *Apodemus sylvaticus*: Forsten (1982);

Kubiak (1982a,b); Sych (1982); Wiszniowska (1982).

Almost 30 per cent of the cave bears at the Bacho Kiro site are individuals older than 10 years with ~18 per cent each for animals of 0.4–1.4 years and 1.4–2.4 years. Other age groups were found in small numbers (Wiszniowska 1982). As in Slovenia and Croatia, the species assemblage implies a warmer, more humid climate. Today, some of those species require deciduous forests and are absent in a typical periglacial fauna. The arctic fox is here the only member of the cold species group; lemmings, mammoths and reindeer are absent or present only in negligible numbers. The faunal association indicates a parkland where steppe areas were interspersed with open deciduous woodland.

Summary: The palaeoclimate of the 4th (Slovenia and Croatia) and 5th (Bulgaria) faunal provinces was quite different from that of the 2nd province (central and southern Moravia and northern Austria) and to a lesser degree the 3rd province (southern Slovakia and the Pannonian basin), because in Slovenia, Croatia and Bulgaria the fauna contains species that in central Europe belonged to the preceding interglacial and possibly early interstadials of the Würm/Weichsel Glaciation. Thermophilic species are present at the same time that in Moravia mammoth and reindeer were widely hunted. Faunal Provinces 4 and 5 are thus considerably warmer than Faunal Province 2.

Regional faunal provinces 2 and 3 themselves differ from each other, the faunal community of Province 2 being considerably colder. Taken together the fauna and flora assemblage from both provinces deviates to some degree from the climate implied by the Greenland ice-core record.

Between 44 and 38 ka BP, several contemporaneous Palaeolithic cultures existed in the study area, i.e. the Bohunician, Mousterian, Szeletian and Micoquian lithic complexes. Their patterns of game use differ regionally: in Moravia species of the cold assemblage dominate, while the Balkans have a suite of thermophilic species, but arctic species such as mammoth and reindeer are scarce or even absent. Clearly, the Neanderthals were adapted to a range of climates.

Mousterian sites without hard dates

Faunal Province 1: southern Poland and northern Moravia

¶ **Čertova díra** (5)
Layers: III–IV; Mousterian.

¶ **Šipka** (6)
Layers: III–IV; (with Neanderthal bones). Mousterian.
Fauna (layer III): *Cervus elaphus, Marmota* sp., *Bison priscus, Bos primigenius, Crocuta spelaea, Equus mosbachensis-abeli, Equus (Asinus) hydruntinus, Coelodonta antiquitatis, Vulpes*

vulpes, Ursus spelaeus, Mammuthus primigenius, Panthera spelaea, Panthera pardus, Canis lupus, Gulo gulo: Musil (1965b).
Fauna (layer IV): *Aquila chrysaetos, Panthera spelaea, Panthera pardus, Cuon alpinus, Vulpes vulpes, Cervus elaphus, Capreolus capreolus, Saiga tatarica, Cervus elaphus maral, Ovibos moschatus, Rupicapra rupicapra, Bison priscus, Bos primigenius, Coelododonta antiquitatis, Ursus spelaeus, Crocuta spelaea, Sus scrofa, Equus mosbachensis-abeli, Equus (Asinus)* cf. *hydruntinus, Vulpes vulpes.*
Fireplace: with *Homo neanderthalensis: Ursus spelaeus, Coelodonta antiquitatis, Bos primigenius, Mammuthus primigenius, Panthera spelaea, Crocuta spelaea, Canis lupus, Equus* sp., *Cervus elaphus*: Musil (1965b).

Faunal Province 2: central and southern Moravia and northern Austria

¶ **Švédův stůl** (4)
Layers: 10–14; Neanderthal find; Mousterian.
Fauna: *Lepus* sp., *Crocuta spelaea, Canis lupus, Vulpes vulpes, Vulpes corsac, Meles meles, Ursus spelaeus, Mammuthus primigenius, Coelodonta antiquitatis, Cervus elaphus maral, Alces alces, Rangifer tarandus, Bos primigenius, Bison priscus, Rupicapra rupicapra, Ovis* sp. or *Capra* sp., *Marmota* sp., *Equus mosbachensis-abeli, Equus germanicus, Equus (Asinus) hydruntinus, Equus* cf. *gmelini*: Musil (1961; 1997a).

Faunal Province 3: southern Slovakia and the Pannonian Basin

¶ **Čertova Pec** (21)
Layers: 2–3; Mousterian.
Fauna: *Castor fiber, Panthera spelaea, Crocuta spelaea, Canis lupus, Vulpes vulpes, Alopex lagopus, Meles meles, Ursus* cf. *taubachensis, Ursus spelaeus, Bos* sp., *Bison* sp.: Musil (1998).

Faunal Province 4: Slovenia and Croatia

¶ **Velika Pećina** (35)
Layer: k; 428 masl; Neanderthal finds; Mousterian.
Fauna: *Canis lupus, Ursus arctos priscus, Gulo gulo, Crocuta spelaea, Panthera spelaea, Megaloceros giganteus, Alces alces, Bison priscus, Ursus spelaeus* (98 per cent of all bones): Malez (1974; 1986).

Interval 38–32 ka BP: Early Cold Phase (rapid cooling; many cold events, only three slightly warmer ones)

Faunal Province 1: southern Poland and northern Moravia

Dzierzyslav I (16)
Upper Layer: 36±5.5 ka BP (TL: 36,000±5500 ka BP). Szeletian.

Spadzista Street (17)
Layer: V; **34±2.3 ka** BP (32,000±2000 uncal b.p.). Aurignacian.
Fauna: *Equus* sp., *Coelodonta antiquitatis*, Cervidae, Bovidae, *Mammuthus primigenius* (Wiszniowski 2000; 2002).
Flora: *Pinus* sp., *Larix* sp., *Abies* sp.: Kozlowski & Sobczyk (1987).

Piekary II (18)
Dates: 34±1.5 ka BP (32,000±500 uncal yrs b.p.). Aurignacian.

¶ Oblazowa (19)
Layer: VIII–X: **33±1.2 ka** BP (30,600±550 uncal yrs b.p.). VII–X: **33±0.9 ka** BP (31.500±550 uncal yrs b.p.). **36±1.3 ka** BP (34,200±650 uncal yrs b.p.). Aurignacian above Szeletian.

Faunal Province 2: central and southern Moravia and northern Austria

Stránská skála (3)
Layer: IIIa/3; **33±0.5 ka** BP (30,980±360 uncal yrs b.p.). Layer IIIb/4; **35±1.9 ka** BP (32,600 1700/–1400 uncal yrs b.p.). Layer IIa/4; **34±1.5 ka** BP (32,350±900 uncal yrs b.p.). All three Aurignacian.

¶ Pod hradem (15)
Layer: 9; **35±1.7 ka** BP (33,300±1100 uncal yrs b.p.). Szeletian. **35±1.2 ka** BP (33,100±530 uncal yrs b.p.). Aurignacian. **29±0.9 ka** BP (26,830±300 uncal yrs b.p.). Gravettian?.
Macrofauna (horizon 8-19, brown earth, single time-layer): *Lepus* sp., *Panthera spelaea*, *Crocuta spelaea*, *Canis lupus*, *Vulpes vulpes*, *Alopex lagopus*, *Martes* sp., *Putorius* cf. *putorius*, *Ursus spelaeus* (dominant), *Ursus arctos*, *Mammuthus primigenius*, *Coelodonta antiquitatis*, *Equus* sp., *Sus scrofa*, *Rangifer tarandus*, *Bison priscus*, *Bos primigenius*, *?Saiga tatarica*, *Rupicapra rupicapra*, *Capra ibex*.
Microfauna: *Citellus citellus*, *Cricetulus* sp., *Clethrionomys glareolus*, *Microtus arvalis-agrestis* group, *Microtus gregalis*, *Microtus oeconomus*, *Microtus nivalis*, *Arvicola terrestris*.
Aves: *Dafila acuta*, *Lyrurus tetrix*, *Lagopus lagopus*, *Lagopus mutus*, *Strix aluco*, *Garrulus glandarius*: Musil (1965a; 1997a).
Macrofauna (horizon 9): *Alopex lagopus*, *Putorius* cf. *putorius*, *Ursus spelaeus*, *Mammuthus primigenius*, *Coelodonta antiquitatis*, *Rangifer tarandus*, *Bison priscus*, *Bos primigenius*, *Capra ibex*.
Microfauna: *Citellus citellus*, *Cricetulus* sp., *Clethrionomys glareolus*, *Microtus arvalis-agrestis*.
Aves: *Lagopus lagopus*, *Lagopus mutus*: Musil (1965a; 1997a).
Flora (charcoal): *Abies alba* (number of specimens 4), *Fagus silvatica* (5), *Alnus* sp. (1), *Pinus* sp. (1), *Pinus* cf. *mugo* (6), *Pinus* cf. *cembra* (1): Opravil (1965).

Faunal Province 3: southern Slovakia and the Pannonian Basin

¶ Istállóskö (28)
Lower Layer: ~38 ka BP (~36,000 uncal yrs b.p.). Upper layer: **32±0.8 ka** BP (30,000±600 uncal yrs b.p.; *Homo sapiens*). Aurignacian.
Macrofauna: *Lepus timidus*, *Canis lupus*, *Vulpes vulpes*, *Ursus spelaeus*, *Ursus arctos*, *Mustela nivalis*, *Mustela erminea*, *Putorius putorius*, *Martes martes*, *Meles meles*, *Crocuta spelaea*, *Lynx lynx*, *Panthera spelaea*, *Mammuthus primigenius*, *Equus* sp., *Cervus elaphus*, *Alces alces*, *Rangifer tarandus*, *Bison priscus*, *Capra ibex*, *Rupicapra rupicapra*, *Sus scrofa*.
Microfauna: *Rana méhelyi*, *Erinaceus* sp., *Sorex araneus*, *Sorex* sp., *Neomys fodiens*, *Talpa europaea*, *Eptesicus nilssonii*, *Citellus citelloides*, *Microtus arvalis-agrestis* group, *Microtus oeconomus*, *Microtus gregalis*, *Arvicola terrestris*, *Clethrionomys glareolus*, *Cricetus cricetus*, *Ochotona pusilla*: Jánossy (1986).
Aves: *Corvus corone*, *Pica pica*, *Coloeus monedula*, *Nucifraga caryocatactes*, *Garrulus glandarius*, *Pyrrhocorax graculus*, *Sturnus* sp.(?), *Coccothraustes coccothraustes*, *Loxia curvirostra*, *Pinicola enucleator*, *Bombycilla garrulus*, *Turdus* cf. *viscivorus*, *Turdus pilaris*, *Turdus iliacus-musicus* group, *Hirundo rustica*, *Apus apus*, *Dryobates major*, *Nyctea* seu *Bubo* sp., *Surnia ulula*, *Aegolius funereus*, *Asio flammeus*, *Falco columbarius aesalon*, *Falco vespertinus*, *Falco tinnunculus*, *Anas* cf. *platyrhyncha*, Charadriidae, *Porzana porzana*, *Crex crex*, *Perdix perdix*, *Lyrurus tetrix*, *Tetrao urogallus*, *Lagopus lagopus*, *Lagopus mutus*.
Flora: *Abies* sp., *Pinus cembra*, *Picea* sp., *Larix* sp., deciduous trees: Jánossy (1955).
Note: Because the lower layer date is too old and both assemblages are identical, the two layers were combined.

Tata (29)
Dates: 36±1.7 ka BP (33,600±1200 uncal yrs b.p.; freshwater limestone). Mousterian.
Fauna: *Cuon alpinus*, *Ursus spelaeus*, *Ursus arctos*, *Mustela nivalis*, *Mustela erminea*, *Putorius putorius*, *Meles meles*, *Crocuta spelaea*, *Mammuthus primigenius*, *Equus* sp., *Equus (Asinus) hydruntinus*, *Sus scrofa*, *Megaloceros giganteus*, *Cervus elaphus*, *Bos* sp. *vel Bison* sp.: Jánossy (1955; 1986), Vertés (1994).

Faunal Province 4: Slovenia and Croatia

¶ Velika Pećina (35)
Layer: g; **33±2.2 ka** BP (31,168±1400 uncal yrs b.p.). Layer i: **36±1.1 ka** BP (33,850±520 uncal yrs b.p.). Aurignacian.
Fauna (layer g): *Canis lupus*, *Alopex* cf. *lagopus*, *Ursus spelaeus* (95 per cent of bones), *Gulo gulo*, *Panthera spelaea*: Malez (1974; 1986); (layer i) - *Canis lupus*, *Alopex* cf. *lagopus*, *Ursus arctos priscus*, *Ursus spelaeus*, *Gulo gulo*, *Panthera spelaea*, *Sus scrofa*, *Megaloceros giganteus*, Bovidae gen. *et* sp. indet.: Malez (1986).

¶ Vindija (34)
Layer: f; **36±1.4 ka** BP (33,850±520 uncal yrs b.p.; *Homo* sp.). Aurignacian. **30±0.5 ka** BP (28,020±360 uncal yrs b.p.); **31±0.6 ka** BP (29,080±100 uncal yrs b.p.). Both fauna.
Macrofauna: *Marmota marmota*, *Alopex lagopus*, *Canis lupus*, *Ursus spelaeus*, *Panthera spelaea*, *Rangifer tarandus*,

Alces alces, Cervus elaphus, Bison priscus, Capra ibex, Rupicapra rupicapra, Saiga tatarica.
Aves: *Lagopus lagopus, Tetrao urogallus, Lyrurus tetrix, Nyctea scandiaca*: Malez & Rukavina (1979); Malez *et al.* (1984).

¶ Krapina (36)
Dates: 32±0.5 ka BP (30,700±750 uncal yrs b.p.; 120 masl). Reaches Mousterian with Neanderthal finds.
Fauna: Dominant species: *Ursus spelaeus, Stephanorhinus kirchbergensis* (320 fragments), *Castor fiber*; Less abundant are: *Bos* or *Bison, Ursus arctos, Canis lupus, Cervus elaphus, Marmota marmota*; Sporadic species: *Capreolus capreolus, Sus scrofa, Equus* sp., *Megaloceros giganteus, Felis catus, Lutra lutra, Mustela foina*: Malez & Malez (1989).

At the time of writing no stratigraphic profile exists for the Krapina site. Layer 1 consist of fluvial sediments with bones of beaver, red deer, auerochs and perhaps rhinoceros. Layers 2–5 yielded rhinoceros, red deer, roe deer and perhaps wild boar. In layer 3 and probably also in 2 two horizons contain *Homo neanderthalensis* with different claims of the numbers of individuals: *c.* 500 specimens from *c.* 50 individuals (Malez 1979c), *c.* 650 specimens with skulls, lower jaws, vertebrae and limbs from more than 70 individuals (Malez 1984) or about 25–30 individuals (Malez & Malez 1989).

Other sites in this faunal region are thought to have assemblages with Sten*orhinus kirchbergensis* that belong to the last glacial): Brodar (in Guenther 1959).

Faunal Province 5: Bulgaria

¶ Temnata (40)
Layer: 4(s); **32±1.6 ka BP** (31,900±1600 ka BP, TL). Layer 4-4: **35±1.4 ka** (33,000 uncal yrs b.p.); **39±1.4 ka BP** (36,900±1300 uncal yrs b.p.); **41±2.3 ka BP** (38,300±1850 uncal yrs b.p.). **Layer:** 4; **34±1.0 ka BP** (31,900±1600 uncal yrs b.p.); **41±1.5 ka BP** (38,200±1500 uncal yrs b.p.); **41±2.8 ka BP** (38,800±1700 uncal yrs b.p.); **42±1.9 ka BP** (39,100±1800 uncal yrs b.p.). All Aurignacian. **Layer:** 3j,3i,3h; (>32,200 uncal yrs b.p.; 3g: >33,100 uncal yrs b.p.). Aurignacian/Gravettian.
Fauna (layer 4, 3h–3j): *Crocuta crocuta spelaea* (MNI% 3.93), *Canis lupus* (4.05), *Ursus spelaeus* (9.11), *Ursus* sp. (15.52), *Cervus elaphus* (20.92), *Megaloceros* sp. or *Alces* sp. (1.23), *Rupicapra rupicapra* (6.52), *Capra ibex* (26.10), *Bison* sp. (2.58), *Bovinae* (19.45), *Equus* sp. (77.63), *Equus (Asinus) hydruntinus* (6.52).
Flora: *Pinus* sp., *Quercus* sp., *Betula* sp., *Artemisia* sp., Asteroideae, Carduaceae, *Centaurea* sp., Chenopodiaceae, Cichorioideae, *Rumex* sp., *Filipendula* sp., Urticaceae, Scrofulariaceae, Rosaceae, Poaceae, Rubiaceae: Marambat (2000).

Faunal Province 6: Romania

Gura Cheii-Rasnov (42)
Dates: 1±2.4 ka BP (28,900 +2400/–1800 uncal yrs b.p.); **32±1.7 ka BP** (29,700 +1700/–1400 uncal yrs b.p.); **32±0.9**

ka BP (30,450±300 uncal yrs b.p.); **35±2.1 ka BP** (33,300 +1900/–1500 uncal yrs b.p.). All Mousterian.

¶ Spurcata (43)
Dates: 32±2.0 ka BP (30,000 +1900/–1500 uncal yrs b.p.). Mousterian.

Aurignacian and Szeletian sites without hard dates

Faunal Province 2: central and southern Moravia and northern Austria

¶ Švédův stůl (4)
Layer: 8; (fireplace). Aurignacian.
Fauna: *Crocuta spelaea, Ursus spelaeus, Ursus arctos priscus, Coelodonta antiquitatis, Rangifer tarandus, Equus mosbachensis-abeli*: Musil (1961; 1997a).

Faunal Province 3: southern Slovakia and the Pannonian Basin

¶ Szeleta (26)
Szeletian.
Fauna: *Ursus spelaeus* (dominant), *Ursus arctos, Canis lupus, Vulpes vulpes, Alopex lagopus, Martes martes, Panthera spelaea, Crocuta spelaea, Cervus elaphus maral, Megaloceros giganteus, Rangifer tarandus* (rare), *Coelodonta antiquitatis, Mammuthus primigenius*: Jánossy (1986).

¶ Bivak (27)
Layers: 1–2: (Szeletian/Aurignacian transition).
Macrofauna (layers 1+2; (xx) = number of specimens): *Lepus timidus* (18), *Canis spelaeus* (1), *Vulpes vulpes* (5), *Ursus arctos* (1), *Ursus spelaeus* (306), *Sus scrofa* (1), *Cervus elaphus* (2), *Rangifer tarandus* (2), *Capra ibex* (14).
Microfauna: *Talpa europaea* (2), *Citellus citellus* (3), *Citellus rufescens* (1), *Cricetiscus songorus* (1), *Arvicola amphibius* (4), *Microtus oeconomus* (1), *Dicrostonyx torquatus* (1).
Aves: *Lyrurus tetrix* (1), *Lagopus mutus* (1), *Lagopus lagopus* (3), *Corvus corax* (1): Jánossy (1986).

Layer 1: *Macrofauna: Lepus timidus* (79/28), *Vulpes vulpes* (2/1), *Alopex lagopus* (1/1), *Ursus spelaeus* (94/33), *Martes martes* (2/1), *Mustela nivalis* (3/1), *Crocuta spelaea* (1/1), *Lynx lynx* (3/1), *Equus* sp. (2/1), *Rangifer tarandus* (4/1), *Capra ibex* (5/2), *Bison* sp. (2/1): Jánossy (1986); **Layer 2:** *Vulpes vulpes* (5/1), *Ursus arctos* (1/1), *Ursus spelaeus* (306/84), *Sus scrofa* (1/1), *Cervus elaphus* (1/1), *Megaloceros giganteus* (1/1), *Rangifer tarandus* (2/1), *Capra ibex* (14/4): Jánossy *et al.* (1957).

Faunal Province 6: Romania

¶ La Adam (41)
Layer: V; upper Aurignacian; VI middle Aurignacian.
Fauna (layer V): *Cervus elaphus, Megaloceros giganteus, Bos primigenius, Castor fiber*: Dumitrescu *et al.* (1963); (layer

VI): *Lepus timidus, Rangifer tarandus, Vulpes corsac, Putorius eversmanni, Equus* sp., *Saiga tatarica, Bison priscus, Crocuta spelaea.*
Microfauna: *Microtus gregalis, Microtus nivalis, Allactaga jaculus, Lagurus lagurus, Lagurus luteus*: Dumitrescu *et al.* (1963).

¶ **La Adam** (41)
Szeletian.
Fauna: *Mammuthus primigenius, Coelodonta antiquitatis, Crocuta spelaea, Cervus elaphus, Bos primigenius, Ursus spelaeus*: Dumitrescu *et al.* (1963).

Summary: The interval 38–32 ka BP is dominated by the Aurignacian except for the Szeletian and a few Romanian sites where Mousterian has been recorded. For game animals the best sites are in Faunal Province 3 (southern Slovakia and the Pannonian Basin) and in particular Province 4 (Slovenia and Croatia). Unfortunately, we have no faunal data from Province 1, but in Provinces 2, 3, and 4 the typical glacial (tundra or arctic) fauna dominates during the 38–32 ka BP interval with only a few thermophilic elements. In Hungary (Province 3) mammoths still roamed, but in Province 4 the absence of mammoths and presence of some thermophilic species suggest a weak climate change towards warmer conditions. The Bulgarian (5) and Romanian (6) Provinces differ from provinces 3 and 4 because of their much more conspicuous assemblage of thermophilic species.

Interval 32–28 ka BP: Early Cold Phase (numerous cold events)

Faunal Province 2: central and southern Moravia and northern Austria

Předmostí (8)
Dates: **27±0.8 ka** BP (25,040±320 uncal yrs b.p.); **28±0.6 ka** BP (26.320±240 uncal yrs b.p.); **29±0.9 ka** BP (26,870±250 uncal yrs b.p.; more than 20 *Homo sapiens* skeletal finds). All three are Pavlovian.
Macrofauna: *Mammuthus primigenius* (MNI/MNI% ± >1.000 of teeth: 72.05), *Canis lupus* (4.143/43.32, 103/7.42), *Alopex lagopus* (2.250/23.47, 96/6.92), *Lepus timidus* (860/8.97, 79/5.62%), *Lepus europaeus* (52/0.54, 8/0.58), *Rangifer tarandus* (890/9.28, 36/2.59), *Gulo gulo* (581/6.06, 12/0.86), *Equus germanicus* (194/2.02, 5/0.36), *Ursus arctos arctos* (233/2.43, 18/0.58), *Ursus spelaeus* (probably *U. arctos arctos*, 82/0.85, 2/0.14), *Coelodonta antiquitatis* (5/0.05, 1/0.07), *Megaloceros giganteus* (13/0.13, 1/0.07), *Alces alces* (13/0.13, 2/0.14), *Castor fiber* (4/0.04, 2/0.14), *Crocuta spelaea* (4/0.04, 1/0.07), *Panthera pardus* (1/0.01, 1/0.07), *Panthera spelaea* (probably *Panthera leo*, 1/0.01, 1/0.07), *Bison priscus* (25/0.25, 2/0.14), *Bos primigenius* (9/0.09, 1/0.07), *Meles meles* (23/0.24, 2/0.14), *Capreolus capreolus* (2/0.02, 1/0.07), *Capra ibex* (2/0.02, 1/0.07), *Ovibos moschatus* (4/0.04, 1/0.07).

Microfauna: *Lemmus lemmus* (12/0.2, 3/0.22), *Dicrostonyx torquatus* (16/0.17, 4/0.29), *Talpa europaea* (25/0.26, 2/0.14): Musil (1955; 1968; 1994; 1997a).
Flora: Conifers (dominant), *Abies* sp., *Corylus* sp., *Quercus* sp., *Carpinus* sp.: Svobodová (2002). The palaeobotanical evidence is scarce, but Předmostí II suggests a decrease in arboreal pollen from 31 per cent to 16 per cent and an increase in sun-loving plants in the layers immediately above the Gravettian occupation: Puchmajerová (1950), Svoboda *et al.* (1994).

The chief prey in this interval was the mammoth; reindeer, hares, foxes and wolves appear to have been taken less often. *All other animals were hunted opportunistically.* Besides species which can survive in a very cold climate, such as the musk-ox (*Ovibos moschatus*), there are several species with contrasting ecological demands, such as roe deer and badgers. Judging by the thickness of the cultural layer together with the number of finds of game, bone and lithic tools, objects of art and human skeletons, Předmostí was clearly one of the largest, if not the largest of all Palaeolithic hunting stations of its time.

Pavlov (9)
Dates: **27±0.8 ka** BP (25,020±150 uncal yrs b.p.); **27±1.1 ka** BP (25,160±170 uncal yrs b.p.); **28±0.5 ka** BP (25,530±110 uncal yrs b.p.); **28±0.6 ka** BP (25,840±290 uncal yrs b.p.); **29±0.6 ka** BP (26,620±230 uncal yrs b.p.); **29±0.7 ka** BP (26,730±250 uncal yrs b.p.); skeleton of *Homo sapiens* and other finds, mostly teeth. All Pavlovian.
Macrofauna (MNI, %): *Rangifer tarandus* (423, 20.6%), *Lepus* sp. (345, 16.8%), *Canis lupus* (318, 15.5%), *Alopex lagopus* (260, 12.6%), *Mammuthus primigenius* (213, 10.4%), *Equus germanicus* (184, 8.9%), Aves (63, 5.0%), *Vulpes vulpes* (100, 4.9%), *Ursus arctos arctos* (50, 2.4%), *Gulo gulo* (31, 1.5%), *Panthera leo* (13, 0.6%), *Panthera pardus* (3, 0.15%), *Lynx lynx* (3, 0.15%), *Capra ibex* (2, 0.1%), *Castor fiber* (2, 0.1%), *Alces alces* (1, 0.05%), Bovidae (1, 0.05%), *Coelodonta antiquitatis* (a few bones), *Cervus elaphus* (a few bones): Musil (1955; 1959a; 1994; 1997a,b).
Flora (charcoal): *Abies alba* (242), cf. *Abies* (1), *Picea excelsa* (100), cf. *Picea* (1), *Picea/Larix* (7), *Pinus sylvestris* (15), *Pinus cembra* (1), *Pinus mugo* (5), *Pinus* sp. (19), *Juniperus communis* (8): Opravil (1994).

The animals hunted were mostly of small to medium size, such as hares, wolves and reindeer and above all foxes. Mammoths, horses and wolverines are less common. Thus the largest animals, the most important ones as a food source, occupy by number only second place. All others are rare, forming an assemblage that appears to be the result of opportunistic hunting: bears, lions, cats, red deer, auerochs and lynx. In addition to species typical for the cold climate of the high glacial, other species normally found in a somewhat milder climate also abound.

Dolní Věstonice (10)
The ¹⁴C dates confirm that Dolní Věstonice was a cluster of sites. Dolní Věstonice I and II were settled between

~31 ka BP (29,000 uncal yrs b.p.) and ~27 ka BP (25,000 uncal yrs b.p.). Dolní Věstonice III (unit 2) existed between 26 ka BP (24,500 uncal yrs b.p.) and (unit 1) at ~28 ka BP (26,200 uncal yrs b.p.).

Individual dates:
Mammoth dumps: **24±0.5 ka** BP (22,250±570 uncal yrs b.p.); **24±0.6 ka** BP (22,368±749 uncal yrs b.p.); **28±0.6 ka** BP (26,100±200 uncal yrs b.p.).
Settlement 3: **25±0.3 ka** BP (22,630±420 uncal yrs b.p.); **29±0.9 ka** BP (27,070±300 uncal yrs b.p.).
Settlement 2: **29±0.9 ka** BP (26,920±250 uncal yrs b.p.). **28±0.5 ka** BP (25,740±210 uncal yrs b.p.).
Settlement 1: **28±0.6 ka** BP (26,390±270 uncal yrs b.p.).
Human burials: **28±1.1 ka** BP (25,570 ±280 uncal yrs b.p.; burial XV). **29±0.6 ka** BP (26,640±110 uncal yrs b.p.; triple burial).
Miscellaneous dates: **30±0.4 ka** BP (27,660±80 uncal yrs b.p.); **28±1.1 ka** BP (25,950 +630/−580 uncal yrs b.p); **28±0.5 ka** BP (25,820 ±170 uncal yrs b.p.); **30±0.4 ka** BP (27,660 ±80 uncal yrs b.p.); **29±0.9 ka** BP (27,250 +590/−550 uncal yrs b.p.; *Homo sapiens* skeleton and other human finds). All are Pavlovian.
Macrofauna: Aves indet., *Lepus timidus* (dominant), *Castor fiber*, *Lynx lynx* (rare), *Panthera leo* (rare), *Canis lupus* (dominant), *Vulpes vulpes* (rare), *Alopex lagopus* (dominant), *Gulo gulo* (abundant), *Ursus arctos arctos* (rare), *Mammuthus primigenius* (abundant), *Coelodonta antiquitatis* (rare), *Equus germanicus*, *Rangifer tarandus*, *Bos* sp. *vel Bison* sp.(rare): Musil (1959b; 1994).
Flora (charcoal): *Picea excelsa* (6–29%), *Pinus sylvestris* (77–94%), *Salix* sp.(3%): Musil (1997a, after Opravil (1994). *Pinus* sp., *Abies alba*, cf. *Ulmus* sp., *Juniperus communis*: Musil (1997 after Opravil 1994). *Pinus* sp. (35 pieces), *Pinus cembra* (29 pieces), *Pinus mugo* (18 pieces), *Picea excelsa* (17 pieces), *Larix decidua* (12 pieces), *Fagus sylvatica* (1 piece): Nečesaný (1951); Kneblová (1954).
Abies alba, cf. *Abies*, *Larix decidua*, cf. *Larix* sp., *Picea excelsa*, *Picea* sp. *vel Larix* sp., *Pinus sylvestris*, *Pinus cembra*, *Pinus mugo*, *Pinus* sp., *Juniperus communis*, *Taxus baccata*, cf. *Ulmus* sp., *Salix* sp., *Fagus sylvatica*, Puchmajerová (1950). *Abies alba*: Opravil (1994), *Taxus baccata* (Mason *et al.* 1994), *Populus* sp.: Damblon (1997). Triple burial – *Abies alba* (1), cf. *Abies* (1), *Larix decidua*, cf. *Larix* (1), *Picea excelsa* (12), cf. *Picea* (1), *Picea/Larix* (10), *Pinus sylvestris* (3): Musil (1997a) after Opravil (1994).
Parts of tree trunks (3000 pieces up to 40 cm in length, diameter to 30 cm, roots of trunks), all from conifers: Klíma (1990).
Flora (pollen): *Pinus* sp. (dominant), *Picea* sp. (dominant), *Fagus* sp., *Quercus* sp., *Pediastrum borynum*, *P. kawraiskyi*, *P. integrum*: Svobodová (1988; 2002).
Flora, non-arboreal pollen: *Artemisia* sp. (4), *Cirsium* sp. (1), Asteracea liguliflorae (7), Asteraceae tubuliflorae (4), Chenopodiaceae (4), Cyperaceae (46), *Plantago major-media* (1), Poaceae (54), Poaceae type. *Glyceria* sp. (2), *Ranunculus t.* (2), *Valeriana* sp. (1), Hydrophyta: *Utriculatria* sp. (1). Pterophyta: *Botrychium* type b (6), *Equisetum* sp. (3), Polypodiaceae (1), *Pteridium aquilinum* (2), Bryales:

Pediastrum duplex, *Pediastrum integrum* (2), Fungi: *Microthyrium* sp. (1), *Rhizopoda* (8): Svobodová (1988; 1991a,b; 2002).
Arboreal pollen: *Pinus* sp. (dominant), *Picea* sp. (dominant), *Betula* sp., *Alnus* sp., *Juniperus* sp., *Tilia* sp., *Carpinus* sp., *Corylus* sp., *Fagus* sp., *Quercus* sp., *Pediastrum borynum*, *P. kawraiskyi*, *P. integrum*: Svobodová (1991a,b; 2002).

The most common prey was the mammoth followed by hares and foxes and numerous wolves and reindeer bones. All other species were hunted only by chance or their presence was unintentional. *Mammuthus primigenius*, *Lepus timidus*, *Alopex lagopus*, *Vulpes vulpes* together amounted to 51 per cent of the total, including 17–25 per cent hares and foxes. At the same time species typical of the Holocene, such as red deer, bovids, elk, lynx, modern lion and brown bear began to appear in south Moravia amongst species of glacial tradition although in small numbers. They appear to have migrated up from southern and southeastern Europe up the main rivers.

The high amount of *Pinus* pollen grains does not prove the presence of pine forest in the vicinity. Pollen of pines preserve much better than pollen of other trees. Recent investigations have indicated that only a few pines in the neighbourhood change the pollen spectrum which has far more pine pollen than corresponds to the number of pines. Pollen spectra from the Pavlovian cultural layer suggest a steppe environment with a heliophilous vegetation and islands of thermophilic deciduous (*Quercus* sp., *Tilia* sp., *Carpinus* sp., *Fagus* sp., *Corylus* sp.) and coniferous (*Pinus* sp., *Picea* sp., *Larix* sp.) trees. *Alnus* sp., *Salix* sp., *Populus* sp. and *Betula* sp. imply a wet environment: Svobodová (1991a,b).

Bulhary (11)
Dates: 28±2.5 ka BP (25,675 +2750/−2045 uncal yrs b.p; clayey peat near Pavlov).
Flora (macroscopic): <u>Aquatic and swamp plants</u>: *Carex rostrata*, *C. limosa*, *C. aquatilis*, *C. lasiocarpa*, *C. pseudocyperus*, *Heleocharis palustris*, *Alopecurus aequali vel geniculatus*, *Myriophyllum spicatum*, *M. verticillatum*, *Hippuris vulgaris*, *Potamogeton* cf. *gramineus*, *P. pusillus*, *P.* cf. *filiformis*, *Potamogeton* sp., *Batrachium* sp., *Zanichellia palustris*. <u>Dry land plants</u>: *Betula* cf. *pubescens*, *Salix* sp., *Filipendula ulmaria*, *Agrostis* sp. (?), *Selaginella selaginoides*. <u>Coniferous forest</u>: *Pinus sylvestris*, *Pinus* cf. *mugo*, *Pinus cembra*, *Picea* sp. (abies?), *Larix europaea*, *Juniperus communis*, *Betula* sp. As a scatter through the area: *Ulmus* sp., *Acer* sp., *Corylus* sp., *Quercus* sp., *Tilia* sp., *Salix* sp., *Alnus* sp. <u>Grass and herbaceous steppe vegetation</u>: *Ephedra cympylopoda*, grasses, *Artemisia* sp., Chenopodiaceae, *Plantago media*, Daucaceae, Ranunculaceae, Asteraceae, Viciaceae (leading plant types). <u>Subalpine vegetation</u>: *Pinus* cf. *mugo*, *Juniperus* sp., Cyperaceae, *Astragalus* sp., *Medicago* sp., *Ephedra* sp., *Juniperus* sp., Cyperaceae, *Pleurospermum* sp., *Melampyrum* sp., *Linum*-type *alpinum*, *Helianthemum* sp., *Silene* sp., *Gypsophyla* sp., *Centaurea* sp. div., *Knautia* sp., *Euphorbia* sp., *Potentilla* sp., *Bupleurum* sp., *Campanula* sp., *Anemone* sp., *Saxifraga* sp., *Primula*

sp., *Gentiana* sp., *Botrychium* sp., *Selaginella selaginoides*, etc. <u>Spring vegetation</u>: *Pinguicula* sp. *(alpina?)*, *Chrysosplenium* sp., *Swertia perennis*, *Montia* sp., *Parnassia* sp., *Pedicularis* sp.

Flora (pollen): *Salix* sp., *Alnus* sp., *Filipendula* sp., *Petasites* sp., *Valeriana officinalis*, *Polygonum bistorta*, *Veratrum* sp., *Urtica* sp., *Lysimachia vulgaris*, *Galium* sp., *Rumex* spec. div., *Caltha* sp., *Polemonium coeruleum*, *Heracleum* sp., *Mentha* sp., *Lycopus* sp., *Thalictrum* sp., *Chamaenerion* sp., Cyperaceae, *Sparganium* sp., *Typha angustifolia*, *Lemna* sp., *Sagittaria* sp., *Utricularia* sp., *Myriophyllum alterniflorum*, *Nymphaea* sp., *Alisma plantago-aquatica*, *Stratiotes* sp., hairs of *Ceratophyllum* sp., rich planctonic algal flora, *Equisetum* sp., *Phragmites* sp., *Menyanthes* sp., *Lycopus* sp., *Galium* sp. etc.: Rybníčková & Rybníček (1991).

Jarošov (12)
Dates: ~27±1.3 ka BP (25,020±600 uncal yrs b.p.); ~28±1.2 ka BP (25,530±600 uncal yrs b.p.); 28±0.6 ka BP (25,780±240–250 uncal yrs b.p.); 26±0.8 ka BP (25.110±230–240 uncal yrs b.p.); 28±0.8 ka BP (26.220±360–390 uncal yrs b.p.); 28±0.6 ka BP (26,340±180 uncal yrs b.p.); 29±0.9 ka BP (26,950±200 uncal yrs b.p.). All Pavlovian.
Fauna: *Rangifer tarandus* (dominant), *Canis lupus*, *Lepus* sp.: Škrdla & Musil (1999). *Rangifer tarandus* (58% of bones), *Mammuthus primigenius* (4.2%), *Vulpes* sp. (16.6%), *Canis lupus* (12.50%), *Equus germanicus* (4.2%), *Gulo gulo* (4.2%) *Lepus* sp. (4.2%): Škrdla & Musil 1999; Škrdla & Kruml (2000).

Willendorf (23)
Layer: I,II/5; 25±3.1 ka BP (23,500±2900 uncal yrs b.p.); 26±2.7 ka BP (24,200±2600 uncal yrs b.p.); 29±0.4 ka BP (27,270±290 uncal yrs b.p.); 32±1.3 ka BP (30,500 +900/ –800 uncal yrs b.p.). Layer 6: 28±0.6 ka BP (26,150±110 uncal yrs b.p.); 28±0.7 ka BP (26,500±480 uncal b.p.); 30±0.5 ka BP (27,600±480 uncal yrs b.p.); 30±0.4 ka BP (27,620±230 uncal yrs b.p.); 27±0.8 ka BP (24,710±180 uncal yrs b.p.); 28±1.3 ka BP (25,800±800 uncal yrs b.p). Layer 6–8: *(Homo sapiens*; evolved Gravettian pre-Pavlovian). Layer 9: 25±0.2 ka BP (23,180±120 uncal yrs b.p.); 26±1.0 ka BP (23,860±270 uncal yrs b.p.); 26±0.9 ka BP (24,370±290 uncal yrs b.p.); 27±0.7 ka BP (24,910±150 uncal yrs b.p.).
Fauna (Willendorf II; number of bones/NMI): *Layer 5:* Aves (2/1), *Lepus* sp. (3/1), *Panthera spelaea* (1/1), *Canis lupus* (1/1), *Vulpes vulpes* (2/1), *Ursus* sp. (1/1), *Mammuthus primigenius* (2/1), *Cervus* sp. (2/1), *Rangifer* sp. (18/2), *Capra ibex* (13/2); *Layer 6:* *Panthera spelaea* (1/1), *Canis lupus* (1/1), *Ursus* sp. (1/1), *Rangifer* sp. (4/1), *Capra ibex* (2/1); *Layer 7:* *Canis lupus* (1/1), *Vulpes vulpes* (1/1), *Ursus* sp. (1/1), *Mammuthus primigenius* (3/1), *Bison* sp. (2/1); *Layer 8:* *Lepus* sp. (1/1), *Panthera spelaea* (4/ 1), *Canis lupus* (1/1), *Vulpes* sp. (1/1), *Mammuthus primigenius* (1/1), *Equus* sp. (4/2), *Cervus* sp. (2/1), *Rangifer* sp. (3/1), *Capra ibex* (20/3); *Layer 9:* *Panthera* sp. (3/1), *Canis lupus* (10/2), *Vulpes vulpes* (38/9), *Alopex lagopus* (48/25), *Gulo gulo* (2/1), *Ursus* cf. *arctos* (3/1), *Mammuthus primigenius* (6/2), *Equus* sp. (9/1), *Cervus* sp. (8/2), *Rangifer* sp. (15/2), *Bison priscus* (2/2), *Capra ibex*

(52/4), Ovicaprinae (1/1).
Fauna (Willendorf I, number of bones/number of bones in %, MNI/MNI in %): *Aves* (1/0.44, 1/2.08), *Lepus* sp. (9/4.00, 2/4.16), *Lynx lynx* (1/0.44, 1/2.08), *Canis lupus* (62/27.55, 7/14.58), *Vulpes vulpes* (7/3.11, 1/2.08), *Alopex lagopus* (2/0.88, 1/2.08), *Ursus* sp. (1/0.44, 1/2.08), *Mammuthus primigenius* (23/10.22, 8/16.66), *Stephanorhinus kirchbergensis* (6/2.66, 1/2.08), *Equus* sp. (7/3.11, 3/ 6.25), *Cervus* sp. (13/5.77, 5/10.71), *Rangifer* sp. (18/8.00, 4/ 8.33), *Bison priscus* (17/7.55, 4/8.33), *Capra ibex* (57/25.33, 8/16.66), Ovicaprinae (1/0.44, 1/2.08): Musil (1994).

The Willendorf I and II faunas are similar in terms of prey, but differ in number of individuals per species. The species *Stephanorhinus kirchbergensis* has probably been wrongly determined or it comes from an older horizon. In the hunted game wolves, foxes and ibex predominate. Red deer, aurochs and bison are also common, indicating a gradual change from the previous, definitely cold environment towards a milder climate. This is supported by the relatively few mammoths, horses and reindeer amongst the prey. Most bears are probably the subspecies *Ursus arctos arctos*. The lions are probably the species still living today.

The Pavlovian or upper Gravettian is the only culture belonging exclusively to the intervals 32–28 and 28–24 ka BP. In that period large settlements were constructed in central and southern Moravia and northern Austria, always along major rivers. They contain large numbers of artefacts which have no parallel in central Europe. The typical glacial fauna has fewer species than previously, but mammoth, reindeer, arctic fox and hare are very common; less common are wolverine, rhinoceros, hyena, musk-ox and oddly, horses. Some thermophilic elements, now appearing in modest numbers, migrated upstream along the rivers from southeastern Europe. The main sites in southern Moravia, Dolní Věstonice and Pavlov, differ distinctly as concerns the local climate.

Faunal Province 3: southern Slovakia and the Pannonian Basin

Bodrogkerestúr (30)
Dates: 31±3.5 ka BP (28,700±3000 uncal yrs b.p.). Gravettian.
Fauna (no. of bones/MNI): *Panthera spelaea* (1), *Equus* sp. (94/9), *Mammuthus primigenius* (7/3), *Alces alces* (59/9), *Bison priscus* (5/1), *Lepus* sp. (1): Jánossy (1986).
Flora: *Larix* sp., *Picea* sp.: Jánossy (1986).

Chronological and stratigraphical correlations suggest that the Gravettian was in part contemporaneous with the Upper Aurignacian. The main prey were horse and elk, but the number of mammoths was sharply reduced relative to central and southern Moravia and northern Austria (Faunal Province 2). A marked climatic diversification of provinces began at this time that is reflected in an abundance of horses and a relative scarcity of mammoths. Pattern and abundance of individual species differs from that in Moravia, again pointing to local environmental differences.

Faunal Province 4: Slovenia and Croatia

¶ **Velika Pećina** (35)
Layer: d; **28±0.6 ka** BP (26,450±300 uncal yrs b.p.). Gravettian.
Fauna: *Marmota marmota, Canis* cf. *aureus, Canis* sp., *Vulpes vulpes, Ursus arctos priscus, Crocuta spelaea, Lynx lynx, Cervus elaphus, Capra ibex, Ursus spelaeus* (95% of bones). Cave bears were not human prey and the arctoid bears were most likely the subspecies *Ursus arctos arctos:* Malez (1974; 1986).

¶ **Vindija** (34)
Layer: e; **29±1.0 ka** BP (26,970±630 uncal yrs b.p.). Gravettian or Aurignacian.
Fauna: *Marmota marmota, Alopex lagopus, Rangifer tarandus, Alces alces, Bison priscus, Capra ibex, Saiga tatarica, Rupicapra rupicapra, Lepus timidus, Canis lupus, Gulo gulo, Coelodonta antiquitatis:* Malez *et al.* (1984).

¶ **Šandalja II** (37)
Layer: d; Aurignacian (72 masl). Layer e: **26±0.2 ka** BP (23,540±180 uncal yrs b.p.). Aurignacian.
Fauna: *Ursus spelaeus, Equus* sp., *Ochotona pusilla, Bovidae* gen. et spec. indet.
Layer: f; **27±1.1 ka** BP (25,340±130 uncal yrs b.p.). Aurignacian.
Fauna: *Ursus spelaeus, Equus* sp., *Sus* sp., *Cervus elaphus, Capreolus capreolus, Bovidae* gen. et spec. indet.
Layer: h; **30±1.0 ka** BP (27,800±850 uncal yrs b.p., Aurignacian): Malez (1972; 1974).

The game suite contains species that cause it to differ markedly from those of the Pannonian Basin and Moravia since they are quite thermophilic. Clearly, the cold glacial species were in retreat and only a few isolated reindeer and arctic foxes were still around. Mammoths and rhinoceros are entirely absent (Malez 1972; 1974).

Faunal Province 5: Bulgaria

¶ **Temnata** (40)
Layer: 3f; **31±1.4 ka** BP (28,900±1400 uncal yrs b.p.). Gravettian.
Flora: *Pinus* sp., Cichorioideae, Poaceae, Asteroideae, *Centaurea* sp., *Artemisia* sp., Chenopodiaceae: Marambat (2000); Ferrier & Laville (2000).

Pavlovian (= upper Gravettian) sites without hard dates

Faunal Province 3: southern Slovakia and the Pannonian Basin

¶ **Čertova Pec** (21)
Layer: 6; Gravettian.
Fauna: *Cricetus cricetus, Martes martes, Bos* sp., *Bison* sp.: Musil (1998).

Faunal Province 4: Slovenia and Croatia

¶ **Ciganska jama** (38)
Gravettian; (17,000–18,000 uncal yrs b.p. date is clearly incorrect).
Fauna (MNI/MNI%): *Castor fiber* (1/1.4), *Marmota marmota* (34/48.6), *Canis lupus* (1/1.4), *Martes martes* (1/1.9), *Ursus arctos priscus* (1/1.4), *Ursus* sp. (1/1.4), *Sus scrofa* (2/2.8), *Alces alces* (4/5.7), *Rangifer tarandus* (13/18.6), *Cervus elaphus* (4/5.7), *Cervus elaphus maral* (1/1.4), *Megaloceros giganteus* (2/2.8), *Bos* sp. vel *Bison* sp. (2/2.8), *Equus* sp. (2/2.8): Pohar (1992).
Flora: *Pinus* sp., *Sorbus* sp., *Carpinus* sp.: Pohar (1992).

This is a cave site and the data must be used with caution. The arctoid bear was probably the subspecies *Ursus arctos arctos*. The abundance of marmots is curious because, although the animal was hunted in the Russian plains, one would not expect it at this site.

Summary: The interval from 32 to 28 ka BP saw the *floruit* of Palaeolithic hunters in Moravia where they were represented by the Pavlovian (= Upper Gravettian) in Moravia, and by Aurignacian and Gravettian in Slovenia and Romania, but Aurignacian sites have been cited for this interval only in the southeast and their presence needs verification. The period is characterized by rapid temporal environmental changes and great regional environmental variation over short distances (Musil 1999; 2000). The game expresses this variability in time and space by showing large differences between the Pannonian basin (Faunal Province 3 and including Slovenia and Croatia) which had a much warmer climate, and in Moravia the climate was no longer so unrelentingly very cold. (deciduous trees!: *Quercus* sp., *Carpinus* sp., *Fagus sylvatica, Alnus* sp., *Betula* sp., *Tilia* sp., *Corylus* sp., *Acer* sp., *Salix* sp., *Populus* sp., *Ulmus* sp.).

Sümegi & Krolopp (2000) believe on the basis of gastropod analysis that between 32 and 25 ka BP, in Faunal Province 3, the climate in the Carpathian (= Pannonian) basin was similar to present July temperatures while on the promontory of the Carpathian Mountains dense woodland existed. Thus conditions between 30 and 20 ka BP would have been optimal for Palaeolithic hunters. However, the thermophilic species found commonly in Faunal Province 3 and 4 only very gradually penetrated Moravia.

Interval 28–20 ka BP (late Early Cold Phase and onset of the Late Glacial Maximum (LGM) followed by sharp cooling) to Last Glacial Maximum

Most sites in intervals 28–24 and 24–20 ka BP cover a long timespan and the exact duration of each occupation is difficult to determine. Therefore sites of the two intervals have been combined.

Faunal Province 1: southern Poland and northern Moravia

Spadzista Street (17)
Layer: III/6b; **23±0.3 ka** BP (21,000±100 uncal. yrs b.p.);
22±0.4 ka BP (20,200±350 uncal yrs b.p.); **26±0.9 ka** BP
(24,380±180 uncal yrs b.p.); **26±1.0 ka** BP (24,040 uncal yrs
b.p.). All Gravettian. **19±0.5 ka** BP (17,400±310 uncal yrs
b.p.).
Fauna: *Mammuthus primigenius, Coelodonta antiquitatis,
Ursus spelaeus, Canis lupus, Lepus timidus, Alopex lagopus.*

The existence of cave bear as late as this is surprising, but it is an open-air site and mixing with an older
layer is therefore very unlikely: Kozlowski & Sobczyk
(1987).

¶ Mamutowa (20)
Layer: upper VI: **22±0.3 ka** BP (20,260±250 uncal yrs b.p.).
Gravettian.
Fauna: *Canis lupus, Alopex lagopus, Vulpes vulpes, Ursus
spelaeus, Mustela nivalis, Mustela erminea, Crocuta spelaea,
Equus* sp., *Rangifer tarandus, Bos* sp. *vel Bison* sp., *Rupicapra
rupicapra, Lepus timidus.*

In this region also the finds of cave bear and cave
hyena are surprising.

Petřkovice (13)
Dates: **23±0.4 ka** BP (20,790±270 uncal yrs b.p.). **25±0.2 ka**
BP (23,370±160 uncal yrs b.p.). Both Gravettian.
Fauna: *Mammuthus primigenius* (teeth only: Klíma (1955).
Bones of all species were decayed under highly unfavourable local conditions.

*Faunal Province 2: central and southern Moravia and
northern Austria*

¶ Kůlna (1)
Layer: 6b; **24±0.2 ka** BP (21,750±140 uncal yrs b.p.; charcoal); **24±0.3 ka** BP (21,630±150 uncal yrs b.p.); **25±0.2 ka**
BP (22,990±170 uncal yrs b.p.); **23±0.3 ka** BP (21,260±140
uncal yrs b.p.). Gravettian.
Fauna: *Alces alces, Cervus elaphus, Equus* sp. (92% of all
species); also *Mammuthus primigenius, Rangifer tarandus*
(6%). Total number of bone fragments about 50 pieces:
Seitl (1988).
Flora (charcoal): *Pinus* sp., *Acer pseudoplatanus.*
Flora (pollen): *Pinus* sp. (7.9%), *Betula* sp. (26%), *Alnus*
sp. (1.6%), tree/herbaceous ratio = 12.3/87.7.

Milovice (14)
Dates: **23±1.0 ka** BP (21,200±1100 uncal yrs b.p.); **24±0.9
ka** BP (22,100±1000 uncal yrs b.p.; mammoth bone dump);
25±0.4 ka BP (22,900±490 uncal yrs b.p.); **27±1.2 ka** BP
(25,220±280 uncal yrs b.p.). Pavlovian.
Fauna (total number of bones/MNI %): *Panthera leo* (7/
0.09), *Canis lupus* (37/0.52), *Vulpes vulpes* (10.9%),
Mammuthus primigenius (6748/95.25), *Equus germanicus*
(145/2.04), *Rangifer tarandus* (135/1.9).

Mammoth bones predominate over horse and reindeer; all other species are isolated finds: Musil (1997a).

Grubgraben (24)
Layer: 1–5; **20±0.4 ka** BP (18,170±300 uncal. yrs b.p.);
20±0.5 ka BP (18,380±130 uncal yrs b.p.); **20±0.5 ka** BP
(18,400±330 uncal yrs b.p.); **21±0.4 ka** BP (18,890±140 uncal yrs b.p.); **21±0.4 ka** BP (18,920±180 uncal yrs b.p.); **21
±0.5 ka** BP (18,960±290 uncal yrs b.p.); **22±0.2 ka** BP (19,380
±90 uncal yrs b.p.). All are Gravettian.
Fauna (MNI %): *Rangifer tarandus* (73, dominant), *Equus
germanicus* (c. 22), *Capra ibex* (2, 4), *Bos* sp. (few teeth),
Alopex lagopus (jaw-bone)?, *Gulo gulo* (one bone),
Mammuthus primigenius (small fragments of tusk), *Ursus
arctos arctos*: Logan (1990); Musil (2001).

Although mammoths declined in the intervals 28–
24 ka BP, they continued to form a small part of the
hunted game. A curious exception is Milovice with an
abundance of mammoth bones unlike all other sites.
This suggests specialized hunting and shows that at this
time it was still possible to find fairly large mammoth
herds. This may not hold for the entire interval, but a
brief immigration into this area of mammoth herds from
the north is certainly conceivable. This agrees with the
gastropod record from the Pannonian basin (Sümegi &
Krolopp 2000) that shows in that area the relatively mild
and humid climate of the period between 23 and 20 ka BP
was subject to significant temperature oscillations.

*Faunal Province 3: southern Slovakia and the
Pannonian Basin*

Moravany Lopata II (22)
Dates: **23±0.6 ka** BP (21,400±610 uncal yrs b.p.); **26±1.1 ka**
BP (24,100±800 uncal yrs b.p.). Late Gravettian post-dating the Moravian Pavlovian.
Fauna (number of bones/MNI): *Ursus arctos* (11/6), *Ursus*
cf. *arctos, Canis lupus* (14/1), *Vulpes vulpes* (18/6.6), cf.
Gulo gulo (7/1), *Mammuthus primigenius* (12/4), *Lepus* sp.
(11/3), *Equus* sp. (1/1), *Rangifer tarandus* (1446/42):
Lipecki & Wojtal (1998); Kozlowski (1998).
Flora (charcoal): *Pinus* sp.: Lityńska-Zajac (1998).

In the interval 28–20 ka BP Moravany Lopata is this
the only site that specialized in reindeer hunting, perhaps because of its location in the foothills of the
Carpathians. The focus on reindeer hunting is obvious,
but other game such as hare, fox, mammoth and wolf are
also fairly abundant. The continuing presence of mammoth, albeit in small numbers, is surprising because at the
same time another arctic indicator, the arctic fox, is completely absent. Rhinoceros are also absent.

Sümegi & Krolopp (2000), using micro-gastropods,
showed that between 25 and 23 ka BP significant cooling
in the Pannonian Basin produced a cool continental
steppe with a low July temperature between 12–15°C,
although forests survived in sheltered places (Willis *et
al.* 2000).

Faunal Province 4: Slovenia and Croatia

¶ **Šandalja II** (37)
Layer: d; **26±0.2 ka** BP (23.540 +/-180 uncal yrs b.p.).
Fauna: *Ursus spelaeus, Equus* sp., Bovidae.
Layer: c; **24±0.5 ka** BP (21,740 ±450 uncal yrs b.p.). Gravettian.
Fauna: *Sorex alpinus, Mustela nivalis, Microtus nivalis, Alopex lagopus, Gulo gulo, Alces alces, Rangifer tarandus, Bison priscus, Equus* sp., *Capra ibex, Marmota marmota, Lepus timidus, Ochotona pusilla, Canis lupus, Ursus spelaeus.* **Aves:** *Lagopus lagopus, Lagopus mutus, Lyrurus tetrix, Bombycilla garrulus, Pyrrhocorax graculus*: Malez (1972; 1974; 1979c).

Layer: b; Date of **12,320 ka** BP (10,830±50 uncal yrs b.p.) is clearly incorrect or industry is not Gravettian).
Macrofauna: *Lepus timidus, Marmota marmota, Castor fiber, Canis lupus, Alopex lagopus, Vulpes vulpes, Ursus arctos priscus, Ursus spelaeus, Martes martes, Putorius putorius, Mustela erminea, Mustela nivalis, Meles meles, Felis catus, Lynx lynx, Panthera spelaea, Equus* sp., *Equus (Asinus) hydruntinus, Sus scrofa, Cervus elaphus, Megaloceros giganteus, Alces alces, Bos primigenius, Capreolus capreolus, Rupicapra* sp., *Bison priscus.*
Microfauna: *Talpa europaea, Sorex araneus, S. alpinus, Crocidura russula, Rhinolophus ferrum-equinum, Rh. hipposideros, Myotis myotis, Eptesicus* sp., *Minipterus schreibersii, Arvicola terrestris, Microtus arvalis-agrestis, Pitymys subterraneus.* **Aves:** Malez (1974).

This game suite differs strongly from that of the same culture in Moravia.

Faunal Province 6: Romania

Giurgiu-Malul Rosu (44)
Dates: **23±0.3 ka** BP (21,140±120 uncal yrs b.p.); **25±0.2 ka** BP (22,790 ±130 uncal yrs b.p.). Both Aurignacian.

Summary: The various provinces differ in their faunal assemblages. This is most obvious in the case of Poland (Province 1), where as late as 24–20 ka BP only the cold glacial fauna existed, while at the same time that central and southern Moravia and northern Austria already had a fauna intermediate between Poland and the Pannonian Basin, Slovenia and Croatia. In Croatia mammoth and rhinoceros no longer appeared, but various other cold species return at this time in varying numbers in all faunal provinces. Of course, some changes in species numbers may simply have been quantitative rather than absolute, but there is no evidence for that.

Medium-sized animals prevailed in the hunting suite, but in isolated cases woolly rhinoceros, arctic fox and wolverine were still taken. Reindeer and mammoth are no longer abundant (with the exception of Milovice and maybe Moravany Lopata), and horses occur only sporadically. It was at this time that specialized hunting of certain species began, for example reindeer at Moravany and mammoth at Milovice. That was a new phenomenon of which the type case is in the Magdalenian.

Interval 20–17 ka BP

Faunal Province 2: central and southern Moravia and northern Austria

Stránská skála IV (3)
Dates: **20±0.2 ka** BP (18,220±120 uncal yrs b.p. and 17,740 uncal yrs b.p.). Epigravettian.
Fauna: *Equus* sp. (dominant), *Rangifer tarandus, ?Bos primigenius, Mammuthus primigenius, Coelodonta antiquitatis*: Svoboda *et al.* (2002).

Brno, Vídeňská str. (45)
Dates: **15±0.6 ka** BP (14,450±90 uncal yrs b.p.): Valoch 1975.

Velké Pavlovice (46)
Dates: **15±0.2 ka** BP (14,460±230 uncal yrs b.p.). Epigravettian.
Fauna: *Equus* sp., *Mammuthus primigenius*: Svoboda *et al.* (2002).

Faunal Province 3: southern Slovakia and the Pannonian Basin

Mende & Tapiosüly (31)
Dates: **19±0.7 ka** BP (16,750±400 uncal yrs b.p.); **23±0.4 ka** BP (20,520±290 uncal yrs b.p.). Both are Epigravettian.

Sagvar (32)
Upper Layer: **19±0.5 ka** BP (17,160±160 uncal yrs b.p.).
Lower Layer: **21±0.4 ka** BP (18,900±100 uncal yrs b.p.). Both are Epigravettian.

Faunal Province 5: Bulgaria

¶ **Temnata** (40)
Layer: 3a; **22±4.0 ka** BP (19,600±3700 uncal yrs b.p.).

Summary: In the interval 20–17 ka BP dated sites are available only for Faunal Province 2 (central and southern Moravia and northern Austria), Province 3 (southern Slovakia and the Pannonian basin) and Province 5 (Bulgaria), and most of them have no description of flora and fauna, thus severely limiting our conclusions. Still, we can see major changes in the mammalian assemblages. This change is well documented in the assemblage of gastropods in the Pannonian Basin (Sümegi & Krolopp 2000) which indicate a strong cooling in the period 22 ka BP to 20 ka BP (c. 20,000–18,000 uncal yrs b.p.) which peaked in the period of 20 ka BP to 18 ka BP (18,000–20,000 uncal yrs b.p.). Most research workers are convinced that around ~20 ka BP a great environmental crisis resulted in depopulation (Mussi *et al.* 1999 *passim*). Reindeer and horse became the dominant animals and arctic fox, wolverine and mammoth, although surviving in small numbers, yielded only one or two finds, implying that they were opportunistic catches. In contrast, and this is most interesting, thermophilic species — which previously had always occurred in small numbers — now disappeared altogether.

Climate variation in time and space and human hunting in central and eastern Europe

This study of the Palaeolithic game suites of eastern central and southeastern Europe has produced some interesting results. The provinces of the region differed from each other in their climatic conditions to various degrees for the whole time between *c.* 32 and 17 ka BP. An extreme example of those differences is provided by the cold climate of southern Poland in Gravettian time where a cold glacial fauna predominated while at the same time milder climates in the Balkan Peninsula, particularly in Slovenia and Croatia, allowed thermophile species of various requirements of warmth to be present. The fauna of Moravia and northern Austria reflected an intermediate climate that tended more to the conditions prevailing in Poland than to the climate of more southerly regions. A special position was occupied by the arid climate of the wide low-lands of the Pannonian basin.

Over the interval of about 45 to 17 ka BP the fauna assemblages of the individual provinces did not remain constant, but the degree of temporal difference between them varied considerably. The faunal pattern in the southern part of Poland appears to have been relatively stable because a cold fauna persisted throughout the whole interval, but the relative abundances of the species may well have varied considerably, but until more quantitative data on species composition become available little more can be said. Less stable were Slovenia and Croatia where rather large changes in species composition occurred in the course of the same period. For the other faunal provinces the data are too sparse to provide sufficient information even for speculation.

Between 50 and 38 ka BP, mammoth, rhinoceros, reindeer and saiga antelope made up a species assemblage in Faunal Province 2 (central and southern Moravia and northern Austria) that consisted entirely of cold-glacial species, with mammoth and reindeer being the dominant ones. Little data is available for the period from 38–28 ka BP despite the occurrence of many Palaeolithic settlements (Chapter 4: van Andel *et al.* 2003b); the hunting preferences and habits of the inhabitants and the animal resources available to them thus remain largely unknown. This situation changes greatly during the period 28–24 ka BP interval. In that period several of the largest settlements in central Europe are found in Moravia. Because three of those, Předmostí, Dolní Věstonice and Pavlov, were close together in a small area, the high human population density of the time suggests favourable environmental conditions and an abun-

dance of animals to be hunted. The fauna consisted of cold-glacial species, above all mammoth, reindeer, arctic fox and wolverines. Woolly rhinoceros appear only rarely and the low number of horses is puzzling for a region where everything suggests a parkland. What is important in this time (28–24 ka BP) and perhaps surprising is the gradual penetration, evidently upstream along the rivers, of thermophilic species originating from southeastern Europe. Subsequently, during the period of 24–20 ka BP, a slow retreat began of some of hitherto still abundant arctic species, especially mammoth, wolverine and arctic fox. At the same time the newly arrived thermophilic species do not seem to increase, either in number of species or in the number of individuals. Then, just before 20 ka BP, a drastic change in climate and/or environment seems to have occurred. Mammoth, wolverine and arctic fox almost disappear while reindeer and horses become the dominant species.

The first finds from Faunal Province 3 (southern Slovakia and the Pannonian basin) date only from the period 38–33 ka BP. They display, besides cold-glacial species, some definitely thermophilic species, a situation that did not change very much until about 28 ka BP. From 28 ka BP onward, however, a fauna unlike that of Moravia came into being that was dominated by horses, mammoths being already very rare, a difference that should be attributed at least to some part to the character of the landscape that supported this fauna so different from that of central and southern Moravia and northern part of Austria.

In Province 4 the faunal pattern of the period 45–38 ka BP reminds one of the penultimate interglacial in central Europe or the first Würm interstadials. Missing are the mammoth and woolly rhinoceros, and all species present require quite high mean annual temperatures. Throughout the whole period of our concern, Slovenia and Croatia (Faunal Province 4) too have climatic and environmental conditions that differ substantially from those prevailing in the second (central and southern Moravia and northern Austria) and third (southern Slovakia and Pannonian basin) faunal provinces, conditions that are incompatible with those of the first faunal province (southern Poland and northern Moravia). For the period 38–32 ka BP there is only very little information, but beginning around 32 ka BP settlements with game reappear. The assemblage reindeer, wolverine, woolly rhinoceros, wolverine, arctic fox and saiga antelope, however, seems normal for the time, but the contrast with the 45–38 ka BP interval requires a substantial evolution in the very period from which we know of so few Palaeolithic settlements. The 32 ka BP faunal assemblage persists more or less until the period 28–24 ka BP, differing from its original composition mainly by the disappearance of many of the cold-glacial species. Totally absent are woolly mammoth and woolly rhinoceros, but reindeer, saiga and arctic fox can still be found. Whether they were abundant or not, we do not know. This gradual change continues until the period of 24–20 ka BP.

The climatic changes taking place between c. 50 and 15 ka BP are well known (Fig. 10.2), but parallel changes in the faunal assemblages are riddled with *lacunae* and the data for Romania are insufficient. Still, a picture is beginning to emerge that differs from the conventional broad generalizations of climatic change across Europe that are expressed mainly in terms of the arctic-to-Mediterranean climate zones. The data presented above suggest that the reality was more complex and strongly influenced by maritime-to-continental gradients. On a regional level those palaeoenvironmental variations in time and space are of the greatest importance in understanding the relations between Palaeolithic human beings and their environments, particularly with regard to the blossoming of the modern humans that is demonstrated by the Gravettian culture groups.

We have made a beginning here with outlining those differences in terms of the natural environment, but only specialists in archaeology can appreciate how they affect human populations and cultures.

Acknowledgements

I wish to express my gratitude to Professor Tjeerd H. van Andel from the Department of Earth Sciences of the University of Cambridge. I thank him also for the editing of my original text in the style nowadays common in the west and for well-meant words of criticism. I thank Dr William Davies for transposing all dates into a chronology in calendar years (calibration of ^{14}C dates) and for preparing Appendix 10.1 which contains the original ^{14}C dates + standard deviation by sites and their calibrated equivalents.

References

Barron, E., T.H. van Andel & D. Pollard, 2003. Glacial environments II: reconstructing the climate of Europe in the Last Glaciation, in *Neanderthals and Modern Humans in the European Landscape during the Last Glaciation*, Chapter 5, eds. T.H. van Andel & W. Davies. (McDonald Institute Monographs.) Cambridge: McDonald Institute for Archaeological Research, 57–78.

Bolus, M. & N.J. Conard, 2001. The late Middle Palaeolithic and earliest Upper Palaeolithic in Central Europe and their relevance for the Out of Africa hypothesis.

Quaternary International 75, 29–40.

Damblon, F., 1997. Anthracology and past vegetation reconstruction. Pavlov I-Northwest. The Upper Palaeolithic burial and its settlement context. *The Dolní Věstonice Studies* 4, 437–42.

Dansgaard, W., S.J. Johnsen, H.B. Clausen, D. Dahl-Jensen, N.S. Gundestrup, C.U. Hammer, C.S. Hvidberg, J.P. Steffensen, H. Sveinbjörnsdottir, J. Jouzel & G. Bond, 1993. Evidence for general instability of past climate from a 250-kyr ice-core record. *Nature* 364, 218–20.

Davies, W. & P. Gollop, 2003. The human presence in Europe during the Last Glacial Period II: climate tolerance and climate preferences of mid- and late glacial hominids, in *Neanderthals and Modern Humans in the European Landscape during the Last Glaciation*, Chapter 8, eds. T.H. van Andel & W. Davies. (McDonald Institute Monographs.) Cambridge: McDonald Institute for Archaeological Research, 131–46.

Dumitrescu, M., P. Samson, E. Terzea, C. Radulescu & M. Ghica, 1963. Pestera „La Adam" station pleistocena. *Lucracile Institutului de Speleologie, Emil Racovitza* 1–2 (1962–1963), 229–91.

Ferrier, C. & H. Laville, 2002. Corrélations stratigraphiques des sédiments interplénieglaciaires, in *Temnata Cave in Karlikovo Karst Area Bulgaria* 2, eds. B. Ginter, J.K. Kozlowski, J. Guadelli & H. Laville. Kraków: Jagellonian University, 31–6.

Forsten, A., 1982. Equidae, in *Excavations in Bacho Kiro Cave*, ed. J.K. Kozlowski. Warsaw: Panstwowe wydawnictwo naukowe, 56–61.

Ginter, B., J.K. Kozlowski, J. Guadelli & H. Laville (eds.), 2002. *Temnata Cave in Karlikovo Karst Area Bulgaria* 2. Kraków: Jagellonian University.

GRIP (Greenland Ice-core Project) Members), 1993. Climate instability during the last interglacial period recorded in the GRIP ice core. *Nature* 364, 203–7.

Guadelli, J.L. & F. Delpech, 2002. Les Grands Mammifères du début du paléolithique supérieur à Temnata, in *Temnata Cave in Karlikovo Karst Area Bulgaria* 2, eds. B. Ginter, J.K. Kozlowski, J. Guadelli & H. Laville. Kraków: Jagellonian University, 54–159.

Guenther, E.W., 1959. Zur Altersdatierung der diluvialen Fundstelle von Krapina in Kroatien. *Bericht über die 6. Tagung der Deutschen Gesellschaft für Anthropologie*, 202–9.

Huntley, B. & J.R.M. Allen, 2003. Glacial environments III: palaeo-vegetation patterns in Late Glacial Europe, in *Neanderthals and Modern Humans in the European Landscape during the Last Glaciation*, Chapter 6, eds. T.H. van Andel & W. Davies. (McDonald Institute Monographs.) Cambridge: McDonald Institute for Archaeological Research, 79–102.

Jánossy, D., 1955. Die Vögel und Säugetierreste der spätpleistozänen Schichten der Höhle von Istállóskö. *Acta Academiae Scientiae Hungaricae* 5, 149–81.

Jánossy, D., 1986. *Pleistocene Vertebrate Faunas of Hungary*. Budapest: Akadémiai Kiadó.

Jánossy, D., S. Kretzoi-Varrok, M. Herrman & L. Vértes, 1957. Forschungen in der Bivakhöhle, Ungarn. *Eiszeitalter und Gegenwart* 8, 18–36.

Jöris, O. & B. Weninger, 1998. Extension of the 14-C calibration curve to *c.* 40,000 cal BC by synchronising Greenland $^{18}O/^{16}O$ ice-core records and North Atlantic Foraminifera profiles: a comparison with U/Th coral data. *Radiocarbon* 40, 495–504.

Jöris, O. & B. Weninger, 1999. Possibilities of calendric conversion of radiocarbon data for the glacial periods. *Actes du Colloque 'C14 Archéologie'* Lyon 1998, 87–92.

Jöris, O. & B. Weninger, 2000. Calendric age-conversion of glacial radiocarbon dates at the transition from the Middle to the Upper Palaeolithic in Europe. *Bulletin de la Société Préhistorique Luxembourgeoise* 18, 43–55.

Klíma, B., 1955. Výsledky archeologického výzkumu na tábořišti lovců mamutů v Ostravě-Petřkovicích. *Časopis Slezského muzea* 4, 1–35.

Klíma, B., 1990. Dřevěné zbytky z paleolitické stanice Dolní Věstonice. *Pravěké a slovanské osídlení Moravy*, 7–14. Musejní a vlastivědný spolek Brno.

Kneblová, V., 1954. Fytopaleontologický rozbor uhlíků z paleolitického sídliště v Dolních Věstonicích. *Anthropozoikum* 3, 297–9.

Kozlowski, J.K. (ed.), 1982. *Excavations in the Bacho Kiro Cave (Bulgaria). Final Report, 1–172.* Warsaw.

Kozlowski, J.K. (ed.) 1998. *Complex of Upper Palaeolithic Sites near Moravany, Western Slovakia*, vol. II: *Moravany - Lopata II (Excavations 1993–1996).* (Jagellonian University Kraków.) Kraków: Slovak Academy of Science Nitra.

Kozlowski, J.K. & K. Sobczyk, 1987. The Upper Palaeolithic site Kraków-Spadzista street C2. *Zeszyty Naukowe UJ. Prace Archeologiczne* 42.

Kubiak, H., 1982a. Rhinocerotidae, in *Excavations in the Bacho Kiro Cave (Bulgaria). Final Report, 1–172*, ed. J.K. Kozlowski. Warsaw, 55–60.

Kubiak, H., 1982b. Artiodactyla, in *Excavations in the Bacho Kiro Cave (Bulgaria). Final Report, 1–172*, ed. J.K. Kozlowski. Warsaw, 61–6.

Kyncl, J., 1984. Vorbericht über die Analyse von Kolzkohlenpartikeln aus dem Würm-Bodenkomplex von Stránská skála (Bez. Brno-město). *Přehled výzkumů Archeologického ústavu AV ČR v Brně* 1982, 13.

Lipecki, G. & P. Wojtal, 1998. Mammal remains, in *Complex of Upper Palaeolithic Sites near Moravany, Western Slovakia*, vol. II: *Moravany - Lopata II (Excavations 1993–1996)*, ed. J.K. Kozlowski. (Jagellonian University Kraków.) Kraków: Slovak Academy of Science Nitra, 103–26.

Lityńska-Zajac, M., 1998. Anthracological analysis, in *Complex of Upper Palaeolithic Sites near Moravany, Western Slovakia*, vol. II: *Moravany - Lopata II (Excavations 1993–1996)*, ed. J.K. Kozlowski. (Jagellonian University Kraków.) Kraków: Slovak Academy of Science Nitra, 97–8.

Logan, B., 1990. The hunted of Grubgraben: an analysis of the faunal remains, in *The Epigravettian Site of Grubgraben, Lower Austria*, ed. A. Montet-White, 65–91. (ERAUL 40.)

Malez, M., 1958. Einige neue Resultate der paläontologischen Erforschung der Höhle Veternica. *Palaeonto-logija Jugoslavica* 1, 5–24. Zagreb.

Malez, M., 1961. Pećina Veternica kao paleolitsko nalazište s tragovima kulta medvjeda, II. *Jugoslavenski speleološki kongres Split* 1958, 123–38.

Malez, M., 1972. Ostaci fosilnog čovjeka iz gorna pleistocena Šandalja kod Pule (Istra) [The remains of the upper Pleistocene man from Šandalja near Pula im Istria/Croatia]. *Palaeontologia Jugoslavica* 12, 1–39.

Malez, M., 1974. Neue Ergebnisse der Paläolithikum-Forschungen in Velika Pećina, Veternica und Šandalja (Kroatien). *Arheološki radovi i razprave* 7, 7–44.

Malez, M., 1979a. Položaj naslaga spilje Vindije u sustavu članjenja kvartara šireg područja Alpa [Lage der Höhlenablagerungen von Vindija im System der quartären Vollgliederung des Alpengebietes]. *RAD, Razred za prirodne znanosti* 18, 187–218.

Malez, M., 1979b. Prilog poznavanju rasprostranjenosti stepske zviždare u gornjem pleistocenu Hrvatske [Beitrag zur Kenntnis der Verbreitung des Pfeifhasen im oberen Pleistozän in Kroatien]. *RAD. Razred za prirodne znanosti* 18, 345–61.

Malez, M., 1979c. Osnovne crte paleolitika i mezolitika u Hrvatskoj [Grundzüge des Paläolithikums und Mesolithikums in Kroatien]. *RAD. Razred za prirodne znanosti* 18, 117–53.

Malez, M., 1984. Kvartarnogeološka, paleontološka i paleolitička istraživanja u spiljama Hrvatske [Quartärgeologische, paläontologische und paläolithische Forschungsarbeiten in den Höhlen Kroatiens]. *The 9th Yugoslavian Congress of Speleology Karlovac 1984, Proceedings*, 169–73.

Malez, M., 1986. Kvarterni sisavci (Mammalia) iz Velike Pećine na Ravnoj Gori (SR Hrvatska, Jugoslavija). *Radovi* 1, 33–139.

Malez, M. & V. Malez, 1989. The Upper Pleistocenne fauna from Neanderthal man site in Krapina (Croatia, Jugoslavia). *Geologik Vjesnik* 42, 41–57.

Malez, M. & D. Rukavina, 1979. Položaj naslaga spilje Vindije u sustavu članjenja kvartara šireg područja Alpa (Lage der Höhlenablagerungen von Vindija im System der quartären Vollgliedeung des Alpengebietes). *RAD, Razred za prirodne znanosti* 18, 187–218.

Malez, M., An. Šimunić & A. Šimunić, 1984. Geološko, sedimentološki i paleoklimatski odnosi Spilje Vindija i bliže okolice [Geologische, sedimentologische und paläoklimatische Verhältnisse der Höhle Vindija und der näheren Umgebung]. *Rad 411, Razred za prirodne znanosti* 20, 231–64.

Marambat, L., 2000. Paléoenvironnements végétaux à Temnata: La séquence aurignacienne, in , in *Temnata Cave in Karlikovo Karst Area Bulgaria* 2, eds. B. Ginter, J.K. Kozlowski, J. Guadelli & H. Laville. Kraków: Jagellonian University, 37–51.

Mason, S.L., J.G. Hather & G.C. Hillman, 1994. Preliminary investigation of the plant macro-remains from Dolní Věstonice II and its implications for the role of plant foods in Palaeolithic and Mesolithic Europe. *Antiquity* 68, 48–57.

Meese, D.A., A.J. Gow, R.B. Alley, G.A. Zielinski, P.M. Grootes, M. Ram, K.C. Taylor, P.A. Mayewski & J.F. Bolzan, 1997. The Greenland Ice-sheet Project 2 depth-age scale: methods and results. *Journal of Geophysical Research* 102, 26,411–23.

Mook, W.G., 1976. Groningen Radiokarbondaten von Bohunice, in *Die altpleistozäne Fundstelle von Brno-Bohunice*. Academia Praha: Studie AÚ ČSAV v Brně 4/1, 76–83.

Mook, W.G., 1988. Radiocarbon - Daten aus der Kůlna-Höhle, in *Die Erforschung der Kůlna-Höhle 1961–1976*, ed. K. Valoch. (Anthropos Studien zur Anthropologie, Paläoethnologie, Paläontologie und Quartärgeologie 24.) Brno: Moravian Museum, 285–6.

Musil, R., 1955. Osteologický materiál z paleolitického sídliště v Pavlově[Das osteologische Material aus der paläolithischen Siedlungsstätte in Pavlov]. *Práce Brněnské základny ČSAV* 27/6, 279–319.

Musil, R., 1959a. Osteologický materiál z paleolitického sídliště v Pavlově. Část II [Das osteologische Material aus der paläolithischen Siedlungsstätte in Pavlov, part II]. *Anthropozoikum* 8 (1958), 83–106.

Musil, R., 1959b. Poznámky k paleontologickému materiálu z Dolních Věstonic [Bemerkungen zum paläontologischen Material aus Dolní Věstonice/ Unter-Wisternitz]. *Anthropozoikum 8*, 73–82.

Musil, R., 1961. Die Höhle Švédův stůlein typischer Höhlenhyänenhorst. *Anthropos Studien zur Anthropologie, Paläoethnologie, Paläontologie und Quartärgeologie 13*, 97–260. Brno: Moravian Museum.

Musil, R., 1965a. Die Bärenhöhle Pod hradem. Die Entwicklung der Höhlenbären im letzten Glazial. *Anthropos Studien zur Anthropologie, Paläoethnologie, Paläontologie und Quartärgeologie 18* (N.F.10), 7–92.

Musil, R., 1965b. Zhodnocení dřívějších paleontologických nálezů ze Šipky [Wertung der früheren paläontologischen Funde aus der Šipka-Höhle]. *Anthropos Studien zur Anthropologie, Paläoethnologie, Paläontologie und Quartärgeologie 17* (N.F.9), 127–34.

Musil, R., 1968. Biostratigraphie des Pleistozäns auf Grund der Faunegesellschaften, in Das Periglazial und das Paläolithikum in der Tschechoslovakei. Brno, 130–37.

Musil, R., 1969. Die Entwicklung der Tiergesellschaft im Laufe der Sedimentation in der Kůlna-Höhle. *Quartär* 20, 8–20.

Musil, R., 1976. Pferdefunde aus der Zeit zwischen dem Alt-und Mittelpleistozän, in Die altsteinzeitliche Fundstelle von Brno-Bohunice, ed. K. Valoch. *Studie AÚ ČSAV v Brně* 4/1, 76–83.

Musil, R., 1986. Palaeobiography of terrestrial communities in Europe during the Last Glacial. *Acta Musei Nationalis Pragae* 41B/1–2, 1–84.

Musil, R., 1988. Ökostratigraphie der Sedimente in de Kůlna-Höhle. *Anthropos Studien zur Anthropologie, Paläoethnologie, Paläontologie und Quartärgeologie* 24 (N.F.16), 215–55.

Musil, R., 1990. Pferdefunde (*Equus taubachensis* und *Equus scythicus*) aus der Kůlna-Höhle in Mähren. *Weimarer Monographien zur Ur- und Frühgeschichte* 26, 1–86.

Musil, R., 1994. The fauna. Hunting game of the culture layer at Pavlov. Pavlov I. Excavations 1952–1953. (ERAUL 66, Études et Recherches archéologiques de l'Université de Liège.) *The Dolní Věstonice Studies* 2, 183–209.

Musil, R., 1997a. Klimatická konfrontace terestrických a marinních pleistocenních sedimentů (Climatic Comparison of Terrestrial and Marine Pleistocene Sediments), in *Dynamika vztahů marinního a kontinentálního prostředí*. Přírod. Brno: Fakulta Masarykovy Univerzity, 93–167.

Musil, R., 1997b. Hunting game analysis, in Pavlov I: Northwest. (ERAUL.) *The Dolní Věstonice Studies* 4, 443–68.

Musil, R., 1998. Čertova pec a její fauna (Die Höhle Čertova pec bei Radošina (Westslowakei) und ihre Fauna). *Slovenský kras 34* (1996), 5–56.

Musil R., 1999. Životní prostředí v posledním glaciálu na území Moravy (The environment in the Last Glacial on the territory of Moravia). *Acta Musei Moraviae, Scientiae geologicae 64*, 161–86.

Musil, R., 2000. Das Studium der Pferde aus der Lokalität Grubgraben. *Acta Musei Moraviae, Scientia geologica* 87, 165–219.

Musil, R., 2001. Natural environment. *Anthropologie* 38, 327–31.

Musil, R., in press. The Early Upper Palaeolithic fauna from Stránská skála, in *Stránská Skála: Origin of the Upper Palaeolithic in the Brno-Basin, Moravia, Czech Republic*, eds. J. Svoboda & J. Bar. (American School of Prehistoric Research Bulletin 47.) Cambridge (MA): Harvard University.

Mussi, M., W. Roebroeks & J. Svoboda, 1999. Hunters of the Golden Age: an introduction, in *Hunters of the Golden Age: the Mid Upper Palaeolithic of Eurasia 30,000–20,000 BP*, eds. M. Mussi, W. Roebroeks & J. Svoboda. Leiden: University of Leiden, 1–11.

Nečesaný, V., 1951. Studie o diluviální flóře Dyjsko-svrateckého úvalu [The diluvian flora of the valleys of Dyje and Svratka]. *Práce Moravsko-slezské Akademie věd přírodních* 14, 291–308.

Opravil, E., 1965. Ergebnisse der Holzkohlenanalyse aus der Grabung der Höhle Pod hradem. *Anthropos Studien zur Anthropologie, Paläoethnologie, Paläontologie und Quartärgeologie 18* (N.F.10), 147–9.

Opravil, E., 1976. Ergebnisse der Holzkohlenanalyse von Brno-Bohunice. *Studie Arch. ústavu ČSAV v Brně* 4/1, 72–4.

Opravil, E., 1988. Ergebnisse der Holzkohlenanalyse aus der Kůlna-Höhle, in Die Erforschung der Kůlna-Höhle 1961–1976, ed. K. Valoch. *Anthropos Studien zur Anthropologie, Paläoethnologie, Paläontologie und Quartärgeologie* 24 (N.F. 16), 211–14.

Opravil, E., 1994. Vegetation, in Pavlov I: Excavation, 1952–1983, ed. J. Svoboda. (Ch. V. ERAUL 66.) *The Dolní Věstonice Studies* 2, 177–80.

Pohar, V., 1992. Mlajšewürmska fauna iz Ciganske jame pri Želnah (Kočevja, južnovihodna Slovenija). *Razprave 4*, 33/6, 147–87.

Puchmajerová, M., 1950. Pylové rozbory spraší a pohřbených půd sídlišt u Dolních Věstonic a Předmostí na Moravě. *Sborník Moravsko-slezské akademie věd přírodních* 3/4, 134–5.

Rink, W.J., H.P. Schwarcz, K. Valoch, L. Seitl & C.B. Stringer, 1996. ESR dating of Micoquian industry and Neanderthal remains at Kůlna Cave, Czech Republic. *Journal of Archaeological Science* 23, 889–901.

Rybníčková, E. & K. Rybníček, 1991. The environment of the Pavlovian: palaeoecological results from Bulhary, South Moravia, in *Palaeovegetational Development in Europe, Proceedings of the Pan-European Palaebotanical Conference 1991*. Vienna: Museum of Natural History, 73–9.

Seitl, L., 1988. Ökologisch-ökonomische Analyse des osteologischen Materials aus dem Magdalénien und aus dem Gravettien in der Kůlna-Höhle. *Anthropos Studien zur Anthropologie, Paläoethnologie, Paläontologie und Quartärgeologie* 24 (N.F.16), 257–60.

Škrdla, P. & O. Kruml, 2000. Uherské Hradiště (k.ú. Jarošov u Uh. Hradiště, okr. Uh. Hradiště). *Přehled výzkumů Archeologického ústavu AV ČR v Brně* 41, 88–92.

Škrdla, P. & R. Musil, 1999. Jarošov II: nová stanice gravettienu na Uherskohradišťsku (tschechisch). *Přehled výzkumů Archeologického ústavu AV ČR v Brně (1995–1996)*, 47–62.

Sümegi, P. & E. Krolopp, 2000. Quaternary malacological analyses from modelling of the Upper Weichselian palaeoenvironmental changes in the Carpathian Basin. *GeoLines* 11, 139–41.

Svoboda, J. & H. Svobodová, 1985, Les industries de type Bohunice dans leur cadre stratigraphique et écologique. *L'Anthropologie* 89/4, 505–14.

Svoboda, J., V. Ložek, H. Svobodová & P. Škrdla, 1994. Předmostí after 110 years. *Journal of Field Archaeology* 21, 457–72.

Svoboda, J., P. Havlíček, V. Ložek, J. Macoun, R. Musil, A. Přichystal, H. Svobodová & E. Vlček, 2002. Paleolit Moravy a Slezska [Palaeolithic of Moravia and Silesia]. *Dolnověstonické studie* 8, 3–303.

Svobodová, H., 1987. Přírodní prostředí. *Studie Arch. ústavu ČSAV v Brně* 4, 18–27.

Svobodová, H., 1988. Pollenanalytische Untersuchung des Schichtkomplex 6-I vor der Kůlna-Höhle, in Die Erforschung der Kůlna-Höhle 1961–1976, by K. Valoch. *Anthropos Studien zur Anthropologie, Paläoethnologie, Paläontologie und Quartärgeologie* 24, 205–10.

Svobodová, H., 1991a. The pollen analysis of Dolní Věstonice II, section No. 1, in Dolní Věstonice II: Western Slope. *ERAUL* 54, 75–88.

Svobodová, H., 1991b. Pollen analysis of the Upper Palaeolithic triple burial at Dolní Věstonice. *Archeologické rozhledy* 43, 505–10.

Svobodová, H., 2002. Vývoj vegetace. *Dolnověstonické studie* 8, 48–51.

Switsur, V.R., 1976. Cambridge Radiocarbondaten von Bohunice: Die altsteinzeitliche Fundstelle von Brno-Bohunice. *Studie Archeologického ústavu ČSAV v Brně* 14/1, 18–27.

Sych, L., 1982. Lagomorpha, in *Excavations in Bacho Kiro Cave*, ed. J.K. Kozlowski. Warsaw: Panstwowe wydawnictwo naukowe, 52.

Valoch, K., 1975. Paleolitická stanice v Koněvově ulici v Brně *Archeologické rozhledy* 27, 3–17.

van Andel, T.H., 2003. Glacial environments I: the Weichselian climate in Europe between the end of the OIS-5 interglacial and the Last Glacial Maximum, in *Neanderthals and Modern Humans in the European Landscape during the Last Glaciation*, Chapter 2, eds. T.H. van Andel & W. Davies. (McDonald Institute Monographs.) Cambridge: McDonald Institute for Archaeological Research, 9–20.

van Andel, T.H., W. Davies, B. Weninger & O. Jöris, 2003a. Archaeological dates as proxies for the spatial and temporal human presence in Europe: a discourse on the method, in *Neanderthals and Modern Humans in the European Landscape during the Last Glaciation*, Chapter 3, eds. T.H. van Andel & W. Davies. (McDonald Institute Monographs.) Cambridge: McDonald Institute for Archaeological Research, 21–30.

van Andel, T.H., W. Davies & B. Weninger, 2003b. The human presence in Europe during the Last Glacial period I: human migrations and the changing climate, in *Neanderthals and Modern Humans in the European Landscape during the Last Glaciation*, Chapter 4, eds. T.H. van Andel & W. Davies. (McDonald Institute Monographs.) Cambridge: McDonald Institute for Archaeological Research, 31–56.

Vertés, L. (ed.), 1994. *Tata: eine mittelpaläolithische Travertinsiedlung in Ungarn*. Budapest.

Willis, K.J., E. Rudner & P. Sümegi, 2000. The full-glacial forests of central and southeastern Europe. *Quaternary Research* 53, 203–13.

Wiszniowska, T., 1982. Carnivora, in *Excavations in Bacho Kiro Cave*, ed. J.K. Kozlowski. Warsaw: Panstwowe wydawnictwo naukowe, 52–5.

Appendix 10A. *Archaeological and mammalian dates used in this chapter; with ^{14}C standard deviations, calibrated ages with their ranges (SD_{cal}) and ages ka BP rounded to the nearest 1000 years. Site numbers are displayed on Figure 10.1. Codes: A - dating method; B - ^{14}C uncalibrated with C and D standard deviation; calBP - age in calendar (calibrated) years; cal sig - calibration range (SD_{cal}); ka BP - age before present in thousands of years.*

Site #	Site name	Layer	A	B	C	D	cal BP	σcal	ka BP	industry
1	Kůlna	layer 6b	14C	21,260	140	140	23,260	293	23	Gravettian
1	Kůlna	layer 6b	14C	21,630	150	150	23,630	253	24	Gravettian
1	Kůlna	layer 6b	14C	21,750	140	140	23,750	234	24	Gravettian
1	Kůlna	layer 6b	14C	22,990	170	170	24,990	205	25	Gravettian
1	Kůlna	layer 6a	14C	38,600	950	800	40,600	1065	41	Micoquian
1	Kůlna	Nean. bone	14C	45,660	2850	2200	47,660	3178	48	Micoquian
1	Kůlna	layer 7a	ESR				50,000	5000	50	Micoquian

Appendix 10.1. *(cont.)*

Site #	Site name	Layer	A	B	C	D	cal BP	σcal	ka BP	industry
2	Bohunice		14C	36,000	1100	1100	38,000	1144	38	Bohunician
2	Bohunice		14C	40,173	1200	1200	42,173	2644	42	Bohunician
2	Bohunice		14C	41,400	1400	1200	43,400	2142	43	Bohunician
2	Bohunice		14C	42,900	1700	1400	44,900	2027	45	Bohunician
3	Stránska skála	IV	14C	18,220	120	120	20,000	210	20	Epigravettian
3	Stránská skála	IIIa, layer 3	14C	30,980	360	360	32,980	520	20	Epigravettian
3	Stránská skála	IIa, layer 4	14C	32,350	900	900	34,350	1511	33	Aurignacian
3	Stránská skála	IIIb, layer 4	14C	32,600	1700	1400	34,600	1932	34	Aurignacian
3	Stránská skála	III-5	14C	38,200	1100	1100	40,200	1275	35	Aurignacian
3	Stránská skála	III-5	14C	38,500	1400	1200	40,500	1900	40	Bohunician?
3	Stránská skála	IIIa-5, layer 4	14C	41,300	3100	2200	43,300	2873	40	Bohunician
3	Švedův stůl		n.d.							Aurignacian
4	Čertova díra		n.d.							
5	Šipka cave		n.d.							Mousterian
6	Vedrovice	V	14C	37,650	550	550	39,650	999	40	Mousterian
7	Vedrovice	V	14C	39,500	1100	1100	41,500	2561	41	Szeletian
7	Předmostí		14C	25,040	320	320	27,040	842	27	Szeletian
8	Předmostí		14C	26,320	240	240	28,320	572	28	Pavlovian
8	Předmostí		14C	26,870	250	250	28,870	868	29	Pavlovian
8	Pavlov		14C	25,020	150	150	27,020	767	27	Pavlovian
9	Pavlov		14C	25,160	170	170	27,160	1148	27	Pavlovian
9	Pavlov		14C	25,530	110	110	27,530	539	28	Pavlovian
9	Pavlov		14C	25,840	290	290	27,840	552	28	Pavlovian
9	Pavlov		14C	26,620	230	230	28,620	628	29	Pavlovian
9	Pavlov		14C	26,730	250	250	28,730	730	29	Pavlovian
9	Dolní Věstonice		14C	22,250	570	570	24,250	495	24	Pavlovian
10	Dolní Věstonice	mammoth dump	14C	22,368	749	749	24,368	654	24	Gravettian
10	Dolní Věstonice	settlement 3	14C	22,630	420	420	24,630	351	25	
10	Dolní Věstonice	burial DV XVI	14C	25,570	280	280	27,570	1077	28	
10	Dolní Věstonice	settlement 1	14C	25,740	210	210	27,740	539	28	
10	Dolní Věstonice		14C	25,820	170	170	27,820	533	28	
10	Dolní Věstonice		14C	25,950	630	570	27,950	1106	28	
10	Dolní Věstonice	mammoth dump	14C	26,100	200	200	28,100	553	28	
10	Dolní Věstonice	settlement 1	14C	26,390	270	270	28,390	584	28	
10	Dolní Věstonice	triple burial	14C	26,640	110	110	28,640	602	29	
10	Dolní Věstonice	settlement 2	14C	26,920	250	250	28,920	884	29	
10	Dolní Věstonice	settlement 3	14C	27,070	300	300	29,070	896	29	
10	Dolní Věstonice		14C	27,250	590	550	29,250	927	29	
10	Dolní Věstonice		14C	27,660	80	80	29,660	377	30	
10	Bulhary	peat nr. Pavlov	14C	25,675	2750	2045	27,675	2507	28	
11	Jarošov		14C	25,020	600	600	27,020	1251	27	
12	Jarošov		14C	25,530	600	600	27,530	1248	28	Pavlovian
12	Jarošov		14C	25,110	230	240	26,070	890	28	Pavlovian
12	Jarošov		14C	25,780	240	230	28,240	679	28	Pavlovian
12	Jarošov		14C	26,320	390	360	27,640	790	28	Pavlovian
12	Jarošov		14C	26,340	180	180	27,890	846	28	
12	Jarošov		14C	26,950	200	200	29,450	200	29	
12	Petřkovice		14C	20,790	270	270	22,790	401	23	
13	Petřkovice		14C	23,370	160	160	25,370	205	25	Gravettian
13	Milovice	mammoth bone	14C	21,200	1100	1100	23,200	1012	23	Gravettian
14	Milovice	mammoth bone	14C	22,100	1000	1000	24,100	932	24	Pavlovian
14	Milovice		14C	22,900	490	490	24,900	398	25	Pavlovian
14	Milovice		14C	25,220	280	280	27,220	1176	27	Pavlovian
14	Pod Hadrem	layer 6	14C	26,830	300	300	28,830	866	29	Pavlovian
15	Pod Hradem	layer 9	14C	33,100	530	530	35,100	1159	35	Aur/Grav?
15	Pod Hradem	layer 9	14C	33,300	1100	1100	35,300	1722	35	Szeletian
15	Dzierzyslav	I, upper layer	TL				36,000	5500	36	Szeletian
16	Spadzista st Krakov	level II/6b	14C	17,400	310	301	19,400	496	19	Szeletian
17	Spadzista st Krakov	level III/6b	14C	20,200	350	350	22,200	396	22	Epigravettian
17	Spadzista st Krakov	level III/6b	14C	21,000	100	100	23,000	303	23	Gravettian
17	Spadzista st Krakov	level III/6b	14C	24,040	200	200	26,040	1009	26	Gravettian
17	Spadzista st Krakov	level III/6b	14C	24,380	180	180	26,380	878	26	Gravettian
17	Spadzista st Krakov	layer V	14C	32,000	2000	2000	34,000	2302	34	Gravettian

Appendix 10.1. *(cont.)*

Site #	Site name	Layer		A	B	C	D	cal BP	σcal	ka BP	industry
17	Piekary	II	14C	32,000	500	500	34,000	1547	34	Aur/Grav?
18	Oblazowa	layer VIII–X	14C	30,600	550	550	32,600	1232	33	Aurignacian
19	Oblazowa	layer VII–X	14C	31,500	500	550	33,500	912	33	Aurignacian
19	Oblazowa	layer VII–X	14C	34,200	650	650	36,200	1289	36	Aurignacian
19	Mamutowa	layer VI top	14C	20,260	250	250	22,260	327	22	Aurignacian
20	Čertova Pec	layer 4	14C	38,300	2480	2480	40,300	2834	40	Gravettian
21	Čertova Pec	layer 4	14C	38,400	2800	2100	40,400	2841	40	Szeletian
21	Moravany Lopata	II	14C	21,400	610	610	23,400	635	23	Szeletian
22	Moravany Lopata	II	14C	24,100	800	800	26,100	1135	26	late Gravettian
22	Willendorf	I&II, layer 9	14C	23,180	120	120	25,180	188	25	late Gravettian
23	Willendorf	I&II, layer 5	14C	23,500	2900	2900	25,500	3094	25	Gravettian
23	Willendorf	I&II, layer 9	14C	23,860	270	270	25,860	1046	26	Gravettian
23	Willendorf	I&II, layer 5	14C	24,200	2600	2600	26,200	2698	26	Gravettian
23	Willendorf	I&II, layer 8	14C	24,710	180	180	26,710	751	27	Gravettian
23	Willendorf	I&II, layer 9	14C	24,370	290	290	26,370	890	26	Gravettian
23	Willendorf	I&II, layer 9	14C	24,910	150	150	26,910	747	27	Gravettian
23	Willendorf	I&II, layer 8	14C	25,800	800	800	27,800	1312	28	Gravettian
23	Willendorf	I&II, layer 6	14C	26,150	110	110	28,150	559	28	Gravettian
23	Willendorf	I&II, layer 6	14C	26,500	480	480	28,500	736	28	Gravettian
23	Willendorf	I&II, layer 5	14C	27,270	290	290	28,600	400	29	Gravettian
23	Willendorf	I&II, layer 6	14C	27,620	230	230	29,620	404	30	Gravettian
23	Willendorf	I&II, layer 6	14C	27,600	480	480	29,600	532	30	Gravettian
23	Willendorf	I&II, layer 5	14C	30,500	900	800	32,500	1310	32	Gravettian
23	Grubgraben	layers 1–5	14C	18,170	300	300	20,170	414	20	Gravettian
24	Grubgraben	layers 1–5	14C	18,360	130	130	20,380	342	20	Gravettian
24	Grubgraben	layers 1–5	14C	18,400	330	330	20,400	508	20	Gravettian
24	Grubgraben	layers 1–5	14C	18,890	140	140	20,890	412	21	Gravettian
24	Grubgraben	layers 1–5	14C	18,920	180	180	20,920	428	21	Gravettian
24	Grubgraben	layers 1–5	14C	18,960	290	290	20,960	478	21	Gravettian
24	Grubgraben	layers 1–5	14C	19,830	90	90	21,830	170	22	Gravettian
24	Érd	layer e	14C	44,300	1400	1400	51,000	7700	51	Gravettian
25	Szeleta		n.d.							Mousterian
26	Bivak		n.d.							Szeletian
27	Istállóskö	upper layer	14C	30,000	600	600	32,000	825	32	Szeletian/Aurig.
28	Istállóskö	lower layer	14C	~36,000			~38,000		~38	Aurignacian
28	Tata		14C	33,600	1200	1200	35,600	1724	36	Aurignacian
29	Bodrogkerestúr		14C	28,700	3000	3000	30,700	3524	31	Mousterian
30	Mende & Tapiosüly		14C	16,750	400	400	18,750	727	19	Gravettian
31	Mende & Tapiosüly		14C	20,520	290	290	22,520	394	23	Epigravettian
31	Sagvar	upper layer	14c	17,160	160	160	19,160	473	19	Epigravettian
32	Sagvar	lower layer	14C	18,900	100	100	20,900	400	21	Epigravettian
32	Veternica	layer i	14C	43,200	200	200	45,200	1933	45	Epigravettian
33	Vindija	layer e	14C	26,970	630	630	28,970	966	29	Mousterian
34	Vindija	layer f	14C	33,850	520	520	35,850	1400	36	Gravettian?
34	Vindija	fauna	14C	28,020	360	360	29,648	478	30	Aurignacian?
34	Vindija	fauna	14C	29,080	100	100	30,987	650	31	
34	Vindija	layer G3 Nean.	14C				>44,000		>44	
34	Vindija	layer G3 Nean.	14C	42,400	4300	4300	44,537	3700	45	Mousterian
34	Velika Pećina	layer d	14C	26,450	300	300	28,450	601	28	Mousterian
35	Velika Pećina	layer g	14C	31,168	1400	1400	33,168	2189	33	Gravettian
35	Velika Pećina	layer i	14C	33,850	250	250	35,850	1130	36	Aurignacian
36	Krapina		14C	30,700	750	750	32,510	1010	32	Aurignacian
37	Šandalja II	II, layer c	14C	21,740	450	450	23,740	465	24	Aurignacian
37	Šandalja II	II, layer d	14C	23,540	180	180	25,540	220	26	Mousterian
37	Šandalja II	II, layer f	14C	25,340	130	130	27,340	1121	27	Aurignacian
37	Šandalja II	II, layer h	14C	27,800	850	850	29,800	1018	30	Aurignacian
37	Ciganska Jama		n.d.							Aurignacian
38	Bacho Kiro	11, 11a	14C	34,800	1200	1200	36,800	1566	37	Gravettian?
39	Bacho Kiro	11, 11a	14C	37,700	1500	1500	39,700	1583	40	Mouster./Aurig.?
39	Bacho Kiro	11, 11a	14C	33,800	900	900	35,800	1579	36	Mouster./Aurig.?
39	Bacho Kiro	11, 11a	14C	38,500	1100	1100	40500	1240	40	Mouster./Aurig.?
39	Bacho Kiro	11, 11a	14C	>43,000			>45,000		>45	Mouster./Aurig.?
40	Temnata	layer 3a	14C	19,600	3700	3700	21,600	4016	22	Mousterian

Appendix 10.1. *(cont.)*

Site #	Site name	Layer	A	B	C	D	cal BP	σcal	ka BP	industry
40	Temnata	layer 4 surf.	TL				31,900	1600	32	Epigravettian
40	Temnata	layer 4b	14C	33,000	900	900	35,000	1399	35	Aurignacian
40	Temnata	layer 4b	14C	36,900	1300	1300	38,900	1371	39	Aurignacian
40	Temnata	layer 4b	14C	38,300	1850	1850	40,575	2312	41	Aurignacian
41	La Adam		n.d.							Aurignacian
41	La Adam		n.d.							Szeletian
42	Gura Cheii-Rasnov		14C	28,900	2400	1800	30,900	2381	31	Aurignacian
42	Gura Cheii-Rasnov		14C	29,700	1700	1400	31,700	1742	32	Mousterian
42	Gura Cheii-Rasnov		14C	30,450	300	300	32,450	942	32	Mousterian
42	Gura Cheii-Rasnov		14C	33,300	1900	1500	35,300	2092	35	Mousterian
43	Spurcata cave		14C	30.000	1900	1500	32,000	2033	32	Mousterian
44	Giurgiu-Malul Rosu		14C	21,140	120	120	23,140	299	23	Mousterian
44	Giurgiu-Malul Rosu		14C	22,790	130	130	24,790	190	25	Aurignacian
45	Brno Videňská str.		14C	14.450	90	90	15,000	390	15	Aurignacian
46	Velké Pavlovice		14C	14,460	23	23	15,400	200	15	Epigravettian

Appendix 10.2. *List of scientific and common names of the fauna cited in this chapter with indication of their ecology where appropriate. Environmental code: 1) tundra; 2) subarctic forest (including taiga); 3) arctic steppe, including loess and cold steppe and open parkland steppe; 4) Mild to summer-warm forest (coniferous and deciduous), meadows and gallery forest.*

Scientific name	Common name	Ecology
Artiodactyla		
Alces alces	Elk	2, 3
Cervus elaphus	Red deer	2, 4
Rangifer tarandus	Reindeer	1, 2
Dama dama	Fallow deer	4
Capreolus capreolus	Roe deer	4
Megaloceros sp./*Megaloceros giganteus*	Giant deer	2, 3
Bos sp./*Bos primigenius*	Aurochs	2, 4
Bison sp./*Bison priscus*	Bison	3
Bos/Bison	Aurochs/Bison	
Rupicapra rupicapra	Chamoix	Mountains today
Ovis sp.	Sheep	
Capra ibex	Ibex	Mountains today
Ovibos moschatus	Musk ox	2, 3
Saiga tatarica	Saiga antelope	3
Sus scrofa	Wild boar	4
Perissodactyla		
Equus (Asinus) hydruntinus	Steppe ass	3
Equus sp./*Equus germanicus*	Horses	3
Stephanorhinus kirchbergensis	Extinct rhino	4
Stephanorhinus hemitoechus	Extinct rhino	3
Stephanorhinus sp.	Extinct rhino	
Coelodonta antiquitatis	Extinct rhino	1, 2, 3
Proboscidea		
Mammuthus primigenius	Mammoth	1, 2, 3
Elephas antiquus/Elephas sp.	Straight-tusked elephant	4
Carnivora		
Panthera leo	Lion	Wide ranging
Panthera spelaea	Cave lion	Wide ranging
Panthera leo/Panthera spelaea	Lion	Wide ranging
Panthera pardus	Leopard	Wide ranging
Panthera sp.		
Felis sylvestris	Wild cat	4
Lynx lynx	Lynx	2, 4
Lynx sp.		2, 4
Crocuta crocuta/Crocuta spelaea	Spotted hyaena	Wide ranging
Canis lupus	Wolf	Wide ranging
Cuon sp.		

Appendix 10.2. *(cont.)*

Scientific name	Common name	Ecology
Carnivora *(cont.)*		
Vulpes vulpes	Red fox	4
Vulpes corsac		
Alopex lagopus	Arctic fox	1, 2, 3
Ursus arctos	Brown bear	Wide ranging
Ursus spelaeus	Cave bear	Wide ranging
Ursus sp.	Bear	
Meles meles	Badger	Wide ranging
Mustela erminea	Stoat	
Mustela nivalis	Weasel	
Mustela sp.		
Putorius putorius	Polecat	
Putorius putorius robusta	Large polecat	
Martes sp.	Marten	
Gulo gulo	Wolverine	1, 2
Lutra lutra	Otter	4
Insectivora		
Talpa europaea	Mole	Mild climate
Crocidura sp. group	White-toothed shrews	
Sorex araneus group	Common shrew group	Wide ranging
Sorex minutus	Pigmy shrew	
Sorex minutissimus	Least shrew	1, 2, 3
Sorex sp.		
Neomys sp.		
Rodentia		
Marmota primigenia		2, 3
Marmota bobak		2, 3
Marmota marmota	Alpine Marmot	2, 3
Apodemus flavicollis	Yellow-necked mouse	
Apodemus sylvaticus	Wood mouse	Wide ranging
Castor fiber	Beaver	2, 4
Cricetus cricetus	Common hamster	3
Cricetulus migratorius	Grey hamster	
Mesocricetus sp.	A hamster	
Arvicola terrestris	Northern water vole	
Arvicola sp.	Water vole	
Pericola (Pitymys) subterraneus		S. open woodland
Ptericola (Pitymys) sp.		S. open woodland
Microtus arvalis	Common vole	
Microtus agrestis	Field vole	
Microtus agrestis/arvalis	Field / common vole	
Microtus gregalis		
Microtus nivalis	Snow vole	
Clethrionomys glareolus	Bank vole	
Lemmus lemmus	Norway lemming	1, 2
Dicrostonyx torquatus	Collared lemming	1, 2
Lagurus lagurus		
Eolagurus luteus		
Hystrix cristata/vinogradovi	Porcupine	
Spermophilus major/Citellus superciliosus		3
Sicista subtilis	Southern birch mouse	
Sicista betulina	Northern birch mouse	
Spalax sp.	Mole rat	
Lagomorpha		
Lepus sp.	Hares	
Lepus europaeus	Brown hare	
Lepus timidus	Mountain hare	1, 2, 3
Ochotona pusilla	Pika	4

Chapter 11

The Human Presence in Europe during the Last Glacial Period III: Site Clusters, Regional Climates and Resource Attractions

William Davies, Paul Valdes, Cheryl Ross & Tjeerd H. van Andel

Mountains, forelands and plains

In preceding chapters we have used the climate of the last glaciation as an independent forcing factor and the human presence in glacial time and across continental space as a dependent variable. However, climate change is only one of the forcing factors that determine the human response to environmental changes. Of equal impact is the regional-scale relief of the European continent which influences the human condition on regional and local scales, but is almost invariant on a human time-scale. In terms of human success and failure, the two forcing factors govern several others that impact human affairs directly such as the availability of animal and vegetal resources or the trafficability of the terrain that influence human moves.

Topography of Europe and Palaeolithic site patterns
Much of Europe has a grand scale, dramatic relief that has a major influence on the European climate, — but is the bane of quantitative demographers whose models work best on a table top (Bocquet-Appel & Demars 2000a,b; Zubrow 1992). The Trans-European mountain ranges rise to two or three thousand metres, forming a climate barrier that shields the Mediterranean zone from the nordic climate (Fig. 11.1). North of the Trans-European mountain barrier lie the hilly forelands of western France, southern Germany and farther east and the North European Plain which in glacial times included the dry bed of the emerged North Sea. This plain gradually widens eastward towards Russia and the Ukraine and in the east extend from the Baltic shores to the Black Sea and eastward to the Urals.

The Fennoscandian ice-sheet, generally thought to have extended to and later beyond the Baltic Sea throughout the last glaciation, actually barely reached its northern shores in OIS-4 and did not again advance so far south until it reached a terminus in the North European Plain during the Last Glacial Maximum (LGM). During OIS-3 the ice-sheet was much reduced in size and may even have been limited for much of that time to small ice-caps on the highest mountains of southern Norway (Arnold *et al.* 2002; Olsen *et al.* 1996; Olsen 1997).

Figure 11.1 shows all dated Palaeolithic sites for the period 70–20 ka BP[1] in our chrono-archaeological data base[2], using lithic technocomplexes as proxies for three human categories, Neanderthals, early Anatomically Modern Humans (AMH) and later AMH. The Neanderthals and early modern humans of the Transitional climate phase (Chapter 4: van Andel *et al.* 2003b, Table 4.3) tended to keep to latitudes below 50°N, but in the Early Cold and LGM Cold Phases humans settled on lands well north of the Trans-European mountain ranges and close to the ice-sheet margin.

The dispersal of Mousterian sites across Europe is the widest of the three technocomplexes but it shows a preference for the coastal zones of Italy and Spain and the Atlantic shores of Portugal, northern Spain and Atlantic France. Less dense but equally well-defined is the broadly west–east trending belt of Mousterian sites that straddles the 50th parallel from Britain to central Russia with extensions there as far north as 55°N. Neanderthal site clusters occur in the Crimea and along the Black Sea shore as well, a region mostly ignored by AMH.

The Aurignacian site distribution is similar to that of the Neanderthals, including the preference for the shores of the central and western Mediterranean and the south coast of the Gulf of Biscay. In northern Europe, Aurignacian sites, though fewer in

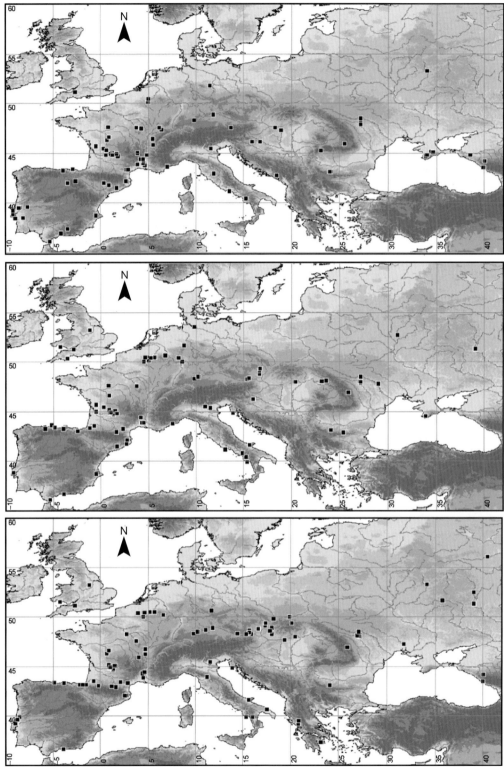

Figure 11.1. *Distribution across Europe of the three main technocomplex sites (dots).*
Top: Mousterian (sites of all ages, 93; visible, 91). Centre: Aurignacian and Early Upper
Palaeolithic (sites of all ages, 103; visible, 103). Bottom: Gravettian and Upper Palaeolithic to
c. 20,000 ka BP (sites of all ages, 114; visible, 76). The deficit of dots relative to the total is due
to the scale; a single dot may cover several sites.

Elevation

0   1000 km

number than those of the Mousterian, occupy the same broad west-to-east zone from Britain to southern Russia including outposts north of 50°N.

In contrast to the Neanderthal and early AMH preference for mid-latitudinal and southern climes, Gravettian and other contemporary Upper Palaeolithic populations appear to have liked life in the foreland and plains north of the Pyrenees and Alps, from the Ardennes eastward to at least the 30°E meridian in the northern European plain, and at several locations beyond 50°N in Russia west of the Ural Mountains.

The Atlantic and Mediterranean shores, especially in the south and west, appear to have attracted all three human populations, in particular where narrow continental shelves exposed at local glacial sea-levels were backed by high rocky coasts (van Andel 1989; van Andel & Shackleton 1982; Shackleton *et al.* 1984). What induced this liking is not clear since evidence for the use of marine resources in the Middle and Upper Palaeolithic is scarce, except for the use of marine shells as personal ornaments by Aurignacian and Gravettian groups (Taborin 1993; 2000). Was it the warm sunshine of southern latitudes or access to coastal resources in the form of the herbivores of the emerged coastal plains? Or is the selectivity merely apparent because the low coasts of the North Sea, the Channel and France border on wide shelves so that Palaeolithic coastal sites are not accessible to archaeologists? During Stage 3 the global sea-level oscillated between –50 and –60 m in the Early Warm Phase, falling slowly to –80 m towards the end of OIS-3 (Lambeck & Chappell 2001).

The high ranges remained, with rare exceptions, uninhabited and the vast plains, hardly occupied during early OIS-3, only gradually acquired an open site scatter that, if plotted on a blank sheet on a continental scale, would appear to be random. In contrast, the occupation of the forelands was marked by vast empty spaces between areas of scattered site that shifted over time. This transitional zone from highlands to lowlands, however, shows a few areas of unusually densely clustered and often long-occupied sites, such as the Ardennes in Belgium, the

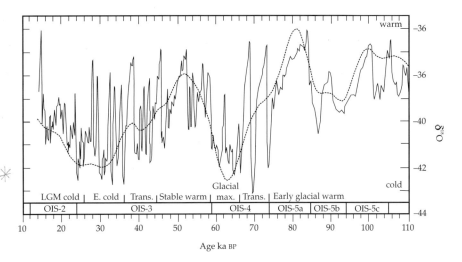

Figure 11.2. *The Greenland GISP2 δ¹⁸O climate record with millennial (Dansgaard/Oeschger) events (Meese et al. 1997; Stuiver & Grootes 2000) and multi-millennial climate phases (Chapter 4:1; Table 4.3). Of special interest in this chapter are the Stable Warm, Transition, Early Cold and LGM Cold Phases.*

Dordogne in southern France and the middle Danube basin in Moravia (Fig. 11.1). What the cause was of this focussed, persistent preference for a few specific regions will occupy us in this chapter.

Long-inhabited site clusters in preferred regions
What made the settings of those few semi-continuously inhabited site clusters so attractive for long-term inhabitation?

To address this question we chose for detailed examination three case-study areas: a) the valley of the Dordogne and its tributary the Vézère in southwestern France; b) the steeply incised antecedent river valleys of the Belgian Ardennes; and c) the site clusters of northwestern Austria, Moravia and southern Poland in the middle Danube basin. The three areas represent climatic archetypes: the Dordogne had an Atlantic maritime climate ameliorated by Mediterranean influence, the Ardennes the cold maritime climate of higher latitudes in western Europe, and the continental climate of the Middle Danube basin. Before we can pronounce on why, throughout OIS-3 and the onset of the LGM, humans should have been consistently present there, we must thoroughly examine the regional climate and related environments of each area to consider living resources and how those, although climate-dependent themselves, might have influenced human presence or absence from an area.

The three case-study areas have been explored archaeologically for well over a century, yielding a treasure of archaeological data and enough local climatic and other environmental information to set the human history in context. Because the chronological resolution of the archaeological and environmental data is limited (Chapter 3: van Andel *et al.* 2003a, we used instead of the D/O climate events the multi-millennial climate phases for the interval 47–18 ka BP (Fig. 11.2), i.e. the Warm Transitional (47–38 ka BP), Early Cold (37–28 ka BP) and Last Glacial Maximum Cold Phases (LGM; 27–18 ka BP).

To describe the palaeoclimatic conditions of the three case-study areas we use the two 'end member' palaeoclimate simulations provided by the Stage 3 Project and discussed in detail in Chapter 5 (Barron *et al.* 2003; see also Alfano *et al.* 2003). The simulations are recorded in two modes: maps and numerical sets of computed climatic variables for each 60 × 60 km output box. The 'OIS-3 warm' simulation stands for warm events in the Transitional Phase (and earlier), the 'LGM (21k) cold' simulation for the two cold phases (see Chapter 5: Barron *et al.* 2003). We confine the description to a small subset of climate variables, temperature, wind-chill, precipitation and snow cover.

The Dordogne case study

The landscape
The Dordogne case-study area is an archetypal site cluster of great antiquity. It lies on the west flank of the Massif Central in southwestern France (Fig. 11.3: top) between the 0°E and 2°E meridians and 44°N and 47°N parallels. The summit of the Massif Central, at 1500 m, is crowned with many Pleistocene volcanoes some of which are still active. During the LGM the summit plateau was partly glaciated (Daugas & Raynal 1989), but was probably ice-free in OIS-3. The ice-covered area was too small to have an impact on the regional climate.

Jurassic, Cretaceous and early Cenozoic deposits onlap onto the western slope of the Massif Central, grading downward into a broad Pleistocene coastal apron traversed by the Garonne river which then drained into the Atlantic across a wide continental shelf that lay fully or partly emerged during the Last Glaciation (Lambeck 1997).

The principal site concentration (Fig. 11.3: top) looks westward across the coastal plain to the Atlantic shore then at –80 m to –120 m. Most sites are in the incised valleys of the Dordogne and Vézère; the caves and abris there were well populated throughout OIS-3 and into the LGM (OIS-2). A smaller number are scattered on the interfluves and a few in

the uplands. A few sites occupy the far field of the Garonne plain such as La Quina (81),[3] Combe Saunière (147) and Barbas III (149). Curiously, the several more southerly tributaries of the Garonne were uninhabited throughout OIS-3.

Palaeoclimate[4]
The Dordogne area is located in the Atlantic maritime climate zone, but except for the few far-field sites, human occupation was limited to the foothills of the Massif Central (Fig. 11.3: top) where elevations of 200–400 m (add –80 to –120 m for the lowered OIS-3 sea-level) caused greater summer–winter temperature, precipitation and snow cover contrasts.

In the warm Transitional Phase, winter temperatures in the core area, 300–500 m above sea-level of the time, ranged from –4°C to 1.5°C. The frost season ended in March or April (mean temperature 1°–3.5°C), but spring began in May with mean temperatures of 7°–9°C, an earlier and warmer spring than the more northerly Ardennes. The summer was long and quite warm (16°–18°C through August). September still had a temperature of 8°–12°C, dropping to 2°–3°C in November and just below freezing in December. The LGM summer was cool with maximum temperatures of 11°–14°C through August, but the January mean of –3° to –5°C was not the bitter glacial winter often invoked. Spring arrived late, 1°–2°C in April.

The wet season of the Transitional Phase lasted from November through May with monthly means from 2.5–5 mm/day, and the driest time was July–September (1–2 mm/day). The same pattern of winter-wet and summer-dry prevailed in the Early Cold and LGM Phases, but with lower means (2.5–3 mm/day) from November to March. The dry summer averaged 1–2 mm/day, a less drastic contrast than today. Throughout OIS-3 the Dordogne summers were sunny, with a cloud cover of less than 50 per cent against a winter cover of 80–90 per cent.

In the Transitional and Early Cold phases the first lasting snow came in November and did not vanish until May. In the core area snow fell on 10–30 days in warmer and 30–40 days in colder times. The resulting cover varied with elevation: downslope 1–8 cm was average in the warm and 2–10 cm in the cold phase with about twice the amount in the uplands.

Climate changes in the Dordogne and mammalian fauna
Early in the Stage 3 Project we recognized that its goals required information on the mammalian fauna,[5] the major food supply for the human population.

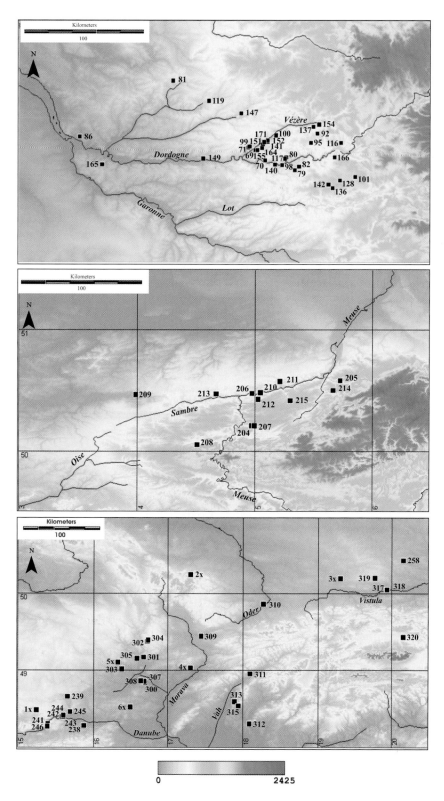

Figure 11.3. *Topography, main rivers and archaeological site locations with numbers for the three case-study areas. For site names see Tables 11.3, 11.5, 11.8. Top: Dordogne area; Centre: Ardennes area; Bottom: Middle Danube Basin area. The scale is elevation above present sea-level in metres. Latitude in °N and longitude in °E marked along the left and bottom sides of each panel. Topography is taken from gtopo30 digital elevation model (USGS 1996); projection is latitude-longitude wgs84.*

Chapter 11

Table 11.1. *Dordogne case study Diversity of mammalian eco-fauna by time-slice, age and type of site for interval of 47 to 18 ka* BP. *Sites with number (xxx). Underneath are dates in [xx] of dated layers. To the right of each date is number of species for each of four eco-suites listed from left to right as follows: Arctic tundra zone; cold-temperate woodland; present-day mountain species; cold steppe species.*

Cave/Abri   **Open-air**   **Cave/Abri**   **Open-air**

Interval 47–38 ka BP

St Césaire (73)
[38]   3   5   0   0

Roc de Combe (79)
[41]   1   2   2   0

Abri Caminade (80)
[40]   1   3   0   0
[38]   1   3   4   0

Chapelle-aux-Saints (116)
[47]   4   4   2   3

Le Moustier (171)
[47–40]   2   4   1   0

Interval 37–28 ka BP

Abri Pataud (69)
[35–31]   5   9   0   2
[30–28]   6   4   2   1

La Flageolet I (70)
[36–35]   1   3   2   1
[29–28]   1   11   2   3

La Ferrassie (71)
[36]   2   5   1   0
[31–28]   4   9   3   3

St Césaire (73)
[37–34]   3   5   0   0

Roc de Combe (79)
[37–29]   5   10   2   5

Abri Caminade (80)
[36]   1   4   0   1

La Quina Y-Z (81)
[37–33]   1   3   4   0

Roc de Marcamps (86)
[28]   4   7   0   4

Abri Mas Viel (101)
[28]   3   5   0   1?

Abri du Facteur (151)
[29]   3   5   3   2
[31]   1   0   0   0

Open-air (Interval 37–28 ka BP):

Jaurens (92)
[35]   6   8   1   2
[35–31]   5   1   0   1

Siréjol (95)
[34–32]   2   7   0   0

Perte de Bramarie (142)
[34]   0   1   0   0

Camiac-St-Denis (157)
[37]   6   3   0   0

Interval 27–18 ka BP

Abri Pataud (69)
[27]   6   5   2   0
[26–25]   5   8   2   0
[24–23]   6   6   2   0
[22–21]   3   7   2   0

Le Flagéolet I (70)
[27]   2   8   3   4
[26]   4   7   3   1
[24]   2   4   1   1
[20]   4   6   0   2

La Ferrassie (71)
[27–26]   5   11   2   3
[24–22]   4   5   1   2

St Césaire (73)
[24]   2   5   0   1

Roc de Combe (79)
[27]   6   10   2   4
[26]   4   5   2   1

La Quina (81)
[26]   1   1   0   0

Laugerie Haute (99)
[23]   2   4   2   1
[22–21]   3   0   0   0

Combe Saunière (147)
[23]   1   3   1   1
[19]   3   6   1   4

Abri du Facteur (151)
[27]   4   6   3   3
[25]   3   5   3   2

Le Raysse (154)
[27–25]   1   2   0   0

Grotte des Fieux (166)
[26]   1   2   0   0

While other faunal data bases had been compiled with taxonomic or stratigraphic objectives, ours was constructed as a function of the mammalian ecological setting (Chapter 7: Stewart *et al.* 2003) and as a main human food resource (Chapter 10: Musil 2003).

North of the Transverse European mountain barrier between the 45° and 55°N parallels, glacial summers were longer, the sun stood higher and growing days were more numerous than in the Arctic now. Therefore, in glacial environments the biomes and resources were not comparable to those of northern Canada, Europe or Siberia today. Many mid-glacial biomes have no modern analogues and species found today in northern latitudes then occurred together with others which now inhabit dry continental areas or more temperate climates. To be able to sustain the herbivore megafauna of the time, the glacial plant cover, the 'mammoth steppe' of Guthrie (1990) must have been quite rich (Lister & Sher 1995; Guthrie 2000; Guthrie & van Kolfschoten 2000). Musil (2003, Chapter 10) states that 'many mammalian species are eurythermic and not narrowly adapted to warm or cold climates. Instead, their environmental preferences are influenced strongly by differences between maritime and continental climates'.

Many species in the mammalian data base are ubiquitous, most large predators for instance, giving little information regarding past environments, while the ecology of others is uncertain or unknown (Chapter 7: Stewart *et al.* 2003). From the faunal data base we have selected a limited number of species whose

Table 11.2. *Dordogne case study. Sites: numbers, names, locations (longitude, latitude), type and lithic industries.*[2]

Site no.	Site name	Long.	Lat.	Type	Lithic industry
69	Abri Pataud	1.000	44.933	abri	Aurignacian/Gravettian
70	Le Flagéolet I	1.083	44.85	abri	Aurignacian/Gravettian
71	La Ferrassie	0.941	44.955	abri	Aurignacian/Gravettian
79	Roc de Combe	1.333	44.767	cave	Châtel./Aurig./Gravet.
80	Abri Caminade	1.257	44.876	abri	Aurignacian
81	La Quina	0.303	45.504	abri	Mousterian
82	Le Piage	1.366	44.797	open	Aurig./Solut./U. Palaeo.
86	Roc de Marcamps	−0.49	45.044	abri	Aurignacian
92	Jaurens	1.523	45.068	open	Fauna
95	Siréjol	1.468	44.997	open	Fauna & Palaeolithic
98	Combe Grenal	1.223	44.808	cave	Mousterian
99	Laugerie-Haute	0.950	44.967	abri	Gravettian/Solutrean
100	Regourdou	1.170	45.055	cave	Mousterian
101	A du Mas Viel	1.848	44.716	abri	Mousterian
116	Chapelle-aux-Saints	1.728	44.995	cave	Mousterian
117	Pech de l'Azé II	1.249	44.867	cave	Mousterian
119	Fonseigner	0.609	45.341	open	Mousterian
128	Gr. de Sanglier	1.717	44.683	cave	Fauna pre-Magdalenian
136	Gr. Pégourié	1.655	44.621	cave	Magdalenian
137	Puy-Jarrige II	1.487	45.125	abri	Gravettian
140	Grotte XVI	1.162	44.816	cave	Châtelperronian
141	A. Castanet	1.099	45.007	abri	Aurignacian
142	Perte de Bramarie	1.617	44.650	open	Fauna
147	A. Combe Saunière	0.883	45.237	abri	Châtelperronian
149	Barbas III	0.558	44.866	open	Mousterian
151	Abri du Facteur	1.062	44.978	abri	Aurignacian/Gravettian
152	La Rochette	1.102	45.016	abri	Aurignacian
154	Le Raysse	1.539	45.143	abri	Aurignacian/Gravettian
155	Vignaud	1.017	44.933	abri	Aurignacian
164	A. Cro-le-Biscop	1.055	44.951	abri	Châtelperronian
165	A. Lespaux	−0.3	44.817	abri	Gravettian
166	Gr. des Fieux	1.673	44.877	cave	Aurignacian
171	Le Moustier	1.07	45.002	abri	Mousterian

presence implies one of four main 'eco-suites' for Europe north of the Transverse ranges (Appendix 11.1): 1) arctic fauna living in various types of tundra and cold steppe; 2) thermophile species that prefer open woodland or meadow, including taïga and deciduous boreal forest; 3) species living today at high elevations; 4) species of the dry cold or dry temperate steppe. Others may disagree with our choices, but the results display consistent trends that parallel broad temporal climate changes. To document environmental changes concisely we express them with a *diversity index*, which for each dated layer gives the number of species in each eco-suite found there.

The mammalian fauna of the Dordogne was rich and its pattern over time quite variable (Table 11.1). At four out of five sites of the Transitional Phase (47–38 ka BP) thermophile species dominate over the arctic ones; only at Chapelle-aux-Saints (116) are both equal (Table 11.1). In the Early Cold Phase (37–28 ka BP) thermophile assemblages surpass to a varying degree the arctic fauna in cave and abri

sites; only at Abri Pataud (69) were arctic species numerous relative to the thermophile eco-suite. The same pattern marks more vaguely the few open-air sites, but only one of those (Camiac-Saint-Denis) has any real archaeology. In the LGM Cold Phase (27–28 ka BP), however, arctic species outnumber the thermophile group, although the latter remain important. Noteworthy is the common presence of desert species that indicates conditions in conflict with the maritime climate.

Human settlement in the Dordogne

The Dordogne area *s.l.* (Fig. 11.3: top; Tables 11.2 & 11.3) has 31 dated archaeological sites (Table 11.2). Neanderthal remains have been recovered from six, including Jaurens (92) which has no associated archaeology. Undated remains from La Ferrassie may also be OIS-3. AMH remains come from three Aurignacian sites (152, 141, 82). We shall use Mousterian and Châtelperronian sites as proxies for Neanderthals, and the Aurignacian and Gravettian sites for AMH. Dated open-air sites are few: only 82, 92, 119, 149 and Camiac-Saint-Denis (157). The number of open-air sites (Table 11.2) does not increase from the Early Cold (37–27 ka BP) to the LGM Phase (27–18 ka BP).

The Dordogne *s.s.* comprises a cluster of sites within the Vézère and Dordogne valleys that outlines the river courses (Fig. 11.3: top). This may in part be due to sampling strategies; Rigaud & Simek (1987, 55) note that Peyrony's work was conducted within one hour's bicycle trip from the Les Eyzies Museum. However, the virtual absence of sites in the Garonne valley cannot be attributed to sampling bias, because research has continued since the days of Peyrony. Scattered outside the core Dordogne cluster (Fig. 11.3: top) are several sites in coastal regions, e.g. La Quina (81), and elsewhere Combe Saunière (147), Mas du Viel (101), Pégourié (136), Lespaux (165), Fonseigner (119) and Marcamps (86).

Only the neighbouring Mousterian sites of Combe-Grenal (98) and Pech de l'Azé II (117) can be directly attributed to OIS-4 (72–60 ka BP), but the number increases in the early Stable Warm Phase

Table 11.3. *Dordogne case study. Sites: elevation (metres), aspect°⁶, type, and dates in ka BP; date sequences of long occupations are shown.[2]*

Site no.	Site name	Elev.	Aspect	Type	Site dates ka BP
69	Abri Pataud	85	286	abri	(22–39)
70	Le Flagéolet I	117	227	abri	(22–38)
71	La Ferrassie	170	240	abri	(24–38)
79	Roc de Combe	123	207	cave	(28–39)
80	Abri Caminade	200	116	abri	31, 36, 38, 40
82	Le Piage	127	241	open	28, 28, 29, 29, 31, 33, 33
86	Roc de Marcamps	?	?	abri	28
92	Jaurens	281	334	open	31, 31, 31, 35
95	Siréjol	278	96	open	31, 32, 34
98	Combe Grenal	91	303	cave	33, 40, 42, 44, 44
99	Laugerie-Haute	135	254	abri	21, 21, 22, 22, 22, 23
101	A du Mas Viel	339	8	abri	28
116	Chapelle-aux-Saints	169	164	cave	47
117	Pech de l'Azé II	195	182	cave	(31–46)
128	Gr. de Sanglier	330	9	cave	31
136	Gr. Pégourié	335	3	cave	18, 18, 18, 26
137	Puy-Jarrige II	179	345	abri	18
140	Grotte XVI	138	36	cave	37, 41
141	A. Castanet	78	347	abri	38
142	Perte de Bramarie	359	31	open	34
149	Barbas III	93	194	open	41
151	Abri du Facteur	183	318	abri	(27–31)
152	La Rochette	79	305	abri	31, 31
154	Le Raysse	166	17	abri	25, 27
155	Vignaud	161	348	abri	26
164	A. Cro-le-Biscop	169	212	abri	20
165	A. Lespaux	?	?	abri	18
166	Gr. des Fieux	251	198	cave	26
171	Le Moustier	?	?	abri	(40–47)

(60–48 ka BP) and two are open-air: Fonseigner (119) and Barbas III (149). Except for (119), all are inside the Dordogne-Vézère cluster. The main Neanderthal expansion occurs near the end of the Stable Warm and during the Transitional Phase (47–38 ka BP); it has been linked to a diversification of their behaviour as evidenced by the earliest Châtelperronian.

During the Transitional Phase the Mousterian and Châtelperronian sites are joined for the first time by Aurignacian ones often at previously-uninhabited locations within the cluster, e.g. Caminade (80) and Castanet (141), and also at Le Flagéolet (70) and Pataud (69) if one believes their earliest dates. Aurignacian sites succeed or interstratify with Châtelperronian ones at La Ferrassie (71) and Roc de Combe (79). Outside the main cluster, the Aurignacian appears only at Combe Saunière (147), preceded there by the Châtelperronian. Neanderthal and AMH sites of this climatic phase cannot be distinguished in environmental terms (Table 11.1); as noted before, all are associated with thermophilous mammalian eco-suites except at La Chapelle-aux-Saints (116) where arctic and thermophile faunas are roughly equal.

During the Early Cold Phase (37–28 ka BP) the number of well-dated sites in the Dordogne region peaks at 14; three may be either earlier or later, but Mousterian and Châtelperronian sites are few. All others have Aurignacian and/or Gravettian assemblages.

Most LGM Cold Phase (27–18 ka BP) sites were previously-occupied, except for Laugerie Haute (99), Le Raysse (154), Vignaud (155), les Fieux (166) and the far-field Pégourié (136). After 27 ka BP the Aurignacian disappeared from the region notwithstanding 'Aurignacian' dates from several LGM sites, and the Gravettian predominated, followed later by the Solutréan. Sites outside the main Dordogne cluster rarely show consistent occupation throughout OIS-3, perhaps owing to changes in landscape use.

The crucial change in Dordogne settlement pattern occurred in the Early Cold Phase (37–28 ka BP) when the scattered distribution of sites became more focused on the Dordogne-Vézère valleys; lower-lying areas show more ephemeral activity. As noted in the previous section, arctic faunas increase in frequency from the Early Cold Phase onward to become dominant in the LGM Cold Phase. Concentrations of sites within a fairly small area have been explained by tracking of herbivore migrations along valleys (White 1980). Until more dated sites are available outside the Dordogne cluster, however, explanations for the settlement pattern there are hard to assess.

The Ardennes case study

Landscape
The Ardennes case-study area (Fig. 11.3: centre) is defined by the 3°E and 6°30'E meridians and the 49°30'N and 51°30'N parallels. The Ardennes massif rises from about 100 m above present sea level in the lowlands of northern and western Belgium to the Hautes Fagnes plateau at 500–700 m. OIS-3 elevations were higher by c. 80 m and LGM ones by 120 m due to the low glacial sea-level. The massif consists mainly of lower and middle Palaeozoic sedimentary rocks with dominant Devonian and early Carboniferous limestones. Middle–late Pleistocene loess and

wind-blown cover-sand, vulnerable to permafrost in the Early Cold and Last Glacial Maximum phases (Antoine *et al.* 1999a,b; van Huissteden *et al.* 2003), blanket the lower slopes of the massif and the lowlands beyond. The uplands are infertile and in many places covered by a blanket of sphagnum peat.

The river Meuse, born in northeastern France, turns northward across the Ardennes massif at about 4°55'E, flowing in an antecedent valley some 200 m deep. Upon reaching the north edge of the massif it is joined from the west by the River Sambre and turns north to northeast in a wide antecedent valley along and inside the northern flank of the Ardennes. The incised valleys date back to an early Mesozoic peneplain since raised to form the present massif and there are many caves up to an elevation of 200–300 m. Palaeolithic habitation was concentrated in the Meuse valley and the lower courses of its dextral tributaries. The sharply V-shaped valleys draining the northeastern, eastern and southeastern flanks to the Moselle and Rhine ones are much younger, and the rugged terrain they have created more difficult to traverse than the gentle relief of the west flank of the massif.

Palaeoclimate[4]

The Ardennes study area is part of the maritime Atlantic palaeo-climate zone (cf. Chapter 5: Barron *et al.* 2003, Fig. 5.11) and has at present damp, quite mild winters and cool, cloudy summers with only a modest contrast between mean winter and summer temperatures.

During warm events of the Transitional Phase (Fig. 11.2: 47–38 ka BP) winter temperatures in the Ardennes ranged from –2° to –4°C, but freezing conditions persisted until spring arrived suddenly in May. Summer temperatures rose from 6°C in May to 16°C in July but slow cooling began in August to a 1°C mean in November; light but persistent frost (–1°C) came in December. The seasonal contrast of 18°C from mid-winter to mid-summer was only a little beyond the 15°C of today and the winter was no colder than *c.* 2°C. Precipitation varied little across the region and through the year from 3–4 mm/day in mid-winter to 3 mm/day in mid-summer. April and September were a little drier. The monthly cloud cover almost never dropped below 60 per cent and in winter hovered at 90–95 per cent.

On the gently rolling plain north of the Meuse annual temperature and precipitation did not differ greatly from those on the higher ground south of the river. Snow fell on average on 30 days/year in striking contrast with the 2–6 days/year of today, but the

Table 11.4. *Ardennes case study. Diversity of eco-mammalian assemblages by time-slice, date and type of site for the interval 47–18 ka BP. Sites with number (xxx). Underneath are dates in [xx] of dated layers. To the right of each date is number of species for each of four eco-suites listed from left to right as follows: Arctic tundra zone; cold-temperate woodland; present-day mountain species; cold steppe species.*

Cave sites					Open-air sites				
Interval 47–38 ka BP									
Trou Magrite (204)									
[44]	3	4	2	0					
[42]	5	9	3	0					
Trou Al'Wesse (215)									
[43]	6	3	0	2					
Sclayn (210)									
[42]	5	9	3	0					
Interval 37–28 ka BP									
Trou Magrite (204)					**L'Hermitage (211)**				
[36]	6	3	1	0	[31]	2	0	0	0
[33]	4	0	3	0	[28]	2	0	0	0
[33]	6	3	1	0	**Maisières-Canal (209)**				
[30]	4	3	4	0	[33]	4	4	0	1
[28]	4	2	0	0	[33]	4	3	0	1
Trou Al'Wesse (215)					[33]	5	3	0	1
[35]	5	2	0	0					
[33]	5	3	0	0					
Trou Walou (205)									
[31]	5	4	0	0					
Trou de l'Abîme (208)									
[28]	1	2	0	0					
Interval 27–18 ka BP									
Trou de l'Abîme (208)					**Maisières-Canal (209)**				
[27]	1	2	0	0	[27]	3	3	0	0
Trou du Renard (207)					[26]	3	3	0	0
[26]	3	2	1	1	[26]	4	3	0	0
Grotte de Spy (213)					**L'Hermitage (211)**				
[27]	5	3	1	0	[26]	3	0	0	0
[23]	3	1	0	0	[24]	3	0	0	0
[23]	3	1	1	0	[19]	5	3	2	0
Grotte Princesse (206)									
[25]	6	0	0	0					
Goyet (212)									
[24]	0	2	0	0					
Trou Magrite (204)									
[24]	5	2	1	0					
[24]	5	2	0	0					
[23]	4	0	2	0					

mean monthly snow depth on the hills (30–45 cm) was more than twice that of the lowlands across the river.

Conditions became more severe during the Early Cold Phase and LGM, but the climate remained maritime. The 'LGM Cold' simulation temperature ranged from –8° in January/February to about 11°C in July, only a little colder than in the Transitional Phase, but the winters were much colder (–8° to –10°C). Precipitation was lower by about half than in the preceding 'Transitional Phase', but the cloud

Table 11.5. *Ardennes case study. Sites: numbers, names, locations (longitude and latitude), type, and lithics industries.[2]*

Site #	Site name	Long.	Lat.	Type	Lithic industry
204	Trou Magrite	4.971	50.214	cave	Mousterian/ E. Aurignacian
205	Trou Walou	5.722	50.589	cave	Magdalenian/Gravettian
206	Gr. de la Princesse	4.972	50.483	cave	Aurignacian
207	Trou du Renard	4.995	50.215	cave	Aurignacian
208	Trou de l'Abîme (Couvin)	4.513	50.060	cave	L.M./E.U./U. Palaeolithic
209	Maisières-Canal	3.987	50.473	open	Gravettian
210	Sclayn	5.046	50.488	cave	Mousterian
211	L'Hermitage	5.205	50.581	open	Gravettian
212	Goyet	5.023	50.433	cave	Aurignacian/Magdalenian
213	Grotte de Spy	4.674	50.478	cave	Aurignacian/Gravettian
214	Grotte du Haleux	5.662	50.507	cave	Aurignacian
215	Trou Al'Wesse	5.294	50.421	cave	Mousterian/Aurignacian

[*]Some sites contain more than one industry.

Table 11.6. *Ardennes case study. Sites: elevation (metres) and aspect° [6], type and available dates in ka BP. Long sequences shown in full.[2]*

Site #	Site name	Elev.	Aspect	Type	Dates (ka BP)
204	Trou Magrite	230	307	cave	19, 23, 24, 30, 33, 33, 36, 44
205	Trou Walou	227	102	cave	22, 24, 26, 28, 30, 31, 36, 44
206	Gr. de la Princesse	110	246	cave	25
207	Trou du Renard	255	171	cave	26
208	Trou de l'Abîme (Couvin)	180	133	cave	27, 55
209	Maisières-Canal	39	56	open	26, 26, 27, 30, 33, 38
210	Sclayn	162	26	cave	42, 42
211	L'Hermitage	181	135	open	24, 26, 28, 31
212	Goyet	178	323	cave	24, 29
213	Grotte de Spy	127	231	cave	23, 23, 27
214	Grotte du Haleux	249	317	cave	27
215	Trou Al'Wesse	240	324	cave	33, 35

cover was similar in both 'warm' and 'cold' simulations.

Climate changes in the Ardennes and the mammalian fauna

Table 11.4 shows the faunal diversity chart for the Ardennes study area; it is clearly much less diverse than the Dordogne fauna where thermophile elements were present throughout the entire interval from 47 to 18 ka BP.

The two oldest sites (47–38 ka BP) have a mixed arctic and thermophile fauna fitting well with the climate of the Transitional Phase. By 37–28 ka BP the arctic signature is dominant as expected for the Early Cold Phase, but some thermophile species persevered. The LGM time-slice (27–18 ka BP), however, leaves no doubt that the arctic eco-suite dominated.

As in the Dordogne area, diversity indices of cave and open-air archaeological sites are subtly different, recalling Musil's (2003, Chapter 10) caution

that archaeological open-air sites preserve the hunting record in contrast with cave sites where hunting and natural fauna alternate or intermingle. The two open-air sites from the region, L'Hermitage (211) and Maisières Canal (209) confirm this.

Human settlement in the Ardennes

With twelve sites (Fig. 11.3: centre; Tables 11.5 & 11.6) the Ardennes have the fewest dated sites of the study regions, although research there extends back to the first half of the nineteenth century. Dated assemblages are mostly from older excavations, and concern is warranted about stratigraphic errors and mislabelling in museums (Otte 1979) because Aurignacian and Gravettian assemblages from nineteenth- and early twentieth-century excavations are difficult to distinguish. Much progress has been made in the late twentieth century, however, and rich data have been obtained from the two open-air Gravettian sites (209) and (211) (Fig.11.3: centre). All of the dozen sheltered sites are within the Sambre-Meuse corridor, but the two open-air sites are in the plain north of the main site concentration, implying differential occupation during the climatic deterioration after 37 ka BP. The Trou de l'Abîme (208) is also located outside the main site concentration, but is not dated convincingly beyond the Stable Warm Phase.

Only two dated Mousterian assemblages may fall into the early Stable Warm Phase (60–48 ka BP): Sclayn cave (210) and Trou de l'Abîme. The latter also has an Early Cold Phase (37–28 ka BP) date, casting doubt on its stratigraphic integrity; it was from this assemblage that a deciduous Neanderthal molar was recovered (Ulrix-Closset *et al.* 1988). Undated Neanderthal remains from the Grotte de Spy (213) may also be from early OIS-3 (*c.* 50–35 ka BP; Bellaire & Otte 2001, 18). A more consistent Mousterian OIS-3 presence is limited to the late Stable Warm Phase (47–38 ka BP) at Sclayn and Trou al'Wesse). By the start of the Early Cold Phase *after* 38 ka BP, the Neanderthals seem to have left the region.

Assemblages from Trou Magrite (204) and pos-

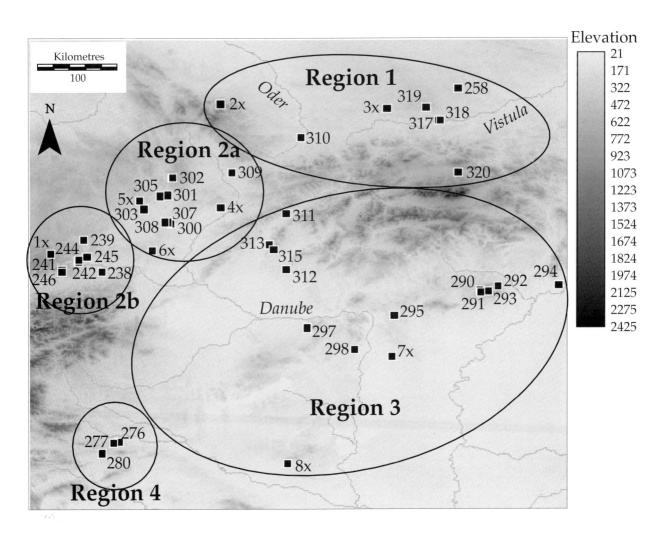

Figure 11.4. *The geography of the Middle Danube Basin case-study area* sensu lato *with dated sites; numbers refer to Table 11.8. Approximate boundaries are shown for the first four ecological regions defined by Musil (2003, Chapter 10). Topography is taken from gtopo30 digital elevation model (USGS 1996); projection is latitude-longitude wgs84.*

sibly Trou al'Wesse (215) imply that the earliest AMH arrived in the Transitional Phase (45–38 ka BP), but the bulk of Aurignacian assemblages derives from the Early Cold Phase, including the Trous Magrite, Walou (205) and Al'Wesse. Dates from Goyet (212) are unreliable, showing the underlying Aurignacian as *younger* than the overlying Magdalenian. The first Gravettian (209, 211) also falls in the Transitional Phase and closely follows the Aurignacian at Trou Walou. The early part of the LGM Cold Phase (27–25 ka BP) shows Gravettian continuity at (205), (209) and (211), but only one site, Grotte de Spy (213), has Aurignacian dates in the Early Cold Phase. Aurignacian dates from Grotte de la Princesse, Haleux and Trou du Renard, based on bulk bone samples, seem too young and probably belong to the Early Cold Phase.

The Gravettian occupations of Walou, l'Hermitage and Maisières-Canal continue into the early LGM Cold Phase, but there is no robust evidence for human occupation in the Ardennes in the later LGM (24–18 ka BP). Conditions may have been too inhospitable or resources too insignificant, because the faunal communities of the Ardennes region are much less diverse than those of the Dordogne and have an increasingly arctic signature from the Early Cold Phase onwards.

The Middle Danube basin case study

Landscape
The Middle Danubian case-study area is bound by the 46°N and 50°N parallels and the 15°E and 21.5°E meridians and covers parts of Austria and Moravia

Table 11.7. *Middle Danube basin case study. Diversity of eco-mammalian by time-slice, date and type of site for the interval of 47 to 18 ka* BP. *Sites with number in (xx). Underneath: [layer date]. Right: diversity indices (number of species/layer) of the four eco-suites: arctic tundra species; cold or cold-temperate woodland species; present-day mountain species; steppe species. Regions: 1) southern Poland and northern Moravia; 2) central and southern Moravia and northern Austria; 3) southern Slovakia and Pannonian Basin; 4) Slovenia and Croatia. (After Musil 2003, Chapter 10.)*

Region 1 Open-air	Region 2 Caves	Region 2 Open-air	Region 3 Caves	Region 3 Open-air	Region 4 Caves
Interval 47–38 ka BP					
	Kůlna [48] 9 5 2 5; [41] 6 3 1 3	Bohunice [45–42] 1 0 0 0; Stránská skála [40] 2 2 0 0	Čertova Pec [41–40] 3 4 1 0	Érd [44] 9 6 2 4	Veternica [45] 3 5 2 0; Vindija [45] 2 2 0 0
Interval 37–28 ka BP					
	Pod hradem [35] 7 6 3 5; [29] 7 6 3 5	Willendorf II [32–29] 3 1 1 5; [30] 5 3 2 6; Předmostí [29–28] 11 4 1 0; [29–28] 5 2 1 0; Pavlov [29–28] 5 2 1 0; Dolní Věstonice [29–28] 6 1 0 0; Jarošov [29–28] 4 0 0 0	Istállóskö [32] 5 9 2 4	Tata [36] 3 6 1 1; Bodrögkeresztúr [31] 1 3 0 0; Moravany Lopata II [26] 4 0 0 0	Velika Pećina [36] 4 3 0 0; [33] 3 0 0 0; [28] 0 1 0 0; Krapina [32] 3 5 1 0; Vindija [31–30] 3 3 3 1
Interval 27–25 ka BP					
Petřkovice [25] 1 0 0 0	Kůlna [25] 2 2 0 0	Willendorf II [27–25] 5 3 2 0; [26] 3 1 1 0; Předmostí [27] 11 4 1 0; Pavlov [27] 11 5 1 0; Milovice [27–25] 2 0 0 0			
Interval 24–18 ka BP					
Spadzista Street [23–19] 5 0 0 0; Petřkovice [23] 1 0 0 0	Kůlna [24] 2 2 0 0	Stránská skála [20] 3 1 0 1; Velké Pavlovice [15] 1 0 0 1; Milovice [24–23] 2 0 0 0; Grubgraben [22–20] 3 1 1 0		Moravany Lopata II [24] 5 2 4 1; [23] 4 0 0 0	Vindija [19] 3 2 3 1

(Czech Republic) and smaller parts of Slovakia and Hungary (Fig. 11.3: bottom; Fig. 11.4). It is located in the rugged Carpathian ranges with elevations up to 1500 m and borders in the north on the North European Plain and in the south on the Pannonian Basin. The southern margin of the Fennoscandian LGM ice sheet was about 500 km away. The case-study area differs from the Dordogne and the Ardennes by being much larger, with a more complex relief and local climate and the existence of several medium site clusters.

The northern Carpathians separate the relatively uniform arctic zone of the North European Plain from the more complex Moravian region and Pannonian basin. The Danube flows eastward at 17°E in a broad valley between the Austrian Alps and the south flank of the Carpathians. Then it enters a lowland basin and is joined by large tributaries such as the Morava, Váh and Tisza rivers that drain the southern Carpathian slopes. The headwaters provide access through passes locally known as 'gates' to the north slopes of the range and from there to the North European Plain (Svoboda *et al.* 1996, ch. 2).

The largest site cluster extends from northeastern Austria to southern Poland (Fig. 11.3: bottom) and forms the heart of the study area. Far-field site clusters are found in southern Poland, Hungary and northwestern Croatia. Unlike the other study areas, the Middle Danubian sites spread across elevations from 200 m to 500 m.

Palaeoclimate[4]

The Middle Danube region is part of the continental climate zone (Chapter 5: Barron *et al.* 2003, Figs. 5.11 & 5.13). It reflects an archetypal regime of warm summers and cold winters, but the mountain ranges and low-level basins traversed by many rivers of the region have local climates that may differ considerably from the regional pattern.

The regional climate of the Transitional Phase had cold winters (–6.5°C to –4°C with wind-chill from –19°C to 16°C) and summer temperatures from 16.5°C to 20.5°C. In the Early Cold and LGM Phases the seasonal contrast differed little from the Transitional Phase except that the winters were colder in January (–12.5°C; –24°C with wind-chill). Mid-summer temperatures were less variable than in the Transitional Phase and somewhat lower (12°C to 14.5°C). On the plain north of the Carpathian range summers were even colder and the seasonal contrast there was less than in the Transitional Phase.

The Middle Danubian case-study area *s.s.* has three main site clusters located at different altitudes on the southern slopes of the western Carpathian range (Fig. 11.3: bottom). The first is the Willendorf (241) — Krems (242–244) — Langenlois (245) cluster in the Danube valley some 100 km upstream of its exit into the Pannonian basin. During the Transitional Phase winter temperatures in the Danube valley there ranged from –2°C in November to –5°C for December to April and even in May the mean value was still just below freezing. Spring was late but in May the summer came swiftly to reach its maximum warmth (17°C) in August. After a fairly warm September the autumn cooled very slowly.

In the Early and LGM Cold Phases the summer high was *c.* 11°C, but September remained mild (*c.* 7°C). Real winter began in November with a –5°C mean and dropped to –11°–12°C in January and February (not counting wind-chill). A –3°C frost persisted into April and spring was not early.

In the Transitional Phase precipitation peaked at 3 mm/day in July and August, then fell to its annual minimum of 1 mm/day in October, staying at an average 1–2 mm/day through winter. The following cold phases had roughly the same rain in summer, but the autumn was a little wetter.

In the semi-enclosed upper Danube basin (Fig. 11.3: bottom; Fig. 11.4) an 80–90 per cent cloud cover prevailed in all three climate phases except for a sunny summer (50 per cent cover) in the Transitional Phase. The fall brought a light ephemeral snow cover already in October that increased to 35 cm in February. The spring thaw came late and not until June was the area snow-free. In the Early Cold and LGM Cold Phases the maximum snow thickness was about the same but it came earlier and peaked at 35 cm in February and March. The area was not snow-free until June. About 30 snow days annually marked all three climate phases.

Upstream on the Morava (Fig. 11.3: bottom) at an altitude of 200–330 m are located the large Dolní Věstonice (308), Milovice (300) and Pavlov (307) site clusters. The climate in all three climate phases was similar to Willendorf (241) for temperature, cloudiness and number of snowy days but a little drier in summer. At Dolní Věstonice the snow was gone in May, bringing a rapid but not particularly early spring.

Farther north, but still on the south flank of the Carpathian range, are the Kůlna (304) and Pod Hradem (302) sites, both on a hillside at *c.* 500 m. Somewhat farther east, Předmostí (309) at 230 m overlooked the broad Morava valley. At these sites the Transitional Phase winters were colder (–6° to –7°C) than at Dolní Věstonice (308) all the way through April (–2°C), but the July maximum of 17°C was as elsewhere. November with –2.5°C (not including wind-chill) was colder than either of the two other clusters. At all upland sites the LGM winter began early with a cold November (–5°C) and December (–0.5°C), but from January onwards to late autumn conditions were much like those at the clusters lower downslope.

Precipitation and cloud cover at the Kůlna, Pod Hradem and Předmostí sites were like those on the lower slopes and so was the snow cover of 32–40 cm but it lasted well into April.

Middle Danube climate changes and the mammalian fauna

The 'ecological' species record (Appendix 11.1) lists 30 species for the Ardennes and 44 for the Dordogne, but 69 for Moravia and adjacent territory. In all eco-suites the diversity in Middle Danubian area is higher than at the other case-study areas. Its continentality is expressed by the steppe eco-suite which has 15 species in the Middle Danube, six in the Dordogne and none in the Ardennes. The continental climate with its longer, hotter summers and an early snow melt is a factor here, as are the rugged topography, the larger area and perhaps the early spring compared to the Atlantic region.

Table 11.7 shows the Middle Danubian faunal diversity by time-slices and regions (Fig. 11.4). In the Transitional Phase arctic and cold temperate species are nearly balanced at six sites while Kůlna cave also has many desert species. The open-air site of Bohunice (305) has only mammoth remains and may have been a special hunting site. Southern Poland and northern Moravia have no data for the Transitional Phase, but much data exist for the early Early Cold Phase. The ratio of arctic and cold-temperate species, however, remains the same and the pattern is complex. Four sites in southern Slovakia and the Pannonian basin and three in Slovenia and Croatia also indicate a climate mosaic during the first part of the Early Cold Phase.

Data from the southern margin of the North European Plain become abundant later in the Early Cold Phase (27–25 ka BP) and in the LGM (24–18 ka BP), showing a simple, purely arctic eco-suite; all are open-air sites and use as hunting camps may well account for the simple assemblages. In the Moravia region, the only cave site (Kůlna 304) has a minimal record, but the five open-air sites have small but non-trivial numbers of cold-temperate species.

Human settlement in the Middle Danube area
Because the Middle Danube basin is so large (Fig. 11.4), it is best discussed in four Regions based upon palaeoclimatic conditions and topography (Chapter 10: Musil 2003). Region 1 (southern Poland and northern Moravia) has yielded eight dated sites of which three are open-air and five sheltered, but only two have adequate faunal analyses. In the 47–38 ka BP time-slice, Nietoperzowa (319) is the only cave and Dzierzyslaw (2x) the only open-air sites with possibly Neanderthal (Szeletian) human occupations. All other sites date to the 37–28 ka BP time-slice and are either Aurignacian or Gravettian, and thus of AMH origin. Sites dated 27–18 ka BP are Gravettian and three out of four are open-air. In Mamutowa cave (258), undated Aurignacian and Szeletian *(s.l.)* assemblages underlie a Gravettian with a 22 ka BP minimal date that puts it in the LGM period.

Region 2 (central and southern Moravia and eastern Austria) has two of the best-researched and archaeologically richest site clusters in late Palaeolithic Europe. Eight of the ten Moravian sites have faunal analyses that allow a detailed examination of the available fauna. In the Transitional Climate Phase (47–38 ka BP) the human presence in Moravia is documented by Kůlna cave (304: Mousterian), Bohunice-kejbaly (305: Bohunician) and Stránská skála (301: Bohunician) and Vedrovice V (303: Szeletian). There is no Bohunician after the Transitional Phase, but at Pod Hradem (302) the Szeletian extends into the Early Cold Phase (37–28 ka BP). Overall this time-slice brought an increase in sites of which two con-

tinued from the Transitional into the Early Cold Phase (Szeletian at Vedrovice and Aurignacian at Stránská skála). At 37–28 ka BP, dated Aurignacian and Gravettian ('Pavlovian' variant) assemblages are present in equal proportions, a ratio that raises doubt about the cultural superiority of the latter technocomplex relative to the former one. By 28 ka BP the Aurignacian has disappeared, however, and the Gravettian dominates but without exploiting new locations.

The Austrian cluster of sites is open-air and centred on the Danube itself; only two out of nine sites have yielded no fauna, probably owing to local preservation conditions. Unlike in Moravia, the Austrian AMH cluster occupies an area without preceding Neanderthal presence (no dated Mousterian sites). The earliest occupation of the Austrian area began in the 37–28 ka BP time-slice with slightly more dated Gravettian than Aurignacian sites. The cluster appears to have collapsed between 27 and 18 ka BP and the LGM period (24–18 ka BP) saw a move to new sites (Langmannersdorf, Horn and Grubgraben), all with typologically unusual assemblages.

Southern Slovakia and the Pannonian basin (Region 3) do not lack dated sites, but they are scattered rather than clustered as in Austria and Moravia. Six out of 16 sites have adequate faunal data, including two caves, Istállóskö (290) and Čertova pec (315). All the sites of the 47–38 ka BP interval are caves, excepting Érd (298), but continuing occupation into the Early Cold Phase (37–28 ka BP) is not evident. Two dated Aurignacian sites, Peskö (291) and Istállóskö (290), are in the northwestern Bükk mountains at 600 m masl but humans were present beyond 47–38 ka BP only at Istállóskö. Contemporary caves, Szeleta (292: not beyond 28 ka BP) and Čertova pec (315) are in the northeastern Bükk mountains and in western Slovakia.

In Region 3, dated Mousterian is represented by two Hungarian open-air sites, Érd (47–38 ka BP) and Tata in the 37–28 ka BP period; then the Neanderthals vanished. The Gravettian is the commonest dated technocomplex with four open-air sites (294, 295, 311, 313) between 37 and 28 ka BP, but no dated occupation is known from between 27 and 25 ka BP. Except for Moravany-Lopata II (314), the Gravettian settlement resumed at *different* sites in 24–18 ka BP. The last Gravettian occupation includes a sheltered site (Balla cave, 293) in the central Bükk mountains, but open-air sites were preferred during the LGM (Mende, 312, Tapiosuly, 313 and Ságvár, 314) in western Slovakia and Hungary.

In Region 4 (Croatia), three cave sites yielded

good faunal data. Vindija (276) may have Aurignacian in the 47–38 ka BP time-slice, but the convincing dates are for the Early Cold Phase (37–28 ka BP) and there is Gravettian until the LGM (24–18 ka BP). Velika Pećina 2 (277) has Aurignacian then Gravettian at 37–28 ka BP. Together the few dated sites suggest a sparse ephemeral human presence. A dated Neanderthal presence exists at Krapina (280) and perhaps at Vindija, but the evidence there is weak.

Summarizing:

- Main site clusters and dated sites occur in the valleys of the Danube, Svitava, Dyje, Vistula, Váh and Nitra, probably good paths of communication. Interfluvial sites are few.
- Neanderthal presence seems to have been generally low or non-existent in the Middle Danube study area.
- Open-air locations are increasingly used after 28 ka BP, notwithstanding the deteriorating climatic conditions. What encouraged people to occupy complex open sites under maximal glacial conditions?
- The Gravettian technocomplex was clearly more successful than the Aurignacian

Humans and climate: a local approach

The Stage 3 simulations have provided much insight in the European glacial climate, but at the scale used in this chapter each cluster only occupies a few grid boxes of 60 × 60 km in size. Therefore the climate simulations can not directly address the impact of climate variations within each cluster, something that would require models on a 1–2 km spatial scale. Such models are currently being developed but are not ready for implementation. However, for some climate variables a

Figure 11.5. *Down-scaled regional climate for the three case study areas. Top: Dordogne; Centre: Ardennes and Bottom: Middle Danube Basin. Wind-chill coldest month mean temperatures (in centigrade) for the LGM simulation, down-scaled to a 30 arc-second resolution using a simple elevation correction for temperature and wind. Projection: longitude-latitude.*

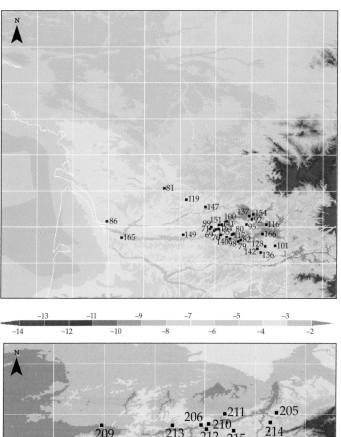

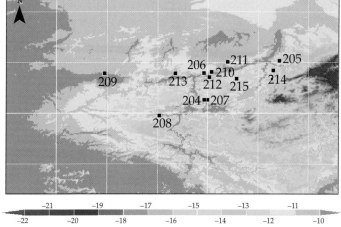

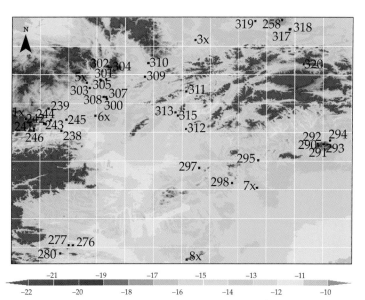

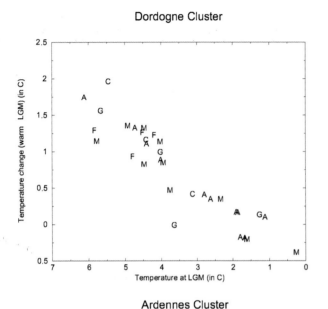

Dordogne Cluster

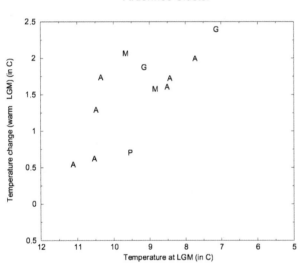

Ardennes Cluster

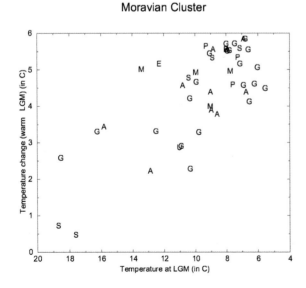

Moravian Cluster

simple down-scaling will predict climate cchanges on the same scale as the topography.[6]

We focus here on the down-scaling of temperature and related variables for two reasons. First, quantitative studies of the impact of climate on human comfort levels suggest that in cold conditions temperature and wind-chill are the most important factors (Terjung 1968), while in warm regions humidity and precipitation are more important. Secondly, small spatial-scale variations of temperature and winds are strongly controlled by the local topography; this enables us to down-scale temperature from local topography only.

The down-scaling technique was used for all model simulations, and for all months, but for brevity we present only the results for the wind-chill temperatures of the coldest month. From the climate zonation (Fig. 11.2) we shall use the Stable Warm, Transitional, Early Cold and LGM Cold phases of which the first two use the 'OIS-3 warm' simulation and the others the 'LGM cold' simulation (Chapter 5).

Figure 11.5 (top) shows wind-chill temperatures down-scaled from the 'LGM cold simulation' for the coldest month in the Dordogne region, the most severe climatic condition in OIS-3. In vast majority, sites are seen to occupy low-lying and relatively sheltered areas within the Dordogne valley in a region with a large area of milder climates. The lower elevation and sheltered conditions resulted in relatively mild temperatures (wind-chill of –2° to –5°C).

By contrast, the Lot valley farther south (Fig. 11.3: top) is almost as warm but, being narrower, the favourable area is smaller. The valley also narrows downstream, impeding access from the west. The model predicts wind speeds twice or three times those in the Dordogne valley with increased wind chill values. This helps to explain why most sites are found along the Dordogne and not farther south where the Garonne, Tarn and Aveyron valleys are also colder than the Dordogne area, again primarily because of a stronger wind-chill factor.

Figure 11.6 (top) shows down-scaling results for wind-chill during the LGM Cold Phase versus

Figure 11.6. *Down-scaled wind-chill coldest month mean temperatures (in centigrade) for the LGM simulation versus the change in wind-chill coldest month mean temperatures (warm simulation – LGM simulation) in the three case-study regions.*
Top: Dordogne; Centre: Ardennes and Bottom: Middle Danube Basin: Letters indicate lithic industries:
A = Aurignacian, C = Châtelperronian, F = fauna only, G = Gravettian and M = Mousterian.

the difference in wind-chill temperatures between the 'OIS-3 warm' and 'LGM cold' simulations. Clearly, most sites are in areas where in the coldest month wind-chill temperatures are less than –5°C. The climatic distribution of the Mousterian compared to the Aurignacian and Gravettian suggests that all seem to have had the same tolerance to cold temperatures.

A point of note is that the climatic differences between the 'OIS-3 warm' and 'LGM cold' simulations are quite small, because changes in temperature are affected by changes in wind chill. In the Dordogne area the 'OIS-3 warm' simulation predicts slightly stronger winds than for the 'LGM cold' simulation, partially compensating for warmer temperatures. The biggest differences in wind-chill between the Stable Warm and Transitional Phases on the one hand and the colder phases on the other occur in areas which were coldest during the LGM and suggest that during warm intervals minimum habitation temperatures would be nearer to –4°C.

The down-scaled temperatures for the 'OIS-3 warm' simulation show that in the Lot and Tarn valleys the temperature during warmer phases was much higher. In the Tarn valley an extensive area had wind-chill values of the coldest month near or above zero. Why then are there no Mousterian and/or Early Upper Palaeolithic open-air sites in the warmest part of OIS-3? Lack of absolute dates prohibits an answer to this question (Rigaud & Simek 1987, 55–6). Dated sites in the Dordogne area are concentrated in the Dordogne and Vézère valleys from the Early Cold Phase (37–28 ka BP) on through the LGM. The La Quina (81), Camiac-Saint-Denis (157) and Saint-Césaire (73) sites outside the cluster were continuously occupied from the Transitional through the Early Cold Phase and, like Roc de Marcamps (86) and Abri du Mas Viel (101), survived maximal wind-chill temperatures of –5° to –4°C, but none was inhabited in the LGM (24–18 ka

Table 11.8. *Middle Danube basin case study. Sites: numbers, names, locations (longitude, latitude), type and lithic industries.[2] Note: site numbers ending in x are not in the chrono-archaeological data base but were provided by Musil (2003, Chapter 10).*

Site no.	Site name	Long.	Lat.	Type	Lithic industry
238	Langmannersdorf	15.879	48.294	open	Gravettian
239	Horn	15.662	48.670	open	Aurignacian
241	Willendorf II	15.399	48.323	open	Gravettian
242	Krems-Hundssteig	15.587	48.413	open	Aurignacian
243	Krems-Wachtberg	15.604	48.415	open	Gravettian
244	Krems-Galgenberg	15.606	48.430	open	Aurignacian
245	Langenlois	15.693	48.475	open	Gravettian
246	Aggsbach	15.395	48.293	open	Gravettian
258	Mamutowa	20.160	50.430	cave	Aurignacian
276	Vindija	16.088	46.305	cave	Aurignacian
277	Velika Pećina	16.016	46.291	cave	Aurignacian
280	Krapina	15.875	46.169	abri	Mousterian
290	Istállóskö	20.418	48.065	cave	Aurignacian
291	Peskö	20.426	48.046	cave	Aurignacian
292	Szeleta	20.631	48.118	cave	Szeletian
293	Balla	20.532	48.051	cave	Gravettian
294	Bodrögkeresztúr	21.386	48.125	cave	Gravettian
295	Püspökhatvan	19.381	47.776	open	Gravettian
297	Tata	18.334	47.633	open	Mousterian
298	Érd	18.917	47.379	open	Mousterian
300	Milovice I	16.699	48.851	open	Aurignacian
301	Stránská skála	16.676	49.184	open	Aurignacian
302	Pod Hradem	16.723	49,389	cave	Environment
303	Vedrovice V	16.381	49.023	open	Szeletian
304	Kůlna	16.737	49.000	cave	Micoquian
305	Bohunice-kejbaly	16.585	49.170	open	Bohunician
307	Pavlov I	16.678	48.872	open	Gravettian
308	Dolní Věstonice	16.634	48.875	open	Gravettian
309	Předmostí	17.439	49.456	open	Gravettian
310	Petřkovice	18.262	49.865	open	Gravettian
311	Nemsova	18.089	48.966	open	Gravettian
312	Nitra-Čermáň	18.084	48.322	open	Gravettian
313	Moravany-Žakovska	17.880	48.609	open	Gravettian
314	Morazany-Lopata II	17.880	48.601	open	Gravettian
315	Čertova pec	17.937	48.459	cave	Szeletian
317	Spadzista Str. A	19.924	50.053	open	Gravettian
318	Kraków	19.930	50.054	open	U. Palaeolithic
319	Nietoperzowa	19.768	50.025	cave	Szeletian
320	Oblazowa 1	20.156	49.443	cave	Gravettian
1x	Grubgraben	15.250	48.500	open	Gravettian
2x	Dzierzyslaw	17.300	50.250	open	Szeletian
3x	Piekary	19.300	50.200	open	Aurig/Szeletian
4x	Jarošov	17.300	49.040	open	Gravettian
5x	Brno Videňská	16.330	49.120	open	Aurignacian
6x	Velké Pavlovice	16.490	48.540	open	Epigravettian
7x	Mende&Tapiosuly	19.350	47.300	open	Gravettian/Epigr.
8x	Ságvár	18.100	46.050	open	Epigravettian

BP). The only sites then occupied were three new ones outside the main cluster, all in areas with maximal wind-chill values of –6° to –3°C.

Figures 11.5 (centre) and 11.6 (centre) show the results of down-scaling temperature for the Ardennes region. Note that because this region was overall substantially colder, the scales on the two figures differ. It is clear that the cold Ardennes plateau acted as a barrier for colonization of this region, while the

Table 11.9. *Middle Danube basin case study. Sites: numbers, names, elevation (metres) and aspect°, type and dates (ka BP) including date sequences of long occupations.[2]*

Site no.	Site name	Elev.	Aspect	Type	Site dates (ka BP)
238	Langmannersdorf	195	29	open	22, 22
239	Horn	303	93	open	24
241	Willendorf II	441	107	open	(23–26), 41, 42, 43
242	Krems-Hundssteig	379	215	open	37
243	Krems-Wachtberg	379		open	29–30
244	Krems-Galgenberg	379		open	(30–35)
245	Langenlois	203	99	open	27, 28, 28, 29
246	Aggsbach	423	199	open	23, 24, 28, 28
258	Mamutowa			cave	22
276	Vindija			cave	28, 29, 30, 30, 31, 45, 46
277	Velika Pećina 2			cave	28, 29, 30
280	Krapina			cave	32
290	Istállóskö	611		cave	33, 34, 35, 43, 45
291	Peskö			cave	38
292	Szeleta			cave	35, 35, 45
293	Balla			cave	21, 24
294	Bodrögkeresztúr			open	31
295	Püspökhatvan			open	29
297	Tata			open	36
298	Érd			open	38, 42
300	Milovice I	178	51	open	31, 23, 23, 24, 27
301	Stránská skála	281	172	open	33, 35, 35, 41, 42, 44
302	Pod Hradem	530	112	cave	23, 28, 30, 31, 35, 36, 36
303	Vedrovice V	261	171	open	31, 38, 40, 40, 43
304	Kůlna	519	232	cave	23, 23, 23, 24, 41
305	Bohunice-kejbaly	280	81	open	38, 43, 44, 45
307	Pavlov I	202	55	open	26, 28, 28, 28, 28, 28, 28, 31
308	Dolní Věstonice	331	308	open	(21–31), 33, 34, 35
309	Předmostí	230	164	open	26, 28, 28, 28
310	Petřkovice			open	22, 25
311	Nemšova	331	140	open	30
312	Nitra-Čermáň	128	162	open	24
313	Moravany-Žakovska	215		open	19
314	Morazany-Lopata II	219	272	open	20, 23
315	Čertova pec	216	162	cave	41, 42
317	Spadzista Str. A			open	18, 22, 22, 22, 24, 26, 33
318	Kraków			open	22
319	Nietoperzowa			cave	41
320	Oblazowa 1			cave	19, 24, 32, 33, 35, 36
1x	Grubgraben			open	20, 20, 20, 21, 21, 21, 22
2x	Dzierzyslaw			open	36
3x	Piekary			open	34
4x	Jarošov			open	27, 28, 28, 28, 28, 28, 29
5x	Brno Videňská			open	15
6x	Velké Pavlovice			open	15
7x	Mende&Tapiosuly			open	19, 23
8x	Ságvár			open	19, 21

LGM sites occupied localities where the mean wind-chill values of coldest month were around –10°C.

The greater cold tolerance in the Ardennes compared to the Dordogne cluster could be due to the nature of the sites. Almost all Ardennes sites are caves; the two open sites, Maisières-Canal (209) and l'Hermitage (211), both late OIS-3 Gravettian, are situated farther away from the plateau. During the LGM, when those sites were probably occupied, they were very cold with wind-chill temperatures of –9°C to –7°C, colder than most sites in the Dordogne area. During warm phases temperatures were several degrees warmer and comparable to those experienced in the Dordogne region.

The only dated sites of the OIS-3 Stable Warm Phase (60–48 ka BP) in the region were Trou de l'Abîme (208), and possibly Sclayn cave (210) in the upper reaches of the Sambre valley, where wind-chill temperatures were colder than –10°C (Figs. 11.5: centre; 11.6: centre). The Mousterian site of Grotte de Spy (213) is undated, leaving the occupation record of the Sambre valley much sparser than the Meuse despite similarities in wind-chill. Exploitation of the resources might have played a role, but two dated sites do not suffice for speculation. Of the four (mostly Aurignacian) sites Trou al'Wesse (215), Trou Walou (205), Trou Magrite (204) and Grotte du Haleux (214), all in areas with a maximal wind-chill of –12° to –10°C (Fig. 11.5: centre), only one has dated occupation in LGM (27–25 ka BP); the others belong to the Early Cold Phase (37–28 ka). No convincingly dated occupation of the Ardennes exists for the full LGM Cold Phase (24–18 ka BP), but this may be due more to a lack of reliability and availability of the resources than directly to the climate.

Figures 11.5 (bottom) and 11.6 (bottom) show the results of downscaling climate for the Middle Danube Basin study area which is much larger and has many more sites than the Ardennes and Dordogne areas. The climate is continental with a seasonal wind-chill temperature range that can be in excess of 25°C. In addition, there is a large variation between cold phases (such as the LGM) and warm ones (such as the Transitional one). However, even during the warmest month, wind-chill temperatures do not exceed 21°C during warm phases and 15°C during cold (LGM) phases. It therefore seems reasonable to suggest that the coldest temperature remained the important climatic control.

Most sites in this region are open and they increase in number towards the LGM. Although the sites are well distributed across the area, more than 80 per cent of them occur where the coldest month's mean wind-chill temperature is below −11°C due to high elevation and strong winds. Those are very cold conditions indeed but broadly comparable to the climatic tolerances of the Ardennes, with a few exceptions. The most extreme site is Mamutowa (258 on Fig. 11.4) whose date of 22 ka BP is a *terminus ante quem* for its Gravettian layer (J. Kozlowski pers. comm.); wind-chill temperatures there dropped to −40°C, but its assemblages are small and suggest short-term occupation.

Sites in Regions 1, 3 and 4 (Fig. 11.4; Table 11.7) are consistently associated with colder areas (Fig. 11.5: bottom; 11.6: bottom), probably due to their higher altitudes. Many of the sites with coldest wind-chill (−14° to −20°C) belong to the LGM (24–18 ka BP), such as Spadzista (317), Horn (239), Grubgraben (1x), Ságvár (8x), all open-air, and the Mamutowa (258) and Balla (293) caves; all are Gravettian. Langmannersdorf (238) is the only one with wind-chill temperatures of less than −11°C. Sites of the late Stable Warm and Transitional Phases (47–38 ka BP) also are common in upland areas, where wind-chill temperatures did not differ much from LGM values (Fig. 11.5: bottom), e.g. at Dzierzyslaw (2x), Szeleta (292), Certova peć (315: Szeletian/Jerzmanowician), Érd (298: Mousterian) and Istállóskö (290) and Peskö (291) caves (Aurignacian). If those sites were occupied in winter (many show evidence for highly specialized, short-term occupation), we must ask what resources attracted Palaeolithic humans.

Table 11.10. *Dordogne technocomplexes by climate phase. (MO = Mousterian; CH = Châtelperronian; AU = Aurignacian; GR = Gravettian; SOL = Solutrean; MAGD = Magdalenian; 'UP' = Upper Palaeolithic; H.n. = Neanderthal remains only; FAU = Faunal site. Dashed lines between columns indicate that the same assemblage/context spans two or more climate phases. Codes in parentheses are of uncertain attribution to a particular climate phase.)*

#	Site name	70–60 ka BP	60–48 ka BP	47–38 ka BP	37–28 ka BP	27–25 ka BP	24–18 ka BP
69	Pataud			(AU)	AU; GR	GR	GR
70	Le Flagéolet I			(AU)	AU; GR	GR	GR
71	La Ferrassie			AU	AU; GR	GR	
73	Saint-Césaire			MO; CH	AU		AU?
79	Roc de Combe			CH; AU	AU; (GR)	GR	
80	Caminade			AU			
81	La Quina			MO	MO; AU	(UP)	SOL
82	Le Piage				AU; 'UP'		
86	Roc de Marcamps				AU		
92	Jaurens				H.n.		
95	Siréjol				FAU		
98	Combe Grenal	MO		MO	MO		
99	Laugerie-Haute					GR	SOL
100	Regourdou		MO				
101	Le Mas Viel				MO		
116	Chapelle-aux-Saints		MO				
117	Pech de l'Azé II	MO	MO	MO	(MO)		
119	Fonseigner		MO				
128	Sanglier				FAU		
136	Pégourié					(MAGD)	MAGD
137	Puy-Jarrige II						GR
140	Grotte XVI			CH			
141	Castanet			AU			
142	Perte de Bramarie				FAU		
147	Combe Saunière			CH; AU			GR; SOL
149	Barbas III			MO--------MO			
151	Le Facteur				AU; GR	GR	
152	La Rochette				AU		
154	Le Raysse				?GR	AU; [GR]	
155	Vignaud					AU	
157	Camiac-St-Denis				CH--------(CH)		
164	Cro le Biscop						CH?
165	Lespaux						GR
166	Les Fieux					AU	
171	Le Moustier	MO	MO				

Table 11.11. *Ardennes technocomplexes by climate phase. (MO = Mousterian; AU = Aurignacian; GR = Gravettian; MAGD = Magdalenian; 'UP' = Upper Palaeolithic. Dashed lines between columns indicate that the same assemblage/context spans two or more climate phases. Codes in parentheses are of uncertain attribution to a particular climate phase.)*

#	Site name	60–48 ka BP	47–38 ka BP	37–28 ka BP	27–25 ka BP	24–18 ka BP
204	Trou Magrite			AU----------AU----------AU?		
205	Trou Walou			AU; GR	GR	(MAGD)
206	La Princesse				AU	
207	Trou du Renard				AU	
208	Trou de l'Abîme	MO		MO/'UP'		
209	Maisières-Canal		?GR	GR	GR	
210	Sclayn	(MO)---------MO				
211	L'Hermitage			GR		
212	Goyet			AU (MAGD)		?MAGD
213	Gr. de Spy			(AU)	AU (GR)	
214	Gr. du Haleux				AU----------AU	
215	Trou Al'Wesse		MO; AU	AU		

209

Table 11.12. *Middle Danube Basin technocomplexes by climate phase. (MO = Mousterian; AU = Aurignacian; SZ = Szletian; BO = Bohunician; GR = Gravettian; EPI-AU = Epi-Aurignacian; EPI-GR = Epi-Gravettian; MAGD = Magdalenian. Dashed lines between columns indicate that the same assemblage/context spans two or more climate phases. Codes in parentheses are of uncertain attribution to a particular climate phase.)*

#	Region 1	47–38 ka BP	37–28 ka BP	27–25 ka BP	24–18 ka BP
258	Mamutowa				GR
310	Petřkovice			GR	GR
317/8	Spadzista St (Kraków)		AU	GR	GR
319	Nietoperzowa	SZ			
320	Oblazowa 1		GR		GR
2x	Dzierzyslaw		SZ		
3x	Piekary		AU/GR		

#	Region 2a	47–38 ka BP	37–28 ka BP	27–25 ka BP	24–18 ka BP
300	Milovice I		AU	GR	GR
301	Stránská skála	BO	AU		
302	Pod Hradem		SZ/AU		
303	Vedrovice V	SZ	SZ		
304	Kůlna	MO		'MAGD'	'MAGD'
305	Bohunice-kejbaly	BO			
307	Pavlov I		GR	GR	
308	Dolní Věstonice		GR	GR	GR
309	Předmostí		GR	GR	
4x	Jarošov		GR	GR	
5x	Brno-videlska				
6x	Velke Pavlovice				

#	Region 2b	47–38 ka BP	37–28 ka BP	27–25 ka BP	24–18 ka BP
238	Langmannersdorf				GR/EPI-AU
239	Horn				GR/EPI-AU
241	Willendorf II		AU; GR	GR	
242	Krems-Hundssteig		AU		
243	Krems-Wachtberg		GR		
244	Krems-Galgenberg		AU		
245	Langenlois		GR-------------GR		
246	Aggsbach		GR----------------------------GR		
1x	Grubgraben				EPI-GR

#	Region 3	47–38 ka BP	37–28 ka BP	27–25 ka BP	24–18 ka BP
290	Istállóskö	AU	AU		
291	Peskö	AU			
292	Szeleta	SZ	SZ		
293	Balla				GR
294	Bodrögkeresztúr		GR		
295	Püspökhatvan		GR		
297	Tata		MO		
298	Érd	MO			
311	Nemšova		GR		
312	Nitra-Čermáň				GR
313	Moravany-Žakovska		GR		GR
314	Moravany-Lopata II		GR		GR
315	Čertova pec	SZ			
7x	Mende & Tapiosuly				GR/EPI-GR
8x	Ságvár				EPI-GR

#	Region 4	47–38 ka BP	37–28 ka BP	27–25 ka BP	24–18 ka BP
276	Vindija	AU?	AU		
277	Velika Pećina 2		AU; GR		GR
280	Krapina		MO		

Humans and resources: a regional approach

'Attractive' environments should display not just the exploitation of their resources (lithic raw materials, local faunas, etc.), but also the latters' subsequent redistribution around the landscape. Material 'resources' can encompass anything derived from plants (wood, vegetable fibres, fruit/edible plants), animals (meat, ivory/antler/teeth, bone hide/fur), shell (perhaps mainly for personal adornment) or stone (for making tools or building (occupation) structures). Such materials have great potential for dictating where humans are in the landscape, whether climatic conditions are harsh or kind.

As mentioned earlier, all three case-study areas contain dissected valley topographies with adjoining lowland areas; thus all accommodate — to varying extents — mixed mosaics of ecotones. Of the three regions, the relatively small Ardennes is the most homogeneous, but the near-absence of dated sites from the North European Plain is puzzling: Maisières-Canal (209) is the closest site, yet it is on the southernmost fringe of the plain, and there is virtually nothing further north, even if one considers undated sites (Otte 1979). We here interpret the thin distribution of sites on the coastal lowlands of the Dordogne *s.l.* as the more ephemeral human occupation of non-dissected landscapes during OIS-3, and not as reflecting the distribution of limestone in Europe (with its associated sheltered sites). Accusations of preservation/excavation bias are certainly falsified by the rich, dated open-air record from the Middle Danube Basin, with evidence of great behavioural complexity particularly in Region 2.

All three case-study areas have a mixture of sites along main river courses and in adjoining side valleys. The Ardennes area probably

Table 11.13. *Estimated faunal resource potential at the three case-study areas and their subdivisions if any. Legend: N is number of sites; SUM is the species frequency in the number of dated faunal samples available at each site.*

| | The Dordogne, SW France | | | Ardennes | Middle Danube and Moravia | | |
| | Lowlands | Uplands | Valleys | | Danube | Moravia | N.Euro plain |
47–38 ka BP	N=3 SUM	N=4 SUM	N=8 SUM	N=4 SUM	N=6 SUM	N=10 SUM	N=3 SUM
Coelodonta antiquitatis		1		2		2	
Alopex lagopus				2	1	1	
Mammuthus primigenius		1	4	5	2	3	
Rangifer tarandus	1	1	11	5	2	2	
Megaloceros giganteus							
Ovibos moschatus		1	4				
Alces alces			7			2	
Bison	1	1	11	5	1		
Sus scrofa		1	3	2			
Cervus elaphus		1	9	2	2		
Capreolus capreolus			5	1			
Rupicapra rupicapra		1	7	1			
Capra ibex		1	7	1	1		
Saiga tartarica		1				1	
Equus hydruntinus			2				
37–28 ka BP							
Coelodonta antiquitatis	6	5	2		1		
Alopex lagopus	1	4	19	7	1	1	1
Mammuthus primigenius	9	4	18	8	4	2	1
Rangifer tarandus	9	5		16	3	4	1
Megaloceros giganteus			5		1	4	
Ovibos moschatus				6		1	
Alces alces							
Bison	9	1		13	2	1	
Sus scrofa	7	5		4		3	
Cervus elaphus	1	3		6	3	1	
Capreolus capreolus	7			2			
Rupicapra rupicapra				2	1		
Capra ibex				5	3	1	
Saiga tartarica	1		7			2	
Equus hydruntinus	1	1				1	
27–18 ka BP							
Coelodonta antiquitatis	1			8	2	1	1
Alopex lagopus				10	5	2	2
Mammuthus primigenius				14	3	6	2
Rangifer tarandus	2			13	5	6	1
Megaloceros giganteus				1		1	
Ovibos moschatus						1	
Alces alces				1		1	
Bison	2			6	2	1	1
Sus scrofa	1			8			
Cervus elaphus	1			2	2	2	
Capreolus capreolus	1					1	
Rupicapra rupicapra				2			1
Capra ibex				6	3	1	
Saiga tartarica				1			
Equus hydruntinus						2	

shows the strongest occupation of side valleys in relation to those directly in the main river corridor, perhaps reflecting concerns for greater shelter from prevailing winds (cf. Fig. 11.5: middle!) or specialized hunting strategies, e.g. ambush in small valleys. Major river corridors, such as those of the Dordogne-Vézère, the Sambre-Meuse and Danube/Morava, were almost certainly (major) migration routes for large herbivores, and thus might explain the clustered presence of sites in, or immediately adjoining, them. The repeated reoccupation of many sites in all three study areas attests to the continuing validity of their strategic/logistical positions and the reliability of their local resources, with successive techno-complexes (representing different human groups) often being present (Tables 11.10–11.12). Uniquely, many of these reoccupied sites in the Middle Danube Basin (especially in Region 2) are open-air. However, at the *chronometric* scale, fewer reoccupied sites show continuous occupation as defined by overlapping dates for their contiguous assemblages, and they all derive from the Dordogne and the Middle Danube (Tables 11.3 & 11.9). The Ardennes has no sites with such consistent use (Table 11.6), perhaps owing to its smaller size and greater susceptibility to climatic and environmental change.

The main exploited species in the Ardennes do not differ greatly from those of the Dordogne and the Middle Danube Basin (Table 11.13; Appendix 11.1), although there were apparent variations in species availability (Table 11.13), doubtless creating subtle variations in hunting strategies. At 47–38 ka BP, the lowland zones of the Dordogne and the Middle Danube Basin are thinly occupied (Tables 11.10 & 11.12) and show a lack of diversity in their hunted faunas (Table 11.13), implying an ephemeral and economically-specialized occupation. The contemporary valley occupations in all three study areas show a much greater diversity of species, with thermophilic ones (and those preferring rocky terrain) commonest in the Dordogne *s.s.* and the Ardennes, together with consistent presence of mammoth and reindeer (Table 11.13). In the Transitional climatic phase (37–28 ka BP), mammoth and associated arctic fauna appear in the hunted assemblages of lowland 'Dordogne' and Middle Danubian Region 1 archaeological sites, perhaps reflecting their increased presence in such areas. Elsewhere, more mixed assemblages of arctic, thermophilous and craggy landscape (persisting mainly in the Ardennes and Austria) faunas coincide with a strong increase in site numbers and densities (Table 11.13). Sites from the LGM Cold Phase (27–18 ka BP), as noted earlier, show a preponderance of arctic fauna, especially mammoth, reindeer and arctic fox (the latter for teeth/furs?), except in the Dordogne, where a more diverse array of woodland/meadow species (boar, red deer and roe deer) was exploited by humans in discrete site clusters (Tables 11.10 & 11.13). Even during the LGM, people evidently had access to both arctic and more thermophilous faunal resources, and exploited them both. Not only does this tell us about the character of the mosaic of faunal eco-suites containing our main archaeological site clusters, but it also helps to explain their attractiveness: the diversity of resources could help to sustain healthy population levels even when arctic herbivores migrated out of the area.

Although people would doubtless have followed any large-scale animal migrations during OIS-3, a wider variety of resources (including plant-derived ones) would have enabled them to occupy territories on a year-round basis. Any apparent concentration upon megafaunal resources is probably misleading because smaller animals tend to be less mobile, and therefore could be more reliably present in the landscape. While mammoth, horse, bison, saiga antelope and reindeer can be assumed to be (long-distance) migrants, with the latter three displaying high potential for seasonal aggregation, species such as *Megaloceros*, elk, musk ox, boar, ibex, chamois and roe deer were perhaps more residential, with a concomitantly low seasonal aggregation potential (Gamble 1986, 105). The scale of herbivore migration would surely have varied in each of our three case-study areas, depending upon topography and the distances between the seasonally-available resources, and therefore we cannot expect the mobility of people in those three areas to have been identical. Mammoth and reindeer may have been hunted primarily by specialized expeditions, or perhaps by the whole group during seasonal aggregation events. The exploitation of ivory and bone from (long-)dead mammoths could further confuse the picture: it would not necessarily have been subject to restrictions of season or of size of exploitation group, and sometimes occurs at sites with additional evidence of mammoth hunting (Christensen 1997; Münzel *et al.* 1994). Further exploration of seasonality in food resource availability must therefore be abandoned here, owing to the scarcity and ambiguity of current data.

Sites of the Middle Danube Basin show the highest group mobility, if one treats the transfer distances of stone raw materials as a proxy for human movement around the landscape (and/or for intergroup exchange), with those from the Dordogne and Ardennes indicating smaller transfers (Féblot-Augustins 1997; 1999). Binford's (1979) concept of Embedded procurement strategies (contrasted with Direct ones) will be used as a working assumption for this discussion, setting as it does the acquisition of stone raw materials within basic or specialized subsistence activities: people combine their hunting with gathering of suitable/desired lithic materials. Direct procurement, i.e. forays with the specific aim of obtaining raw materials, is rare in the modern ethnographic record (Binford 1979), and is in any case difficult to distinguish from embedded strategies in the archaeological record, as faunal material may be lacking for taphonomic, rather than economic, reasons.

Most of the lithic raw materials used in OIS-3 archaeological sites are of local (perhaps <5 km away) provenance, but transfer distances begin to increase in the late Mousterian, and expand substantially in the Upper Palaeolithic (especially in the Aurignacian and Gravettian) (e.g. Miller 2001; Féblot-Augustins 1997; 1999). The test of whether such materials are an attraction should be whether they were transported to other sites, rather than simply used on-site. Materials which underwent longer-distance transfers tended to be the higher-quality siliceous ones (flint, chert, jasper, etc.), and they were normally tested and worked before transport (e.g. Miller 2001, 62), presumably to avoid transporting unwanted mass. All three regions show this pattern of movement, although the Dordogne *s.l.* and Ardennes display shorter transport distances — generally <80 km in the Upper Palaeolithic (Dordogne maximum = <120 km; Ardennes maximum = <250 km, if acquisitions from the Paris Basin are genuine) (Féblot-Augustins 1997; 1999) — than do the sites of the Middle Danube Basin, where the Aurignacian site of Krems-Hundssteig (242) contains materials from Hungary, *c.* 400 km to the east (Féblot-Augustins 1999, 242). Such differences between these maritime and continental areas, respectively, might be attributed to different regional patterns of seasonal movement (Féblot-Augustins 1999, 233).

The transfer patterns of marine shells, whether 'fresh' or fossil, often mirror those of the lithic materials. The Aurignacian of Spy (213) contains fossil shells which could come from the Paris Basin and from the Red Crags deposit in eastern England (Otte 1979). Middle Danubian sites extensively exploited fossil shells from Miocene outcrops, necessitating their transport over long distances (Taborin 2000). Mediterranean shells, e.g. at Krems-Hundssteig, were also used on occasion (Hahn 1997, 609), perhaps indicating (exchange) connections with that zone. Western Europe, exemplified by the Dordogne, preferred to use fresh marine shells during the Upper Palaeolithic, although the number of preferred species seems to contract between the Aurignacian and the Gravettian (Taborin 2000). Castanet (141) is unique among the dated sites in having both Atlantic and Mediterranean species (Taborin 1993), but that might be explained by its apparent specialization as a 'factory' for the production of beads and pendants (White 1989). The remaining Dordogne sites have Atlantic fresh shells, if they have any at all

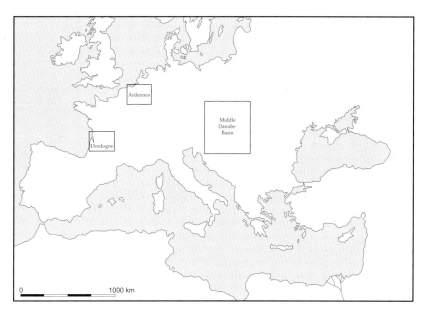

Figure 11.7. *Case-study areas in the continental context.*

(Taborin 1993); presumably the original collection sites are now under-water in the Bay of Biscay. Therefore, while we can easily conclude that marine shells were desirable commodities in the early Upper Palaeolithic of the Dordogne, we cannot attempt to reconstruct settlement patterns near their sources: was there exchange with coastal groups, or something more akin to direct procurement?

Caution is always advisable when trying to infer site functions from their artefactual quantities and diversity, faunal remains, spatial organization and transfer distances of their commodities; attempts to reconstruct settlement systems will always be vulnerable to any augmentation in our excavated data. Residential sites of varying complexity and duration can be picked out fairly easily (e.g. Trou Magrite, Goyet, Spy, Maisières-Canal, Dolní Věstonice, Pavlov, Předmostí, Willendorf, Krems-Hundssteig, Kraków-Spadzista St, Castanet, Pataud, Laugerie-Haute, Grubgraben), but trying to identify specialised hunting sites (perhaps Istállóskö and Trou de l'Abîme) is more difficult (Miller 2001; Svoboda *et al.* 1996; White 1989; Sonneville-Bordes 1960; Otte 1981; Neugebauer-Maresch 1995; Kozlowski 1983; 1998). Many sites have a lithic workshop component (e.g. Stránská skála, Bohunice-kejbaly, Vedrovice V, Moravany-Lopata, Dzierzyslaw, Bodrögkeresztúr, Püspökhatvan, Grubgraben), but it is accompanied by other (subsistence/butchery, domestic) activities (Vértes 1966; Dobosi 2000; Hromada & Kozlowski 1995, 87), supporting the assertion that raw material acquisitions were embedded in other activities.

213

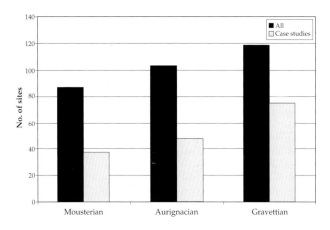

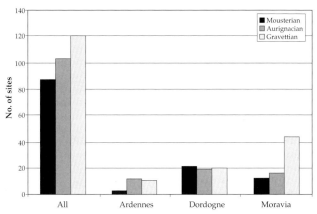

Figure 11.8. *The number of sites per technocomplex for the total study area compared to the number of sites found in the total area of the case studies.*

Figure 11.9. *The total number of sites in each of the case-study areas according to broad technocomplex.*

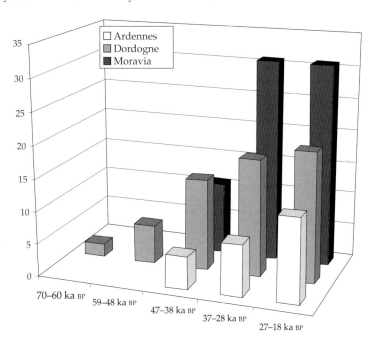

Figure 11.10. *Number of sites for each case-study area according to OIS-3 climate phases.*

The long-distance materials found in significant numbers at large residential sites, such as Castanet, Dolní Věstonice and Pavlov, imply that these loci were convergence points in the landscape. The lithic raw material provenances in the Middle Danubian Upper Palaeolithic suggest two separate systems: one in the western half, reaching as far as the Váh valley (eastern Slovakia) and north onto the North European Plain through the Moravian Gate, and the other stretching eastwards from the Váh valley into Hungary (Féblot-Augustins 1999, 241–2). The Ardennes has no such convergence sites, per-

haps owing to its more restricted and isolated catchment area. Unlike the other two case-study areas, its occupation withers towards the LGM, with no convincing occupation after 25 ka BP. The relatively late sites of Maisières-Canal and l'Hermitage probably represent ephemeral small-scale occupations with a strong lithic workshop component. The groups responsible could have lived outside the area for much of the year, susceptible to food resource availabilities determined by local climatic/environmental conditions (Fig. 11.5: middle).

The Middle Danube Basin witnesses a behavioural change also with the onset of the LGM Cold Phase, certainly after 25 ka BP (Table 11.12). While the Gravettian (Pavlovian variant) shows increasing sedentism in its occupation of the landscape, combined with elaborate resource exploitation systems, and technological and ideological ones, it is concentrated in the Early Cold Phase (down to 28 ka BP) (Svoboda *et al.* 2000). With the onset of the LGM Cold Phase (27–18 ka BP), this elaborate social system could not be sustained, and the Gravettian apparently became simpler. We might speculate that the exploitation of plant (with possible grinding stones in Moravia and Germany: Svoboda *et al.* 2000; Bosinski 2000) and other reliable food resources may have been a crucial underpinning of Pavlovian occupational stability, which became unsustainable with environmental changes towards the LGM.

Economic necessity would not necessarily have dictated which materials in the landscape were identified as 'resources', e.g. the essentially Upper Palaeolithic (especially Aurignacian and Gravettian) use of marine shells and animal teeth for personal adornment and also the exploitation of bone, antler and ivory. Middle Palaeolithic use of these materials, by contrast, is seldom recorded. Additionally, the *combinations* of these new resource elements could have changed or augmented over time in response to climatic change, perhaps necessitating different patterns and modes of exploitation. Evidence for heterogeneity of resource exploitation strategies within an area would encourage us to define it as 'attractive'. We believe that approaches which emphasize the attractiveness of available resources to explain human spatio-temporal distributions ('aggregation regions'?) should be preferred over ones which view any clusters as relict, 'refugial' populations.

The study areas in a continental context

Three discrete clusters of archaeological sites have been described in terms of climate, spatio-temporal distribution patterns and resources (Fig. 11.7), and general conclusions about them and their continental context can now be drawn. Figure 11.8 shows that of the dated sites in our chrono-archaeological data base,[4] 32 per cent of the Mousterian ones, 49 per cent of the Aurignacian ones and 89 per cent of the Gravettian ones are distributed within the three case-study clusters. This is significant in that the combined area of the studied regions amounts to less than 25 per cent of the European landmass we are considering (Fig. 11.7). Therefore, the archaeological richness of our three study areas cannot simply be attributed to sampling biases, for if that were the case, we should expect roughly equivalent site concentrations for each of our three technocomplexes. While there appears to be a uniform increase in the number of sites overall, the Mousterian and Aurignacian ones are more evenly dispersed than those of the Gravettian where the concentration of sites towards/at the LGM is almost completely within the case-study areas. Almost half of the Gravettian sites are Middle Danubian (Fig. 11.9), perhaps reflecting new social and economic adaptations. These changing spatio-temporal distributions can be taken to support the notion that (AMH) people increasingly show a clear preference for (rather than tolerance of) areas in climatically inhospitable regions (Chapter 8: Davies & Gollop 2003).

Figure 11.10 shows the site distributions according to the OIS-3 climate phases. Although the Ardennes have significantly fewer sites than the Dordogne, both regions share a similar maritime climate (cf. Chapter 5: Barron *et al.* 2003, Fig. 5.11) and site distribution pattern. Both show a steady increase in the number of sites over time, while exhibiting consistency and stability in the spatial distribution of the three technocomplexes (Figs. 11.1, 11.3 & 11.9). On the other hand, the more continental Middle Danube basin (cf. Chapter 5: Barron *et al.* 2003, Fig. 5.11) shows a different pattern of site distribution: the sharp increase in the number of Gravettian sites reaches a stable peak in the OIS-3 Early and LGM Cold Phases (Fig. 11.10). Middle Danubian sites in the latter cold phases are dominated by the large open-air Gravettian occupations in the vicinity of Aggsbach-Willendorf-Krems-Langenlois (on the northern slope of the Danube valley) and at Dolní Věstonice-Pavlov-Milovice (with Předmostí a significant northern outlier). Differences in site occupation patterns between the Middle Danube Basin (dominated by complex open sites) and the Ardennes and Dordogne (dominated by sheltered sites) might be attributed to climatic and topographic influences, which in turn affect faunal and other resources. Conversely, in the Dordogne the number of sites remains stable regardless of climate changes, although the sites do become more localized within the Dordogne and Vézère valleys during the Early Cold Phase.

If human settlement distributes spatio-temporally in response to resource distributions, the assemblages from the sites within our three major clusters should be compared to those in between. The more equitable spatial distributions of the Mousterian and Aurignacian sites across mid- and lower-latitude Europe demand a 'continental-scale' comparative approach: we should expect to find equally 'complex' sites in the interstitial (non-case-study) areas with similar resources. However, the Gravettian spatio-temporal distribution would be the truly interesting test of our resource-based assumptions, as almost 90 per cent of them are found inside our three study areas. What attracted people to our case-study areas? Are there any similarly 'complex' sites in the interstitial areas, in addition to the ephemeral ones, which might falsify the assumption that the records of the three regions are unique responses?

Raw material analyses perhaps lend the most support to the integrity of our case-study areas (Féblot-Augustins 1997; 1999; Scheer 2000), although some goods were imported from outside our drawn boundaries (especially marine shells: Taborin 1993; 2000). Whether all such exotic pieces were obtained through exchange or through Direct procurement (implying high mobility of [sub]groups) is difficult to establish at present. It is possible that northern European groups traded ivory for marine shells with groups in the north of Italy (Mussi *et al.* 2000), but as evidence this is ambiguous; one could alternatively argue that such northerly groups made ephemeral forays into Mediterranean zones carrying ivory and collecting shells themselves. Interstitial regions with consistent human presence throughout OIS-3 are southern Germany, with its rich and significant Aurignacian sites (e.g. Conard & Bolus 2003), the Rhône basin, Cantabrian Spain and the Russian Plain.

Except for parts of the Middle Danube basin, we have virtually no seasonality data from OIS-3 sites. Yet it is axiomatic that the various human socio-economic strategies must cope with a critical dependence on seasonal food supplies, almost everywhere to some degree, but especially in glacial Europe. The reality of arctic and subarctic living is that the climate, even today, forces seasonal transhumance from southerly winter sites to northerly summer sites to maximize benefits from the most abundant animal resources. The arctic/subarctic mode has existed in northern Europe and Eurasia since the early Holocene and still survives in many places, but it may date back much farther in time. We believe that it would be very peculiar indeed if the Gravettians who have left much evidence of an advanced mode of living in a rather small part of north and central Europe had not developed the subarctic mode as a reasonably reliable way to extract maximum benefit from tundra animals such as mammoth and reindeer. In fact, their highly original idea of responding to the deteriorating climate by going north rather than southwest and southeast as Aurignacians did, seems to point strongly in the direction of annual tundra transhumance. To test this possibility, however, we really need seasonality data from areas other than the Middle Danubian region or at least more from the Ardennes and the Dordogne.

We have argued that the site clusters comprising our three case-study areas repesent aggregates of attractive loci ('magnets') in the landscape, distributed in a relatively fine-grained spatial mosaic that might wax and wane in productivity as a result of localised climatic conditions, but which would comprise a stable resource area at the regional scale. Attempts to define areas with attractive resources are better at taking changing human preferences into account than are ones which assume they represent 'refugia'. What is held to be 'attractive' by human groups could easily change over time.

We can conclude that climate, topography and environmental resources are capable of strongly influencing human choices about subsistence and location of settlement. Those choices may be partly or entirely controlled by other social, technical or other specialized factors, such as a desire to explore, or a need for inter- and intra-group relationships, even in the face of fluctuating climatic and environmental conditions.

Acknowledgements

For this chapter we needed more detailed climate, landscape and archaeological information than for the continental-scale problems addressed in most other chapters. These required an advanced analysis of data sets, in particular the numerical data available from the climate simulations. For this data processing we owe much to Piers Gollop who compiled the required extracts from the palaeoclimate simulations. We are deeply grateful to Jon Jolly and family, Mark Bearwaldt and Mies van Andel for the generosity that enabled us to make full use of the material. We also thank Janusz Kozlowski for clarifying the stratigraphic position of several dates used in this paper, John Stewart for his help in our use of the mammalian data and Clive Gamble for his review of the manuscript.

Notes

1. The notation ka BP stands for calendar dates (ice-core, U-Th, TL, OSL and ESR) and calibrated ^{14}C dates in millennia (ka) before present (BP). The CalPal ^{14}C calibration programme used is available at its website: http://www.calpal.de.
2. The chrono-archaeological data base is described in detail in Chapter 3 (van Andel *et al.* 2003a) and in the Preamble of the Stage 3 Project data bases which can be found at the Stage 3 website: http://www.esc.esc.cam.ac.uk/oistage3/Details/Homepage.html.
3. Numbers in brackets are the site numbers of the chrono-archaeological data base; also shown on Figure 11.3. Name lists of sites used for the study case clusters are Tables 11.2 (Dordogne), 11.5 (Ardennes) and 11.8 (Middle Danube basin).
4. Palaeoclimate: The climate phases are shown in Figure 11.2. Unless noted otherwise, the palaeoclimate data in this chapter derive from the Stage 3 climate simulations available on the Stage 3 Project website: http://www.esc.esc.cam.ac.uk/oistage3/Details/Homepage.html.
5. The mammalian data base is available at the Stage 3 Project website. For details see Chapter 7 (Stewart *et*

al. 2003) and the Preamble to the Stage 3 data bases; both can be consulted at: http://www.esc.esc.cam.ac.uk/oistage3/Details/Homepage.html.

6. The downscaling procedure rests on the assumption that small-scale spatial climate variations are controlled mainly by local topography; small-scale circulation changes are of less importance. The predicted temperatures are modified to account for the difference between the model topography of the regional model grid and relatively smooth) and a high resolution (30 arc second) topography. It has four steps:

(1) All climate models have regional biases when their present day climate simulation is compared to observations (Chapter 5: Barron *et al.* 2003) which tend to be small. To minimize associated problems we use the climate changes predicted by the regional model, adding it to a high-resolution ($0.5° \times 0.5°$) observed modern climatology (New *et al.* 1999) as widely done in impact studies; this eases the comparison of climates in different cluster regions.

(2) All climate and climate model data are interpolated with a simple bi-linear interpolator from the various input grids onto a common grid of very high-resolution (30 arc second or 1/120 of a degree) of about 0.7 km at 45°N. The resolution was chosen since it corresponds to the highest resolution orographic data set (ETOP30) available.

(3) A lapse rate-based correction is applied to the temperatures so that they reflect the temperatures at the height of the local topography rather than the height within the PSU model. Because predicted lapse rates are remarkably uniform, we decided to use a constant lapse rate of 7.8°C/km for all simulations.

(4) Because wind-chill is a function of temperature and wind the wind speed was corrected to take into account the local elevation. However, because inspection of the results showed that the wind shear varied across the region, we used the wind shear from the model directly. This final downscaling modification has two small weaknesses. The local wind is affected by small-scale circulation around local topography but we neglect this in our calculations. Secondly, we calculate the wind-chill temperature from the local monthly mean temperature and monthly mean winds. Due to the non-linearity of the calculation, this is not the same as calculating wind-chill using daily data and then calculating the monthly mean.

The 4-step down-scale procedure is available from: http://edcdaac.usgs.gov/gtopo30/gtopo30.html (GTOPO30 dataset).

References

Alfano, M.J., E.J. Barron, D. Pollard, B. Huntley & J.R.M. Allen, 2003. Comparison of climate model results with European vegetation and permafrost during Oxygen Isotope Stage Three. *Quaternary Research* 59, 97–107.

Antoine, P., J.L. de Beaulieu, P. Bintz, J.P. Brugal, M. Girard, J.L. Guadelli, M.T. Morzadec-Kerfourn, J. Renault-Miskovsky, A. Roblin-Jouvé, B. van Vliet-Lanoë & J.D. Vigne, 1999a. *La France dans les deuz derniers extrèmes climatiques (Cartes 1:1,000,000)*. Châtenay-Malabry, France: ANDRA.

Antoine, P., D.D. Rousseau, J.-P. Lautridou & C. Hatté, 1999b. Last interglacial-glacial climatic cycle in loess-palacosol successions of north-western France. *Boreas* 28, 551–63.

Arnold, N.S., T.H. van Andel & V. Valen, 2002. Extent and dynamics of the Scandinavian ice sheet during Oxygen Isotope Stage 3 (60,000 to 30,000 ka BP). *Quaternary Research* 57, 38–48.

Barron, E., T.H. van Andel & D. Pollard, 2003. Glacial environments II: reconstructing the climate of Europe in the Last Glaciation, in *Neanderthals and Modern Humans in the European Landscape during the Last Glaciation*, Chapter 5, eds. T.H. van Andel & W. Davies. (McDonald Institute Monographs.) Cambridge: McDonald Institute for Archaeological Research, 57–78.

Bellaire, C. & M. Otte, 2001. La Grotte de Spy (Province de Namur), in *Guide des sites préhistoriques et proto-historiques de Wallonie (Vie Archéologique)*, eds. C. Bellaire, J. Moulin & A. Cahen-Delhaye. (Bulletin de la Fédération des Archéologues de Wallonie asbl, Numéro spécial 2001.) Namur, Belgium, 18–21.

Binford, L.R., 1979. Organization and formation processes: looking at curated technologies. *Journal of Anthropological Research* 35, 255–73.

Bocquet-Appel, J.-R. & P.-Y. Demars, 2000a. Neanderthal contraction and modern human colonization of Europe. *Antiquity* 74, 544–52.

Bocquet-Appel, J.-R. & P.-Y. Demars, 2000b. Population kinetics in the Upper Palaeolithic in Western Europe. *Journal of Archaeological Science* 27, 551–70.

Bosinski, G., 2000. The period 30,000-20,000 bp in the Rhineland, in *Hunters of the Golden Age: the Mid Upper Palaeolithic of Eurasia 30,000–20,000 BP*, eds. W. Roebroeks, M. Mussi, J. Svoboda & K. Fennema. Leiden: University of Leiden, 271–80.

Christensen, M., 1997. The depositional formation process of usewear polish and its implications for the identification of worked material: tools used in ivory working in the Upper Paleolithic. *Poroilo o raziskovanju paleolitika, neolitika in eneolitika v Sloveniji (Ljubljana)* XXIV, 99–112.

Conard, N.J. & M. Bolus, 2003. Radiocarbon dating the appearance of modern humans and timing of cultural innovations in Europe: new results and new challenges. *Journal of Human Evolution* 44, 331–71.

Daugas, J.-P. & J.-P. Raynal, 1989. Quelques tapes du peuplement du Massif Central français dans leur contexte paléoclimatique et palégéographique, in *Variations de Paléomilieux et Peuplement Préhistorique*, ed. H. Laville. (Cahiers du Quaternaire 13.) Paris: CNRS Université de Bordeaux, 67–95.

Davies, W. & P. Gollop, 2003. The human presence in

Europe during the Last Glacial Period II: climate tolerance and climate preferences of mid- and late glacial hominids, in *Neanderthals and Modern Humans in the European Landscape during the Last Glaciation*, Chapter 8, eds. T.H. van Andel & W. Davies. (McDonald Institute Monographs.) Cambridge: McDonald Institute for Archaeological Research, 131–46.

Dobosi, V., 2000. Interior parts of the Carpathian Basin between 30,000 and 20,000 bp, in *Hunters of the Golden Age: the Mid Upper Palaeolithic of Eurasia 30,000–20,000 BP*, eds. W. Roebroeks, M. Mussi, J. Svoboda & K. Fennema. Leiden: University of Leiden, 231–9.

Féblot-Augustins, J., 1997. Middle and Upper Paleolithic raw material transfers in western and Central Europe: assessing the pace of change. *Journal of Middle Atlantic Archaeology* 13, 57–90.

Féblot-Augustins, J., 1999. La mobilité des groupes paléolithiques. *Bulletin et Mémoire de la Société d'Anthropologie de Paris* 11, 219–60.

Gamble, C.S., 1986. *The Palaeolithic Settlement of Europe*. Cambridge: Cambridge University Press.

Guthrie, R.D., 1990. *Frozen Fauna of the Mammoth Steppe: the Story of Blue Babe*. London: University of Chicago Press.

Guthrie, R.D., 2000. Origin and cause of the mammoth steppe: a story of cloud cover, woolly mammal tooth pits, buckles and inside-out Beringia. *Quaternary Science Reviews* 20, 549–74.

Guthrie, D. & T. van Kolfschoten 2000. Neither warm and moist, nor cold and arid, the ecology of the Mid Upper Palaeolithic, in *Hunters of the Golden Age: the Mid Upper Palaeolithic of Eurasia 30,000–20,000 BP*, eds. W. Roebroeks, M. Mussi, J. Svoboda & K. Fennema. Leiden: University of Leiden, 13–20.

Hahn, J., 1997. Krems-Hundssteig, in *Dictionnaire de la Préhistoire*, ed. A. Leroi-Gourhan. Paris: Quadrige/ Presses Universitaires de France, 609–10.

Hromada, J. & J.K. Kozlowski (eds.), 1995. *Complex of Upper Palaeolithic Sites near Moravany, Western Slovakia*, vol. I: *Moravany-Žakovska (Excavations 1991–1992)*. Kraków: Jagellonian University Press.

Kozlowski, J.K., 1983. Le Paléolithique supérieur en Pologne. *L'Anthropologie* 87, 49–82.

Kozlowski, J.K., 1998. Function of the site and its place in seasonal activities of reindeer and mammoth hunters, in *Complex of Upper Palaeolithic Sites near Moravany, Western Slovakia*, vol. II: *Moravany-Lopata II (excavations 1993–1996)*, ed. J.K. Kozlowski. Kraków: Jagellonian University Press, 128–30.

Lambeck, K., 1997. Sea-level change along the French Atlantic and Channel coasts since the time of the Last Glacial Maximum. *Palaeogeography, Palaeoclimatology, Palaeoecology* 129, 1–22.

Lambeck, K. & J. Chappell, 2001. Sea level change through the last glacial cycle. *Science* 292, 679–86.

Lister, A.M. & A.V. Sher 1995. Ice cores and mammoth extinction. *Nature* 378, 23–4.

Meese, D.A., A.J. Gow, R.B. Alley, G.A. Zielinsky, P.M.

Grootes, M. Ram, K.C. Taylor, P.A. Mayewski & J.F. Bolzan, 1997. The Greenland Ice Sheet Project 2 depth-age scale: methods and results. *Journal of Geophysical Research* 102, 26,411–23.

Miller, R., 2001. *Lithic Resource Management during the Belgian Early Upper Paleolithic: Effects of Variable Raw Material Context on Lithic Economy*. (ERAUL 91.) Liège: Études et Recherches Archéologiques de l'Université de Liège.

Münzel, S., P. Morel & J. Hahn, 1994. Jungpleistozäne Tierreste aus der Geißenklösterle-Hohle bei Blaubeuren. *Fundberichte aus Baden-Württemberg* 19, 63–93.

Musil, R., 2003. The Middle and Upper Palaeolithic game suite in central and southeastern Europe, in *Neanderthals and Modern Humans in the European Landscape during the Last Glaciation*, Chapter 10, eds. T.H. van Andel & W. Davies. (McDonald Institute Monographs.) Cambridge: McDonald Institute for Archaeological Research, 167–90.

Mussi, M., J. Cinq-Mars & P. Bolduc, 2000. Echoes from the mammoth steppe: the case of the Balzi Rossi, in *Hunters of the Golden Age: the Mid Upper Palaeolithic of Eurasia 30,000–20,000 BP*, eds. W. Roebroeks, M. Mussi, J. Svoboda & K. Fennema. Leiden: University of Leiden, 105–24.

Neugebauer-Maresch, C., 1995. Altsteinzeitforschung im Kremser Raum, in *Perspektiven zum Werdegang von Krems und Stein (Archäologie Österreichs, Sonderausgabe 1995)*, ed. A. Krenn-Leeb, 14–25.

New, M., M. Hulme & P.D. Jones, 1999. Representing twentieth century space-time climate variability, part 1: Development of a 1961–90 mean monthly terrestrial climatology. *Journal of Climate* 12, 829–56.

Olsen, L., 1997. Rapid shifts in glacial extension characterise a new conceptual model for glacial variations during the Mid and Late Weichselian in Norway. *Norges Geologiske Undersøgelse Bulletin* 433, 54–5.

Olsen, L., V. Mejdahl & S.F. Selvik, 1996. Middle and late Pleistocene stratigraphy, chronology and glacial history in Finnmarken, North Norway. *Norges Geologisk Undersøgelse Bulletin* 429.

Otte, M., 1979. *Le paléolithique supérieur ancien en Belgique*. Brussels: Monographies d'Archéologie Nationale 5.

Otte, M., 1981. *Le Gravettien en Europe Centrale (2 volumes)*. Brugge: De Tempel (Dissertationes Archaeologicae Gandenses 20).

Rigaud, J.-P. & J.F. Simek, 1987. 'Arms too short to box with God': problems and prospects for Paleolithic prehistory in Dordogne, France, in *The Pleistocene Old World: Regional Perspectives*, ed. O. Soffer. New York (NY): Plenum Press, 47–61.

Roebroeks, W., M. Mussi, J. Svoboda & K. Fennema (eds.), 2000. *Hunters of the Golden Age: the Mid Upper Palaeolithic of Eurasia 30,000–20,000 BP*. Leiden: University of Leiden.

Scheer, A., 2000. The Gravettian in Southwest Germany: stylistic features, raw material resources and settlement patterns, in *Hunters of the Golden Age: the Mid Upper Palaeolithic of Eurasia 30,000–20,000 BP*, eds. W.

Roebroeks, M. Mussi, J. Svoboda & K. Fennema. Leiden: University of Leiden, 257–70.

Shackleton, J.C., T.H. van Andel & C.N. Runnels, 1984. Coastal paleogeography of the central and western Mediterranean during the last 125,000 years and its archaeological implications. *Journal of Field Archaeology* 11, 307–14.

Sonneville-Bordes, D. de, 1960. *Le Paléolithique Supérieur en Périgord.* Bordeaux: Delmas.

Stewart, J.R., T. van Kolfschoten, A. Markova & R. Musil, 2003. The mammalian faunas of Europe during Oxygen Isotope Stage Three, in *Neanderthals and Modern Humans in the European Landscape during the Last Glaciation,* Chapter 7, eds. T.H. van Andel & W. Davies. (McDonald Institute Monographs.) Cambridge: McDonald Institute for Archaeological Research, 103–30.

Stuiver, M. & P. Grootes, 2000. GISP2 oxygen isotope ratios. *Quaternary Research* 53, 277–84.

Svoboda, J., V. Ložel & E. Vlček, 1996. *Hunters between East and West: the Paleolithic of Moravia.* New York (NY): Plenum Press.

Svoboda, J., B. Klíma, L. Jarošova & P. Škrdla, 2000. The Gravettian in Moravia: climate, behaviour and technological complexity, in *Hunters of the Golden Age: the Mid Upper Palaeolithic of Eurasia 30,000–20,000 BP,* eds. W. Roebroeks, M. Mussi, J. Svoboda & K. Fennema. Leiden: University of Leiden, 197–217.

Taborin, Y., 1993. Shells of the French Aurignacian and Périgordian, in *Before Lascaux: the Complex Record of the Early Upper Paleolithic,* eds. H. Knecht, A. Pike-Tay & R. White. Boca Raton (LA): CRC Press, 211–27.

Taborin, Y., 2000. Gravettian body ornaments in Western and Central Europe, in *Hunters of the Golden Age: the Mid Upper Palaeolithic of Eurasia 30,000–20,000 BP,* eds. W. Roebroeks, M. Mussi, J. Svoboda & K. Fennema. Leiden: University of Leiden, 135–41.

Terjung, W.H., 1968. World patterns of the distribution of the monthly comfort index. *International Journal of Biometeorology* 12, 119–51.

Ulrix-Closset, M., M. Otte & P. Cattelain, 1988. Le 'Trou de l'Abîme' à Couvin (Province de Namur, Belgique), in *L'Homme de Neandertal,* vol. 8: *La Mutation,* ed. J. Kozlowski. (ERAUL 35.) Liège: Études et Recherches Archéologiques de l'Université de Liège, 225–39.

USGS EROS DATA CENTER, 1996. *GTOPO30 Digital Elevation Model.* EROS Data Center Distributed Active Archive Center, URL: http://edcwww.cr.usgs.gov/landaac/.

van Andel, T.H., 1989. Late Quaternary sea level changes and archaeology. *Antiquity* 63, 733–45.

van Andel, T.H. & J.C. Shackleton 1982. Late Paleolithic and Mesolithic coast lines of Greece and the Aegean. *Journal of Field Archaeology* 9, 445–54.

van Andel, T.H., W. Davies, B. Weninger & O. Jöris, 2003a. Archaeological dates as proxies for the spatial and temporal human presence in Europe: a discourse on the method, in *Neanderthals and Modern Humans in the European Landscape during the Last Glaciation,* Chapter 3, eds. T.H. van Andel & W. Davies. (McDonald Institute Monographs.) Cambridge: McDonald Institute for Archaeological Research, 21–30.

van Andel, T.H., W. Davies & B. Weninger, 2003b. The human presence in Europe during the Last Glacial period, I: human migrations and the changing climate, in *Neanderthals and Modern Humans in the European Landscape during the Last Glaciation,* Chapter 4, eds. T.H. van Andel & W. Davies. (McDonald Institute Monographs.) Cambridge: McDonald Institute for Archaeological Research, 31–56.

van Huissteden, J., D. Pollard & J. Vandenberghe, 2003. Paleotemperature reconstructions of the European permafrost zone during Oxygen Isotope Stage 3 compared with climate model results. *Journal of Quaternary Research* 18, 453–64.

Vértes, L., 1966. The Upper Palaeolithic site on Mt Henye at Bodrögkeresztúr. *Acta Archaeologica Hungaricae* 18, 3–14.

White, R., 1980. The Upper Paleolithic Occupation of the Perigord: a Topographic Approach to Subsistence and Settlement. Unpublished PhD dissertation, Toronto: University of Toronto.

White, R., 1989. Production complexity and standardisation in Early Aurignacian bead and pendant manufacture: evolutionary implications, in *The Human Revolution: Behavioural and Biological Perspectives in the Origins of Modern Humans,* eds. P.A. Mellars & C. Stringer. Edinburgh: Edinburgh University Press, 366–90.

Zubrow, E.B.W., 1992. An interactive growth model applied to the expansion of Upper Palaeolithic populations, in *The Origins of Human Behaviour,* ed. R.A. Foley. Cambridge: Cambridge University Press, 82–96.

Appendix 11.1. *Mammals hunted for food, fur, bone, antler and ivory that are present at the three case-study areas ('ubiquitous' species have not been listed).*

Arctic: tundra and steppe		Ardennes	Dordogne	Mid. Danube
Coelodonta antiquitatis	woolly rhinoceros	x	x	x
Alopex lagopus	arctic fox	x	x	x
Dicrostonyx torquatus	collared lemming	x	-	x
Lepus timidus	mountain hare	x	x	x
Lemmus lemmus	Norway lemming	x	-	x
Megaloceros giganteus	giant deer	x	x	x
Mammuthus primigenius	mammoth	x	x	x
Ovibos moschatus	musk ox	-	x	x
Rangifer tarandus	reindeer/caribou	x	x	x

Woodland, meadows, cold to temperate				
Alces alces	elk (moose)	-	-	x
Bison priscus	wood bison	x	x	x
Bos primigenius	aurochs	x	x	x
Capreolus capreolus	roe deer	x	x	x
Cervus elaphus	red deer	x	x	x
Lepus europaeus/capensis	brown hare	-	x	x
Megaloceros giganteus	giant deer	x	x	x
Sus scrofa	wild boar	x	x	x

Living in mountains at present				
Capra ibex	ibex	x	x	x
Marmota marmota	alpine marmot	x	x	x
Oryctolagus cunniculus	rabbit	x	x	x
Ovis aries	mountain sheep	x	x	---
Rupicapra rupicapra	chamois	x	x	x

Steppe, cold or warm				
Equus hydruntinus	steppe ass	-	x	x
Equus hemionus	hemione	-	-	x
Ochotona pusilla	pika	x	-	x
Spalax sp.	mole rat	-	-	x
Saiga tartarica	Saiga antelope	x	x	x

Hunted for furs, skins				
Gulo gulo	wolverine	-	x	x
Ursus spelaeus	cave bear	x	x	x
Putorius putorius	polecat	-	x	x
Alopex lagopus	arctic fox	x	x	x
Meles meles	badger	-	-	-

Chapter 12

Neanderthals as Part of the Broader Late Pleistocene Megafaunal Extinctions?

John R. Stewart, Thijs van Kolfschoten, Anastasia Markova &
Rudolf Musil

The fate of the Neanderthals remains controversial despite more than a century of research. Most researchers agree that the Neanderthals disappeared during Oxygen Isotope Stage 3 (OIS-3), but there is no consensus regarding what happened to them (e.g. Aiello 1993; Stringer & Gamble 1993; Stringer & McKie 1996; Krings *et al*. 1997; 1999; Wolpoff & Caspari 1997; Kramer *et al*. 2001; Wolpoff *et al*. 2001). Two models dominate the dispute: the Out of Africa model and the Multiregional Evolution model, both of which have 'extreme' and 'moderate' formulations. The 'extreme' Out of Africa model contends that the Neanderthals were replaced with little or no interbreeding by migrating Anatomically Modern Humans (AMH) whose origins lay in Africa around 150–200 thousand years before present (ka BP) (Stringer & Andrews 1988). The 'extreme' Multiregional Evolution model asserts that the disappearance of the Neanderthals is illusory, and discounts migration of African modern humans as a significant factor in the appearance of the modern humans (Wolpoff 1989). It asserts instead that the Neanderthals simply evolved into the modern humans.

Within the framework of the Out of Africa model, most explanations for the disappearance of the Neanderthals have focused on their inability to compete effectively with the migrating modern humans (e.g. Mellars 1989; Mithen 1996; Klein 2000). Authors differ regarding the source of the modern human's competitive advantage. For example, some suggest that it was technological (Pettitt 1999), while others contend that it was linguistic (Lieberman 1989), cognitive (Mellars 1989; Mithen 1996; Klein 2000) or demographic (Bocquet-Appel & Demars 2000). But there is widespread agreement that competition with the modern humans was the major or perhaps sole factor in the demise of the Neanderthals. Recently,

however, Finlayson *et al*. (2000a,b) have criticized the assumption that the modern humans played a role in the extinction of the Neanderthals. Drawing on the results of a computer simulation study, these authors contend that explanations requiring competitive interactions between the Neanderthals and modern humans are unnecessary. They assert instead that the decline and extinction of the Neanderthals and the growth and dispersal of modern human populations are best explained as independent, climate-linked events. In Finlayson *et al*.'s (2000a) model, Neanderthals and modern humans had been separated ecologically for a long period, and had consequently developed different resource acquisition strategies. The Neanderthals' strategy focused on local exploitation of a variety of resources, whereas the modern humans' strategy was oriented towards long-range hunting of large herbivores. Between 60 and 20 ka BP, deterioration in climatic conditions caused changes in ecology, most notably the expansion of the so-called steppe tundra and its associated fauna, which favoured the resource acquisition strategy of modern humans and made the Neanderthals' resource acquisition strategy ineffective. Accordingly, the size of the modern human population increased leading to their migration from Africa, while the Neanderthal population declined and eventually became extinct around 30 ka.

A further reason for rejecting modern human involvement in Neanderthal extinction is the lengthy co-occurrence, up to 10 ka BP, of the two species in Europe in areas such as Cantabria. This lengthy sympatry between modern humans and Neanderthals is contested by some authors and d'Errico *et al*. (1998) and Zilhão & d'Errico (1999) reject any early dates for modern humans in Europe on the grounds of poor stratigraphy and dating. Klein (2000) dismisses

Table 12.1. *Large mammal taxa in the Stage Three Project Mammalian data base.[1] Categories: 1a) Taxa whose distribution remained the same in Europe after the Pleistocene; 1b) Taxa whose distribution contracted northwards and/or into montane areas in Europe after the Pleistocene; 1c) Taxa whose distributions contracted eastwards into Southwest Asia and Central Asia after the Pleistocene; 2a) Taxa that went extinct at the termination of the Pleistocene; 2b) Taxa that went extinct at the approach of the Glacial Maximum.*

Taxon no.	Taxon Latin	English	Status	Status category (see text)
2	*Alces alces*	Elk	Extant	1b
3	*Cervus elaphus*	Red deer	Extant	1a
4	*Rangifer tarandus*	Reindeer	Extant	1b
5	*Dama dama*	Fallow deer	Extant but reintroduced to northern Europe	2b*
6	*Capreolus capreolus*	Roe deer	Extant	1a
7	*Megaloceros giganteus*	Giant deer	Extinct[1, 2]	2a‡
8	*Bos / Bison*	Auroch / Bison	Effectively extinct in Europe	1a/2a
9	*Rupicapra rupicapra*	Chamoix	Extant	1b
10	*Capra ibex / pyrenaica*	Ibex	Extant	1b
11	*Ovibos moschatus*	Musk ox	Extant	1b
12	*Saiga tartarica*	Saiga antelope	Extinct in Europe[1]	1c
13	*Sus scrofa*	Wild boar	Extant	1a
14	*Equus ferus*	Horse	Extant	1a
15	*Equus hydruntinus*	European wild ass	Extinct[1]	1c§
16	*Equus asinus*	Donkey	Extant	1c**
17	*Equus latipes*	Extinct horse	Extinct	?
18	*Stephanorhinus kirchbergensis*	Merck's rhino	Extinct[1]	2b
19	*Stephanorhinus hemitoechus*	Narrow-nosed rhino	Extinct in Europe[1]	2a/b†
20	*Coelodonta antiquitatis*	Wooly rhino	Extinct[1]	2a
21	*Mammuthus primigenius*	Mammoth	Extinct[1]	2a
22	*Elephas antiquus*	Straight-tusked elephant	Extinct[1]	2b
23	*Panthera leo*	Lion	Extinct in Europe[1]	1c
24	*Panthera pardus*	Leopard	Extinct in Europe[1]	1c
25	*Lynx lynx*	Lynx	Extant	1a
26	*Lynx pardina*	Iberian lynx	Extant	1a Iberian endemic
27	*Felis sylvestris*	Wild cat	Extant	1a
28	*Crocuta crocuta*	Spotted hyaena	Extinct in Europe[1]	1c
29	*Canis lupus*	Wolf	Extant	1a
30	*Cuon*	Dhole	Extant	1c?
31	*Vulpes vulpes*	Red fox	Extant	1a
32	*Alopex lagopus*	Arctic fox	Extant	1b
33	*Ursus arctos*	Brown bear	Extant	1a
34	*Ursus spelaea*	Cave bear	Extinct[1]	2a
35	*Meles meles*	Badger	Extant	1a
36	*Mustela*	Stoat / Weasel	Extant	1a
37	*Putorius putorius*	Polecat	Extant	1a
38	*Martes* spp.	Marten	Extant	1a
39	*Gulo gulo*	Wolverine	Extant	1b
40	*Lutra lutra*	Otter	Extant	1a
41	*Homo neanderthalensis*	Neanderthal human	Extinct[4]	2b
42	*Homo sapiens*	Modern human	Extant	1a

Note: Extant/extinct status of taxa has been derived from literature sources. Modern distribution (extant taxa) in Europe comes from Mitchell-Jones *et al.* (1999) and Stuart (1982). Sources for extinction dates: 1 - Stuart (1991); 2 - Gonzalez *et al.* (2000); 3 - Markova *et al.* (1995); 4 - Smith *et al.* (1999).

* Fallow deer not extinct but was very much reduced in distribution to the southern and southeastern Europe by the last glaciation. They were subsequently reintroduced by Romans to Northern Europe (Mitchell-Jones *et al.* 1999).

‡ Giant deer extinct in most areas of Europe by the end of the Pleistocene. However, dates for the early Holocene exist in the Isle on Man and Scotland (Gonzalez *et al.* 2000).

** *Equus asinus* may well be a mistakenly identified taxon as it is generally believed to have arrived from Africa during the Holocene as a domesticate (Musil pers. comm.).

§ *E. hydruntinus* survived the Pleistocene/Holocene boundary but became extinct some time during the Holocene (Benecke 1999).

† *S. hemitoechus* is traditionally an 'interglacial survivor' (Stuart 1991) but current analysis questions this.

the late surviving (post 30 ka BP) Neanderthals in Europe on similar grounds. It is clear, however, that many European early modern human sites have dates that are older than 36.5 ka, and in some cases the dates are based on more than one absolute dating technique (Davies 2001). Likewise, recent dates from Vindija demonstrate that the Neanderthals survived until at least 29 ^{14}C ka BP in some parts of Europe (Smith *et al.* 1999).

It is noteworthy that the debate between those anthropologists and archaeologists who contend that the modern humans played an important role in the demise of the Neanderthals and those who view the Neanderthal's extinction as primarily the consequence of climatic deterioration-induced ecological change parallels the debate in vertebrate palaeontology regarding the Late Pleistocene megafaunal extinctions. In the latter debate two main hypotheses dominate, one involving humans and often called the prehistoric overkill theory (Martin 1967; 1984; Schuster & Schüle 2000), and one where climate is the dominant cause (Graham & Lundelius 1984; Guthrie 1984). The large size of the vertebrates that became extinct with associated relatively small population numbers and slow-breeding strategies is emphasized in both explanations. However, Guthrie (1990) contends that too much focus on the extinctions themselves has obscured the fact that other events had taken place, and that the Late Pleistocene megafaunal extinctions should be seen as part of a Late Pleistocene, climate-driven 'faunal revolution', which involved rapid [micro-] evolutionary changes, fractionation of biotic communities and significant reductions in distributional ranges.

With few exceptions (e.g. Stewart 2000), the disappearance of the Neanderthals has rarely been considered in the light of other Late Pleistocene extinctions, and the Late Pleistocene extinctions have not been viewed in the light of Neanderthal extinctions. Given Guthrie's more holistic perspective (Guthrie 1990) it is clear that such a consideration is overdue. In this paper, we therefore attempt to shed light on the role of climate change and associated ecological developments in the disappearance of the Neanderthals. We do so by analyzing the temporal and geographic distribution of large mammal faunas from European archaeological and palaeontological sites that have been absolutely dated to between 60 and 20 ka BP. This includes an analysis of some of the animals thought to have become extinct during this time frame as well as ones that went extinct after this period and ones that survived until the present.

Materials and methods

In order to investigate the role of climate change and associated ecological developments in the disappearance of the Neanderthals the Stage Three Mammalian Data Base[1] was analyzed in such a way as to test whether there was any synchrony between the demise of Neanderthals and other mammalian ecological occurrences. The data base is described in some detail in Chapter 7 (Stewart *et al.* 2003).

To better understand the Neanderthals as part of the evolving fauna of OIS-3 a historical biogeographical approach was taken. This entailed a consideration of the extent to which each taxon or taxonomic grouping listed in Table 12.1 was distributed in Europe and how that may have changed between 60 and 20 ka BP. Only the large mammal taxa were included in the analysis as they are more consistently reported.

First of all, however, the large mammals were divided into five historical biogeographical categories based on previous knowledge of their extant versus extinct status, what the modern distribution of extant taxa is and when the extinct taxa are believed to have become extinct (for references see Table 12.1):

1. *Extant taxa.* These fall into three categories consisting of:
   a) taxa whose distribution remained the same in Europe after the Pleistocene;
   b) taxa whose distribution contracted northwards and/or into montane areas in Europe after the Pleistocene;
   c) taxa whose distributions contracted eastwards into Southwest Asia and central Asia after the Pleistocene.
2. *Extinct taxa.* These fall into two categories consisting of:
   a) taxa that went extinct near the termination of the Pleistocene;
   b) taxa that went extinct at the approach of the LGM.

Table 12.1 gives the categories for each of the 41 large mammalian taxa (including the Neanderthals and modern humans) in the data base. The categories created suggested differences in tolerances of the mammals to prevailing environmental conditions and hence led one to expect different responses to the changing environments of OIS-3.

The temporal resolution of the analysis was guided by two factors regarding the practical potential of making meaningful comparisons between the absolutely dated mammalian faunas and the global

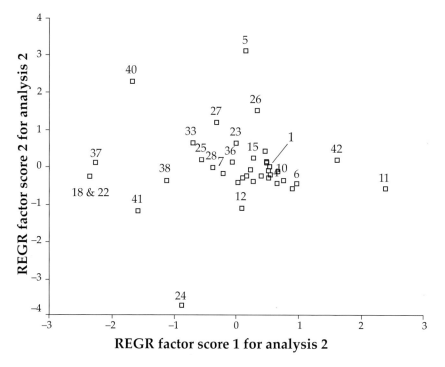

Figure 12.1. *Principal component analysis of the percentage frequency analysis through time. The numbers corresponding with each taxon are listed in Table 12.1. No. 1 represents all taxa taken together.*

spread of dates associated with all the taxa together, which gave reason for concern. The possibility existed that the overall pattern may be driving the individual patterns. This was also problematic as the length of each time zone was not equal (Early = 23 ka BP, Middle = 9 ka BP, Late = 8 ka BP) although it was hoped that the disproportionate length of the early time zone would be countered by the comparatively lack of dates falling within it. Another possibility is that their preferred association with one or other human species may dominate patterns, because the number of dates associated with Upper Palaeolithic (modern human) and Middle Palaeolithic (Neanderthal) are not identical (Table 12.2). There are, for instance, about twice as many dated faunas from Upper Palaeolithic sites than Middle Palaeolithic ones (Table 12.2). Therefore to ascertain the reliability of the results of the initial analysis percentages were calculated for the large mammals, but divided into those from Upper Palaeolithic and Middle Palaeolithic sites respectively (Table 12.3). If the trends in percentages filling each time zone observed in the undifferentiated analysis proved to be the same for the Middle Palaeolithic and the Upper Palaeolithic then the trend could be taken to be real rather than a product of their predominant archaeological association.

The analysis of taxon geographic distribution was achieved in two ways. First, the maximum and minimum latitude for the absolutely dated large mammal taxa in Europe for each of the three time zones were calculated. Then all the large mammal taxa were mapped for each of the time zones to better investigate any changes in geographical distribution through time.

A final part of the geographical analysis was an investigation into the percentage number of dated faunas associated with the Neanderthals or the Middle Palaeolithic that fall into the latitudinal and longitudinal bands represented on the maps. These bands are 5° deep for the latitude and 10° wide for the longitude. The results of this analysis are shown in Table 12.4.

climate signature. The first involved the precision of the dates in relation to the amplitude of the climatic cycles during OIS-3, while the second simply involved the problems of turning the ^{14}C dates into calendrical years. Three time zones were defined, an early, middle and late phase of which the early and late phases can certainly be compared climatically as they represent relatively warmer and colder conditions respectively.

The analysis of the faunal data had two separate parts. These were:
1. analysis of taxon frequency of occurrence in the three time periods;
2. analysis of taxon geographic distribution.
The analysis of the taxon frequency of occurrence in each of the three time periods was done by calculating the percentage number of dated faunas with a taxon in each time zone as part of the total number of dated faunas with the taxon. These percentage data were subjected to Principal Components Analysis (PCA) with and without arcsine transformation (Sokal & Rohl 1995) (Fig. 12.1).

A further analysis had to be designed, however, in order to test the reliability of the results of the percentage analysis. This is because the spread of percentages of dated faunas in each time zone with many individual taxa was similar to the overall

Table 12.2. *Breakdown of mammalian data base by archaeological association. Percentage number of dated faunas associated with each broad archaeological industry.*

Industry	No.	%
OIS-3 dated assemblages with large mammals	468	100
OIS-3 dated assemblages with Upper Palaeolithic industries and large mammals	224	47.86
OIS-3 dated assemblages with Middle Palaeolithic industries and large mammals	107	22.86
OIS-3 dated assemblages with transitional industry and large mammals	24	4.94
OIS-3 dated assemblages with large mammals and no industry	90	19.23
OIS-3 dated assemblages with large mammals and undefined industry	23	4.91

Upper Palaeolithic includes Aurignacian, Gravettian, Solutréan, Magdalenian and unspecified Upper Palaeolithic industries; Middle Palaeolithic includes Mousterian and unspecified Middle Palaeolithic industries; transitional industries includes Châtelperronian, Bohunician, Uluzzian, Szeletian, Bachokirian and Micoquian.

Table 12.3. *Percentage number of dated assemblages with large mammal taxa falling in each of the three time zones but broken down into from horizons with Middle Palaeolithic (Neanderthal) and Upper Palaeolithic (modern human) archaeology.*

Taxon	No. of dated faunas with taxon	% no. of dated faunas with taxon per time zone					
		>37 ka		37–28 ka		<28 ka	
		MP	UP	MP	UP	MP	UP
All taxa	322	48.19	19.24686	39.75904	43.51464	12.04819	37.23849
Alces alces	15	50	23.07692	50	46.15385	0	30.76923
Cervus elaphus	186	50	20	37.5	43.84615	12.5	36.15385
Rangifer tarandus	204	52.5	17.68293	42.5	40.85366	5	41.46341
Dama dama	10	66.66667	0	33.33333	75	0	25
Capreolus capreolus	71	7.692308	15.51724	61.53846	41.37931	30.76923	43.10345
Megaloceros giganteus	32	63.63636	9.52381	27.27273	61.90476	9.090909	28.57143
Bos/Bison	219	56.14035	20.98765	40.35088	44.44444	3.508772	34.5679
Rupicapra rupicapra	125	46.875	19.35484	34.375	41.93548	18.75	38.70968
Capra	149	45.45455	20.68966	36.36364	42.24138	18.18182	37.06897
Ovibos moschatus	7	33.33333	0	33.33333	50	33.33333	50
Saiga tartarica	15	100	35.71429	0	42.85714	0	21.42857
Sus scrofa	81	50	22.80702	25	49.12281	25	28.07018
Equus ferus	201	51.85185	17.68707	40.74074	42.17687	7.407407	40.13605
Equus hydruntinus	52	70	11.90476	20	47.61905	10	40.47619
Equus asinus	3	None	33.33333	None	33.33333	None	33.33333
Equus latipes	4	None	50	None	50	None	0
Stephanorhinus kirchbergensis	5	25	100	50	0	25	0
Stephanorhinus hemitoechus	9	50	0	50	33.33333	0	66.66667
Coelodonta antiquitatis	67	62.5	21.56863	31.25	45.09804	6.25	33.33333
Mammuthus primigenius	111	55	20.87912	35	46.15385	10	32.96703
Elephas antiquus	5	50	66.66667	50	33.33333	0	0
Panthera leo	54	46.15385	21.95122	46.15385	51.21951	7.692308	26.82927
Panthera pardus	14	70	50	10	25	20	25
Lynx lynx	20	66.66667	23.52941	33.33333	52.94118	0	23.52941
Lynx pardina	13	28.57143	16.66667	57.14286	50	14.28571	33.33333
Felis sylvestris	32	26.66667	35.29412	53.33333	47.05882	20	17.64706
Crocuta crocuta	94	57.14286	23.72881	37.14286	52.54237	5.714286	23.72881
Canis lupus	148	42.85714	22.12389	45.71429	43.36283	11.42857	34.51327
Cuon	10	40	0	40	60	20	40
Vulpes vulpes	156	45.45455	17.88618	42.42424	43.90244	12.12121	38.21138
Alopex lagopus	87	46.15385	20.27027	30.76923	48.64865	23.07692	31.08108
Ursus arctos	42	41.17647	32	47.05882	56	11.76471	12
Ursus spelaea	96	51.72414	23.8806	37.93103	46.26866	10.34483	29.85075
Meles meles	27	66.66667	19.04762	16.66667	47.61905	16.66667	33.33333
Mustela	28	33.33333	24	66.66667	48	0	28
Putorius putorius	11	100	44.44444	0	55.55556	0	0
Martes	14	50	30	50	60	0	10
Gulo gulo	36	0	29.41176	50	47.05882	50	23.52941
Lutra lutra	3	0	50	100	50	0	0

Table 12.4. *Percentage number of dated assemblages with Neanderthals or the Middle Palaeolithic found in latitudinal and longitudinal bands in the three time OIS-3 time zones.*

Latitude

	35–40	40–45	45–50	50–55
Early	7.87	29.21	57.3	5.62
Middle	15.38	30.77	48.08	5.77
Late	17.65	29.41	47.06	5.88

Longitude

	–10–0	0–10	10–20	20–30	30–40	40–50
Early	17.98	44.94	21.35	8.99	4.49	2.25
Middle	21.15	53.85	15.38	3.85	1.92	3.85
Late	29.41	52.94	11.76	5.88	0	0

Results

The two-part analysis of the OIS-3 large mammal fossil data was designed to distinguish any changes in temporal distribution of taxon frequency and any changes of geographical range of the large mammalian taxa of OIS-3.

The percentages derived from the number of dated faunas with each taxon in the three time zones defined were then subjected to principle component analysis (PCA). Figure 12.1 presents the two principal components extracted from the percentage data in the PCA. Principal component 1 (PC1) accounted for 59.4 per cent of the variance in the data set, and PC2 for 40.6 per cent. The PCA results shown in Figure 12.1 have a PC1 which describes the relationship between the early and late time zones of OIS-3. As PC1 increases so the percentage for the early time zone decreases while the late time zone increases. PC2, on the other hand, describes the middle time zone percentage in relation to the early and late time zones. As PC2 increases so the middle time zone percentage increases while the early and late time zones decrease. Together, Figure 12.1 and Table 12.3 indicate that there are three main patterns of temporal distribution among the percentage data. The first, are those mammals that do not appear to change their occurrence through time and comprise the largest group. These mammals can only approximate an even spread though time. This is because the percentage occurrence of all mammals in this study shows a 'humped' distribution through time, with 29.29 per cent in the early zone, 41.21 per cent for the middle one and 29.5 per cent for the late zone. The mammals falling in this first cluster include extant taxa whose distribution did not change at the end of the Pleistocene, those that retreated north or east and some that became extinct around the end of the Pleistocene. They include herbivorous taxa such as

reindeer *Rangifer tarandus*, horse *Equus ferus* and mammoth *Mammuthus primigenius* as well as carnivorous mammals including wolf *Canis lupus*, red fox *Vulpes vulpes* and arctic fox *Alopex lagopus*. These taxa can be seen to cluster around the point representing all taxa in Figure 12.1 and have a relatively 'humped' distribution, although this may be a function of the humped data set.

The second group includes the taxa that decreased significantly over OIS-3 and includes those that became extinct in the first phase of the Late Pleistocene megafaunal extinctions together with an assortment of other mammals. This group includes mostly carnivores like the leopard *Panthera pardus*, the polecat *Putorius putorius* and the martens *Martes* spp., larger herbivores such as *Stephanorhinus kirchbergensis*, *Elephas* (*Palaeoloxodon*) *antiquus* as well as the Neanderthal. In Figure 12.1 they fall to the left of the main central cluster, signifying that they have low percentage values for the late time zone relative to the early zone. Among these the carnivores are all extant while the herbivores and Neanderthals are now extinct or absent from Europe. Most important in this category are two of the 'interglacial survivors'; a term first used by Stuart (1991), i.e. *Stephanorhinus kirchbergensis* and *Elephas* (*Palaeoloxodon*) *antiquus*. These last two species belong to the group in Figure 12.1 plotting furthest in the negative for PC1 (Nos. 18 and 22 respectively) to have values of 0 per cent in the last time zone which distinguishes them from the cluster with Neanderthals that falls to around 10 per cent (No. 42). It appears, however, that one should add the Neanderthals to the extinct group as illustrated by their percentage decrease which goes from 56.6 per cent to 32.7 per cent and finally to 10.69 per cent. This decrease is not solely because the decrease in the number of sites with Neanderthal fossils, as dates associated with Middle Palaeolithic archaeology are also included which augment the sample of dates for these hominids. It is not surprising that the 'interglacial survivors' should decrease in the percentage number of dates associated with them through OIS-3, as they are believed to have become extinct towards the end of this time. The temporal frequency pattern for *Stephanorhinus hemitoechus* was unexpected as the species is often said to have become extinct in Europe during the first phase of megafaunal extinctions (Stuart 1991). The results of the analysis, however, show that they appreciably increase in percentage through time. This may signify that this species survived into the Late Glacial after all. The fact that of the two *Stephanorhinus* species it was the specialist browser *S. kirchbergensis* rather than the grazer *S. hemitoechus* (Loose 1975) that became extinct in the early phase may be of significance as the megafaunal grazers of the Late Pleistocene mostly died out at the end of the Pleistocene when the grasslands were particularly affected.

The other taxa whose percentages appear to decrease over OIS-3 are more surprising, such as the leopard *Panthera pardus*, the otter *Lutra lutra*, the polecat *Putorius putorius* and the martens *Martes* spp. It should be pointed out that carnivores form the major group that have a negative deviation relative to the pattern for all other mammalian taxa. There is also a slight decrease in all large mammal taxa towards the LGM, if we consider the two later time zones and that may be a further reflection of general environmental impoverishment. The latter may not be the case, however, as the pattern of decline in percentage numbers of sites between the intermediary time zone and the late one mirrors that for all the sites with fossil mammals taken together.

The only taxa that reliably increased their percentages markedly are the musk ox *Ovibos moschatus* and the modern humans *Homo sapiens*. The increase in percentage in the musk ox may be explained by their modern northern distribution and tolerance of cold temperatures (Mitchell-Jones *et al.* 1999). In Figure 12.1 these taxa have the highest values for PC1. It is interesting that none of the other taxa that might be considered to be cold adapted, such as the mammoth *Mammuthus primigenius*, show a similar trend and this may be because none were as cold tolerant as the musk ox. The final pattern to note is where taxa increased and then decreased. These include the fallow deer *Dama dama*, and the otter *Lutra lutra* can perhaps be disregarded, as the number of sites with these species is low, making it dangerous to draw reliable conclusions from them.

It became apparent, however, that a possible bias may have occurred due to the association of specific archaeological industries with specific large mammal taxa. This is the reason behind Table 12.3 where the percentage number of dated assemblages with species showing changes in each time zone was calculated, but divided into those from either Upper or Middle Palaeolithic archaeological sites. The results in Table 12.3 confirm that taxa such as the red deer *Cervus elaphus*, reindeer *Rangifer tarandus*, horse *Equus ferus* and mammoth *Mammuthus primigenius* had percentage patterns for the Middle and Upper Palaeolithic that correspond reasonably well with the patterns of all the mammals associated with each industry (Table 12.3). Of those taxa that showed a

decrease through the time zones in the archaeologically undifferentiated analysis the extinct rhino *Stephanorhinus kirchbirgensis*, the straight-tusked elephant *Elephas antiquus*, the otter *Lutra lutra*, the polecat *Putorius putorius* and the martens *Martes* spp. show the same pattern in the percentages from Upper Palaeolithic as Middle Palaeolithic sites (Table 12.3). It should be added, however, that the percentage number of faunas with wild cat *Felis sylvestris* and brown bear showed a marked decrease towards the LGM on Upper Palaeolithic sites suggesting that they too suffered a decline with time. Finally, the increase in dated faunas with musk ox *Ovibos moschatus* cannot be proved. The result of the test in Table 12.3 shows that, by and large, the picture revealed by the archaeologically undifferentiated analysis was substantiated.

If the results of the temporal distribution analysis in Figure 12.1 and Table 12.3 are associated with climatic and vegetational deterioration through OIS-3, it is likely that the geographical distribution of these mammals in Europe would have changed. Therefore the maximum and minimum latitudes for each taxon in the three time zones of OIS-3 were calculated. The main observations to be made from this analysis was that on the whole very few taxa appear to change either their maximum or minimum latitude. The changes that can be observed are difficult to interpret without knowing more exactly where all the sites are distributed in each time zone. The most interesting observation to be made from the maximum and minimum latitudes over-all is that some species never reach very far north while others never reach far to the south. The taxa which generally have more northern latitudes are those such as the mammoth *M. primigenius*, the reindeer *R. tarandus* and the arctic fox *A. lagopus* which never reached further south than 42° and approximates with northern Spain, northern Italy and the more northern parts of the Balkans (see Fig. 12.2 for latitude lines). These taxa are ones that belong in category 1b, extant taxa whose ranges today are northern and montane (although the montane species ibex *Capra* spp. and chamoix *Rupicapra* do not follow this geographical pattern) and ones from category 2a, taxa that became extinct at the end of the Pleistocene. Then there are those taxa whose distribution never reached further north of 46° or the latitude of the northern parts of the Dordogne. These include the three 'interglacial survivors', *Stephanorhinus kirchbergensis*, *S. hemitoechus* and *Elephas antiquus* as well as the endemic Iberian lynx *Lynx pardina* and the fallow deer *Dama dama*. Finally, there are taxa that

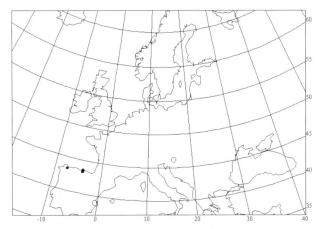

Figure 12.2. *Map of taxon* Stephanorhinus kirchbirgensis *that went extinct at the approach of the Late Glacial Maximum. Small solid circle - Early time zone; medium circle - Intermediate time zone; large circle - Late time zone.*

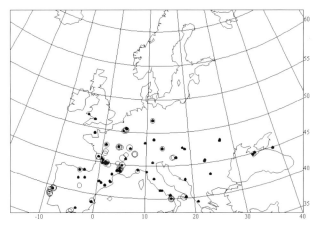

Figure 12.3. *Map of taxon* Homo neanderthalensis *that went extinct at the approach of the Late Glacial Maximum. Small solid circle - Early time zone; medium circle - Intermediate time zone; large circle - Late time zone.*

are distributed throughout Europe such as the red deer *C. elaphus*; the horse *E. ferus* and the wolf *Canis lupus*, all taxa that fall within category 1a. The northern and southern taxa have a latitudinal band of overlap between 42° and 46°. This band of overlap encompasses important areas of dense mammalian fossil and archaeological site occurrence such as the Dordogne and the Rhone Valley (Davies *et al.* in press).

In order to further investigate any geographical change in occurrence of taxa, maps were constructed for all the large mammals in the Stage Three Data Base. The results show that the red deer *Cervus elaphus* (Category 1a), the reindeer *Rangifer tarandus* (Category 1b) and the saiga *Saiga tatarica* (Category 1c) do not appear to change their distributions through OIS-3.

There are also extinct taxa such as the mammoth (Category 2a) whose distribution is restricted in the south, although it is found in Scandinavia in greater numbers than any other mammal in the Stage Three Project Mammalian Data Base. There are no apparent changes in the distributions of the mammoth through OIS-3 although recent evidence shows that they were absent from Europe during the LGM (Stuart *et al.* 2002).

The final category of mammals contains those which are at the heart of the main argument in this paper and are those which became extinct at the end of OIS-3. They are the straight-tusked elephant, the Merck's rhino and the Neanderthals. As can be seen from Figures 12.2 and 12.3 they all appear either to be retreating or to have retreated to the south and to some extent to the west and south in the case of the

Neanderthals. The straight-tusked elephant and the Merck's rhino are absent by the last time zone while Neanderthals appear to survive, although in fewer numbers. The Neanderthals also seem to be retreating into refugia in more northern areas such as the Belgian Ardennes, their most northern haunt at the end of OIS-3. This may be explained by the refugium effect described recently by Stewart & Lister (2001). These taxa seem to be contracting towards warmer climes as the climate deteriorates.

Table 12.4 was constructed to better test the southwards and westwards retreat of Neanderthals through OIS-3. It shows the percentage number of dated assemblages with Neanderthals or their Middle Palaeolithic industries in each latitudinal and longitudinal band. The results for the longitude are clearer than those for the latitude as there is a clear and progressive retreat from the east towards the west through the three time zones (Chapter 4: van Andel *et al.* 2003). The latitude analysis shows that while there is a clear northward decline with fewer sites in the north the Neanderthals manage to persist in specific areas to the last time zone. This could be explained by the northern refugium effect described above. There is some indication that the Neanderthals retreated south as the 35°–40° latitudinal band can be seen to increase though time at the expense of the 45°–50° band.

Conclusion

The larger mammals in the Stage Three Project Data Base were classified at the outset into a number of historical biogeographical categories. These categories are first based on whether the taxa are globally extinct today as opposed to extant. Within the extinct category there are two groups including taxa that became extinct towards the Glacial Maximum earlier than *c*. 20 ka BP, and ones that became extinct around the end of the Pleistocene *c*. 10 ka BP. Within the extant category there are those taxa whose distributions have not significantly changed since the Late Pleistocene as a whole, those whose distributions contracted north or up into more montane regions such as the Alps, and taxa whose distributions retreated eastwards and toward Southwest Asia. These large mammals were then subjected to two analyses to further investigate their temporal and geographical patterns during OIS-3. This led to the following main conclusions:

1. The Neanderthals are most similar to the 'interglacial survivors' in terms of their frequency and geographic distributions throughout Oxygen Isotope Stage 3.
2. The results of the analysis also suggest that the biomass declined towards the LGM.

As we hope we have demonstrated there is much merit in viewing the extinction of the Neanderthals from the perspective of an evolving biosphere with particular emphasis on their fellow mammals. An *a-priori* classification of the large mammals of OIS-3 immediately suggests that the Neanderthals belong with the megafauna that became extinct in the first part of the Late Pleistocene extinction event, known as the extinction of the 'interglacial survivors'. Historical biogeographical analyses conducted further reveal that Neanderthals decrease in numbers towards the LGM and retreat westwards and possibly southwards throughout OIS-3 (for practical purposes in this paper between 60 and 20 ka BP). This conclusion appears to support a case for the extinction of the Neanderthals, *Elephas antiquus* and *Stephanorhinus kirchbergensis* as a result of the increasingly cold global temperatures. The analysis of the other mammals yielded the unexpected result that many of the carnivores, in particular, were decreasing in numbers towards the LGM. This is further supported by evidence that the mammoth was doing likewise (Stuart *et al*. 2002), suggesting a general decrease in the environment's carrying capacity at this time. The latter adds a further layer of complexity to the argument regarding the extinction of the 'interglacial survivors', including the Neanderthals, as it may be that it was not simply cold temperatures that were to blame but a series of ecological repercussions caused by the increasing cold of the LGM.

It should be stated that it may not be necessary to invoke a role for modern humans in the demise of the Neanderthal and that a solely climatically driven extinction mechanism may be sufficient. Neanderthals appear to be a European endemic species that evolved on the western fringes of the Eurasian landmass during the Middle Pleistocene, presumably from an African ancestor that may have been *Homo heidelbergensis* (Hublin 1998). The period during which they evolved seems to have been dominated by continental phases of climate during which the Atlantic Ocean was influencing the European landmass to a far lesser extent than it is today. In relation to today's climate Western Europe was more continental, but in relation to more central parts of Eurasia at the time it must have been relatively oceanic. Therefore it may be that the *Homo heidelbergensis* / Neanderthal lineage became particularly well-adapted to the western fringes of the Eurasian landmass which we would term Western Eurasian continental-adapted rather than simply continental-adapted. These requisite conditions began to unravel towards the LGM, immediately before the rest of the animals that had become adapted to such dominant conditions became extinct or retreated to their present ranges at the end of the Pleistocene.

Acknowledgements

We wish to acknowledge the Leverhulme Trust and the AHRB Centre for the Evolutionary Analysis of Cultural Behaviour, University College London, both of which funded JS's contribution to this paper. Tjeerd van Andel is thanked for providing the fertile environment of the Stage Three Project that led to the central idea in this paper. William Davies is thanked for helping collect the dates in the mammalian fossil data base as is the McDonald Institute, Cambridge University that partially funded him. Bernie Weninger and Olaf Jöris are thanked for providing their CALPAL program that calibrated the ^{14}C dates. Mark Collard is acknowledged for suggesting the uses of percentages in the analysis of the temporal distribution of taxa through OIS-3. We would also like to acknowledge the help of Margaret Clegg in performing the Principle Component Analysis. Finally the following are also thanked for numerous helpful discussions: T. van Andel, W. Davies, C.B. Stringer, L. Aiello, P. Gollop, C. Gamble, B. Huntley, A. Lister, A. Stuart, S. Parfitt, J.-L. de Beaulieu, J. Allen and K. van Huissteden.

Note

1. Stewart, J.R., M. van Kolfschoten, A. Markova & R. Musil, 2001. *Stage Three Project Mammalian Data Base.* http://www.esc.esc.cam.ac.uk/oistage3/Details/Homepage.html.

References

Aiello, L.C., 1993. The fossil evidence for the origins of modern humans. *American Anthropology* 95, 73–96.

Benecke, N. (ed.), 1999. *The Holocene History of the European Vertebrate Fauna: Modern Aspects of Research.* (Workshop, 6–9th April 1998.) (Archaeologie in Eurasien; Band 6.) Berlin: Rahden/Westf., Leidorf.

Bocquet-Appel, J.-P. & P.Y. Demars, 2000. Neanderthal colonization and modern human colonization of Europe. *Antiquity* 74, 544–52.

Davies, S.W.G., 2001. A very model of a modern human industry: new perspectives on the origin and spread of the Aurignacian in Europe. *Proceedings of the Prehistoric Society* 67, 195–217.

Davies, S.W.G., T.H. van Andel & J.R. Stewart, in press. *An Overview of the Results of the Stage Three Project: Hominids in the Landscape.* (CALPE 2001.) Chicago (IL): Chicago University Press.

d'Errico, F., J. Zilhão, M. Julien, D. Baffier, & J. Pelegrin, 1998. Neanderthal acculturation in western Europe? A critical review of the evidence and its interpretation. *Current Anthropology* 39 (Suppl.), S1–S44.

Finlayson, C., R.N.E. Barton, F. Giles Pacheco, G. Finlayson, D.A. Fa, A.P. Currant & C.B. Stringer, 2000a. Human occupation of Gibraltar during Oxygen Isotope Stages 2 and 3 and a comment on the late survival of Neanderthals in Southern Iberian Peninsula, in *Paleolitico da Peninsula Iberrica.* (Actas do 3.º Congresso de Arqueologia Peninsular 2.) Porto: ADECAP 2000, 277–86.

Finlayson, C., D.A. Fa & G. Finlayson, 2000b. Biogeography of human colonisations and extinctions in the Pleistocene. *Memoirs Gibcemed* 1, 1–69.

Gonzalez, S., A.C. Kitchener & A.M. Lister, 2000. Survival of the Irish elk into the Holocene. *Nature* 405, 753–4.

Graham, R.W. & E.L. Lundelius Jr, 1984. Co-evolutionary disequilibrium and Pleistocene extinctions, in Martin & Klein (eds.), 223–49.

Guthrie, R.D., 1984. Mosaics, allelochemics and Nutrients: an ecological theory of Late Pleistocene Megafaunal extinctions, in *Quaternary Extinctions: a Prehistoric Revolution*, eds. P.S. Martin & R.G. Klein. Tucson (AZ): University of Arizona Press, 259–98.

Guthrie, R.D., 1990. Late Pleistocene faunal revolution – a new perspective on the extinction debate, in *Megafauna and Man: Discovery of America's Heartland*, eds. L.D. Agenbroad, J.I. Mead & L.W. Nelson. Flagstaff (AZ): North Arizona Press, 42–53.

Hublin, J.-J., 1998. Climatic changes, paleogeography, and the evolution of the Neanderthals, in *Neanderthals and Modern Humans in Western Asia*, eds. T. Akazawa, K. Aoki & O. Bar-Yosef. New York (NY): Plenum Press, 295–310.

Klein, R.G., 2000. Archaeology and the evolution of human behavior. *Evolutionary Anthropology* 9, 17–36.

Kramer, A., T.L. Crummett & M.H. Wolpoff, 2001. Out of Africa and into the Levant: replacement or admixture in Western Asia. *Quaternary International* 75, 51–63.

Krings, M., A. Stone, R.W. Schmitz, H. Krainitzki, M. Stoneking & S. Pääbo, 1997. Neanderthal DNA sequences and the origin of modern humans. *Cell* 90, 19–30.

Krings, M., H. Geisert, R.W. Schmitz, H. Krainitzki & S. Pääbo, 1999. DNA sequence of the mitochondrial hypervariable region II from the Neanderthal type specimen. *Proceedings of the National Academy of Science of the USA* 96, 5581–5.

Lieberman, P., 1989. The origins of some aspects of human language and cognition, in *The Human Revolution: Behavioural and Biological Perspectives in the Origins of Modern Humans*, eds. P. Mellars & C.B. Stringer. Edinburgh: Edinburgh University Press, 391–414.

Loose, H., 1975. Pleistocene Rhinocerotidae of W. Europe with reference to the recent two-horned species of Africa and S.E. Asia. *Scripta Geologica* 33, 1–59.

Martin, P.S., 1967. Pleistocene overkill, in *Pleistocene Extinction: the Search for a Cause*, eds. P.S. Martin & H.E. Wright. Yale (CT): Yale University Press, 75–120.

Martin, P.S., 1984. Pleistocene overkill: a global model, in Martin & Klein (eds.), 354–403.

Markova, A.K., N.G. Smirnov, A.V. Kozharinov, N.E. Kazantseva, A.N. Simakova & L.M. Kitaev, 1995. Late Pleistocene distribution and diversity of mammals in northern Eurasia. *Paleontologia I Evolucio* 28–9, 5–143.

Mellars, P., 1989. Technological change at the Middle-Upper Palaeolithic transition: economic, social and cognitive perspectives, in *The Human Revolution: Behavioural and Biological Perspectives in the Origins of Modern Humans*, eds. P. Mellars & C.B. Stringer. Edinburgh: Edinburgh University Press, 338–65.

Mitchell-Jones, A.J., G. Armori, W. Bogdanowicz, B. Kryštufek, P.J.H. Reijnders, F. Spitzenberger, M. Stubbe, J.B.M. Thissen, V. Vohraľík, & J. Zima, 1999. *The Atlas of European Mammals.* London: The Academic Press.

Mithen, S., 1996. *Prehistory of the Mind.* London: Thames & Hudson Ltd.

Pettitt, P.B., 1999. Disappearing from the world: an archaeological perspective on Neanderthal extinction. *Oxford Journal of Archaeology* 18, 217–40.

Schuster, S. & W. Schüle, 2000. Anthropogenic causes, mechanisms and effects of Upper Pliocene and Quaternary extinctions of large vertebrates. *Oxford Journal of Archaeology* 19, 223–39.

Smith, F.H., E. Trinkaus, P.B. Pettitt, I. Karavanic & M. Paunic, 1999. Direct radiocarbon dates for Vindija and Velika Pecina Late Pleistocene hominid remains.

Proceedings of the National Academy of Science of the USA 96, 12,281–6.

Sokal, R.R. & F.J. Rohl, 1995. *Biometry.* 3rd Edition. New York (NY): Freeman.

Stewart, J.R., 2000. The Fate of the Neanderthals – a special case or simply part of the broader Late Pleistocene megafaunal extinctions? *Final Program and Abstracts. 6ᵗʰ Annual Meeting. European Association of Archaeologists.* Lisbon.

Stewart, J.R. & A.M. Lister, 2001. Cryptic northern refugia and the origins of modern biota. *Trends in Ecology and Evolution* 16, 608–13.

Stewart, J.R., T. van Kolfschoten, A. Markova & R. Musil, 2003. The mammalian faunas of Europe during Oxygen Isotope Stage Three, in *Neanderthals and Modern Humans in the European Landscape during the Last Glaciation,* Chapter 7, eds. T.H. van Andel & W. Davies. (McDonald Institute Monographs.) Cambridge: McDonald Institute for Archaeological Research, 103–30.

Stringer, C. & P. Andrews, 1988. Genetic and fossil evidence for the evolution of modern humans. *Science* 239, 1263–8.

Stringer, C. & C. Gamble, 1993. *In Search of the Neanderthals.* New York (NY): Thames & Hudson.

Stringer, C. & R. McKie, 1996. *African Exodus: the Origins of Modern Humanity.* London: Pimlico.

Stuart, A.J., 1982. *Pleistocene Vertebrates in the British Isles.* London: Longman.

Stuart, A.J., 1991. Mammalian extinction in the Late Pleistocene of Northern Eurasia and North America. *Biological Review* 66, 453–562.

Stuart, A.J., L.D. Sulerzhitsky, L.A. Orlova, Y.V. Kuzmin & A.M. Lister, 2002. The latest woolly mammoths (*Mammuthus primigenius* Blumenbach) in Europe and Asia: a review of the current evidence. *Quaternary Science Reviews* 21, 1559–69.

van Andel, T.H., W. Davies & B. Weninger, 2003. The human presence in Europe during the Last Glacial period, I: human migrations and the changing climate, in *Neanderthals and Modern Humans in the European Landscape during the Last Glaciation,* Chapter 4, eds. T.H. van Andel & W. Davies. (McDonald Institute Monographs.) Cambridge: McDonald Institute for Archaeological Research, 31–56.

Wolpoff, M., 1989. Multiregional evolution: the fossil alternative to Eden, in *The Human Revolution: Behavioural and Biological Perspectives in the Origins of Modern Humans,* eds. P. Mellars & C.B. Stringer. Edinburgh: Edinburgh University Press, 62–108.

Wolpoff, M.H. & R. Caspari, 1997. *Race and Human Evolution: a Fatal Attraction.* New York (NY): Simon & Schuster.

Wolpoff, M.H., J. Hawks, D.W. Frayer & K. Hunley, 2001. Modern human ancestry at the peripheries: a test of the replacement theory. *Science* 291, 293–7.

Zilhão, J. & F. d'Errico, 1999. The chronology and taphonomy of the earliest Aurignacian and its implications for the understanding of Neanderthal extinction. *Journal of World Archaeology* 13, 1–68.

Chapter 13

Climatic Stress and the Extinction of the Neanderthals

Chris Stringer, Heiko Pälike, Tjeerd H. van Andel, Brian Huntley, Paul Valdes & Judy R.M. Allen

The Neanderthals evolved in Europe during the Middle Pleistocene (Stringer & Gamble 1993; Arsuaga *et al.* 1997; Hublin 1998; Stringer 1998), and appear to have been the sole human occupants of Europe from more than 200 ka until about 40 ka, when archaeological and fossil evidence of the additional presence of modern humans begins to appear. The advent and wide application of accelerator radiocarbon dating has provided a working framework for study of the period between about 30–40 ka BP, when the last Neanderthals and early modern humans (Cro-Magnons) may have co-existed in western Eurasia (Stringer & Grün 1991). Accelerator radiocarbon dating gives good precision, but has serious problems of accuracy compared with calendar years over this period (Beck *et al.* 2001; Jöris & Weninger 1998; 2000; CalPal 2003). Nevertheless, several Neanderthal fossils and sites have been dated to between about 28–30,000 ^{14}C uncal. bp. These include two Neanderthal fossils from Vindija, Croatia (29,080±400 ^{14}C uncal. bp, 28,020±360 ^{14}C uncal. bp: Smith *et al.* 1999) and a child's skeleton from Mezmaiskaya, Caucasus (29,195±965: Ovchinnikov *et al.* 2000). There are also accelerator or conventional dates on associated charcoal or fauna for presumed late Neanderthal occupation sites such as Zafarraya Spain (29,800±600, 31,800±550: Hublin *et al.* 1995) and Buran-Kaya III, Crimea (28,520±60 ^{14}C uncal. bp, 28,840±460 ^{14}C uncal. bp: Pettitt 1999). Despite problems of accuracy and possible contamination (Pettitt & Pike 2001), these dates suggest that the Neanderthals were still widespread in western Eurasia until about 30 ka BP, yet they apparently disappeared shortly afterwards.

Over the last few years, many explanations for the disappearance of the Neanderthals have been proposed. Some of these ideas are reworkings of old hypotheses, but novel suggestions based on palaeoclimatic or palaeoecological factors have also emerged. For some researchers, the Neanderthals never became completely extinct. They either evolved into the Cro-Magnons (Brace 1995) or were genetically absorbed into their populations through interbreeding (Smith 1994; Zilhão & Trinkaus 2003). However, for those workers who accept that Neanderthal extinction did occur (give or take a minimal amount of gene flow), a variety of possible scenarios have been proposed. These include single-cause theories, and others that postulate multiple factors. Of the single-cause models, several focus on Neanderthal–Cro-Magnon interaction or competition, or Cro-Magnon behavioural advantage (Stringer & Gamble 1993; Mellars 1999). Zubrow (1989) demonstrated how a small demographic advantage of modern humans over Neanderthals in factors such as birth rate, life span or mortality rate could have led to Neanderthal extinction in a time period as short as a millennium. Skinner (1997) showed how even a difference in weaning time could have produced such a reproductive advantage for the Cro-Magnons. Flores (1998) presented a competitive exclusion model for Neanderthal extinction, with a small difference in mortality rate. Even in the face of additional migration Neanderthal extinction would, according to him, have been mathematically inevitable. Gat (1999) revived the idea that inter-population conflict may have been responsible for Neanderthal extinction. He argued that the larger group sizes of the Cro-Magnons might have given them a decisive advantage over the smaller social groupings of the Neanderthals in territorial warfare. Differences in social structure were also implicated by Soffer (1994) in Neanderthal replacement, while Richards *et al.* (2001) found evidence for increased dietary breadth in the Cro-Magnons. Van Blerkom & McGowan (1999) suggested that incoming modern humans may have introduced infectious diseases to which the Neanderthals had little resistance. As an example of multifactored models, Stringer & Grün (1991) ar-

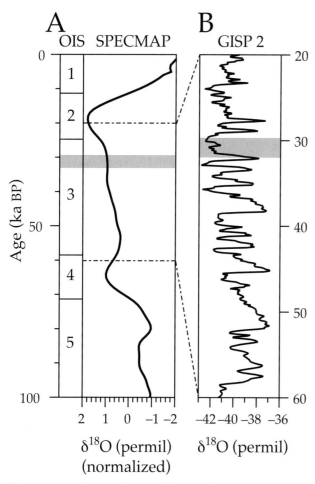

A OIS SPECMAP B GISP 2

δ¹⁸O (permil) (normalized)   δ¹⁸O (permil)

Figure 13.1. *Oxygen isotope history of the last glacial-interglacial cycle (after van Andel 2002).*
A) SPECMAP marine record. B) Millennial scale fluctuations recorded by the GISP2 Greenland ice core for the interval 20–60 ka BP. Note that 1A indicates that Neanderthal extinction (~30 ka BP, marked in grey) preceded the last glacial maximum by several millennia, and apparently occurred during a time of relative climatic stability, whereas the more detailed record of 1B shows much greater complexity around this time.

gued that Neanderthals suffered gradual population attrition in the face of displacement and fluctuating climates and food supplies. Bocquet-Appel & Demars (2000) attempted to model the phased replacement of Neanderthals by Cro-Magnons through inferred archaeological proxies, using the Stage 3 data base, but the approach was considered overambitious by Pettitt & Pike (2001).

Until recently, few models of Neanderthal extinction have implicated environmental or climatic change as major factors in their demise. This is partly because it has long been recognized that the Neanderthals evolved and survived successfully through the

fluctuating environments of later Pleistocene Europe, and their extinction around 30 ka BP preceded the glacial maximum of Oxygen Isotope Stage 2 (OIS-2) by several millennia (Fig. 13.1a). Moreover, the Neanderthals are generally regarded as having been physically cold adapted (Steegman *et al.* 2002), and thus well able to have survived the impact of glacial conditions, as they had previously (Chapter 9: Aiello & Wheeler 2003). But with the recent advent of detailed palaeoclimatic records from terrestrial and marine sediments, other models have introduced more palaeoenvironmental or palaeoecological considerations into discussions of Neanderthal extinction (Stringer & Davies 2001). Faria (2000) applied economic analysis to Neanderthal–Cro-Magnon interaction by modelling how incoming modern humans impacted Neanderthal exploitation of the renewable resources of their habitats. Neanderthals could have co-existed successfully in a 'survival trap', but this state of equilibrium would have been unstable and subject to perturbation e.g. by climatic change. Zilhão (e.g. Zilhão 2001; Zilhão & Trinkaus 2003) has argued that replacement was a complex process where Neanderthals could have been outbred by Cro-Magnons in less favourable environments such as central Europe, where they ceded territory, but in core areas such as southern Iberia they maintained their numbers until further climatic deterioration led to Cro-Magnon introgression, followed by a merging of the populations. Stewart and colleagues (Chapter 12, Stewart *et al.* 2003) have argued that Neanderthal extinction should not be treated as a unique event, but instead should be seen in the context of other megafaunal extinctions occurring during the later Pleistocene. The Neanderthals went the way of other 'interglacial survivors' such as *Elephas antiquus* and *Stephanorhinus kirchbergensis* during OIS-2–4, with no necessary influence from the presence of the Cro-Magnons. Finlayson *et al.* (2000) proposed that increasing fragmentation of the landscape due to cyclical post-OIS-5 climatic deterioration led to Neanderthal population decline and extinction, and this extinction was independent of economic competition from the Cro-Magnons. Indeed, the Cro-Magnons may only have entered parts of Europe *after* the Neanderthals had become extinct, but they were perhaps preadapted to the increasingly arid and open environments by their social systems and subsistence preferences. Langbroek (2001) rather similarly argued that rapid rates of environmental change during OIS-3 fortuitously favoured the Cro-Magnons, not through any inherent 'superiority' on their part, but because their subsistence systems happened to suit the prevailing conditions of that particular time better than those of the Neanderthals. A contrary view was taken by d'Errico & Sánchez-Goñi (2003), who argued that increasing aridity approaching the OIS-3/2 transition may have served to exclude early modern humans from southern Iberia, allowing the Neanderthals to persist longer. These views form part of a general re-evaluation of Neanderthals in the light of evidence that they displayed some, at least, of the behavioural repertoire previously thought to be exclusive to modern humans (Hublin *et al.* 1996; Zilhão 2001; Zilhão & Trinkaus 2003).

Modelling climatic stress

In this paper we focus on the possible role of 'climatic stress', in order to assess whether this could have contributed to an extinction event prior to the Last Glacial Maximum. 'Climatic stress' as we model it here is not the direct physiological effect of extreme climate on an organism, but rather the indirect effects caused

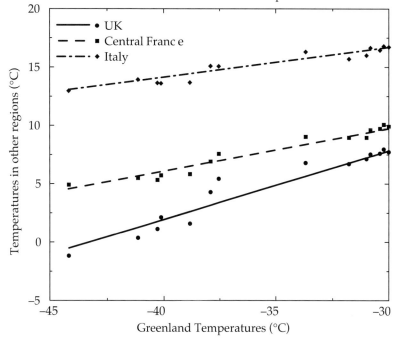

Figure 13.2. *Model of predicted relationship between Greenland temperatures and the temperature in various parts of Europe, based on a set of climate model simulations for the last 21 ka at 1 ka intervals (except 20 ka BP, 19 ka BP and 2 ka BP which lack data). Example grid points are in the UK, Central France and Italy.*

by environmental change and perturbation. Our climatic stress hypothesis considers a very low order model with only two contributing factors: changes in the absolute local temperature, and the rate of change of temperature. The detailed processes that link these two environmental factors to the distribution of a given species are likely to be complicated, as climatic change influences a variety of ecological processes (Stenseth *et al.* 2002). For example, absolute temperatures will exert direct evolutionary pressure during the lifetime of an organism, e.g. by influencing the base-line conditions for the fauna and flora that form the local habitat and resources. The situation is more complex for rates of temperature change. Pressure will nevertheless be exerted through changes in the supporting flora and habitat, as rates of temperature change will control how quickly adaptations to new food and habitat resources can be made. Periods of rapid change can also destabilize local ecologies, reducing the carrying capacity of the environment until stability is re-established. In particular, rapidly changing and less productive glacial/stadial environments will be less

able to support specialized organisms at the top of the food chain. At present there is no detailed knowledge that helps to assess the relative importance of the two considered contributions to climatic stress, and we treat both factors equally, acknowledging that this could require modification in the light of new data.

Here, we use two types of data to model possible climatic stress: oxygen isotope data from Greenland, and the terrestrial pollen record from Lago Grande di Monticchio, Italy. We have chosen these records as they are of high quality but quite different character, as well as being located many thousands of kilometres apart. The most complete and highest resolution temperature proxy data available come from oxygen stable isotope measurements of air bubbles trapped in snow from Greenland, and we have used the GISP2 data (Grootes *et al.* 1993; The Greenland Icesheet Project 2 website 2003). The oxygen isotope data have been calibrated to absolute temperature with the help of borehole temperature measurements (Cuffey *et al.* 1992), and correspond closely to variations in global ice volume, as inferred from

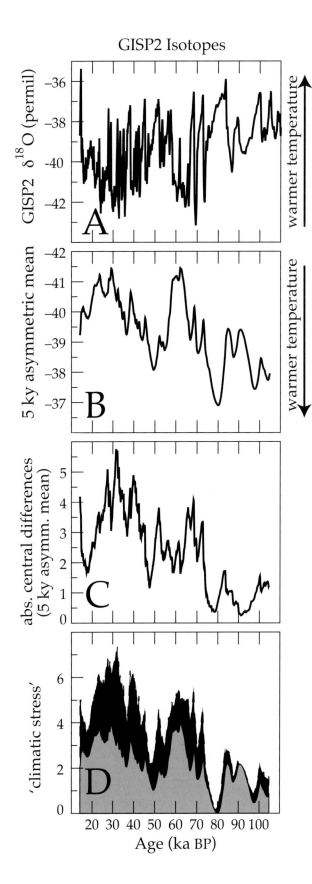

GISP2 Isotopes

variations in $\delta^{18}O$ of $CaCO_3$ in deep-sea sediments (Shackleton & Pisias 1985). In this study we consider GISP2 data back to 110 ka BP.

While the oxygen isotope data from Greenland arguably form the most direct and highest resolution temperature proxy for conditions in Europe, it is important to assess whether they are likely to represent conditions across Europe. One of us (PV) ran climate model simulations for the past 21 ka at approximately 1 ka intervals to evaluate the predicted relationship between Greenland temperatures and trans-European temperatures (grid points in the UK, Central France, and Italy), shown in Figure 13.2. All three locations show a reasonable correlation coefficient (~0.6) with conditions in Greenland, but the amplitude of variability is reduced with distance from Greenland. Italy shows a clear correlation, although absolute temperature changes are only about one-third of those in Greenland. Based on model results, the Greenland record is a useful indicator for the basic temperature conditions across Europe.

In order to further assess the validity of using Greenland data to model climatic stress in continental Europe we have also used a second data set, namely the relative proportion of pollen of woody taxa in the sediments from Lago Grande di Monticchio, Italy (Allen & Huntley 2000; Allen *et al.* 2000). Pollen data can give detailed information about local climate conditions, although the vegetation signal reflects a more complex combination of temperature and moisture than the Greenland data. Here we are only considering first order variations in the pollen record in order to evaluate whether our findings from the Greenland record are also reflected in different and geographically distant proxy measurements.

Figure 13.3. *Panel (A) shows the raw GISP2 oxygen isotope data, plotted with warm equivalent temperatures up. Panel (B) shows a smoothed version of (A), which was obtained by integrating asymmetrically over a 5 ky long window, where the largest weight is set at a given age, with linearly decreasing weights down to zero weight 5 ky before each point. Panel (B) is inverted with respect to panel (A). Panel (C) shows the absolute values of the central differences of panel (A), smoothed as in panel (B). Panel (D) shows the sum of panels (B) and (C), after the curves were normalized by setting the highest absolute value and the lowest central difference value to zero, and dividing by their respective standard deviations. Different shades of black and grey indicate the relative contribution from panels (B) and (C) to create the final stress curve in (D).*

The results and method of our study are illustrated in Figure 13.3 (oxygen isotope data) and Figure 13.4 (pollen data). In both figures, panel A illustrates the raw data. Oxygen isotopes and the percentage of woody taxa show a clear decreasing trend over the period 110 to 10 ka BP, as well as more rapid and pronounced variations towards the younger half of the records. The first component of our modelled climatic stress is shown in panel B, which is a smoothed absolute component of isotope and pollen data, respectively. Biologically, one might expect that 'cumulative stress' would be important in generating environmental and ecological pressures. For this reason, we calculated the smoothed absolute values of panel B by integrating the raw data over a 5 ka long window, and, on the basis that future climate change can have no current effect on an environment, we calculated an 'asymmetric mean' by assigning the largest weight to the youngest point on the 5 ka long window, decreasing linearly to a zero weight 5 ka earlier. Tests reveal that the exact shape of these 'asymmetric running means' is robust to small variations of the window length. The second component of our 'climatic stress' curve is contributed by the rate of change of data values. This component is illustrated in panel C and was computed by taking the absolute values of the central differences (as rapid warming presumably exerts pressure as well as rapid cooling) and then smoothing, using the same method described for panel B.

To generate the final 'climatic stress curve' in panel D, both components (absolute mean from panel B, and rate of change from panel C) were first normalised (using their standard deviation) and then added in a 1:1 ratio. Different shadings show the contributions from each component in the stacked plots.

Neanderthal extinction: concluding remarks

Figure 13.5a shows the two stress curves superimposed. Each shows clear peaks or plateaus at around 30 ka BP and 65 ka BP, with stress minima during OIS-5a (about 80 ka BP) and within OIS-3 (around 50 ka BP). However, the stress curves show less agreement beyond 70 ka BP, and this may reflect different regional conditions during OIS-5, or correlation problems between the two data sets. The two periods of maximum stress at about 30 and 65 ka BP probably affected the human populations of Europe in similar ways. But in the earlier stress peak the Neanderthals could have survived in refugia from which they recovered when conditions ameliorated during the ear-

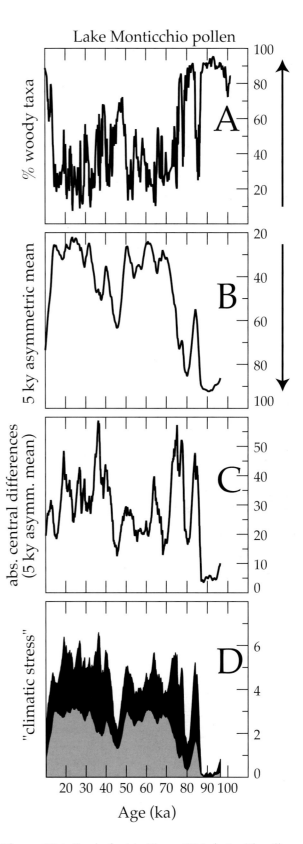

Figure 13.4. *Equivalent to Figure 13.3, but with pollen data from Lago Grande di Monticchio, Italy.*

237

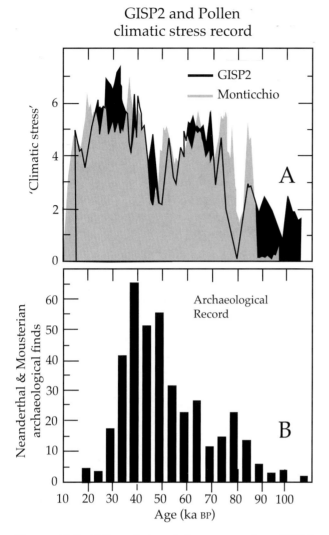

Figure 13.5. *A) Superimposed stress curves from GISP2 oxygen isotope data and Lago Grande di Monticchio pollen data. B) Histogram of radiometric dates for Middle Palaeolithic sites and fossils from the Stage 3 data base.*

lier part of OIS-3. However, the prolongation and accumulation of climatic stress towards the OIS-3/2 transition appears to have exceeded that of the earlier peak. If Neanderthal population numbers were already small, the ensuing environmental stress may well have led to their extinction. Circumstantial evidence may also implicate the Cro-Magnons in this process, as they were by then sharing the European continent with the Neanderthals, potentially competing with them for its rapidly changing and stressed resources, in environments that must, at times, have had considerably reduced carrying capacities. The populations with the most innovative and dynamic adaptive strategies, the best cultural buffering, and

the best-developed social networks would have had the greatest chances of survival at this time, and while many Cro-Magnon populations must also have gone extinct, along with the Neanderthals, others survived to form the basis for succeeding European populations (Stringer & Gamble 1993; Richards & Macaulay 2000).

We believe that this new approach to modelling late Pleistocene climatic stress not only helps to explain why the Neanderthals became extinct at about 30 ka BP but may also account for their greater archaeological visibility during parts of OIS-5 and the earlier part of OIS-3, around 50 ka BP. Figure 13.5b shows a histogram of radiometric dates for Neanderthals sites and fossils from the Stage 3 data base, and this shows a reasonable inverse relationship to the modelled stress curve, particularly bearing in mind the paucity of dates beyond the effective range of radiocarbon. In the future it would be valuable to extend such comparisons to early Upper Palaeolithic radiometric dates to see if this pattern also applies to the Cro-Magnons, as we would expect it to, and to other terrestrial evidence such as faunal records. It may also be possible to widen this approach to earlier time periods, and to regions of interest beyond Europe.

Acknowledgements

The National Snow and Ice Data Center, University of Colorado at Boulder, and the WDC-A for Paleoclimatology, National Geophysical Data Center, Boulder, Colorado provided GISP2 data. We thank the McDonald Institute for Archaeological Research, the Leverhulme Trust, and private donors, for their financial support of the Stage 3 Project, from which this paper has emanated. We would also like to thank members of the Stage 3 project, particularly John Stewart, for their comments on the approach adopted in this paper, and to Andy Currant for his review of the manuscript.

References

Aiello, L.C. & P. Wheeler, 2003. Neanderthal thermoregulation and the glacial climate, in *Neanderthals and Modern Humans in the European Landscape during the Last Glaciation,* Chapter 9, eds. T.H. van Andel & W. Davies. (McDonald Institute Monographs.) Cambridge: McDonald Institute for Archaeological Research, 147–66.

Akazawa, T., K. Aoki & O. Bar-Yosef (eds.), 1998. *Neandertals and Modern Humans in Western Asia.* New York (NY): Plenum.

Allen, J.R.M. & B. Huntley, 2000. Weichselian palynological records from southern Europe: correlation and chronology. *Quaternary International* 73/74, 111–25.

Allen, J.R.M., W.A. Watts & B. Huntley, 2000. Weichselian palynostratigraphy, palaeovegetation and palaeoenvironment; the record from Lago Grande di Monticchio, southern Italy. *Quaternary International* 73/74, 91–110.

Arsuaga, J.L., J.M. Bermúdez de Castro & E. Carbonell (eds.), 1997. The Sima de los Huesos Hominid site. *Journal of Human Evolution* 33, 105–421 (whole volume).

Beck, J.W., D.A. Richards, R.L. Edwards, B.W. Silverman, P.L. Smart, L. Donahue, S. Hererra-Osterheld, G.S. Burr, L. Calsoyas, A.J.T. Jull & D. Biddulph, 2001. Extremely large variations of atmospheric ^{14}C concentration during the last glacial period. *Science* 292, 2453–8.

Bocquet-Appel, J.-P. & P.-Y. Demars, 2000. Neanderthal contraction and modern human colonization of Europe. *Antiquity* 74, 544–52.

Brace, C.L., 1995. *The Stages of Human Evolution*. Englewood Cliffs (NJ): Prentice Hall.

CalPal, 2003. Cologne Radiocarbon Calibration & Palaeoclimate Research Package, available online from http://www.calpal.de/.

Cuffey, K., R. Alley, M. Grootes & S. Anandakrishnan, 1992. Toward using borehole temperatures to calibrate an isotopic paleothermometer in central Greenland, 1992. *Palaeogeography, Palaeoclimatology and Palaeoecology* 98, 265–8.

d'Errico, F. & M.F. Sánchez-Goñi, 2003. Neandertal extinction and the millennial scale climatic variability of OIS 3. *Quaternary Science Reviews* 22, 769–88.

Faria, J., 2000. What happened to the Neanderthals? – The survival trap. *KYKLOS* 53, 161–72.

Finlayson, C., D. Fa & G. Finlayson, 2000. Biogeography of human colonizations and extinctions in the Pleistocene. *Memoir GIBCEMED* 1, 1–69.

Flores, J., 1998. A mathematical model for Neanderthal extinction. *Journal of Theoretical Biology* 191, 295–8.

Gat, A., 1999. Social organisation, group conflict and the demise of Neanderthals. *The Mankind Quarterly* 39, 437–54.

The Greenland Icesheet Project 2, 2003. Summary web site, http://www.gisp2.sr.unh.edu/GISP2/.

Grootes, P., M. Stuiver, J. White, S. Johnsen & J. Jouzel, 1993. Comparison of oxygen isotope records from GISP2 and GRIP Greenland ice cores. *Nature* 366, 552–4.

Hublin, J.-J., 1998. Climatic changes, paleogeography, and the evolution of the Neandertals, in *Neandertals and Modern Humans in Western Asia*, eds. T. Akazawa, K. Aoki & O. Bar-Yosef. New York (NY): Plenum, 295–310.

Hublin, J.-J., C. Barroso Ruiz, P. Medina Lara, M. Fontugne & J.-L. Reyss, 1995. The Mousterian site of Zafarraya (Andalucia, Spain): dating and implications on the Palaeolithic peopling processes of Western Europe. *Comptes rendus de l'Académie des sciences Paris* IIa, 321, 931–7.

Hublin, J.-J., F. Spoor M. Braun, F. Zonneveld & S. Condemi, 1996. A late Neanderthal from Arcy-sur-Cure associated with Upper Palaeolithic artefacts. *Nature* 381, 224–6.

Jöris, O. & B. Weninger, 1998. Extension of the 14-C calibration curve to c. 40,000 cal BC by synchronising Greenland $^{18}O/^{16}O$ ice-core records and North Atlantic *Foraminifera* profiles: a comparison with U/Th coral data. *Radiocarbon* 40, 495–504.

Jöris, O. & B. Weninger, 2000. Calendric age-conversion of glacial radiocarbon dates at the transition from the Middle to the Upper Palaeolithic in Europe. *Bulletin de la Société Préhistorique Luxembourgeoise* 18, 43–55.

Langbroek, M., 2001. The trouble with Neandertals. *Archaeological Dialogues* 8, 123–35.

Mellars, P., 1999. The Neanderthal problem continued. *Current Anthropology* 40, 341–50.

Nitecki, M. & D. Nitecki (eds.), 1994. *Origins of Anatomically Modern Humans*. New York (NY): Plenum Press.

Ovchinnikov, I.V., A. Götherström , G.P. Romanova, V.M. Kharitonov, K. Lidén & W. Goodwin, 2000. Molecular analysis of Neanderthal DNA from the northern Caucasus. *Nature* 404, 490–93.

Pettitt, P., 1999. Middle and Early Upper Palaeolithic Crimea: the radiocarbon chronology, in *Préhistoire d'Anatolie: Genèse de deux mondes*, ed. M. Otte. Liège: Études et Recherches Archéologiques de l'Université de Liège (ERAUL) 85, 329–38.

Pettitt, P. & A. Pike, 2001. Blind in a cloud of data: problems with the chronology of Neanderthal extinction and anatomically modern human expansion, with comment by J.P. Bocquet-Appel & P.Y. Demars. *Antiquity* 75, 415–20.

Richards, M. & V. Macaulay, 2000. Genetic data and the colonization of Europe: genealogies and founders, in *Archaeogenetics*, eds. C. Renfrew & K. Boyle. (McDonald Institute Monographs.) Cambridge: McDonald Institute for Arcaeological Research, 139–51.

Richards, M., P. Pettitt, M. Stiner & E. Trinkaus, 2001. Stable isotope evidence for increasing dietary breadth in the European mid-Upper Paleolithic. *Proceeedings of the National Academy of Sciences of the USA* 98, 6528–32.

Shackleton, N. & N. Pisias, 1985. Atmospheric carbon dioxide, orbital forcing and climate, in *National Geophysics Monographs Series 32*, eds. E. Sundquist & W. Broecker. Washington (DC): American Geophysical Union, 303–17.

Skinner, M., 1997. Dental wear in immature Late Pleistocene hominines. *Journal of Archaeological Science* 24, 677–700.

Smith, F.H., 1994. Samples, species, and speculations in the study of modern human origins, in *Origins of Anatomically Modern Humans*, eds. M. Nitecki & D. Nitecki. New York (NY): Plenum Press, 228–49.

Smith, F.H., E. Trinkaus, P.B. Pettitt, I. Karavanić & M. Paunović, 1999. Direct radiocarbon dates for Vindija

G1 and Velika Pećina Late Pleistocene hominid remains. *Proceedings of the National Academy of Sciences of the USA* 96, 12,281–6.

Soffer, O., 1994. Ancestral lifeways in Eurasia: the Middle and Upper Paleolithic records, in *Origins of Anatomically Modern Humans*, eds. M. Nitecki & D. Nitecki. New York (NY): Plenum Press, 101–19.

Steegmann, A.T. Jr, F.J. Cerny & T.W. Holliday, 2002. Neandertal cold adaptation: physiological and energetic factors. *American Journal of Human Biology* 14, 566–83.

Stenseth, N., A. Mysterud, G. Ottersen, J. Hurrell, K.-S. Chan & M. Lima, 2002. Ecological effects of climate fluctuations *Science* 297, 1292–6.

Stewart, J.R., T. van Kolfschoten, A. Markova & R. Musil, 2003. Neanderthals as part of the broader Late Pleistocene megafaunal extinctions?, in *Neanderthals and Modern Humans in the European Landscape during the Last Glaciation,* Chapter 12, eds. T.H. van Andel & W. Davies. (McDonald Institute Monographs.) Cambridge: McDonald Institute for Archaeological Research, 221–32.

Stringer, C., 1998. Chronological and biogeographic perspectives on later human evolution, in *Neandertals and Modern Humans in Western Asia*, eds. T. Akazawa, K. Aoki & O. Bar-Yosef. New York (NY): Plenum, 29–37.

Stringer, C. & W. Davies, 2001. Those elusive Neanderthals. *Nature* 410, 791–2.

Stringer, C. & C. Gamble, 1993. *In Search of the Neanderthals*. London: Thames & Hudson.

Stringer, C. & R. Grün, 1991. Time for the Last Neanderthals. *Nature* 351, 701–2.

van Andel, T., 2002. Climate and landscape of the middle part of the Weichselian glaciation in Europe - The Stage 3 Project. *Quaternary Research* 57, 2–8.

Van Blerkom, L. & N. McGowan, 1999. Disease exchange and Neandertal extinction. *Abstract Human Origins Meeting, Cold Spring Harbor Laboratory, April 1999* (Abstracts, 3).

Zilhão, J., 2001. *Anatomically Archaic, Behaviourally Modern: the Last Neanderthals and their Destiny* (23rd Kroon Lecture). Amsterdam: Nederlands Museum voor Anthropologie en Praehistorie.

Zilhão, J. & E. Trinkaus (eds.), 2003. Portrait of the artist as a child. The Gravettian human skeleton from the Abrigo do Lagar Velho and its archeological context. *Trabalhos de Arqueologia* 22 (whole volume).

Zubrow, E., 1989. The demographic modelling of Neanderthal extinction, in *The Human Revolution*, eds. P. Mellars & C. Stringer. Edinburgh: Edinburgh University Press, 212–31.

Chapter 14

Demography, Dispersal and Human Evolution in the Last Glacial Period

Marta Mirazón Lahr & Robert A. Foley

The Stage 3 Project represents a collaborative effort to unravel the complexities of European climate and environment during the long interstadial (~60,000 to 25,000 years) of the last glacial cycle. From interdisciplinary approaches and sophisticated climatic modelling, the main outcome of the project is the confirmation that Stage 3 was not just a slightly warmer phase of the last glaciation, but a climatically variable period that at moments approached nearly interglacial conditions, while at others experienced abrupt and profound changes towards glacial climates (van Andel 2002).

This focus on understanding Stage 3 — trends, variability, frequency of climatic change, the amplitude of those changes, and their effects on the fauna and flora of Europe — is important for the development of temporal biogeographic models of the Pleistocene, in particular the later part of the period. Such models tend to generalize change on the basis of the extremes within a glacial cycle — i.e. between interglacial and peak glacial conditions (Lahr & Foley 1998). Parameters derived from these models are important for interpreting the order and directionality of demographic change, both population expansion and contraction, and explaining such changes in the context of resource availability. These approaches have provided a general model for understanding the population history of Europe on a glacial-interglacial scale (Gamble 1993; Foley & Lahr in press).

However, from a phylogeographic perspective small-scale changes can have significant demographic and genetic effects. Recent research, such as that developed in other chapters of this book, highlights the degree of smaller-scale climatic changes between interglacials and peak glaciations (see Chapter 5: Barron *et al.* 2003; also Barron & Pollard 2002). This change largely reflects medium-term fluctuations in temperature and ice-sheet extent, but also short-term abrupt changes identified as Dansgaard/Oeschger oscillations. The role of such interstadial fluctuations on evolutionary and biogeographic patterns is only just beginning to be explored.

These aspects are important at least partly because they provide the context for events following the emergence of modern humans. This period, encompassing approximately the last 160,000 years, witnessed the expansion of a small African population throughout the world and the subsequent disappearance of all other hominin species, leading to the existence of a single species of hominin throughout the world for the first time in five million years. How many other hominin species existed throughout the world in the early Upper Pleistocene is a matter of much controversy. However, for the focus of this book there is no doubt that the hominin species that had occupied Europe for a considerable period of time was replaced by expanding modern humans. This major demographic event, the extinction of the Neanderthals and the establishment of a modern human population in Europe, took place during Stage 3 (Stringer & Gamble 1993; Stringer 2002).

The fact that Stage 3 was the time when this happened is particularly significant. Modern humans evolved in Africa at some point during the previous glacial period (Stage 6) and experienced at least phases of moderate geographical expansion in the subsequent interglacial (Stage 5). However, these early modern human groups did not disperse further into Europe at that time. They did so later on, around, or soon after, approximately 60,000 years ago, before the peak of the last glaciation (Chapter 4: van Andel *et al.* 2003b; Lahr & Foley 1994; 1998). These two facts contradict predictions based on broad glacial–interglacial biogeographic models, and raise a major question — was the expansion of modern humans into Europe, the event that led to the extinction of Neanderthals, climatically driven?

In order to tackle this question, we will first discuss the broad biogeographical models of glacial–interglacial faunal exchange between Europe and Africa. Secondly, we will discuss the climatic context of Stage 3 and the implications for such biogeographic models. Thirdly, we will review the palaeoanthropological and genetic evidence for the timing and extent of modern human expansion during Stages 5, 4 and 3. Finally, we will explore analytically whether the fluctuating site demography of the Middle and Upper Palaeolithic in Europe during Stage 3 (using the Stage 3 Project Archaeological Data Base; see Chapter 3: van Andel *et al.* 2003a) can be linked to climatic change and discuss the implications of the findings.

Biogeographical models: evolutionary geography and climatic change

Evolutionary change in a dynamic climatic context

While it is a truism that climatic and environmental change is a major influence on evolution, it is less clear exactly how this relationship operates. At one extreme it has been argued that climate is a necessary and sufficient cause of evolutionary change, and that in its absence speciation and directional changes are unlikely to occur (Vrba 1993; 1996). Under this model, often referred to as 'pulse-turnover', evolutionary change is expected to be synchronized with climatic change. At the other extreme, continuous co-evolutionary interactions have been seen as the main driver of change, and that this occurs independent of climatic shifts (van Valen 1973). Under this model, a tight relationship between evolution, such as the appearance of new species, and climatic change, is not expected. Between these two extremes lie a number of more intermediate positions, which recognize that while climate might prompt particular bursts of change, they occur in particular competitive contexts.

Elsewhere we have argued for one of those more intermediate positions, focusing on the role of geographical processes in shaping evolutionary patterns (Lahr & Foley 1998). In particular, we have argued that the populational response to climatic change is most directly distributional. When climates change, so does habitat distribution; this is the essence of the climatic simulations and reconstructions outlined in this book. In response to this, animal populations will either contract or disperse. It is these changed distributions that provide the new conditions under which selection may bring about evolu-

tionary change, or indeed extinction. However, the relationship between climate and these evolutionary processes is much less direct than is the case with dispersals or contractions (Foley 1999).

In terms of human evolution and Stage 3, one of the opportunities provided by the fine-grained analyses of climatic change is to see the extent to which changes in hominin distributions are a response to changes in habitat distribution, and whether this provides the microevolutionary basis for understanding later Pleistocene human evolution. We can consider this by first looking at how Pleistocene climatic change could provide the framework for understanding major dispersal patterns in a biogeographical context.

Afro-European biogeography and glacial cycles

Glacial cycles provide the primary framework for Pleistocene biogeography. Although cooling of the Earth's climate began at the end of the Pliocene (Denys 1985; Loubere 1988), cyclical glacial fluctuations date from approximately 800 ka BP (Shackleton 1987). The first full glacial stage is recognized in the marine sequences as Stage 22, within the Matuyama subchron and before the Jaramillo event, and therefore between 900 and 790 ka BP. The succeeding glacial cycles each show the slow build-up of continental ice sheets, a rapid period of deglaciation, followed by a relatively short warm interglacial stage, although each cycle also shows unique aspects in terms of duration and extent (Shackleton 1987; 1996; Malatesta & Zarlenga 1988). These cyclical events influenced equatorial and northern latitudes in predictably different ways. Maximum cold in the north led to a southward shift of climatic belts, reduced temperatures globally, low sea-level stands, and aridity in many parts of the tropics. Warming is followed by the release of water trapped in the northern hemisphere glaciers, which is responsible for the global rise in sea level and for pluvial and high lake level short episodes in equatorial regions.

These changes have effects on African and Eurasian faunal distributions (Fig. 14.1) (Tchernov 1992a,b; Lahr & Foley 1998). Large mammalian faunas in both continents undergo periods of range expansion during early interglacials, while only Eurasian faunas seem to shift ranges during glacial build-up (in Africa extensive aridity causes the contraction of available ranges and a level of isolation and endemism). In the case of Europe, interglacial faunal expansions were associated with the retreat of ice and tundra along the Eurasiatic plains to the northeast, and occasionally towards the southeast, reaching the

Middle East (although the Taurus-Zagros mountain range and the interglacial forests of the Greek and Turkish peninsulas acted as important barriers to movement in this direction). During glacial periods, European animal ranges shifted southwards as continental areas became covered by ice sheets and permafrost terrain. At these times, the Middle East acted as a cul-de-sac, for these northern elements could not overcome the Saharan barrier at its maximum extent during glacial stages. Therefore, the main direction of Palaearctic expansions was East–West, as reflected by past and present animal distributions. In the case of Africa, population expansions were associated with increased moisture occurring particularly during the early phases of interglacials. These expansions were also directional, as forests expanded equatorially and savannas in a northerly direction across the Sahara. During these episodes, the Ethiopian faunal range also encompassed the Sahara, northern Africa and the Levant, which shows indications of savanna conditions, while movement into Europe would reflect a subsequent dispersal if the Taurus-Zagros barrier was transcended. Therefore, the main direction of non-forest Ethiopian expansions was north–south, reaching into the Middle East through the Sinai Peninsula.

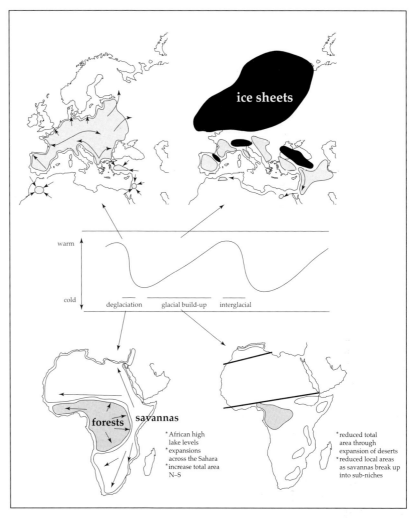

Figure 14.1. *Schematic diagram of the effect of the glacial cycles on habitat distribution and biogeographical relationships between Europe and Africa. (From Lahr & Foley 1998.)*

Hominin dispersals in relation to the climatic cycles
The implication of this general model is that there should be a predictable relationship between climatic change and hominin dispersals between Africa and Europe. In particular, it can be expected that during periods of glacial tropical aridity, the Sahara would be a barrier to movements out of Africa, and European populations are likely to have experienced contractions as habitats become latitudinally compressed. Conversely, during early interglacials, the northward spread of savanna environments would open up routes of dispersal from Africa to Eurasia, while Eurasian populations are likely to have dispersed northwards in response to habitat expansion due to warmer climates.

This model fits well the pattern of dispersals

into Europe during the lower and middle Pleistocene, and the apparent depopulation that occurs in glacial stages. It also conforms to what is observed for the extra-African dispersals that occur in Stage 5, although they do not appear to have penetrated Europe at this time (Lahr & Foley 1998). However, this general dispersal model does not allow us to estimate the role of smaller-scale climatic change in promoting multiple or relatively few such dispersals, or what the outcomes of such dispersals could have been for the various hominin populations that lived during this time period.

Flux & fragility as a demographic model for the Upper Pleistocene
A further implication of the evolutionary model described here is that hominin populations may have

243

been highly unstable in numbers and distribution. The standard anthropological model of human demography has been one of overall stability, leading to a gradual increase in human population. Underlying this view, was the idea that as the human species — and its ancestral forms — became more sophisticated, intelligent or simply better adapted, they experienced population growth, or at least maintained local demographic stability. However, this stable model is inconsistent with the genetic evidence and the emerging archaeological evidence. This evidence suggests that human and hominin populations experienced major fluctuations in population size, and that rather than being characterized by stability, they were in fact in a permanent condition of flux and fragility (Foley & Lahr in press). The local level would be the starting point of this, as communities responded to immediate adverse challenges (or failed to), ranging from resource depletion, inter-group conflict, and disease. Where such factors extended regionally, then local effects would be amplified. The converse of this situation is expansion, leading to dispersals and geographical expansions of existing ranges; as resources recover, or alternatively, when new and successful adaptive strategies are developed, then populations may be expected to expand very rapidly. In the latter case, there may be interactive effects between population collapse and population expansion.

A 'flux and fragility' model of prehistoric demography has implications for human evolution, and the high resolution of the Stage 3 climatic and environmental data provides the context for testing some of these ideas.

Stage 3 as a climatic context for human evolutionary: events and models

The pattern of climatic change in Stage 3
Figure 14.2 is a simplification and representation of the climatic sequence of the period from 64,000 to 15,000 years ago, based on the Atlantic Ocean core MD95-2042, using $\delta^{18}O$ values derived from planktonic Foraminifera (Shackleton *et al.* 2000). The period has been divided into 1000-year units. Although this is more finely resolved than might be warranted, the purpose is to summarize the main climatic trends. The climate quality for each 1000-year period was estimated, using a four-level scale: a) extreme cold phases, when the planktonic values for $\delta^{18}O$ were no more than 20 per cent less than during the last glacial maximum (LGM); b) cold phases, when $\delta^{18}O$ values were between 20–40 per cent less than those

at the LGM; c) cool phases, when $\delta^{18}O$ values were between 40–60 per cent less than LGM ones; and d) warm phases, when $\delta^{18}O$ values were at least 60 per cent less than at the LGM. For the most part, each of the 1000-year units could be ascribed as having a modal value within each of these categories; where there were aberrant warm or cold surges or spikes, these are indicated in the figure.

A number of points can be made about the climate of the Stage 3 period (which are further elaborated in other chapters of this book: Chapter 2: van Andel 2003; Chapter 5: Barron *et al.* 2003 & Chapter 6: Huntley & Allen 2003):

1. *Comparison with LGM and Holocene*: Stage 3 is clearly considerably warmer than LGM (only two 1000-year blocks approach LGM inferred temperatures — 43–44 ka BP, and 63–64 ka BP). Although at no point does the temperature approach that of the Holocene, the period can be considered not just intermediate between extreme glacial and interglacial conditions, but composed of significant lengths of time of relatively warm temperatures. If we look at Stage 3 as a single period, 13,000 years of its 35,000 year length (37 per cent) are what we have defined here as warm (as opposed to extreme cold, cold, and cool); 8000 years (23 per cent) are cool; 12,000 years (34 per cent) are cold; and 2000 years (6 per cent) are extremely cold.

2. *Overall trends*: As can be seen in Figure 14.2A, there is a clear trend towards warming that stabilizes between 50–35 ka BP (i.e. the warming phase compared to Stage 4), followed by a general deterioration into the LGM. As noted elsewhere in this book, Stage 3 is really composed of two periods, an early warm phase and a colder late phase. After 33 ka BP, there are no further warm phases.

3. *Climatic variability*: However, these trends mask what is perhaps the most intriguing aspect of Stage 3, which is its high variability. Figure 14.2B shows the warmest and coldest phases for each 5000-year block of Stage 3. It clearly shows that the period from 40–45 ka BP has the widest range; it also has the largest number of different climatic phases (Fig. 14.2C), and the most changes (Fig. 14.2D). Although no other part of Stage 3 matches these 5000 years, the period between 40–30 ka BP is also relatively variable, certainly compared to Stage 2.

4. *Stability and change*: Beyond overall variability is the rate of change, and by inference the degree of stability. Using the simple classification of major shifts shown in Figure 14.2, there are 22 phase

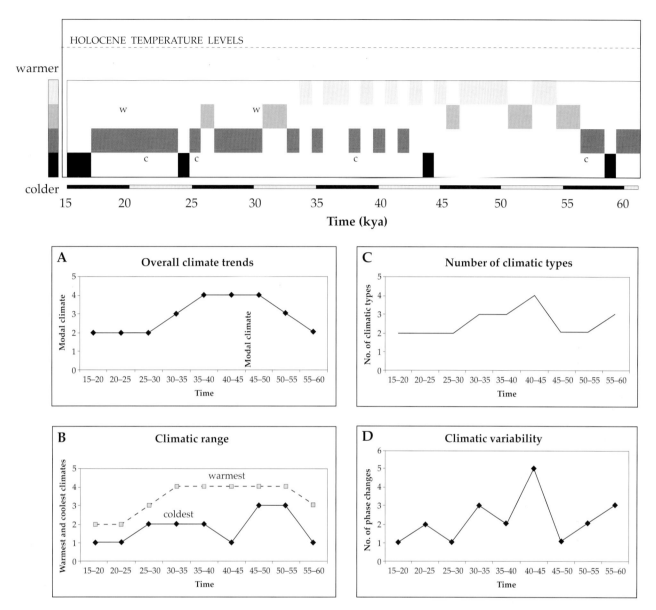

Figure 14.2. *Simplified scheme of climatic change across Stage 3. The vertical scale in the upper diagram shows a four-part categorization of temperatures derived from the oxygen isotope record of MD95-2042, in 1000-year units. The darker shading shows increasingly cold conditions (see text for details). A–D in the figure shows various parameters of Stage 3 by 5000-year units: A) overall climatic trends based on modal* ^{18}O *values; B) climatic range based on maximum and minimum* ^{18}O *values; C) the number of each of climatic category found in each 5000-year period; and D) climatic variability based on the number of changes occurring in each 5000-year period.*

changes across the 35,000-year period; in other words, the climate seldom remains within a single phase for more than 1000 years. Indeed, the longest period of stability is 4000 years, which occurs with a warm phase between 46–50 ka BP, and a cold phase between 26–30 ka BP.

5. *Interactive effects*: One further point is that the period between 46–50 ka BP has been characterized by the most prolonged phase of climatic sta-

bility, while the following 5000 years (45–40 ka BP) have the greatest levels of instability. The interactive effect of this combination may have been of significance for human evolutionary patterns.

Relationship between Stage 3 scale of climatic change and the glacial–interglacial cycle

The question is how can these characteristics of an interstadial, in this case Stage 3, be integrated into the biogeographic model described before so as to provide more sophisticated or detailed predictions of hominin demographic change?

Some simple expectations can be outlined. First, if full interglacial conditions are necessary to prompt 'out of Africa' hominin dispersals, then these do not occur in Stage 3, even though it is a relatively warm period. Stage 3 might thus offer a test for aridity/ humidity thresholds for hominin dispersals across the Sahara. Second, it is clear, as shown elsewhere in this book, that the earlier part of Stage 3 is considerably warmer than the later part. If contractions and extinctions are an expectation of colder conditions, then the slide into more glacial conditions during the course of Stage 3 might provide insights into how sensitive to such changes hominin populations are. Third, the high level of variability and oscillation between relatively warm and relatively cold conditions, particularly in the middle parts of Stage 3, should provide an indication of the sensitivity of hominin populations to such changes — for example, whether they respond on a scale of 40 generations (i.e. a thousand years), or whether they are sufficiently buffered to damp out the effects of such rapid change. These general questions will be addressed in the remainder of this chapter, first by considering the overall evidence for hominin dispersals and demography, and then by considering the quantitative approaches that might indicate the effect of climate on later human evolution.

Palaeoanthropological and genetic evidence for modern human expansion: tracing early modern human demographic change

Evidence for the pattern of later human evolution
The fossil evidence suggests that the evolutionary lineage leading to modern humans is a uniquely African one, and that although the fossil record is relatively poor, it is possible to trace a trajectory from larger brained archaic forms such as Florisbad or Ngaloba (which we would refer to as *Homo helmei*) to early modern forms represented by the Idaltu and Omo Kibbish fossils from Ethiopia (Stringer & Andrews 1988; Lahr & Foley 1994; Stringer 2002; White *et al.* 2003). Although the date of the transition between archaic and morphologically modern forms continues to be a matter of debate, it is likely, on the basis of the Ethiopian material, that it took place before 150–160 ka BP, during Stage 6. No similar evidence for such a transitional process is found elsewhere in the world, and this is supported by abundant palaeoanthropological and genetic evidence.

Two observations about the genetic evidence are paramount. The first is that humans display relatively little genetic variation in fast-mutating loci, giving rise to the model that human diversity is a recent phenomenon derived from a very small population. Chronological estimates for when this occurred vary, but there is general consensus, based on a number of gene systems, that it was approximately 150,000 years ago. This accords well with the first evidence for generalized anatomically modern humans. The second is that there is greater genetic diversity in African populations than in non-African ones, and that the latter are a subset of the former. This, together with the directionality of genetic change observed in uniquely African lineages, supports the hypothesis that modern human origins lie in Africa, on the basis that the amount of diversity in populations is, in part, a reflection of the time over which they have persisted. This conclusion is derived from both unique phylogenies of particular loci, such as the Y chromosome and the mitochondrial genome, as well as by the geographical distribution of such genes in particular populations (Watson *et al.* 1997; Quintana-Murci *et al.* 1999; Ingman *et al.* 2000; Underhill *et al.* 2001).

During Stage 5 there is more widespread evidence for modern humans (Foley & Lahr 1997; Klein 2000). Within Africa, the fossil evidence from Klasies River Mouth shows that if the original population was in northeastern Africa, by this time descendent groups had expanded to the Cape. Archaeological evidence in the form of the MSA industries scattered over much of Africa would perhaps support this view. Furthermore, the presence of modern humans from at least 100,000 years ago in the Levant shows that dispersals across the Sahara or through the Nile had also occurred, as predicted by the biogeographical model for inter-glacial conditions. However, at that time, no evidence for modern humans beyond what can be considered an essentially African biogeographical zone has been found (Lahr & Foley 1998).

This is consistent with the genetic data. The rare and yet widespread distribution of Y-chromosome haplogroups I and II in Africa has been interpreted as evidence for these early dispersals. Today, these lineages are found mostly in small, outlier populations of hunter-gatherers throughout Africa (Underhill *et al.* 2001). The mtDNA evidence also reflects this pattern (Quintana-Murci *et al.* 1999).

From the perspective of demographic change, it is clear that African hominin groups consisted of small, isolated populations, and that these were subjected to stresses that caused significant demographic contractions. Estimates of the size of the ancestral population of anatomically modern humans in Africa at the end of the Middle Pleistocene have been of the order of tens of thousands of individuals. If the populations were sub-structured, as is likely, then total numbers may have been even smaller. Fluctuating demographies are thus a key element in the process of late human evolution, as implied by the general evolutionary models discussed above. Archaeological and chronological evidence, and even ancient DNA results, increasingly suggest that such fluctuating demographies probably characterized all Pleistocene hominin groups.

Major human dispersals beyond Africa

While the earlier phases of modern human origins and dispersal seem to fit the general models presented earlier, it is clear that expansion beyond the African biogeographic zone had not occurred prior to the onset of Stage 4. Neither genetic evidence based on coalescence estimates, nor archaeological evidence, would suggest a greater geographical range for the earliest modern humans. Some caution has to be expressed on this point, as it is possible that later extinctions of modern human populations outside Africa may have erased the genetic evidence, and little work has been done in key areas such as Saudi Arabia and India to provide good dated sequences for the Middle Stone Age.

The earliest evidence for modern humans beyond Africa and the Levant comes from Australia. Although there is some controversy concerning dates, most researchers accept that humans reached Australia by at least 50,000 years ago, and possibly more than 60,000 years ago. We have argued elsewhere that this evidence, in conjunction with genetic data, shows that there was a separate 'Southern Dispersal' independent of the Upper Palaeolithic dispersals into and across Eurasia (Lahr & Foley 1994; 1998; Foley & Lahr 1997). Depending on a more precise chronology of such a Southern Dispersal, it may have taken place during the warming phase associated with the onset of Stage 3. If this is the case, then such dispersals would fit the general expectations of the biogeographical model, and would suggest that the conditions were sufficiently ameliorated to allow extra-Africa population expansion, presumably through the Horn of Africa. This would have been a coastal, sub-tropical expansion, in turn suggesting that it may have been either too cold to allow hominin populations to spread to higher latitudes, or too dry to allow crossing the Sahara. On the other hand, if

the very earliest proposed Australian dates prove to be correct (Roberts *et al.* 1994), then it would mean that the Southern Dispersal occurred against the directionality of the model. As we shall discuss later, this contrast between biogeographical expectation and archaeological evidence may indicate over-riding human behavioural capacities.

The spread of modern humans into Eurasia (Fig. 14.3) is generally associated with the Upper Palaeolithic. Although there are some indications of sites with Upper Palaeolithic industries occurring very early, most of the evidence suggests that major expansion occurred after 45,000 years ago, with dispersal across Europe taking place up to 30,000 years ago, and across central and eastern Asia somewhat later (Torroni *et al.* 2000; Ke *et al.* 2001; Underhill *et al.* 2001).

The key point to stress is that these dispersals occur against the trend of climatic change. Although the period concerned is relatively warm compared to the LGM, nonetheless the overall direction of climatic change is towards colder conditions. In other words, as noticed by many before, the Upper Palaeolithic dispersals do not conform to the predictions of the general biogeographic model.

It is possible that the apparent lack of fit between the evidence for human Eurasian dispersals and the climatic evidence is the result of a mismatch of scale. As mentioned before, Stage 3 was characterized by high variability, with periods of relative warmth oscillating with colder phases on a scale of one thousand years or even less. It could therefore be that modern human behaviour allows Upper Palaeolithic populations to respond sufficiently fast to allow geographical expansion after very short-term climatic amelioration, and thus to disperse into Europe in a more interrupted manner, with bursts of expansion during the short warmer phases, interrupted by colder contractions or even extinctions. The alternative to this model is that the pattern of modern human dispersal into Europe is not influenced by climate, but either by the behavioural characteristics of modern humans, or the nature of the competitive interactions with the local European Neanderthals.

In the next section we will consider these possibilities by using the archaeological evidence compiled by William Davies as part of the Stage 3 Project (Chapters 3 & 4: van Andel *et al.* 2003a,b) as a proxy for human dispersals at a fine-grained chronological level, and by exploring whether it is possible to measure the interaction between human demography and climate during Stage 3.

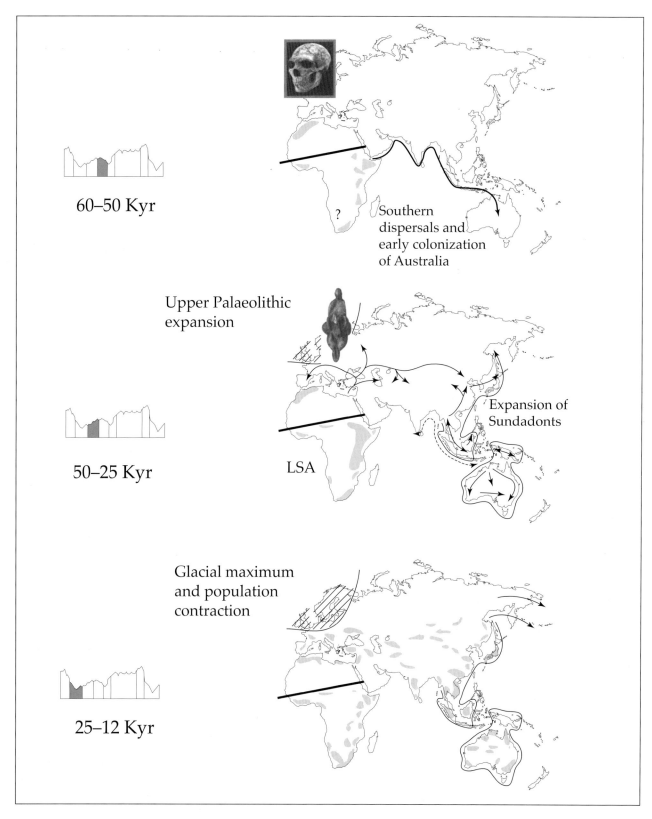

60–50 Kyr

Southern
dispersals and
early colonization
of Australia

?

Upper Palaeolithic
expansion

50–25 Kyr

LSA

Expansion of
Sundadonts

Glacial maximum
and population
contraction

25–12 Kyr

Figure 14.3. *Multiple dispersals and contractions of early modern humans. The top figure shows the early southern dispersals in Stage 4; the middle figure shows the Stage 3 Eurasian dispersals; in Stage 2 (bottom figure) the contraction and fragmentation of the human population is shown. See text for full discussion.*

Measuring the impact of climate on human dispersals

In order to examine the relationship between climatic variables and the number of Middle and Upper Palaeolithic sites in Europe, the raw $\delta^{18}O$ data from the marine core MD95-2042 (Shackleton *et al.* 2000) were used.

Although the Stage 3 Archaeological Data Base contains dated occurrences of Upper Palaeolithic (UPal) sites in Europe as early as between 65–60 ka BP, the first clear evidence of UPal sites occurs after a period of stable and relatively warm climate (between 55 and 48 ka BP), when the number of Middle Palaeolithic (MPal) sites decreases considerably. The following period, between 45–40 ka BP, witnesses the growth of UPal sites, but also of MPal ones (Chapters 3, 4: van Andel *et al.* 2003a,b & Chapter 8: Davies & Gollop 2003) (Fig. 14.4).

The extent to which European hominin populations were sensitive to some aspect of climatic conditions (temperature, variability, etc.) was addressed at three levels: 1) through the overall number of sites, regardless of whether they were Middle or Upper Palaeolithic; 2) Middle Palaeolithic (and presumably Neanderthal); and c) Upper Palaeolithic (and presumably modern human). The Châtelperronian and other 'intermediate' sites were treated as Middle Palaeolithic. A number of climatic variables were generated from the $\delta^{18}O$ data, namely: average $\delta^{18}O$ values for 1000-, 3000- and 5000-year periods; maximum and minimum, as well as amplitude of $\delta^{18}O$ values in the same periods; absolute and relative difference of $\delta^{18}O$ values between units (1000-, 3000- or 5000-year long); or transformed into categorical measures of relative climatic quality, or stability of rate of change. Available evidence from Stage 4 (as existent in the Stage 3 Archaeological Data Base) was included in the analyses. The approach we adopted was to treat the archaeological and climatic data statistically rather than historically, and thus explore hypotheses relating to the measurable influence of climate on hominin populations.

Middle and Upper Palaeolithic sites in Europe during Stage 3

Average ^{18}O values/1000 years, extreme negative and positive $\delta^{18}O$ values/1000 years, or amplitude of ^{18}O values/1000 years do not account for the variation in the number of archaeological sites in Europe between 73 and 23 ka BP (insignificant cubic regressions). However, a cubic regression of number of archaeological sites through time is extremely sig-

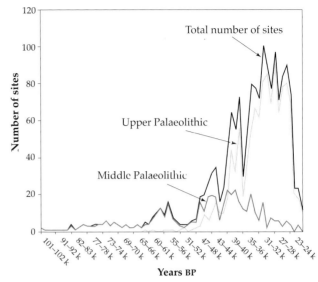

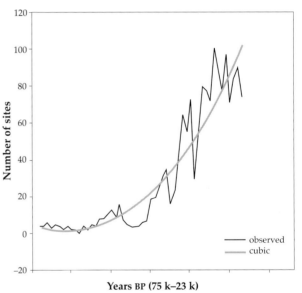

Figure 14.4. *Number of archaeological sites from the end of Stage 5 to the beginnings of Stage 2. The number of sites attributable to Middle and Upper Palaeolithic are also shown (top). Number of sites over time, and the cubic regression derived from these data are shown in the bottom graph.*

nificant ($F = 181.19936$, $p < 0.001$), with an r^2 of 0.88089 (Fig. 14.4). In other words, the main predictor of the number of archaeological sites is time itself; between 73–23 ka BP, one can predict with 88 per cent accuracy the number of archaeological sites in Europe (both Middle and Upper Palaeolithic) for each 1000 year time period. At this time scale, there is no relationship between climate and number of archaeological sites.

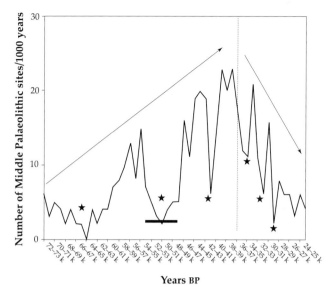

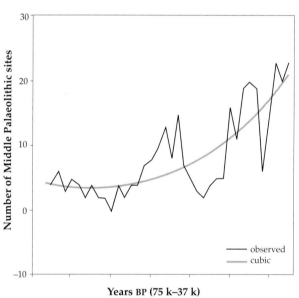

Figure 14.5. *Number of Middle Palaeolithic sites over time (top). There is a marked trend towards increased numbers prior to 38 ka BP, and a decline after that time. Major departures from that trend are indicated by stars. The pattern indicates that Neanderthal populations may have been subject to fluctuations within the general trends. The increase in number of sites prior to 38 ka BP is best described by a cubic relationship (shown in bottom graph, see text for details).*

The role of time in explaining the number of archaeological sites highlights the historical nature of the observed change — there is a cumulative addition of sites. However, it could be that climate is affecting the rate of accumulation, and that the 12 per cent variation not explained by time itself can be accounted for by climatic change. In order to test this, the residuals of the cubic regression between number of sites and time were tested against the climatic variables. No significant relationship or pattern was found between these, i.e. climate does not seem to explain the overall increase in number of archaeological sites in Stage 3 Europe, nor the observed fluctuations in numbers relative to the general trend over time.

Middle and Upper Palaeolithic sites in Europe during Stage 3

It may be that the observed relationship with time conflates different patterns for the Middle and Upper Palaeolithic. The number of Middle Palaeolithic sites in Europe during Stage 3 is very variable, but can be described overall as progressively larger up to ~37 ka BP, and declining subsequently (Fig. 14.5). These generalizations, however, are marked by pronounced departures, particularly in the intervals of 66–65 ka BP, 54–49 ka BP, 44–43 ka BP, 38–36 ka BP, 34–33 ka BP, and 32–31 ka BP. The most significant of these periods in terms of the duration of the effect is clearly the 54–49 ka BP interval.

However, again, no strong statistical relationship between climatic variables and Middle Palaeolithic sites can be demonstrated. An analysis of the number of Middle Palaeolithic sites in Europe between 73–37 ka BP shows a strong relationship with time (Fig. 14.5) ($F = 29.74072$, $p < 0.001$, $r^2 = 0.60839$ (cubic regression). In other words, the more recent the period (between 73 and 37 ka BP), the larger the number of Middle Palaeolithic sites that can be expected (with a 61 per cent accuracy).

We also considered whether, once time was taken into account, there was a relationship between the number of Middle Palaeolithic sites and climatic variables. Using the same approach as above, the residuals of the cubic regression were tested against the climatic variables. The results obtained show that there is no relationship between $\delta^{18}O$ values (1000-year averages, maximum or minimum). There is, however, a weak relationship between amplitude of ^{18}O values in 1000-year intervals and whether there are too few or too many Middle Palaeolithic sites in 1000-year intervals as predicted by age (between 73 and 37 ka BP). The smaller the amplitude of climatic change, the smaller the residual values (i.e. the closer to expected by age). In other words, the greater the degree of climatic variability in 1000-year periods, the greater the probability of affecting the historical trend of increase in number of sites ($r^2 = 0.17214$, $F = 3.64$, $p < 0.05$). These data therefore show some evidence that Neanderthal populations were adversely affected by the amount of climatic change.

In the period between 37 and 23 ka BP, the temporal distribution of Middle Palaeolithic sites is best described by a linear decrease ($F = 9.83$, $p <0.01$; $r^2 = 0.472$). Although this historical trend can only account for 47 per cent of the variation in number of Middle Palaeolithic sites observed (and indeed, the majority of 1000-year periods have either too many or too few sites in relation to the number predicted by the historical trend), none of the climatic variables correlates significantly with the residual data (with $\delta^{18}O$: $r^2 = 0.392$, $p >0.05$; neg$\delta^{18}O$: $r^2 = 0.293$, $p >0.05$; pos$\delta^{18}O$: $r^2 = 0.4.22$, $p >0.05$; amp^{18}O: $r^2 = -0.133$, $p >0.05$).

We also examined whether the residual variance in number of sites during this period was related to the increase in the number of Upper Palaeolithic sites, but no such relationship was found ($r^2 = 0.088$, $p >0.05$). Multiple regression analyses including both climatic variables and number of Upper Palaeolithic sites also failed to explain residual Middle Palaeolithic temporal distributions

Upper Palaeolithic sites in Europe during Stage 3
We carried out the same set of analyses on the Upper Palaeolithic data set. As mentioned before, although there are a few archaeological sites in Europe described as containing Upper Palaeolithic material culture before 48 ka BP, these really become an actual demonstrable demographic phenomenon after this date (Fig. 14.6). Again, cumulative increase through time is the primary relationship. A cubic positive relationship best explains mathematically the progressive increase in the numbers of Upper Palaeolithic sites in the period between 57 ka BP (when they are absent) and 24 ka. This is a highly significant relationship ($F = 120.17894$, $p <0.0001$) that explains more than 88 per cent of the variance in the data ($r^2 = 0.88164$).

The analysis of the residual data fails to find any correlation with either the climatic variables (^{18}O: $r^2 = 0.03$, $p >0.05$; neg^{18}O: $r^2 = 0.053$, $p >0.05$; pos^{18}O: $r^2 = 0.114$, $p >0.05$; amp^{18}O: $r^2 = 10.193$, $p >0.05$), or with the total or partial residual data of number of Middle Palaeolithic sites through time (with Middle Palaeolithic residuals: $r^2 = 0.104$, $p >0.05$; with Middle Palaeolithic residuals up to 37 ka BP: $r^2 = 0.245$, $p >0.05$; with Middle Palaeolithic residuals from 37 ka BP: $r^2 = 0.111$, $p >0.05$).

We also examined the data to see whether any climatic effects are lagged due to the time it takes a population to respond. Neither the actual number of

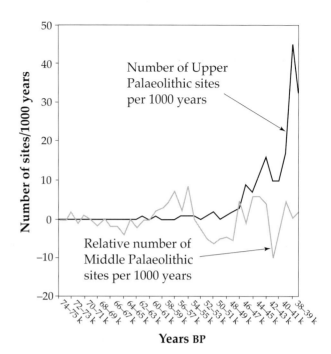

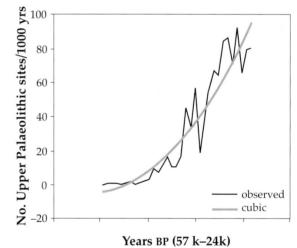

Figure 14.6. *Number of Upper Palaeolithic sites over time, compared to the residual values of Middle Palaeolithic sites (top). Upper Palaeolithic sites increase over time after 57 following a cubic pattern (bottom).*

sites, however, nor residual number of sites through time, correlate with lagged ^{18}O values by 1000 years, or with the difference in ^{18}O values from 1000-year period to the next. There is, however, a weak relationship between categories of difference between $\delta^{18}O$ values (i.e. whether the difference in ^{18}O values from one 1000-year period to the next was less than -0.5, between -0.5 and $+0.5$, and greater than 0.5) and the total ($F = 3.613$, $p <0.05$) and residual ($F = 3.610$, $p <0.05$) number of Middle Palaeolithic sites.

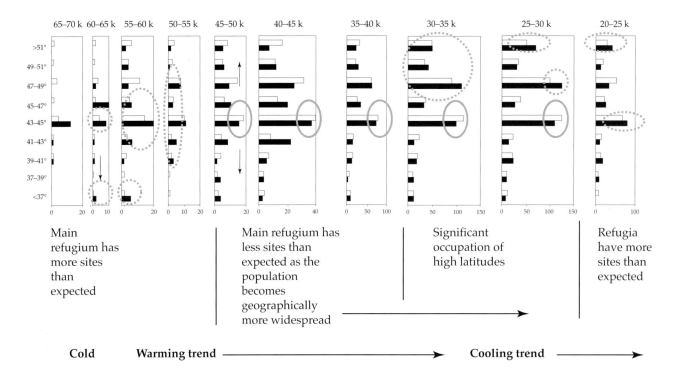

Figure 14.7. *Expected and observed numbers of archaeological sites for different phases of Stage 3. The vertical axes on the graphs show latitudinal bands, and the horizontal axes the percentage of archaeological sites at each of those bands. The graphs show successive chronological periods, from earlier to younger. The white bars on the histograms are expected numbers of sites based on the total archaeological distribution; the black bars are the observed number. According to the hypothesis (see text), observed sites should exceed expected in the preferred refugia during cold periods, and expected should exceed observed in the expansion zones during cold periods. In warmer periods this should be reversed. The data shown support this hypothesis.*

In other words, a very high degree of climatic change did have some, albeit very small, effect on numbers of Middle Palaeolithic sites. As before, we observed that the number of Middle Palaeolithic sites in Europe was weakly affected by departures from climatic stability.

All analyses were repeated at 3000-year units (not shown), and a similar lack of association between archaeological demography and climatic change was observed.

Discussion and conclusions

The role of climate in modern human Eurasian dispersals
The analyses presented above, as well as those in other parts of this book (Chapter 4: van Andel *et al.* 2003b; Chapter 8: Davies & Gollop 2003) suggest that although it is clearly the case that Europe was colonized by modern humans during a period that was warmer than either the preceding Stage 4, or the succeeding Stage 2, and therefore broadly consistent with a model of Eurasian expansions associated with

warmer climates, nonetheless the data do not fit the general evolutionary model when applied at this high resolution level.

There are a number of methodological reasons why this may be the observed result. The obvious ones relate to the nature of the archaeological record and date-based data sets (Housley *et al.* 1997; Chapter 3: van Andel *et al.* 2003a). The archaeological record is far from perfect, with taphonomic distortions occurring at all levels. It may therefore be that the results described are a sampling artefact. Furthermore, the early periods of Stage 3 are notoriously difficult to date, and the later periods are subject to calibration problems that have yet to be resolved. Again, therefore, the results may be an artefact. In addition it could also be the case that the resolution of the archaeological data set simply does not meet that of the climatic record.

More subtly, it could be that there are strong geographical patterns in the data which have not been explored here, and these are masked by the total sample (see Chapter 8: Davies & Gollop 2003).

Finally, the analyses here use raw data, rather than smoothed time series, which may be a means of revealing hidden patterns.

While all these may be factors that explain the results, and are certainly worth further investigation, nonetheless if the observed patterns are not artefacts of the data and analyses, it is necessary to consider what their interpretation could be in terms of the early European dispersals of modern humans. Why might both the expansion of modern humans and the decline of the Neanderthals not be associated strongly with the fine-grained climatic changes of Stage 3?

The answer may lie in the interaction between the behaviour of the hominins and the geographical patterns. Let us suppose that in warmer periods hominin populations increase and so expand, resulting in dispersals into Europe, and further within Europe dispersals. When conditions become more difficult, those populations contract and either survive in a few refugia, or else become extinct. When conditions become better then, there will again be expansions out of any refugia. This, in effect is the evolutionary geography model described at the outset. However, the archaeological signature of this process will not be simple. These areas where there were refugia will show a relatively constant presence of humans, while only in those areas where humans could survive only in relatively warm conditions would there be a signal of flux and fragile, temporary populations. The overall signal of such flux would therefore be smothered by the 'noise' from the refugia.

We tested this hypothesis by considering the distribution of archaeological sites in latitudinal sections in Western Europe (Fig. 14.7). Using the archaeological data base (up to +10° longitude), the expected number of sites per latitudinal unit (2° units) for 5000-year periods between 70–20 ka BP was calculated. This was then compared to the observed number of sites. The data support the proposal that there is indeed a refugium where site numbers are greater than expected during colder phases, and lower than expected during warmer phases. Secondary refugia can also be identified. Conversely, areas that have few or no sites during colder periods have higher than expected numbers during the warmer periods. A X^2 test showed that there is a significant difference between expected and observed sites at each latitude during 5000-year intervals ($X^2 = 236.45$, $p < 0.001$). This was the case whether all sites were treated together, or whether modern humans and Neanderthals were treated separately. The refugium

zone lies in the latitude 43–45°, which is consistent with the view that the southwest of France was an area where populations could be sustained during colder periods.

The link between human population history and the dynamism of the Stage 3 environment is thus complex, and is at least partially explicable in the context of a model of flux and fragility of hominin demography, and in the context of populations being confined to refugia in colder periods and expanding in warmer periods, thus giving rise to separate archaeological signatures. The various periods of Stage 3 can thus be categorized in terms of the percentage of archaeological sites confined to particular latitudes, and this can be related to time and temperature. Figure 14.8 shows that the end of Stage 4 has a high percentage of sites in the preferred latitudinal zones of the refugia; the bulk of Stage 3 shows a decline in that percentage, with a return towards a tighter latitudinal banding in Stage 2. The fact that it did not return to levels seen in Stage 4, however, may indicate the difference in adaptive tolerance of modern humans compared to Neanderthals.

To this more complex model, with its recognition of the difference between refugia and expansion zones, a further important idea can be added. When looked at in the broad context of human evolution in the Pleistocene, it appears more probable that hominins did not survive in many refugia during glacial periods, and therefore the signal of the Pleistocene dispersals is stronger. Two things, however, make the period of Stage 3 more complex. The first is that the climate itself seems to have been less extreme, probably allowing more refugia to retain populations. And second, the adaptations of both modern humans and Neanderthals may have been such as to allow them to be more buffered to these changes than was the case for earlier hominins such a *Homo heidelbergensis* and *Homo erectus/ergaster* (see Chapter 13: Stringer *et al.* 2003 for a discussion). Much has been made of differences between Neanderthals and modern humans, and indeed there must have been major adaptive differences, but these analyses show that there is also an overarching similarity in direction if not intensity of demographic response among all later Pleistocene hominins in Europe.

Integrating genetic evidence

Given the difficulties in applying quantitative techniques to the analysis of archaeological data, it might be thought that the evolutionary genetics evidence

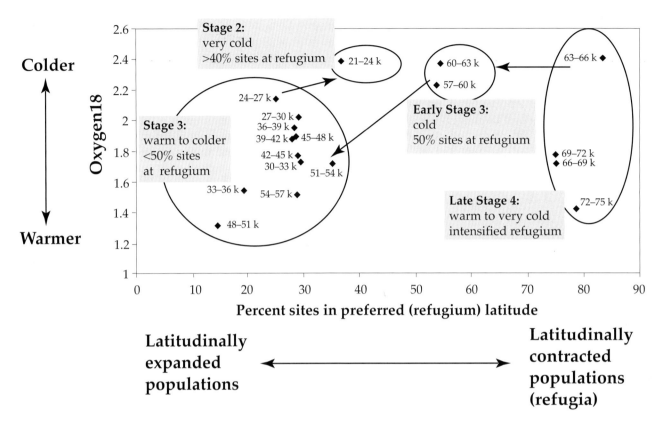

Figure 14.8. *Percentage of sites in latitude 43–45°, the refugial band. In late Stage 4 the Neanderthals are confined to this latitude to a very large extent; both warming during Stage 3 and the arrival of modern humans reduces this compression effect. Climatic cooling in at the end of Stage 3 reverses this trend, presumably associated with changes in modern human distribution and the extinction of the Neanderthals. Numbers (k) refer to thousand-year time intervals.*

may provide a better approach. Certainly recent research in this field has provided key insights into patterns of human dispersals, and the chronological and geographical resolution of these is improving all the time. At this stage, however, it is unlikely that genetic data can resolve the fine-grained questions about human–climate relationships prompted by the results of the Stage 3 Project. While the resolution on chronometric estimates of archaeological and fossil data during Stage 3 might typically have confidence limits of approximately ±1500 years, the confidence limits for coalescence estimates and for the age of demographic expansions might be in excess of 5000 years. In other words, while we might be comfortable in placing the earliest genetically-observable demographic expansion of humans in Europe at around 50,000 years ago (as a maximum estimate), in practice this is actually an estimate lying between 60,000 and 40,000 years ago. Phylogeographic methods can generate important hypotheses about possible correlations with the archaeological record (see for example Underhill *et al.* 2001), but they cannot be

used to test detailed chronological hypotheses.

Conclusions
In recent years both palaeoanthropologists and evolutionary geneticists have converged on a model of later human evolution in which dispersals, initially from Africa, but also more generally, are seen as a major mechanism for evolutionary change. As we have discussed here, this model is also consistent with the way in which evolutionary biologists have been thinking about processes of change. While genetics offers important insights into the history of human distribution and demography, at present it does not have the resolution to link events to the fine-grained environmental record available.

Multi-disciplinary approaches, such as the Stage 3 Project, offer a way forward. We have shown here that the relatively warmer environments of the Stage 3 phase of the last glacial fit a general model of hominin population expansions in relation to climatic change. Nonetheless it is currently not possible to show that such expansions were fine-tuned to the climatic variability of the time. In general terms, it is possible to show that both modern humans and Neanderthals had the capacity to cope better with colder environments than their predecessors, and thus appear to be buffered in some ways. This supports the ideas developed elsewhere in this volume of the importance of refugia in maintaining population continuity through time and promoting intense competitive circumstances. Further work, however, is necessary to be able to discriminate statistically the demography of refuge areas and expansion zones.

Finally, we have attempted in this paper to exploit the high level of inter-disciplinarity and the development of quantified data bases that have been promoted by the Stage 3 Project. As palaeoanthropology matures, we can expect not just better and better resolution in the environmental, chronological and archaeological data, but also a response on the part of palaeoanthropologists to use quantitative techniques and formal models to analyze evolutionary issues.

Acknowledgements

We thank Tjeerd van Andel for his invitation to be part of the Stage 3 Project, the wonderful dinners and discussions that it fostered, and the members of the project for stimulating our interest in this multidisciplinary endeavour. We particularly thank William Davies for access to the archaeological data base, and for advice and help with its analysis. Clive Gamble read and commented on an earlier draft of this paper.

References

Barron, E. & D. Pollard, 2002. High-resolution climate simulations of Oxygen Isotope Stage 3 in Europe. *Quaternary Research* 58, 296–309.

Barron, E., T.H. van Andel & D. Pollard, 2003. Glacial environments II: reconstructing the climate of Europe in the Last Glaciation, in *Neanderthals and Modern Humans in the European Landscape during the Last Glaciation*, Chapter 5, eds. T.H. van Andel & W. Davies. (McDonald Institute Monographs.) Cambridge: McDonald Institute for Archaeological Research, 57–78.

Davies, W. & P. Gollop, 2003. The human presence in Europe during the Last Glacial Period II: climate tolerance and climate preferences of mid- and late glacial hominids, in *Neanderthals and Modern Humans in the European Landscape during the Last Glaciation*, Chapter 8, eds. T.H. van Andel & W. Davies. (McDonald Institute Monographs.) Cambridge: McDonald Institute for Archaeological Research, 131–46.

Denys, C., 1985. Paleoenvironmental and paleobiogeographical significance of the fossil rodent assemblages of Laetoli (Pliocene, Tanzania). *Palaeogeography Palaeoclimatology Palaeoecology* 52, 77–97.

Foley, R.A., 1999. The evolutionary geography of Pliocene hominids, in *African Biogeography, Climatic Change, and Hominid Evolution*, eds. T. Bromage & F. Schrenk. Oxford: Oxford University Press, 328–48.

Foley, R.A. & M.M. Lahr, 1997. Mode 3 technologies and the evolution of modern humans. *Cambridge Archaeological Journal* 7(1), 3–36.

Foley, R.A. & M.M. Lahr, in press. Flux and fragility: demographic models and the relationships between Eurasian and African Late Pleistocene hominins, in *Biogeography and Neanderthal Evolution*, ed. C. Finlayson.

Gamble, C., 1993. *Timewalkers*. London: Allen Lane.

Housley, R.A., C. Gamble, M. Street & P. Pettitt, 1997. Radiocarbon evidence for the late glacial recolonisation of northern Europe. *Proceedings of the Prehistoric Society* 63, 25–54.

Huntley, B. & J.R. Allen, 2003. Glacial environments III: palaeo-vegetation patterns in Late Glacial Europe, in *Neanderthals and Modern Humans in the European Landscape during the Last Glaciation*, Chapter 6, eds. T.H. van Andel & W. Davies. (McDonald Institute Monographs.) Cambridge: McDonald Institute for Archaeological Research, 79–102.

Ingman, M., H. Kaessmann, S. Paäbo & U. Gyllensten, 2000. Mitochondrial genome variation and the origin of modern humans. *Nature* 408, 708–13.

Ke, Y.H., B. Su, X.F. Song, D.R. Lu, L.F. Chen, H.Y. Li, C.J. Qi, S. Marzuki, R. Deka, P. Underhill, C.J. Xiao, M. Shriver, J. Lell, D. Wallace, R.S. Wells, M. Seielstad, P. Oefner, D.L. Zhu, J.Z. Jin, W. Huang, R. Chakraborty, Z. Chen & L. Jin, 2001. African origin of modern humans in East Asia: A tale of 12,000 Y chromosomes. *Science* 292, 1151–3.

Klein, R.G., 2000. Archeology and the evolution of human behavior. *Evolutionary Anthropology* 9, 17–36.

Lahr, M.M. & R.A. Foley, 1994. Multiple dispersals and modern human origins. *Evolutionary Anthropology* 3, 48–60.

Lahr, M.M. & R.A. Foley, 1998. Towards a theory of modern human origins: geography, demography, and diversity in recent human evolution. *Yearbook of Physical Anthropology* 41, 137–76.

Loubere, P., 1988. Gradual Late Pliocene onset of glaciation: a deep-sea record from the northeast Atlantic. *Palaeogeography, Palaeoclimatology, Palaeoecology* 63, 327–34.

Malatesta, A. & F. Zarlenga, 1988. Evidence of Middle Pleistocene marine transgressions along the Mediterranean coast. *Palaeogeography, Palaeoclimatology, Palaeoecology* 68, 311–15.

Quintana-Murci, L., R. Veitia, S. Santachiara-Benerecetti, K. McElreavey, M. Fellous & T. Bourgeron, 1999. Mitochondrial DNA, Y chromosome and human population history. *M S-Medicine Sciences* 15, 974–82.

Roberts, R.G., R. Jones, N.A. Spooner, M.J. Head, A.S. Murray & M.A. Smith, 1994. The human colonisation of Australia: optical dates of 53,000 and 60,000 years bracket human arrival at Deaf Adder Gorge, Northern Territory. *Quaternary Science Reviews* 13, 575–86.

Shackleton, N.J., 1987. Oxygen isotopes, ice volumes, and sea level. *Quaternary Science Review* 6, 183–90.

Shackleton, N.J., 1996. New data on the evolution of Pliocene climatic variability, in *Palaeoclimate and Neogene Evolution*, ed. E. Vrba. New Haven (CT): Yale University Press, 282–90.

Shackleton, N.J., M.A. Hall & E. Vincent, 2000. Phase relationships between millennial scale events 64,000 to 24,000 years ago. *Paleoceanography* 15, 565–9.

Stringer, C., 2002. Modern human origins: progress and prospects. *Philosophical Transactions of the Royal Society of London Series B-Biological Sciences* 357, 563–79.

Stringer, C. & P. Andrews, 1988. Genetic and fossil evidence for the origin of modern humans. *Science* 239, 1263–8.

Stringer, C. & C. Gamble, 1993. *In Search of the Neanderthals*. London: Thames and Hudson.

Stringer, C., H. Pälike, T.H. van Andel, B. Huntley, P. Valdes & J.R.M. Allen, 2003. Climatic stress and the extinction of the Neanderthals, in *Neanderthals and Modern Humans in the European Landscape during the Last Glaciation*, Chapter 13, eds. T.H. van Andel & W. Davies. (McDonald Institute Monographs.) Cambridge: McDonald Institute for Archaeological Research, 233–40.

Tchernov, E., 1992a. Biochronology, paleoecology and dispersal events of hominids in the southern Levant, in *The Evolution and Dispersal of Modern Humans in Asia*, eds. T. Akazawa, K. Aoki & T. Kimura. Tokyo: Hokusen-sha Publ. Co., 149–88.

Tchernov, E., 1992b. Eurasian-African biotic exchanges through the Levantine corridor during the Neogene and Quaternary. *Courier Forschungsinstitut Senckenberg* 153, 103–23.

Torroni, A., M. Richards, V. Macaulay, P. Forster, R. Villems, S. Norby, M.L. Savontaus, K. Huoponen, R. Scozzari & H.-J. Bandelt, 2000. MtDNA haplogroups and frequency patterns in Europe. *American Journal of Human Genetics* 66, 1173–7.

Underhill, P.A., G. Passarino, A.A. Lin, P. Shen, M.M. Lahr, R.A. Foley, P.J. Ofner & L.L. Cavalli-Sforza, 2001. The phylogeography of Y chromosome binary haplotypes and the origins of modern human populations. *Annals of Human Genetics* 65, 43–62.

van Andel, T.H., 2002. The climate and landscape of the middle part of the Weichselian glaciation in Europe: The Stage 3 Project. *Quaternary Research* 57, 2–8.

van Andel, T.H., 2003. Glacial environments I: the Weichselian climate in Europe between the end of the OIS-5 interglacial and the Last Glacial Maximum, in *Neanderthals and Modern Humans in the European Landscape during the Last Glaciation*, Chapter 2, eds. T.H. van Andel & W. Davies. (McDonald Institute Monographs.) Cambridge: McDonald Institute for Archaeological Research, 9–20.

van Andel, T.H., W. Davies, B. Weninger & O. Jöris, 2003a. Archaeological dates as proxies for the spatial and temporal human presence in Europe: a discourse on the method, in *Neanderthals and Modern Humans in the European Landscape during the Last Glaciation*, Chapter 3, eds. T.H. van Andel & W. Davies. (McDonald Institute Monographs.) Cambridge: McDonald Institute for Archaeological Research, 21–30.

van Andel, T.H., W. Davies & B. Weninger, 2003b. The human presence in Europe during the Last Glacial period I: human migrations and the changing climate, in *Neanderthals and Modern Humans in the European Landscape during the Last Glaciation*, Chapter 4, eds. T.H. van Andel & W. Davies. (McDonald Institute Monographs.) Cambridge: McDonald Institute for Archaeological Research, 31–56.

van Valen, L., 1973. A new evolutionary law. *Evolutionary Theory* 1, 1–30.

Vrba, E., 1993. Turnover-Pulses, the Red Queen, and related topics. *American Journal of Science* 293A, 418–52.

Vrba, E., 1996. *Palaeoclimate and Neogene Evolution*. New Haven (CT): Yale University Press.

Watson, E., P. Forster, M. Richards, & H.-J. Bandelt, 1997. Mitochondrial footprints of human expansions in Africa. *American Journal of Human Genetics* 61, 691–704.

White, T.D., B. Asfaw, D. DeGusta, H. Gilbert, G.D. Richards, G. Suwa & F.C. Howell, 2003. Pleistocene *Homo sapiens* from Middle Awash, Ethiopia. *Nature* 423, 742–7.

Epilogue

Humans in an Ice Age - The Stage 3 Project: Overture or Finale?

Tjeerd H. van Andel

The Stage 3 Project was an exercise in the use of palaeoclimate studies in research on human evolution research; it can best be described by one general (A), two specific (B) and one alternative question (C). A. 'Was the extinction of the Neanderthals due to major changes in the glacial environment, being contemporaneous with and analogous to the demise of relict mammals of the preceding interglaciation?' B.1. 'Was extinction a direct consequence of the Neanderthal inability to adapt to a severely cold climate?' or B.2. 'Was it due to their failure to exploit cold-adapted food animals of the high glacial instead of the dwindling survivors of the interglacial?' An alternative question — popular with archaeologists — is C. 'Did the Neanderthals succumb to the technical, cultural, social or warfare superiority of Anatomically Modern Humans?'.

When we conceived the Stage 3 Project it was, and probably still is, the most far-reaching attempt to consider Old Stone Age humans and their cultures in an environmental context, because of the vast territory (all of Europe) and long time-span (60–30 ka BP, later extended to 18 ka BP) that covered nearly all of the last glacial period. It was unique, because we rejected the usual way of assembling all handily available kinds of environmental data and to apply those to an equally heterogeneous set of all kinds of archaeological evidence, a common approach that often allows so many options for interpretation that any conclusion becomes arbitrary (e.g. d'Errico & Sánchez-Goñi 2003).

Instead, we chose a deliberately limited library of existing homogeneous and quantitative data sets that were internally consistent and externally compatible to serve as input for a rigorously defined, computerized palaeoclimate model. Similarly chosen and equally robust sets of palaeoenvironmental data were set aside to test all climate simulations.

We intended to trace all clear lines of reasoning that led from the input data to environmental simulations and from those to human responses to the changing glacial environment. Therefore we inferred those responses from human movements in time and space derived from a chrono-archaeological data base containing only quantitative dates.

In design and execution our approach thus differed fundamentally from many other studies that exploited simultaneously diverse kinds of information, anecdotal as well as systematic, and used qualitative with quantitative data on diverse time-scales with varying levels of relevance to the objectives.

The reader must decide whether we have succeeded in making clear the structure of our reasoning, but we hope that the comparisons of palaeoclimate reconstructions with human behaviour chosen for this book will permit readers to follow our arguments, while drawing their attention to the much larger number of deductions (such as those in Chapter 4: van Andel *et al.* 2003b) that deserve further analysis.

The project, now finished, was deliberately exploratory and its outcome was therefore not predictable. The initial phase of building quantitative (or semi-quantitative!) images of climate, fauna and flora, landscape and human life during OIS-3 did increase step-by-step our understanding of what the data had to say, what the power of our processing methods was and what were the limits of both. But when in the second phase we set out to explore palaeoenvironmental images as a tool to analyze the human distribution in time and space, such a diverse range of windows was opened on so many aspects of the theme 'humans in an ice age' that a comprehensive review of our gains was simply not feasible. And so we abandoned the idea of a complete and systematic report and left the authors free to select their topics.

Reconstructing the OIS-3 environment

The simulation of the mid-glacial (OIS-3) palaeo-climate was the key to the entire enterprise, but even the supercomputer at Pennsylvania State University could not model the continuously changing climatic history of OIS-3 as an ongoing process at the meso-scale resolution that we needed. Therefore we chose two, later three snapshots of the end members of the climatic range displayed by the Greenland GISP2 ice-core record for OIS-3. They were: 1) a mid-glacial warm event from the interval 45–38 ka BP, the 'OIS-3 warm' event, typical also of the warm period of 60–45 ka BP; 2) a cold event from about 30 ka BP, the 'OIS-3 cold' event which preceded the Last Glacial Maximum (LGM) by at least 5000 years; and 3) the 'LGM cold' event itself (Chapter 5 passim; Barron et al. 2003). The first and third simulations agreed quite well with our expectations and have been used as reliable guides for the Stable Warm, Transitional, Early Cold and LGM Cold Phases (Chapter 4: van Andel et al. 2003b, Table 4.3), all widely used in this book.

The 'OIS-3 cold' simulation, however, seemed too warm to reflect the many cold spikes of the interval 37–25 ka BP (Chapter 5: Barron et al. 2003; Barron & Pollard 2002; Pollard & Barron 2003). Worse, the open woodland/taïga simulated by coupled meso-scale RegCM2/BIOME 3.5 models (Chapter 6: Huntley & Allen 2003; Huntley et al. 2003) was received sceptically by Quaternary geologists and palynologists alike. To the geologists it was the evidence for wide-spread permafrost deformation north of 50°N (Alfano et al. 2003; van Huissteden et al. 2003) that rendered the 'OIS-3 cold' simulation unsafe. To palynologists the ubiquitous occurrence of coniferous and even deciduous taïga across so much of Europe was incompatible with the tree-less OIS-3 landscape north of the Trans-European mountain barrier that has been accepted by many palynologists on the basis of long pollen cores (Chapter 6: Huntley et al. 2003).

Persuaded by their critics the modellers carried out a series of tests with a large range of dummy input data to determine what might have caused the discrepancy between simulations and evidence and what climate conditions could have created a tree-less tundra (Alfano et al. 2003). The experiments showed that the tree-less tundra demanded a climate far too extreme to be plausible; either the input data were at fault or the climate dynamics of the models were inappropriate or both. The 'OIS-3 cold' simulation was duly rejected, leaving us with only

two snapshots, the 'OIS-3 warm' event, also suitable for the warm interval from c. 59–45 ka BP, and the 'LGM cold' simulation. The latter has served most studies in this book for the period from c. 37 ka BP to the start of the deglaciation.

Was this justified? At the time many project members thought so, but now, a few years later, there are reasons to reconsider the issue. For a start, permafrost data are difficult to pin down by age, by climate of origin and by the duration and geographic extent of the cold spell. In present permafrost belts the hummocky nature of the periglacial zone generates a mosaic of environments that relates to soil morphology and porosity and other local properties. A recent study by Plug & Werner (2002) shows that brief, very cold events on a time-scale of centuries can leave behind permanent frost wedges of the same dimensions as those of much longer stadials. Van Huissteden et al. (2003) cite mismatches between ice wedges and palaeoclimates, for example for the LGM (Kageyama et al. 2001) and the Younger Dryas (Isarin & Renssen 1997; 1999) that are quite similar to the Stage 3 situation. So our unexpectedly warm 'OIS-3 cold event' might after all be compatible with frost wedges created by brief, very cold Dansgaard/Oeschger (D/O) events.

The pollen-based 'barren tundra' concept has been questioned by mammalian palaeontologists too (e.g. Lister & Sher 1995; Guthrie 2000; Guthrie & van Kolfschoten 1999; Yurtsev 2001), because the carrying capacity adequate to feed numerous large herbivores such as mammoth (Mammuthus primigenius), woolly rhinoceros (Coelodonta antiquus) or giant deer (Megaloceros giganteus) implies a productive environment known as the 'steppe-tundra' or 'mammoth steppe' (Guthrie 1990). Analysis of the Stage 3 Mammalian Data Base (Chapter 7: Stewart et al. 2003a) and abundant central European data made accessible by Musil (Chapter 10: Musil 2003; Willis et al. 2000; Liivrand 1990; 1991), have supplied evidence for wide-spread, patchy open woodland (Chapter 6: Huntley & Allen 2003, Fig. 6.12). The same domain would also have suited the more thermophile herbivores (Chapter 10: Musil 2003) of the later glacial period. In many areas such as southern France (Delpech 1993), central and eastern Europe (Chapter 10: Musil 2003), those herbivores, roe deer (Capreolus capreolus) for instance, red deer (Cervus elaphus) and wild boar (Sus scrofus), have been found together with a pure arctic fauna (for details see Chapter 11: Davies et al. 2003).

Experiments with the BIOME 3.5 model have shown that during the warm D/O events the so-called 'barren' tundra may have been less barren than generally believed (Chapter 6: Huntley & Allen 2003, Fig. 6.12). Instead, a mosaic of parkland with a savanna-like vegetation may have been prominent across much of Europe during warm D/O events, providing the primary production to feed the herbivores of the palaeontological record (Chapter 7: Stewart *et al.* 2003a).

The 'OIS-3 cold' simulation of the Early Cold Phase remains contentious, however. Palynological data suggest that *c.* 40 ka BP an increased frequency of cold D/O events gradually reduced the trees and encouraged a predominantly steppic herbaceous vegetation of less productivity (Chapter 6: Huntley & Allen 2003, Fig. 6.12). Still, human site patterns show that during the Early Cold Phase the vegetation and herbivore fauna sufficed to feed a widespread human population in Europe north of the Trans-European mountain barrier (Chapter 4: van Andel *et al.* 2003b). The settlement patterns are supported by faunal, pollen and charcoal data for France (Antoine *et al.* 1999) and central Europe (Chapter 10: Musil 2003; Willis *et al.* 2000 and elsewhere (e.g. Stewart & Lister 2001).

New marine cores from along the entire European Atlantic margin (PAGES cores) may soon provide better sea-surface temperature and ice-margin position data (Michael Sarnthein pers. comm. 2003), but it is too early to continue the discussion here. It is obvious, however, that a substantial amount of information, some old but not recognized and some new, ought to be taken into consideration.

The severity of the glacial climate has often been exaggerated, especially by archaeologists. Phrases like 'When the glacial maximum approached, the cold became so intolerable that the Neanderthals perished, and even modern humans struggled to avoid the same fate', are not uncommon in even the recent literature but they are nonsense as Chapters 5 (Barron *et al.* 2003), 6 (Huntley & Allen 2003, Fig. 6.12), 7 (Stewart *et al.* 2003a), 10 (Musil 2003) and 11 (Davies *et al.* 2003) clearly show. Even during the LGM, when winter conditions north of the Trans-European mountain barrier were indeed harsh, in summer the climate was no more life-threatening than it is today in, for example, northern Canada around Hudson Bay (Chapter 5: Barron *et al.* 2003).

Early in the project but too late to affect the modelling experiments, serendipitous events suggested that the thick ice sheet assumed to cover most of Scandinavia throughout the interval OIS-4/OIS-2 might not have existed or only in a much-reduced size than commonly thought (Arnold *et al.* 2002; Olsen

1997; Olsen *et al.* 2001a, b). Radiocarbon dates from the interval 45–25 ka BP that implied a largely ice-free Scandinavia in OIS-3 had been rejected (Donner 1995). Thinking this premature, Arnold *et al.* (2002), using much new data of Olsen and others (1997; Olsen *et al.* 1996; 2001a,b) for dynamic ice-sheet modelling, concluded that the OIS-3 ice sheet was probably limited to the high mountains of southwestern Norway and possibly the distant north. Pavlov *et al.* (2001; Pavlov & Indrelid 1999) on archaeological grounds and Markova *et al.* (2002) on mammalian fauna data came to a similar conclusion for northern Russia. The concept of an ice-free Scandinavia in OIS-3 is not universally popular but the much strengthened supporting evidence now demands that it be taken very seriously indeed. But would it have made much difference for the climate south of Scandinavia and south of Baltic? Probably not: sensitivity tests suggest that the impact would have been felt mainly in Russia east of Scandinavia.

Humans in a glacial climate

For the second phase of the project the chrono-archaeological data base (Chapter 3: van Andel *et al.* 2003a) was our key to linking climate changes recorded in the Greenland ice cores to human responses to those changes. Because 4 ka or 5 ka time-slices are the shortest units allowed by the confidence limits of the dated human record (Chapter 3: van Andel *et al.* 2003a), attempts to track millennial D/O oscillations must fail; instead we used a subdivision of the GISP2 record based on major climate changes on multi-millennial time-scales (Chapter 4: van Andel *et al.* 2003b, Fig. 4.1, Table 4.3).

Our confidence in this approach, derived from the success of earlier studies using ^{14}C data for a similar purpose (Ammerman & Cavalli Sforza 1984), turned out to be justified when meaningful, non-random patterns of human movement appeared on the palaeogeographic maps in Chapter 4 (van Andel *et al.* 2003b). The maps raise many interesting points such as the clear parallel between Neanderthal migrations and climate changes across the northern Mediterranean and throughout Europe south of 50°N as soon as the OIS-4 ice sheet had melted away. Similar is the two-pronged withdrawal westward to the Atlantic shore and southeastward to the Black Sea, probably induced by the final climate deterioration after 37 ka BP. Another example is the close similarity of the migration paths of Neanderthals and early Anatomically Modern Humans (AMH) in the 'Transitional' climate phase (*c.* 48–38 ka BP). Together these and other patterns raise the question whether Neanderthals and early AMH were both best adapted to temperate or at worst boreal conditions, living on sedentary animal resources and therefore both equally handicapped when an arctic mode of living with its seasonal mobility became the better option.

The earliest dated Gravettian sites are thinly-scattered across Europe, giving no hint to their point of origin. Beginning in the 33–30 ka BP time-slice the number of sites north of the Trans-European mountain barrier increased greatly especially in central Europe, while dated Upper Palaeolithic sites became rare in southeastern and southwestern Europe and the Mediterranean. Was this Gravettian expansion due to local technical and social adaptations to the coldest period of the entire glaciation or were they introduced by immigrants from the East?

For some time a debate has raged between those who, on the basis of skeletal features, see the Neanderthals as intrinsically cold-adapted (e.g. Trinkaus 1981; Trinkaus *et al.* 1991), and those who regard them as typical inhabitants of the middle latitudes for whom much of northern Europe was too cold. These and related issues are discussed in Chapters 8 (Davies & Gollop 2003) and 9 (Aiello & Wheeler 2003).

Davies & Gollop plotted all Mousterian, Aurignacian+EUP and Gravettian+UP sites on simulations of winter and summer temperatures, wind chill and snow cover, variables thought to be of key importance to Neanderthals and AMH alike (Chapter 5: Barron *et al.* 2003). Histograms (Chapter 8: Davies & Gollop 2003, Figs. 8.5–8.10) display the frequency of sites across the full range of each variable for each of the climate phases. They document two main characteristics of the two human species relative to all climate variables: *tolerance* refers to the full range of acceptable climatic variation while *preference*, displayed by the concentration of sites in one or more narrow segments of the tolerance range, indicates the main zone of comfortable living. Davies & Gollop (2003) concluded that the temperature, wind-chill and snow tolerances of the three major techno-complexes were remarkably similar (Chapter 8: Davies & Gollop 2003, Table 8.3) and fairly consistent through time.

In terms of preferred temperatures the Neanderthal and early AMH site patterns are close; they consistently preferred the warmer parts of Europe throughout the year, a preference that helped them to expand across large parts of Europe during the Stable Warm Phase and to select mainly areas that permitted permanent occupation.

During the Transitional and Early Cold Phases, Mousterian and early AMH sites show a modest preference for the –20°C to –8°C zone, while the Gravettian site pattern marks a strong, even dominant preference for extremely cold sites. However, this apparent choice may be due to an ambiguity in the analysis. Therefore we do not know whether all or even any sites were occupied all year round or only in summer or winter. Hominid preferences for the coldest winter sites are *a priori* implausible, and it seems very likely that the high concentrations in the coldest areas merely represents aestival occupation by people who preferred the benefits of the local summer but lived elsewhere in winter.

Aiello & Wheeler (Chapter 9: 2003) approach the question of cold adaptation from a physiological angle (Steegmann *et al.* 2002). The Neanderthals have often been described as being cold-adapted because their body form corresponds to that found in modern cold-adapted peoples and is consistent with our expectation for ice-age humans. Relative to modern and fossil people who live in warmer climates, Neanderthals had large rib-cages, barrel-shaped chests and short extremities (e.g. Holliday 1997a,b) and therefore have been described as having an arctic or even hyperarctic body form. However, Aiello & Wheeler show that this body form would have provided only a modest advantage over the AMH as far as the lowest critical and sustainable minimum temperatures are concerned. This would be the case even if we allow for the insulating effect of their greater muscle mass and for a dietary-related elevated BMR (Basal Metabolic Rate). They would have had little difficulty surviving all Stage 3 summers without additional insulation, but wind-chill simulations suggest that Mousterian sites were preferentially located in areas with warmer winter temperatures than those accepted by Aurignacians or Gravettians. Elsewhere the cost of maintaining internal heat production at the required winter level would have been bearable only if the Neanderthals were able to sustain a suitably high level of dietary energy intake. It thus seems that they could not have survived local cold phase winters without additional artificial insulation, although they would not have had any real problem with warm phase winters. This may have been a significant factor in their decline, especially if they were in competition with better-insulated AMH for favourable climate conditions, refugia and dietary resources.

Two chapters exploit the potential of the Stage 3 Project with a quite different perspective. In Chapter 14 Lahr & Foley (2003) consider the methods and results of the Stage 3 Project in a geographically wider and chronologically much longer anthropological context. They explore the period that witnessed the migration of a small African population throughout the world and the disappearance of all others, leaving a single human species in control for the first time in 5 million years. Offering alternatives to interpretations in other chapters, they discuss the ever-present possibility of artefacts in the interpretation of data, an issue optimistically dealt with in Chapters 3 and 4 (van Andel *et al.* 2003a,b). Their thoughtful demographic models regarding glacial cycles as a primary framework for the biogeography of the Pleistocene will, applied to various issues discussed throughout the Stage 3 book, be certain to generate fruitful ideas of where one might go in the future with palaeoenvironmental research relevant to human evolution.

In Chapter 11 Davies *et al.* (2003) look in the opposite direction when they draw attention to events scaled down well below continental size and for time-scales in thousands of years. Noting that the predominant OIS-3 pattern of loosely scattered sites is replaced in a very few places by dense, long-lasting site clusters, they ask what the attraction of those locations might have been of those few special places in Europe. Generally the assumption has been that, given an extremely harsh glacial climate, local settlement clusters were refugia, dwelling sites of last resort. Davies *et al.* argue that the clustering, rather than being a suitable shelter to wait out the hardships of winter, might well have chosen areas endowed with year-round favourable qualities that were much superior to those of the surrounding regions. To address the question what those attractions might have been and what purpose they served, Davies *et al.* selected three well-studied areas, the Dordogne, the Ardennes, and Moravia in the mid-Danube basin. They assembled much information on all aspects, archaeological as well as environmental, for all three study cases, and consider in detail their topographic features, the local climate, their occupation history and the evidence for economic, cultural and social activities.

However, the areas, each occupying only a few of the 60 × 60 km output boxes of the climate simulations, are too small to provide an adequately detailed understanding of their climatic advantages and disadvantages. Here recent advances in high-resolution palaeoclimate modelling have made it possible to down-scale, if only for a small area and a few variables at the time, the Stage 3 meso-scale

simulations to a local resolution of a few kilometres. This enables us, for example, to distinguish between winter temperature levels and wind-chill impact *within* the densely occupied Dordogne river valley bottom with conditions in the similar but for unknown reasons largely unoccupied adjacent Lot valley. The down-scale analysis shows clearly that the Lot valley, mainly because of its orientation, had very much harsher winter conditions than the Dordogne valley. Similar high-resolution climate insights were obtained for the Ardennes and mid-Danube basin.

Davies *et al.* (2003), after evaluating the copious information on landscape, climate, resources and archaeology, conclude that, far from being shelters of last resort, all three clusters had their origin as particularly well-endowed settings with a large variety of resources, good climate conditions and, perhaps above all, very good long-distance access along rivers and through passes.

Where mesoscale climate simulations are available, the down-scaling technique might be applied to other parameters than temperature or wind chill. Being relatively undemanding in computer terms, it provides a useful link between the continental scale of the Stage 3 Project and the regional or local studies that provide the daily bread of most archaeologists for which Roebroeks *et al.* (1999) provide several examples, such as Bosinski *et al.* (1999), Hahn (1999), Rigaud (1999), Scheer (1999) and Street & Terberger (1999), all in that rich volume *Hunters of the Golden Age* (Roebroeks *et al.* 1999).

The Neanderthal extinction

So what, if anything, did we uncover that allows us to speculate beyond the current views on the demise of the Neanderthals? Two chapters address this question, so much argued about throughout the life-time of the Stage 3 Project: Chapter 12 (Stewart *et al.* 2003b) and Chapter 13 (Stringer *et al.* 2003).

Stewart's explanation is based on an analysis of the Stage 3 Mammalian Fauna Data Base and its several categories of mammals, each with different late Pleistocene histories. From the perspective of an evolving biosphere they draw attention to the close similarity between the Neanderthal extinction and that of a mammalian group known as the 'interglacial survivors' (perhaps not the most accurate name?), both around 30 ka BP, but they do not comment in detail on the causes of the extinction. Nor do they dwell on the accepted view that the Neanderthals were continuously present in Europe for more than

250,000 years that included the long and severe penultimate glaciation of OIS-6 (Gamble 1999, 175 *sqq.*). Yet why should the Neanderthals, having already survived at least one major glaciation, have succumbed to the relatively mild conditions of OIS-3. Do I dare to propose that they, like the clan of 'interglacial survivors', became periodically extinct some time after the end of each glaciation, only to re-immigrate from the Near East with the same faunal group as companions of the Thames hippopotamus or the straight-tusked elephant? Quite likely, one would say, but then why did they not survive the last glaciation? Perhaps we must turn, after all that has been said and done, to looking modern humans straight in the eye and say 'What did you do?'.

Chapter 13 (Stringer *et al.* 2003) presents a plausible argument not only for the cause but also for the timing of the exit of the 'interglacial survivors'. They show that the deteriorating glacial climate generated major environmental stresses, major in large part because the high-frequency D/O oscillations (Chapter 1: van Andel 2003) demanded rapid adaptation to the very different conditions on the downward limb of each D/O event compared to those of the rising limb. They document this plausible suggestion with a 'stress curve' derived from the Greenland GISP2 ice-core record (Chapter 13: Stringer *et al.* 2003, Fig. 13.3) and for comparison another stress curve calculated from a Monticchio lake core in Italy (Allen *et al.* 2000) which gave very similar results. On both stress curves two instances of major stress stand out beyond all others, one in OIS-4 and the other around 30 ka BP, a time of major mammalian extinction. To the best of my knowledge this is the first time that glaciogene climatic stress does not merely appear as a non-specific *deus ex machina*, but is derived quantitatively from the environmental record. In addition the chapter provides a helpful, long list of proposed causes for the Neanderthal extinction which, if we had seen it laid out in front of us earlier, might have caused us to give the whole subject a much lower priority or at least a different label.

As we expected, we have come up with few answers but, as we hoped, we have generated many questions, some old but now seen in a new light, some new that look exiting and ready to be tackled. Some have large dimensions and demand a whole new project, but others are suited to a more modest enterprise. To one of them three chapters refer, the shift in the Early Cold Phase to the coldest locations on record. A sensible explanation would be that they are *summer* sites unhabitable in winter, and that sug-

gests seasonal migration like the Sami in Lappland or the Inuit and northern Indians in Canada. Our climate simulations are ambiguous but there are other ways for a sensible start, both requiring a return to the publications that underpin our data base. Perhaps a modest problem, but a challenging one.

A word of caution, however, is in order here. The Greenland ice cores are widely seen as a reliable climatic record not only for Stage 3 but also for later times, such as the deglaciation period and the early Holocene, splendid archaeological targets. As Chapter 2 (van Andel 2003) shows, however, their usefulness has so far been demonstrated only for the maritime climate zone of western Europe and the western Mediterranean. Major climate changes in central, eastern and southeastern Europe do exist (Chapter 10: Musil 2003), but their nature is not well understood nor do we know whether they are synchronous with the ice-core record. The farther east and southeast we extend our interest the more likely is it that other major regional climate systems than that of the North Atlantic play a dominant role. In fact, for the interval from *c.* 15 to 5 ka BP the main forcing factor in the Middle and Near East and probably in southeastern Europe is the Monsoon system of the northwestern Indian Ocean (van Andel 2000). Thus, tempting as such a project might seem, a large amount of expert study of the climatic setting is still required before it is ready to be used as background for a regional archaeological study

Here we arrive at the point where application by others must pick up the burden, application of methods, application of insights, application of inter-disciplinary collaboration, all the while coping with the many curious misunderstandings that tend to mark inter-disciplinary studies (van Andel 1994). Important in this regard is a large dose of optimistic scepticism that recognizes that not all that is novel may be worth further study, and accepts that waiting for the advances made by others may slow progress but must be borne. If others judge our efforts to be worthwhile and amenable to further pursuit, the *Epilogue* will be an *Overture*. If circumstances should prevent those as yet anonymous others from following where we are pointing, the Epilogue will be the Finale but that would be a pity indeed.

References

Aiello, L.C. & P. Wheeler, 2003. Neanderthal thermoregulation and the glacial climate, in *Neanderthals and Modern Humans in the European Landscape during the Last Glaciation,* Chapter 9, eds. T.H. van Andel & W. Davies. (McDonald Institute Monographs.) Cambridge: McDonald Institute for Archaeological Research, 147–66.

Alfano, M.J., E.J. Barron, D. Pollard, B. Huntley & J.R.M. Allen, 2003. Comparison of climate model results with European vegetation and permafrost during Oxygen Isotope Stage 3. *Quaternary Research* 59, 97–107.

Allen, J.R.M., W.A. Watts & B. Huntley, 2000. Weichselian palynostratigraphy, palaeovegetation and palaeoenvironment; the record from Lago Grande di Monticchio, southern Italy. *Quaternary International* 73/74, 91–110.

Ammerman, A.J. & L.L. Cavalli-Sforza, 1984. *The Neolithic Transition and the Genetics of Population in Europe.* Princeton (NJ): Princeton University Press.

Antoine, P., J.J. de Beaulieu, P. Bintz, J.P. Brugal, M. Girard, J.L. Guadelli, M.T. Morzadec-Kerfourn, J. Renault-Miskowsky, A. Roblin-Jouvé, B. van Vliet-Lanoë & J.D. Vigne, 1999. *La France pendant les deux derniers extrèmes climatiques - Variabilité des environnements.* Châtenay-Malabry: Andra, Agence Nationale pour la Gestion des Radioactifs.

Arnold, N.S., T.H. van Andel & V. Valen, 2002. Extent and dynamics of the Scandinavian ice sheet during Oxygen Isotope Stage 3 (60,000–30,000 yr BP). *Quaternary Research* 57, 38–48.

Barron, E.J. & D. Pollard, 2002. High-resolution climate simulations of Oxygen Isotope Stage 3 in Europe. *Quaternary Research* 58, 296–309.

Barron, E., T.H. van Andel & D. Pollard, 2003. Glacial environments II: reconstructing the climate of Europe in the Last Glaciation, in *Neanderthals and Modern Humans in the European Landscape during the Last Glaciation*, Chapter 5, eds. T.H. van Andel & W. Davies. (McDonald Institute Monographs.) Cambridge: McDonald Institute for Archaeological Research, 57–78.

Bosinski, G., 1999. The period 30,000–20,000 bp in the Rhineland, in *Hunters of the Golden Age: the Mid Upper Palaeolithic of Eurasia 30,000–20,000 BP*, eds. W. Roebroeks, M. Mussi, J. Svoboda & K. Fennema. Leiden: *Analecta Praehistorica Leidensia* 31, 271–80.

Davies, W. & P. Gollop, 2003. The human presence in Europe during the Last Glacial Period II: climate tolerance and climate preferences of mid- and late glacial hominids, in *Neanderthals and Modern Humans in the European Landscape during the Last Glaciation*, Chapter 8, eds. T.H. van Andel & W. Davies. (McDonald Institute Monographs.) Cambridge: McDonald Institute for Archaeological Research, 131–46.

Davies, W., P. Valdes, C. Ross & T.H. van Andel, 2003. The human presence in Europe during the Last Glacial Period III: site clusters, regional climates and resource attractions, in *Neanderthals and Modern Humans in the European Landscape during the Last Glaciation*, Chapter 11, eds. T.H. van Andel & W. Davies. (McDonald Institute Monographs.) Cambridge: McDonald Institute for Archaeological Research,

191–220.

Delpech, F., 1993. The fauna of the Early Upper Paleolithic: Biostratigraphy of large mammals and current problems in chronology, in *Before Lascaux: the Complex Record of the Early Upper Paleolithic*, eds. H. Knecht, A. Pike-Tay & R.White. Ann Arbor (MI): CRC Press, 71–84.

d'Errico, F. & M.F. Sánchez-Goñi, 2003. Neanderthal extinction and the millennial scale climatic variability of OIS 3. *Quaternary Science Reviews* 22, 769–88.

Donner, J., 1995. *The Quaternary of Scandinavia.* Cambridge: Cambridge University Press.

Gamble, C., 1999. *The Palaeolitic Societies of Europe.* Cambridge: Cambridge University Press.

Guthrie, R.D., 1990. *Frozen Fauna of the Mammoth Steppe: The Story of Blue Babe.* London: University of Chicago Press.

Guthrie, R.D., 2000. Origin and cause of the mammoth steppe: a story of cloud cover, woolly mammal tooth pits, buckles, and inside-out Beringia. *Quaternary Science Reviews* 20, 549–74.

Guthrie, R.D. & T. van Kolfschoten, 1999. Neither warm and moist nor cold and arid: the ecology of the Mid Upper Palaeolithic, in *Hunters of the Golden Age: the Mid Upper Palaeolithic of Eurasia 30,000–20,000 BP*, eds. W. Roebroeks, M. Mussi, J. Svoboda & K. Fennema. Leiden: *Analecta Praehistorica Leidensia* 31, 13–20.

Hahn, J., 1999. The Gravettian in southern Germany - environment and economy, in *Hunters of the Golden Age: the Mid Upper Palaeolithic of Eurasia 30,000–20,000 BP*, eds. W. Roebroeks, M. Mussi, J. Svoboda & K. Fennema. Leiden: *Analecta Praehistorica Leidensia* 31, 249–346.

Holliday, T.W., 1997a. Postcranial evidence of cold adaptation in European Neandertals. *American Journal of Physical Anthropology* 104, 245–58.

Holliday, T.W., 1997b. Body proportions in late Pleistocene Europe and modern human origins. *Journal of Human Evolution* 32, 423–48.

Huntley, B. & J.R.M. Allen, 2003. Glacial environments III: palaeo-vegetation patterns in Late Glacial Europe, in *Neanderthals and Modern Humans in the European Landscape during the Last Glaciation*, Chapter 6, eds. T.H. van Andel & W. Davies. (McDonald Institute Monographs.) Cambridge: McDonald Institute for Archaeological Research, 79–102.

Huntley, B., M.J. Alfano, J. Allen, D. Pollard, P. Tzedakis, J.-L. de Beaulieu, E. Grüger & W. Watts, 2003. European vegetation during Marine Oxygen Isotope Stage 3. *Quaternary Research* 59, 195–212.

Isarin, R.F.B. & H. Renssen, 1997. Surface temperatures in northwestern Europe during the Younger Dryas: AGCM simulation compared with temperature reconstructions, in *The Climate in Northwestern Europe during the Younger Dryas: a Comparison of Multi-proxy Climate Reconstructions with Simulation Experiments*, ed. R.F.B. Isarin. Amsterdam: Thesis, Vrije Universiteit, 75–92.

Isarin, R.F.B. & H. Renssen, 1999. Reconstructing and modelling Late Weichselian climates: the Younger Dryas in Europe as a case study. *Earth Science Reviews* 48, 1–38.

Kageyama, M., O. Peyron, S. Pinot, P. Tarasov, J. Guiot, S. Joussaume & G. Ramstein, 2001. The Last Glacial Maximum climate over Europe and western Siberia: a PMIP comparison between models and data. *Climate Dynamics* 17, 23–43.

Lahr, M.M. & R.A. Foley, 2003. Demography, dispersal and human evolution in the last glacial period, in *Neanderthals and Modern Humans in the European Landscape during the Last Glaciation*, Chapter 14, eds. T.H. van Andel & W. Davies. (McDonald Institute Monographs.) Cambridge: McDonald Institute for Archaeological Research, 241–56.

Liivrand, E., 1990. Type section of the lower and middle-Valdaian interstadial deposits at Töravere in southeast Estonia. *Proceedings Estonian Academy of Sciences* 39, 12–17.

Liivrand, E., 1991. *Biostratigraphy of the Pleistocene Deposits in Estonia and Correlations in the Baltic Region.* (Department of Quaternary Research Report 19.) Stockholm: Stockholm University.

Lister, A.M. & A.V. Sher, 1995. Ice cores and Mammoth extinction. *Nature* 378, 23–4.

Markova, A.K., A.N. Simakova, A.Y. Puzachenko & L.M. Kitaev, 2002. Environments of the Russian Plain during the Middle Valdai Briansk Interstade (33,000–24,000 yr BP) indicated by fossil mammals and plants. *Quaternary Research* 57, 391–400.

Musil, R., 2003. The Middle and Upper Palaeolithic game suite in central and southeastern Europe, in *Neanderthals and Modern Humans in the European Landscape during the Last Glaciation*, Chapter 10, eds. T.H. van Andel & W. Davies. (McDonald Institute Monographs.) Cambridge: McDonald Institute for Archaeological Research, 167–90.

Olsen, L., 1997. Rapid shifts in glacial extension characterise a new conceptual model for glacial variations during the Mid- and Late Weichselian in Norway. *Norges Geologiske Undersøgelse Bulletin* 433, 54–5.

Olsen, L., V. Mejdahl & S.F. Selvik, 1996. Middle and Late Pleistocene stratigraphy, chronology and glacial history in Finnmark, North Norway. *Norges Geologisk Undersøgelse Bulletin* 429, 111 pp.

Olsen, L., K. van der Borg, B. Bergstrøm, H. Svein, S.-E. Lauritzen & G. Hansen, 2001a. AMS radiocarbon dating of glacigenic sediments with low organic content: an important tool for reconstructing the history of glacial variations in Norway. *Norsk Geologisk Tidsskrift* 81, 59–92.

Olsen, L., H. Svein & B. Bergstrøm, 2001b. Rapid adjustments of the western part of the Scandinavian ice-sheet during the mid- and Late Weichselian: a new model. *Norsk Geologisk Tidsskrift* 81, 93–118.

Pavlov, P. & S. Indrelid, 1999. Human occupation in northeastern Europe during the period 35,000–18,000 BP, in *Hunters of the Golden Age: the Mid Upper Palaeolithic of Eurasia 30,000–20,000 BP*, eds. W. Roebroeks, M.

Mussi, J. Svoboda & K. Fennema. Leiden: *Analecta Praehistorica Leidensia* 31, 165–72.

Pavlov, P., J.I. Svendsen & S. Indrelid, 2001. Human presence in the European Arctic nearly 40,000 years ago. *Nature* 413, 64–7.

Plug, L.J. & B.T. Werner, 2002. Non-linear dynamics of ice-wedge networks and resulting sensitivity to severe cooling events. *Nature* 417, 929–33.

Pollard, D. & E.J. Barron, 2003. Causes of model-data discrepancies in European climates during Oxygen Isotope Stage 3 with insights from the Last Glacial Maximum. *Quaternary Research* 59, 108–13.

Rigaud, J.-P., 1999. Human adaptation to the climatic deterioration of the last Pleniglacial in southwestern France (30,000–20,000 bp), in *Hunters of the Golden Age: the Mid Upper Palaeolithic of Eurasia 30,000–20,000 BP*, eds. W. Roebroeks, M. Mussi, J. Svoboda & K. Fennema. Leiden: *Analecta Praehistorica Leidensia* 31, 325–36.

Roebroeks, W., M. Mussi, J. Svoboda & K. Fennema (eds.), 1999. *Hunters of the Golden Age: the Mid Upper Palaeolithic of Eurasia 30,000–20,000 BP.* Leiden: *Analecta Praehistorica Leidensia* 31.

Scheer, A., 1999. The Gravettian in Southwest Germany, stylistic features, raw material resources and settlement patterns, in *Hunters of the Golden Age: the Mid Upper Palaeolithic of Eurasia 30,000–20,000 BP*, eds. W. Roebroeks, M. Mussi, J. Svoboda & K. Fennema. Leiden: *Analecta Praehistorica Leidensia* 31, 257–70.

Steegmann, A.T., F.J. Cerny & T.W. Holliday, 2002. Neandertal cold adaptation: physiological and energetic factors. *American Journal of Human Biology* 14, 566–83.

Stewart, J.R. & A.M. Lister, 2001. Cryptic northen refugia and the origins of modern biota. *Trends in Ecology and Evolution* 16, 608–13.

Stewart, J.R., T. van Kolfschoten, A. Markova & R. Musil, 2003a. The mammalian faunas of Europe during Oxygen Isotope Stage Three, in *Neanderthals and Modern Humans in the European Landscape during the Last Glaciation*, Chapter 7, eds. T.H. van Andel & W. Davies. (McDonald Institute Monographs.) Cambridge: McDonald Institute for Archaeological Research, 103–30.

Stewart, J.R., T. van Kolfschoten, A. Markova & R. Musil, 2003. Neanderthals as part of the broader Late Pleistocene megafaunal extinctions?, in *Neanderthals and Modern Humans in the European Landscape during the Last Glaciation*, Chapter 12, eds. T.H. van Andel & W. Davies. (McDonald Institute Monographs.) Cambridge: McDonald Institute for Archaeological Research, 221–32.

Street, M. & T. Terberger, 1999. The German Upper Palaeolithic, 35,000–15,000 bp. New dates and insights with emphasis on the Rhineland, in *Hunters of the Golden Age: the Mid Upper Palaeolithic of Eurasia 30,000–20,000 BP*, eds. W. Roebroeks, M. Mussi, J. Svoboda & K. Fennema. Leiden: *Analecta Praehistorica Leidensia* 31, 280–91.

Stringer, C., H. Pälike, T.H. van Andel, B. Huntley, P. Valdes & J.R.M. Allen, 2003. Climatic stress and the extinction of the Neanderthals, in *Neanderthals and Modern Humans in the European Landscape during the Last Glaciation,* Chapter 13, eds. T.H. van Andel & W. Davies. (McDonald Institute Monographs.) Cambridge: McDonald Institute for Archaeological Research, 233–40.

Trinkaus, E., 1981. Neanderthal limb proportions and cold adaptation, in *Aspects of Human Evolution,* ed. C.B. Stringer. London: Taylor & Francis, 187–224.

Trinkaus, E., S.E. Churchill, I. Villemeur, K.G. Riley, J.A. Heller & C.B. Ruff, 1991. Robusticity versus shape - the functional interpretation of Neanderthal appendicular morphology. *Journal of the Anthropological Society of Nippon* 99, 257–78.

van Andel, T.H., 1994. Geo-archaeology and archaeological science: a personal view, in *Beyond the Site: Regional Studies in the Aegean Area,* ed. P.N. Kardulias. New York (NY): University Press of America, 24–43.

van Andel, T.H., 2000. Where received wisdom fails: the mid-Palaeolithic and Early Neolithic climates, in *Archaeogenetics: DNA and the Population Prehistory of Europe,* eds. C. Renfrew & K. Boyle. (McDonald Institute Monographs.) Cambridge: McDonald Institute for Archaeological Research, 31–9.

van Andel, T.H., 2003. The Stage 3 Project: initiation, objectives, approaches, in *Neanderthals and Modern Humans in the European Landscape during the Last Glaciation,* Chapter 1, eds. T.H. van Andel & W. Davies. (McDonald Institute Monographs.) Cambridge: McDonald Institute for Archaeological Research, 1–8.

van Andel, T.H., W. Davies, B. Weninger & O. Jöris, 2003a. Archaeological dates as proxies for the spatial and temporal human presence in Europe: a discourse on the method, in *Neanderthals and Modern Humans in the European Landscape during the Last Glaciation,* Chapter 3, eds. T.H. van Andel & W. Davies. (McDonald Institute Monographs.) Cambridge: McDonald Institute for Archaeological Research, 21–30.

van Andel, T.H., W. Davies & B. Weninger, 2003b. The human presence in Europe during the Last Glacial period I: human migrations and the changing climate, in *Neanderthals and Modern Humans in the European Landscape during the Last Glaciation,* Chapter 4, eds. T.H. van Andel & W. Davies. (McDonald Institute Monographs.) Cambridge: McDonald Institute for Archaeological Research, 31–56.

van Huissteden, K., J. Vandenberghe & D. Pollard, 2003. Palaeotemperature reconstructions of the European permafrost zone during marine oxygen isotope Stage 3 compared with climate model results. *Journal of Quaternary Science* 18, 453–64.

Willis, K.J., E. Rudner & P. Sümegi, 2000. The full-glacial forests of central and southeastern Europe. *Quaternary Research* 53, 203–13.

Yurtsev, B.A., 2001. The Pleistocene 'Tundra-Steppe' and the productivity paradox: the landscape approach, in Beringean Paleoenvironments: Festschrift in Honour of D.M Hopkins, eds. S.A. Elias & J. Brigham-Grette. *Quaternary Science Reviews* 20, 165–74.